Direct Methods for Solving
Macromolecular Structures

NATO ASI Series

Advanced Science Institutes Series

A Series presenting the results of activities sponsored by the NATO Science Committee, which aims at the dissemination of advanced scientific and technological knowledge, with a view to strengthening links between scientific communities.

The Series is published by an international board of publishers in conjunction with the NATO Scientific Affairs Division

A Life Sciences **B Physics**	Plenum Publishing Corporation London and New York
C Mathematical and Physical Sciences **D Behavioural and Social Sciences** **E Applied Sciences**	Kluwer Academic Publishers Dordrecht, Boston and London
F Computer and Systems Sciences **G Ecological Sciences** **H Cell Biology** **I Global Environmental Change**	Springer-Verlag Berlin, Heidelberg, New York, London, Paris and Tokyo

PARTNERSHIP SUB-SERIES

1. Disarmament Technologies	Kluwer Academic Publishers
2. Environment	Springer-Verlag / Kluwer Academic Publishers
3. High Technology	Kluwer Academic Publishers
4. Science and Technology Policy	Kluwer Academic Publishers
5. Computer Networking	Kluwer Academic Publishers

The Partnership Sub-Series incorporates activities undertaken in collaboration with NATO's Cooperation Partners, the countries of the CIS and Central and Eastern Europe, in Priority Areas of concern to those countries.

NATO-PCO-DATA BASE

The electronic index to the NATO ASI Series provides full bibliographical references (with keywords and/or abstracts) to more than 50000 contributions from international scientists published in all sections of the NATO ASI Series.
Access to the NATO-PCO-DATA BASE is possible in two ways:

– via online FILE 128 (NATO-PCO-DATA BASE) hosted by ESRIN, Via Galileo Galilei, I-00044 Frascati, Italy.

– via CD-ROM "NATO-PCO-DATA BASE" with user-friendly retrieval software in English, French and German (© WTV GmbH and DATAWARE Technologies Inc. 1989).

The CD-ROM can be ordered through any member of the Board of Publishers or through NATO-PCO, Overijse, Belgium.

Series C: Mathematical and Physical Sciences – Vol. 507

Direct Methods for Solving Macromolecular Structures

edited by

Suzanne Fortier

Departments of Chemistry and Computing and Information Science,
Queen's University,
Kingston, Ontario, Canada

Kluwer Academic Publishers

Dordrecht / Boston / London

Published in cooperation with NATO Scientific Affairs Division

Proceedings of the NATO Advanced Study Institute on
Direct Methods for Solving Macromolecular Structures
Erice, Italy
May 22–June 2, 1997

A C.I.P. Catalogue record for this book is available from the Library of Congress.

ISBN 0-7923-4949-0

Published by Kluwer Academic Publishers,
P.O. Box 17, 3300 AA Dordrecht, The Netherlands.

Sold and distributed in the U.S.A. and Canada
by Kluwer Academic Publishers,
101 Philip Drive, Norwell, MA 02061, U.S.A.

In all other countries, sold and distributed
by Kluwer Academic Publishers,
P.O. Box 322, 3300 AH Dordrecht, The Netherlands.

Printed on acid-free paper

Printed in the Netherlands

TABLE OF CONTENTS

PREFACE

This book of the proceedings of the 1997 NATO Advanced Study Institute (ASI) on *Direct Methods for Solving Macromolecular Structures* was assembled from the lecturers' contributions and represents a comprehensive and in-depth overview of crystallographic structure determination methods for macromolecules. While having a focus based on the direct methods, the Institute adopted an inclusive and broad perspective. Thus, both direct and experimental phasing techniques are presented in this book, highlighting their complementarities and synergies. As well, methodologies spanning the full crystallographic image reconstruction process - from low resolution envelope definition to high resolution atomic refinement- are discussed.

The first part of the book introduces the array of tools currently used in structure determination, whether originating from a mathematical, computational or experimental framework. This section of the book displays the variety and ingenuity of old and new phasing approaches developed to solve increasingly complex structures. Some of the contributions focus on recent developments and/or implementations that have given older approaches a new life. A case in point is the re-implementation of Buerger's superposition approach, which is now solving protein structures. Another beautiful example is found in the introduction to the traditional multiple isomorphous replacement approach where new techniques, such as site-directed mutagenesis and the use of inert gases in the preparation of heavy atom derivatives, are described. Equally impressive are the presentations of newer approaches, which take advantage of advances on the experimental front (e.g., MAD and solvent contrast variation) the mathematical front (e.g., minimal function and maximum entropy) and the computational front (e.g., simulated annealing and density modification).

The second part of the book focuses on applications to macromolecular structures. At the Institute, participants witnessed, through multiple examples presented in the lectures and hands-on workshops, the incredible progress achieved in recent years. Perhaps most enlightening is the realisation that progress is being made on all fronts. MAD, for example, is solving larger and larger structures. The use of phase information from electron microscopy images within the ME approach can attack such complex structures as membrane proteins. Shake-and-bake and the SHELX "half-baked" approaches continue to dazzle us with their successes, the culmination of which was the structure determination of a 1,001 atom protein structure on site at the Institute by SHELX! We see also that the range of problems being considered is getting richer and more daring. Exceedingly large structures, such as ribosomal particles and virus structures, are being investigated using tailored-made methods.

The book concludes with a series of presentations on latest developments, which provides a view of the newest phasing methodologies and offers the promise of yet many more achievements to come. Two main ideas emerged from the Institute, which I hope will also have a central presence in this book. First, the time has passed for looking at different phasing approaches as "competing" approaches. Much can be gained by combining the various techniques, exploiting their complementarities and synergies . Second, direct methods have often set their sights high, sometimes against the prevailing opinions of the day. This has provided us with the inspiration and motivation needed to continually re-energize and re-vitalize our field. It has also resulted in unexpected achievements, such as the *ab initio* structure determination of lysozyme achieved in Erice during the Institute.

It has been a great pleasure for me to act as Director of the Institute and I wish to express my most sincere thanks to Professors Tom Blundell, Paola Spadon and Lodovico Riva di Sanseverino for providing me such a wonderful opportunity. I also want to acknowledge the assistance of Professors Chris Gilmore and Henk Schenk who, as members of the Organizing Committee, contributed so generously not only to the design of the scientific programme but also to its execution. I am most grateful to all of the lecturers who contributed so much to the Institute, sharing their expertise with enthusiasm and generosity, and all of the participants who, through their contributions, made it such an intellectually alive event. Special thanks are given as well to John Irwin, who organized and supervised all of the computing facilities, making possible for the participants to have hands-on experience with the latest tools in the field. Sandra Crocker devoted much of her time and talents to the Institute both on-site in Erice and at home in Kingston. Her assistance in the preparation of the lecture notes and, subsequently, in the editing of the Proceedings deserve special thanks. Finally, on behalf of all of the participants, I wish to acknowledge the superb and flawless local organization, led by Paola and Lodovico. They created an extraordinary environment for the exchange of ideas, the formation of new collaborations and new friendships and Erice 1997 will stay in our minds and lives for many years to come.

The Institute was sponsored and generously financed by several national and international bodies and corporations. First, I would like to acknowledge the generous contribution of NATO, Scientific Affairs Division, Brussels, Belgium, which provided the most significant single financial contribution to the Institute. Assistance is gratefully acknowledged also from the following organizations: Azienda Provinciale Turismo, Trapani, Italy; European Commission, Direction General XII; International Union of Crystallography, Chester, UK; Office for Central and Eastern Europe, Brussels, Belgium and the National Science Foundation, Washington, USA; and the following corporations: Merck and Company, Rahway, NJ, USA and Silicon Graphics Europe, Basel, Switzerland.

Suzanne Fortier
Director, Advanced Study Institute on
Direct Methods for Solving Macromolecular Structures.

HALF TITLE PAGE
1. INTRODUCTION

THE PHASE PROBLEM OF X-RAY CRYSTALLOGRAPHY: OVERVIEW

HERBERT A. HAUPTMAN
Hauptman-Woodward Medical Research Institute, Inc., 73 High St., Buffalo, NY 14203 USA

Abstract

The intensities of a sufficient number of X-ray diffraction maxima determine the structure of a crystal, that is, the positions of the atoms in the unit cell of the crystal. The available intensities usually exceed the number of parameters needed to describe the structure. From these intensities a set of numbers $|E_{\mathbf{H}}|$ can be derived, one corresponding to each intensity. However, the elucidation of the crystal structure also requires a knowledge of the complex numbers $E_{\mathbf{H}} = |E_{\mathbf{H}}| exp(i\phi_{\mathbf{H}})$, the normalized structure factors, of which only the magnitudes $|E_{\mathbf{H}}|$ can be determined from experiment. Thus, a "phase" $\phi_{\mathbf{H}}$, unobtainable from the diffraction experiment, must be assigned to each $|E_{\mathbf{H}}|$, and the problem of determining the phases when only the magnitudes $|E_{\mathbf{H}}|$ are known is called the "phase problem". Owing to the known atomicity of crystal structures and the redundancy of observed magnitudes $|E_{\mathbf{H}}|$, the phase problem is solvable in principle.

Probabilistic methods have traditionally played a key role in the solution of this problem. They have led, in particular, to the so-called tangent formula which, in turn, has played the central role in the development of methods for the solution of the phase problem. A number of computer programs, stressing different aspects of the central theme, have proven to be particularly effective.

Finally, the phase problem may be formulated as one in constrained global optimization. A method for avoiding the countless local minima in order to arrive at the constrained global minimum leads to the *Shake-and-Bake* algorithm, a completely automatic solution of the phase problem for structures containing as many as 600 atoms when data are available to atomic resolution.

1. THE PHASE PROBLEM

The electron density function $\rho(\mathbf{r})$ in a crystal is a three-dimensional periodic function of the position vector $\mathbf{r}$ and may therefore be represented by the three-dimensional Fourier series:

$$\rho(\mathbf{r}) = \frac{1}{V}\sum_{\mathbf{H}} F_{\mathbf{H}} \exp(-2\pi i\mathbf{H}\cdot\mathbf{r}) \tag{1}$$

where V is the volume of the fundamental parallelepiped, the so-called unit cell of the crystal, and the three components of the vector $\mathbf{H}$ range over all the integers. The Fourier coefficient $F_{\mathbf{H}}$, said to be the structure factor corresponding to the reciprocal lattice vector $\mathbf{H}$, may then be calculated in the usual way:

$$F_{\mathbf{H}} = \int_V \rho(\mathbf{r}) \exp(2\pi i\mathbf{H}\cdot\mathbf{r}) dV \tag{2}$$

S. Fortier (ed.), Direct Methods for Solving Macromolecular Structures, 3–10.

where the integration is carried out over the unit cell V. Clearly, each $F_{\mathbf{H}}$ is a complex number that may be written in polar form

$$F_{\mathbf{H}} = |F_{\mathbf{H}}|\exp(i\phi_{\mathbf{H}}) \tag{3}$$

where $\phi_{\mathbf{H}}$ is the phase of the structure factor $F_{\mathbf{H}}$.

When a beam of monochromatic X-rays is incident on a crystal, the radiation is scattered in discrete directions determined by the crystal lattice and labeled by the reciprocal lattice vectors **H**. Both the amplitude and phase of each scattered ray, or reflection, depend on the crystal structure. The amplitude, or intensity, of a reflection leads in a straightforward way to the magnitude $|F_{\mathbf{H}}|$ of the complex structure factor $F_{\mathbf{H}}$. However, the phases $\phi_{\mathbf{H}}$, which are also needed if one is to determine $\rho(r)$ from Eq. 1, are lost in the diffraction experiment. If one uses the known values of the magnitudes $|F_{\mathbf{H}}|$ but arbitrary values for the phases $\phi_{\mathbf{H}}$ in Eq. 1, then density functions $\rho(\mathbf{r})$ consistent with the observed values of the diffraction intensities are obtained. Thus, diffraction intensities alone do not determine a unique density function $\rho(\mathbf{r})$. Even if the known non-negativity of $\rho(\mathbf{r})$ is assumed, thus greatly restricting the values of the phases [1,2] the observed diffraction intensities are, in general, still not sufficient to determine $\rho(\mathbf{r})$ uniquely. It follows that the phase problem, to determine the values of the phases $\phi_{\mathbf{H}}$ of the structure factors $F_{\mathbf{H}}$ when only the magnitudes $|F_{\mathbf{H}}|$ are given, is, in principle, unsolvable when formulated in these terms. It was this argument that led the crystallographic community, prior to 1950, to believe also that crystal structures could not, even in principle, be determined from the diffraction intensities alone. However, by invoking the prior structural knowledge that crystals consist of discrete atoms, one readily refutes this argument, as shown below.

If one replaces the real crystal, with continuous electron density $\rho(\mathbf{r})$, by an idealized one, the unit cell of which consists of N discrete, non-vibrating point atoms, then the structure factor $F_{\mathbf{H}}$ is replaced by the normalized structure factor $E_{\mathbf{H}}$ and Eqs. 3, 2, and 1 are replaced by

$$E_{\mathbf{H}} = |E_{\mathbf{H}}|\exp(i\phi_{\mathbf{H}}) \tag{4}$$

$$E_{\mathbf{H}} = \frac{1}{\sigma_2^{1/2}} \sum_{j=1}^{N} Z_j \exp(2\pi i \mathbf{H} \cdot \mathbf{r}_j) \tag{5}$$

$$\left\langle E_{\mathbf{H}} \exp(-2\pi i \mathbf{H} \cdot \mathbf{r}) \right\rangle_{\mathbf{H}} = \frac{1}{\sigma_2^{1/2}} \left\langle \sum_{j=1}^{N} Z_j \exp\left[2\pi i \mathbf{H} \cdot (\mathbf{r}_j - \mathbf{r})\right] \right\rangle_{\mathbf{H}}$$

$$\left.\begin{aligned} &= \frac{Z_j}{\sigma_2^{1/2}} \text{ if } \mathbf{r} = \mathbf{r}_j \\ &= 0 \text{ if } \mathbf{r} \neq \mathbf{r}_j \end{aligned}\right\} \tag{6}$$

respectively, where Z_j is the atomic number and $\mathbf{r}_j$ is the position vector of the atom labeled j, and

$$\sigma_n = \sum_{j=1}^{N} Z_j^n, \; n = 1,2,3,\ldots \tag{7}$$

In practice, the magnitudes $|E_{\mathbf{H}}|$ of the normalized structure factors $E_{\mathbf{H}}$ are obtainable (at least approximately) from the observed magnitudes $|F_{\mathbf{H}}|$, while the phases $\phi_{\mathbf{H}}$, as defined by Eqs. 4 and 5, cannot be determined experimentally. Since one now requires only the $3N$ components of the N position vectors $\mathbf{r}_j$ rather than the much more complicated electron density function $\rho(\mathbf{r})$, it turns out that, in general, the known magnitudes are more than sufficient. This is most readily seen if we equate the magnitudes of both sides of Eq. 5, thus eliminating the unknown phases $\phi_{\mathbf{H}}$, in order to obtain

$$|E_{\mathbf{H}}| = \frac{1}{\sigma_2^{1/2}} \left| \sum_{j=1}^{N} Z_j \exp(2\pi i \mathbf{H} \cdot \mathbf{r}_j) \right| \tag{8}$$

a system of equations in which the only unknowns are the $3N$ components of the position vectors $\mathbf{r}_j$. Since the number of Eq. 8, equal to the number of reciprocal lattice vectors $\mathbf{H}$ for which the magnitudes $|E_{\mathbf{H}}|$ are observed, usually exceeds the number of unknowns, $3N$, by far, the system in Eq. 8 is redundant. Thus the phase problem is, in principle, solvable when reformulated in terms of fixed point atoms, as reference to Eq. 5 shows.

The system of Eq. 5 implies the existence of relationships among the normalized structure factors $E_{\mathbf{H}}$ since the (relatively few) unknown position vectors $\mathbf{r}_j$ may, at least in principle, be eliminated. In this way we obtain the system of equations among the complex normalized structure factors $E_{\mathbf{H}}$:

$$F(E_{\mathbf{H}}) = F(|E_{\mathbf{H}}|, \phi_{\mathbf{H}}) = 0. \tag{9}$$

Since the magnitudes $|E_{\mathbf{H}}|$ are obtainable from the diffraction experiment, the system (9) leads to the set of identities, dependent on the known magnitudes $|E_{\mathbf{H}}|$, which the phases must of necessity satisfy

$$\mathrm{G}(\phi_{\mathbf{H}} \| E_{\mathbf{H}}|) = 0. \tag{10}$$

By the term "direct methods" is meant that class of methods that exploits relationships among the normalized structure factors in order to go directly from the observed magnitudes $|E|$ to the needed phases ϕ.

An explicit system of identities among the structure factors, valid in the case that the structure consists of N identical atoms in the unit cell and that diffraction data are available at atomic resolution, was first found by Sayre [3]:

$$\mathrm{F_H} = \frac{\mathrm{f_H}}{\mathrm{g_H V}} \sum_{\mathbf{K}} \mathrm{F_K F_{H-K}} \tag{11}$$

where V is the volume of the unit cell and $\mathrm{f_H}$ and $\mathrm{g_H}$ are the scattering factors for the true equal atoms of the structure and for the "squared" structure. When expressed in terms of the normalized structure factors E the Sayre equations retain approximate validity:

$$\mathrm{E_H} = \frac{\mathrm{C}}{\mathrm{g_h}} \sum_{\mathbf{K}} \mathrm{E_K E_{H-K}} \tag{12}$$

where C is an overall scale factor.

2. THE STRUCTURE INVARIANTS

Equation 6 implies that the normalized structure factors $E_{\mathbf{H}}$ determine the crystal structure. However, Eq. 5 does not imply that, conversely, the crystal structure determines the values of the normalized structure factors $E_{\mathbf{H}}$ since the position vectors $\mathbf{r}_j$ depend not only on the structure but on the choice of origin as well. It turns out, nevertheless, that the magnitudes $|E_{\mathbf{H}}|$ of the normalized structure factors are in fact uniquely determined by the crystal structure and are independent of the choice of origin, but that the values of the phases $\phi_{\mathbf{H}}$ depend also on the choice of origin. Although the values of the individual phases depend on the structure and the choice of origin, there exist certain linear combinations of the phases, the so-called structure invariants, whose values are determined by the structure alone and are independent of the choice of origin. The most important class of structure invariants, and the only one to be considered here, consists of the three-phase structure invariants (triplets),

$$\phi_{\mathbf{HK}} = \phi_{\mathbf{H}} + \phi_{\mathbf{K}} + \phi_{-\mathbf{H}-\mathbf{K}}, \tag{13}$$

where H and K are arbitrary reciprocal lattice vectors.

3. THE PROBABILISTIC BACKGROUND

The techniques of modern probability theory lead to the joint probability distributions of arbitrary collections of diffraction intensities and their corresponding phases. These distributions constitute the foundation on which direct methods are based. They have provided the unifying thread from the beginning, *ca* 1950, until the present time. They have led, in particular to the tangent formula Eq. (19), the basis of almost all computer programs for the direct determination of phase, and to the (first) minimal principle.

While the tangent based computer programs have steadily become more powerful over the years, they have rarely solved structures having more than some 200 independent non-hydrogen atoms *ab initio*. However, when combined with other techniques, e.g. solvent flattening and histogram matching, they appear likely to be useful for phase refinement and extrapolation even in the macromolecular range.

The minimal principle [5], on the other hand, which has found expression in the *Shake-and-Bake* formalism [6,7], does provide a completely automatic solution to the phase problem, *ab initio*, even for structures having as many as 600 non-H atoms, provided that diffraction data to atomic resolution are available. Its ultimate potential is still unknown.

It is assumed that (i) A crystal structure is specified, (ii) Three non-negative numbers R_1, R_2, R_3 are also specified; and (iii) The reciprocal lattice vectors H and K are the primitive random variables which are assumed to be uniformly and independently distributed in the subset of reciprocal space defined by

$$|E_{\mathbf{H}}| = R_1,\ |E_{\mathbf{K}}| = R_2,\ |E_{\mathbf{H+K}}| = R_3 \tag{14}$$

where the magnitudes |E| are defined by (5). Then the structure invariant ϕ_{HK} (Eq. (13)), as a function of the primitive random variables **H** and **K** (Eq. (5)), is itself a random variable.

4. THE CONDITIONAL PROBABILITY DISTRIBUTION OF $\phi_{\mathbf{HK}}$, GIVEN $|E_{\mathbf{H}}|$, $|E_{\mathbf{K}}|$, $|E_{\mathbf{H+K}}|$

For simplicity it will be assumed henceforth that all N atoms in the unit cell are identical. Under the three assumptions of §3. the conditional probability distribution of the triplet $\phi_{\mathbf{HK}}$ (Eq. (13)), where $|E_{\mathbf{H}}|$, $|E_{\mathbf{K}}|$, and $|E_{\mathbf{H+K}}|$ are given by (14), is known to be

$$P(\Phi / R_1, R_2, R_3) = \frac{1}{2\pi I_0(A_{\mathbf{HK}})} \exp(A_{\mathbf{HK}} \cos\Phi) \tag{15}$$

where Φ represents the triplet $\phi_{\mathbf{HK}}$,

$$A_{\mathbf{HK}} = \frac{2}{N^{1/2}} R_1 R_2 R_3 = \frac{2}{N^{1/2}} |E_{\mathbf{H}} E_{\mathbf{K}} E_{\mathbf{H+K}}| \tag{16}$$

and I_0 is the Modified Bessel Function. Equation (15) implies that the mode of $\phi_{\mathbf{HK}}$ is zero, and the conditional expectation value (or average) of cos $\phi_{\mathbf{HK}}$, given $A_{\mathbf{HK}}$, is

$$\mathcal{E}(\cos\phi_{\mathbf{HK}} / A_{\mathbf{HK}}) = \frac{I_1(A_{\mathbf{HK}})}{I_0(A_{\mathbf{HK}})} > 0, \tag{17}$$

where I_1 is the Modified Bessel Function. It is also readily confirmed that the larger the value of $A_{\mathbf{HK}}$ the smaller is the conditional variance of $\cos\phi_{\mathbf{HK}}$, given $A_{\mathbf{HK}}$. It is to be stressed that the conditional expected value of the cosine, Eq. (15), is always positive since $A_{\mathbf{HK}}>0$.

5. THE TANGENT FORMULA

It is assumed that a crystal structure consisting of N identical atoms in the unit cell is fixed, but unknown, that the magnitudes |E| of the normalized structure factors E are known, and that a sufficiently large base of phases, corresponding to the largest magnitudes |E|, is specified.

The mode of the triplet distribution (Eq. (15)) is zero and the variance of the cosine is small if $A_{\mathbf{HK}}$ (Eq. (16)) is large. In this way one obtains the estimate for the triplet $\phi_{\mathbf{HK}}$ (Eq. (13)):

$$\phi_{\mathbf{HK}} = \phi_{\mathbf{H}} + \phi_{\mathbf{K}} + \phi_{-\mathbf{H}-\mathbf{K}} \approx 0 \tag{18}$$

which is particularly good in the favorable case that $A_{\mathbf{HK}}$, (Eq. (16)), is large, i.e. that $|E_{\mathbf{H}}|$, $|E_{\mathbf{K}}|$, and $|E_{\mathbf{H+K}}|$ are all large. The estimate given by Eq. (18) is one of the cornerstones of the traditional techniques of direct methods. It is surprising how useful Eq. (18) has proven to be in the applications especially since it yields only the zero estimate of the triplet, and only those estimates are reliable for which $|E_{\mathbf{H}}|$, $|E_{\mathbf{K}}|$, and $|E_{\mathbf{H+K}}|$ are all large. Clearly the coefficient $2/N^{1/2}$ in Eq. (16), and therefore $A_{\mathbf{HK}}$ as well, both decrease with increasing N, i.e. with increasing structural complexity. Hence the relationship (Eq. (18)) becomes increasingly unreliable for larger structures, and the traditional step-by-step sequential direct methods procedures based on Eq. (18) eventually fail. (However, see §6 for recent advances.)

An important supplement to the relationship (18), and an immediate consequence of the distribution (15), is the so-called tangent formula [4];

$$\tan\phi_H = \frac{\sum_K |E_K E_{H-K}| \sin(\phi_K + \phi_{H-K})}{\sum_K |E_K E_{H-K}| \cos(\phi_K + \phi_{H-K})} \tag{19}$$

in which the summations are taken over the same (arbitrary) set of reciprocal lattice vectors K. The tangent formula, in one form or another, plays a major role in the traditional techniques of direct methods.

6. SOME COMPUTER PROGRAMS

Probably the most widely used tangent based computer programs are the various versions of MULTAN, SHELXS, SIR, and MITHRIL. Only two of the most popular of these will be briefly described here.

The first is the RANTAN (Random MULTAN) program devised by Yao Jia-xing [8]. In this procedure initial values of all the desired phases are assigned at random, (rather than only a few as had been done in earlier versions of MULTAN) and all relationships (Eq. (18)) which link them are used from the beginning. A suitably weighted version of the tangent formula (Eq. (19)) is then employed to refine the initial values of the individual phases. An important aspect of the program is the selection of the weights which change during the course of the refinement process. Because of the large number of phases involved at the very beginning, the method is relatively insensitive to the failure of one or several relationships (Eq. (18)) which may occur early on in the process, and this aspect of RANTAN undoubtedly contributes to its success in the applications.

Another recent advance is the development of the Sayre Equation Tangent Formula as expressed in the computer program SAYTAN [9]. This variant of the tangent formula tends to develop phases which satisfy the Sayre Equation (Eq. (12)) for both large and small magnitudes |E| and not only exploits the triplet relationships (Eq. (18)) but involves the analogous quartet relationships as well. It has proven to be a particularly effective direct methods technique especially for larger structures [10].

7. THE MINIMAL PRINCIPLE

Finally, referring to Eqs. (16) and (17) one may formulate the phase problem as one in constrained global minimization [5,6,7]. First, the minimal function $m(\phi)$ is defined by

$$m(\phi) = \frac{1}{\sum_{\mathbf{H,K}} A_{\mathbf{HK}}} \sum_{\mathbf{H,K}} A_{\mathbf{HK}} \left\{ \cos\phi_{\mathbf{HK}} - \frac{I_1(A_{\mathbf{HK}})}{I_0(A_{\mathbf{HK}})} \right\}^2 \tag{20}$$

where $\phi_{\mathbf{HK}}$ and $A_{\mathbf{HK}}$ are defined by Eqs. (18) and (16), respectively. Then, in view of Eq. (17), one infers that the constrained global minimum of $m(\phi)$, where the phases are constrained to satisfy the identities which they must satisfy, Eq. (10), yields the correct values of all the phases appearing in $m(\phi)$ (the minimal principle). The computer program *Shake-and-Bake*, described in the paper in these Proceedings by Miller and Weeks [11], shows how the constrained global minimum of $m(\phi)$ is reached and identified. *Shake-and-Bake* has solved, *ab initio*, six small protein structures ranging in size from about 200 to 600 non-H atoms. Its major limitation is the requirement that diffraction data to a resolution of 1.1Å be available. Its ultimate potential is still unknown.

8. PROGRESS SINCE 1950

It seems fair to state that direct methods had their beginning in the Harker-Kasper inequalities [1]. Progress since then has been spectacular. In the decade of the fifties some eight to ten structures, mostly containing perhaps some 30 non-H atoms, were solved by direct methods. Since then the number of structures solved by direct methods has increased dramatically and, of equal importance, the complexity of solved structures has approximately doubled in each decade (Fig. 1). Thus in the present decade structures having as many as 500 non-H atoms have been routinely solved. In view of Figure 1 it seems safe to predict that in the decade beginning with the year 2000 structures having as many as 1000 non-H atoms will be routinely solvable by direct methods.

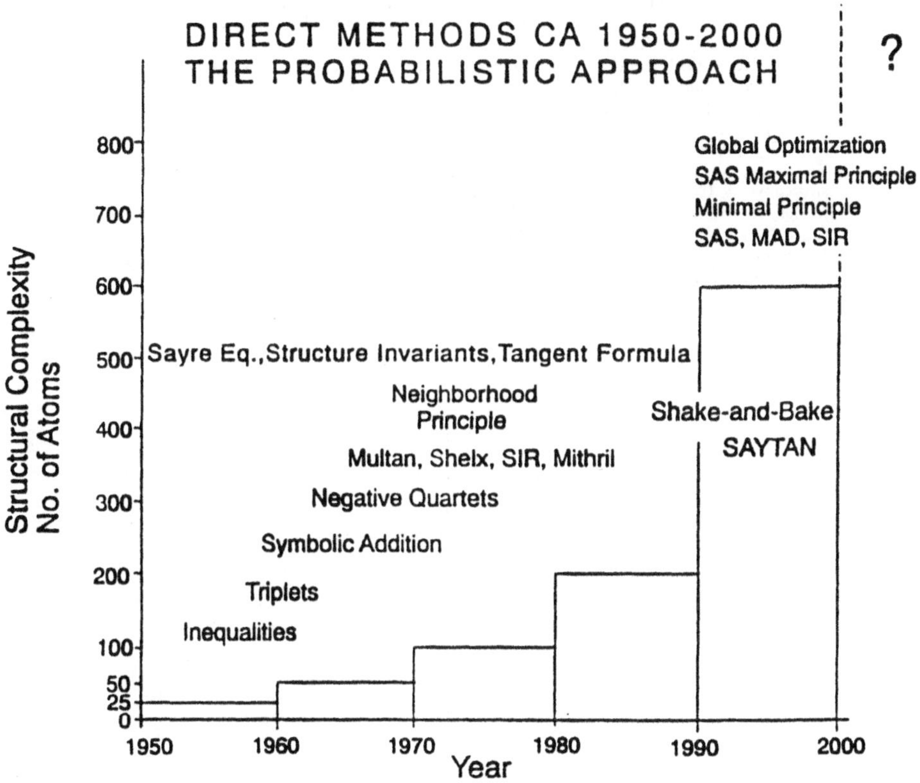

Figure 1. Complexity of structures solved by Direct Methods has doubled in each decade since 1950.

9. ACKNOWLEDGMENTS

The *Shake-and-Bake* algorithm and the *SnB* program have been made possible by the financial support of grants GM-46733 from NIH and IRI-9412415 from NSF.

10. REFERENCES

[1] Harker, D. and Kasper, J.S. (1948) Phases of Fourier coefficients directly from crystal diffraction data, *Acta Cryst.* **1**, 70-75.

[2] Karle, J. and Hauptman, H. (1950) the phases and magnitudes of the structure factors, *Acta Cryst.* **3**, 181-187.

[3] Sayre, D. (1952) The squaring method: a new method for phase determination, *Acta Cryst.* **5**, 60-65.

[4] Karle, J. and Hauptman, H. (1956) A theory of phase determination for the four types of non-centrosymmetric space groups 1P222, 2P22, $3P_12$, $3P_22$, *Acta Cryst.* **9**, 635-651.

[5] Hauptman, H.A. (1991) A Minimal Principle in the Phase Problem, in D. Moras, A.D. Podnarny & J.C. Thierry (eds.), *Crystallographic Computing 5: From Chemistry to Biology*, IUCr Oxford Univ. Press, pp. 324-332.

[6] DeTitta, G.T., Weeks, C.M., Thuman, P., Miller, R. and Hauptman, H.A. (1994) Structure solution by minimal function phase refinement and Fourier filtering: Theoretical basis, *Acta Cryst.* **A50**, 203-210.

[7] Weeks, C.M., DeTitta, G.T., Hauptman, H.A., Thuman, P., and Miller, R., (1994) Structure solution by minimal function phase refinement and Fourier filtering: II. Implementation and applications, *Acta Cryst.* **A50**, 210-220.

[8] Yao, J.-X. (1981) On the application of phase relationships to complex structures. XVIII. RANTAN - random MULTAN, *Acta Cryst.* **A37**, 642-664.

[9] Debaerdemaeker, T., Tate, C. and Woolfson, M.M. (1988) On the application of phase relationships to complex structures. XXVI. Developments of the Sayre-Equation tangent formula, *Acta Cryst.* **A44**, 353-357.

[10] Woolfson, M.M. and Yao Jia-xing (1980) On the application of phase relationships to complex structures. XXX. *Ab initio* solution of a small protein by SAYTAN, *Acta Cryst.* **A46**, 409-413.

[11] Miller, R. and Weeks, C.M. (1997) *Shake-and-Bake*: Applications and advances, in S. Fortier (ed.), *Direct Methods for Solving Macromolecular Structures*, Kluwer Academic Publishers, Dordrecht, The Netherlands, 25th Course, Intl. School of Crystallography, NATO ASI Series, Erice, Sicily, May (1997).

A SIMPLE INTRODUCTION TO DIRECT METHODS

PAUL T. BEURSKENS
Crystallography Laboratory, University of Nijmegen,
Toernooiveld 1, 6525 ED Nijmegen, The Netherlands

Direct methods for solving the phase problem in X-ray crystallography *were* not supposed to work for macromolecules, or on reflection intensity data which do not reach atomic resolution. As a consequence, young protein crystallographers may not have been exposed to direct methods as much as colleagues in the field of "small molecules".

Direct methods *are* applicable to macromolecules, as will be shown in this summerschool (Erice 1997) on "Direct methods for solving macromolecules". But -maybe?- some modern methods may be more complicated and -maybe?- some lecturers expect that the students of the present school have a reasonable knowledge about classical direct methods. This lecture is planned to fill the gap between knowing nothing about direct methods and a reasonable level from where to start lectures on modern developments in this specialized field of direct methods applications. After the lecture there will be ample time for taylor-made tutorial sessions.

Literature references

Proper literature references will be given with each topic in the present summerschool. In the present introduction we will refrain from doing so.

In the early days of direct methods (1950 - 1960 ?) there were no textbooks on direct methods. The first one was a most important book by M.M.Woolfson: "Direct methods in Crystallography" (1961).

Although out of date, it still gives good insight in the basic principles and in many practical aspects of direct methods for structure solution. A good advice for students is to read the original structural publications of I. Karle *et al.* in *Acta Cryst.* (1960 - 1970): these papers include summaries of methods and show how they work in practice. Today, the book edited by C. Giaccovazzo: "Fundamentals of Crystallography" (1992), is a good book for studying the basic theory and many practical aspects of direct methods. In the present simple introduction I will discuss many important items without giving any reference: the interested reader can look up the proper references in Giaccovazzo (*lit.cit.*).

What are direct methods?

The term "direct methods" is used for methods, procedures, strategies, equations, algorithms, *et cetera,* which lead to the determination of phases or the refinement of (a set of) phases, and which generally make use of mathematical tools (algebraic methods, statistics or probability methods) which are (nowadays) computorized, and which do NOT depend on any knowledge of the molecular structure at hand. Neither the direct determination of the scale and overall temperature factor by Wilson's statistics nor finding the position of a heavy atom by an automated Patterson interpretation procedure are considered to be "direct methods".

S. Fortier (ed.), Direct Methods for Solving Macromolecular Structures, 11–18.

A large number of different "direct methods" have been developed and as many computer programs have been designed (just to name a few terms: symbolic addition, magic integers, random phases, sigma-2 relation, triplet formula, quartet, quadruplet, tangent formula, convergence mapping, neighbourhood, seminvariants, MULTAN, YZARC, DIRDIF, SIMPLE, SIR ....). Only a few of these terms and methods will be discussed here.

Notation

$\mathbf{r}_j$ position vector of atom j in the unit cell
Σ_j summation over all (N) atoms in the entire unit cell
F short-hand notation for $F_\mathbf{h} = F(\mathbf{h})$, $\mathbf{h}$ = reflection h,k,l
$|F_o|$ observed structure factor on an absolute scale
$\phi_\mathbf{h} = \phi(\mathbf{h})$ phase of F: $F = |F| \exp(i\phi_\mathbf{h})$
B_j temperature factor parameter for atom j
B_{ov} overall temperature factor parameter
s $\sin\theta / \lambda$
$f_j^o(s)$ scattering factor of atom j at rest
$f_j(s)$ scattering factor of atom j: $f_j(s) = f_j^o(s) \exp(-B_j s)$
$|E|$ observed normalized structure factor
$\Sigma_\mathbf{h}$ summation over all or a selected set of reflections $\mathbf{h}$

The structure factor formula is: $F_\mathbf{h} = \Sigma_j\ f_j(s)\ \exp(2\pi i\ \mathbf{h}.\mathbf{r}_j)$

The Wilson plot

The intensity of a reflection, $I_\mathbf{h}$, is the square of the structure amplitude, here assumed to be on absolute scale: $I_\mathbf{h} = |F_\mathbf{h}|^2$. The expectation value of the intensity of a reflection, or the expected average of intensities for an arbitrary set of reflections is given by (with F^* is complex conjugate F)

$$<I> = <F F^*> = <\Sigma_j f_j^2(s) \quad + \Sigma_j\Sigma_{j'}\ jj'\text{-cross-terms} >$$

For so-called "random" structures the cross-terms cancel. The first use of this result is the Wilson plot for the determination of the scale and overall temperature factor, and will not be discussed here! The second use leads to the definition of E-values.

Normalization

For a given scattering angle the expectation value for the intensity of a reflection is:

$$<I(s)> \ = \ \Sigma_j f_j^2(s) \ \approx \ \Sigma_j [f_j^o(s)]^2 . \exp(-2B_{ov}\, s)$$

where f_j includes the individual B_j value. The averaging <I(s)> is done over all reflections within a short range around s. A reflection of which the actual intensity I is larger than < I(s) > is said to be a strong reflection. If I/<I(s)> is greater than 2, it is a very strong reflection. At rather large scattering angles all reflections are weak, and if I/<I(s)> is greater than 2, the reflection only is relatively very strong. Strong reflections have high information contents: high order reflections which are stronger than expected for the given scattering angle appear to be very important in direct methods! Therefore reflection data are normalized and E-values are introduced by:

$$|E_{\mathbf{h}}|^2 = I_{\mathbf{h}} / [\,\epsilon < I(s) >\,] \qquad (1)$$

or: $$E = F / [\,\sqrt{\epsilon\, \Sigma_j f_j^2}\,] \qquad (\epsilon = \text{symmetry enhancement factor}) \qquad (2)$$

or: $$E = [1/\sqrt{N}]\ \Sigma_j \exp(2\pi i\, \mathbf{h}.\mathbf{r}_j) \qquad \text{for an equal atom structure.} \qquad (3)$$

So reflections having $|E| > 1$ are the strong reflections. In principle, the introduction of E-values means that all direct method equations look better. In practice, such and similar definitions have led to various discussions about the proper way of normalization. One does not know whether or not (1) is the best way of doing if a given structure is not a collection of randomly distributed atoms. And (2) assumes a straight Wilson-plot which particularly need not be the case in macromolecular crystallography. And, of course, (3) is used only for theoretical purposes.

The triplet phase sum or sigma-2 relation

It was known already (at least) in the 1930's that phases of strong reflections are correlated, which could be of help in trying to solve the phase problem by 'trial and error' methods. A reflection triplet is defined as a set of three reflections denoted **h**, **k** and **l** with the condition that **h**+**k**+**l**=0 .
Or written in full: for **h** = (h,k,l) and **k** = (h',k',l'), **l** = (-h-h',-k-k',-l-l') . If all three reflections are very strong, then the sum of their phases is probably approximately zero:

$$\phi(\mathbf{h}) + \phi(\mathbf{k}) + \phi(\mathbf{l}) \approx 0 \qquad \text{or:} \qquad \phi(\mathbf{h}) \approx \phi(\mathbf{k}) + \phi(\mathbf{h-k})$$

Note that in the righthand form use has been made of Friedel's law: $\phi(\mathbf{h}) = -\phi(-\mathbf{h})$.
For centrosymmetric structures the product of the signs of the reflections probably is 'plus':

$$S(\mathbf{h})\, S(\mathbf{k})\, S(\mathbf{l}) \approx + \qquad \text{or:} \qquad S(\mathbf{h}) \approx S(\mathbf{k})\, S(\mathbf{h-k})$$

S(x) means 'sign of x' which can be expressed as both [+ or -] and [+1 or -1].
In the 1950's this triplet phase or sign relation became known as the sigma-2 relation, written as:

$$\phi(\mathbf{h+k}) \approx \phi(\mathbf{h}) + \phi(\mathbf{k}) \qquad \text{and} \qquad S(\mathbf{h+k}) \approx S(\mathbf{h})\, S(\mathbf{k}),$$

respectively, and it was shown that the reliability increases for larger magnitudes of

$$|\,(E(\mathbf{h})\, E(\mathbf{k})\, E(\mathbf{h+k})\,|\,.$$

[Note that the meaning of **h**, **k** and **l** depends on their use in the equations !]

If several phases are known, then the probable phase $\phi(\mathbf{h})$ can sometimes be estimated by two or more pairs of known phases $\phi(\mathbf{k})$ and $\phi(\mathbf{h-k})$, and the resulting $\phi(\mathbf{h})$ values have to be averaged using the participating E-values as weighting factors: this is the *tangent formula*:

$$\phi(\mathbf{h}) \approx \text{phase of } \Sigma_{\mathbf{k}} \; |E(\mathbf{k})\, E(\mathbf{h-k})| \exp(2\pi i\, [\phi(\mathbf{k}) + \phi(\mathbf{h-k})])$$

The reliability of this result increases with $| E(\mathbf{h})\, \Sigma_{\mathbf{k}} E(\mathbf{k})\, E(\mathbf{h-k}) |$.
Note that indeed contradicting terms lead to less reliable phases.

When many phases are known approximately, then the tangent formula approaches the *Sayre equation* which is valid for an *equal atom structure* provided all phases are known:

$$F(\mathbf{h}) = c(s)\, \Sigma_{\mathbf{k}} F(\mathbf{k})\, F(\mathbf{h-k})$$

where c(s) is a known function. The Sayre equation is used for phase refinement and phase expansion (also for incommensurate structures) and plays an important role in theoretical derivations.

Origin and enantiomorph definition

If all atoms are shifted simultaneously over a vector **t**, the phases of $F_{\mathbf{h}}$ are modulated by

$$F_{\mathbf{h}}(\text{new}) = F_{\mathbf{h}}(\text{old}) \exp(2\pi i\, \mathbf{h.t}) .$$

The phase of the reflection 100 varies over 360° when **t** ranges from [0,0,0] to [1,0,0]. Conversely, any arbitrary choice of the phase of this reflection fixes the origin along the **a** direction. Similarly, reflections 010 and 001 can be used to specify the origin in other directions. Other reflections defining a primitive unit cell in reciprocal space may be better choices. [What would happen if we use the reflection 200 as one of the origin defining reflections?] For space groups other than P1 the symmetry leads to (a) more possibilities for selecting origin choices and (b) restrictions to the 'type' of reflections and to allowed numerical values.

The triplet, *i.e.* the phase sum $\phi(\mathbf{h})+\phi(\mathbf{k})+\phi(\mathbf{-h-k})$, or the product $E(\mathbf{h})E(\mathbf{k})E(\mathbf{-h-k})$ is independent of the choice of the origin in the structure. The sum $\phi(\mathbf{h})+\phi(\mathbf{k})+\phi(\mathbf{-h-k})$ can have any value (it tends to be close to zero for strong reflections) but its value will not change if all the atoms in the unit cell are shifted in the same way. This phase sum is called a *structure invariant.* (The term structure invariant is misleading: of course the value of the phase sum depends on the contents of the unit cell, but for a given structure the phase sum is invariant under a shift of origin.) In general:

$$\Sigma_{\mathbf{h}}\, \phi(\mathbf{h}) \quad \text{is a structure invariant if} \quad \Sigma_{\mathbf{h}}\, \mathbf{h} = \mathbf{0} .$$

A structure invariant is a physical quantity and can in principle be determined from experimental data. A phase of an individual reflection is not a physical quantity, it can never be determined from experiment, and determining one single phase always involves some arbitrary choice.

Example: structure CsCl, space group F43m, two possiblilities, (1) and (2):

	Cs at:	Cl at:	F(100)	ϕ(100)	F(200)	ϕ(200)
(1):	0,0,0	1/2,1/2,1/2	+f(Cs)-f(Cl)	0	f(Cs)+f(Cl)	0
(2):	1/2,1/2,1/2	0,0,0	-f(Cs)+f(Cl)	180 degr	f(Cs)+f(Cl)	0

The origin is fixed by choosing one of the two possibilities (either by defining the position of one of the atoms, or by defining the phase of one (just one!) carefully chosen reflection. For the present space group no other atomic positions or reflection phases are possible if we wish to stick to the usual convenient settings of the symmetry elements for this space group. The structure invariant:

$\phi(100) + \phi(100) - \phi(200)$ is equal to 0 for both origin choices.

For non-centrosymmetric space groups there is an enantiomer ambiguity: there is no *a priori* way to tell whether a reflection is **h** or -**h**, or -alternatively- whether we have chosen the correct handedness of the axial system. In direct methods the anomalous scattering is ignored (except, of course, when Bijvoet pairs are measured, or multiple wavelength experiments are used). This usually leads to one more *a priori* arbitrary choices to be made.

Rules for specifying the origin and enantiomorph for all space groups are well established and correctly handled by all computer programs.

The symbolic addition method

The use of the triple-product sign relationship will be illustrated by a very simple example in space group P-1. Table I tabulates the 25 strongest reflections of a 'small' hypothetical structure. It is assumed that this small number of reflections is sufficient to solve the phase problem. The structure is not an existing structure: reflection intensities are not defined. The 'measured' reflections are supposed to be normalized and sorted in decreasing magnitudes of the |E| value, and only the top 25 are given in the table. The data of the example and the consecutive actions presented in Table I are discussed stepwise.

- Column 1 gives the sequence number (#) of the reflection in this table, for reference only.
- Column 2 gives the reflection indices. So, 0-11 is the reflection with the largest |E| value.

All triplets among the listed reflections are traced ('calculated') and presented in column 4. While doing so, all symmetry is used: for P-1 it means that when **h** is listed, also -**h** is used. Usually a convergence map is calculated: it represents an analysis of participating triples and a method to select the best set of reflections to start with, both for origin definition and for assignment of symbols. To keep things simple, we just start from the top, and we refrain from defining the origin at the start of the analysis, and do so at the end.

- The first reflection, #1 , is assigned symbol A.
  This means, the sign of this reflection is A by definition;
  we will find out later whether A=+ or A=- (*i.e.*: A=0 or A=180 in degrees).
  All symbol definitions are given in column 3.
- Reflection #2 is assigned symbol B.

Table I. Illustration of the symbolic addition method.
Underscored digits represent negative indices.

column: 1 refl. #	2 hkl	3 symbol def.	4 triplet relation	5 resulting symbol	6 symbol-relation	7 nr	8 accepted sign / symbol	9	10
1	$0\underline{1}1$	A					+	+	+
2	022	B					B	B	-
3	$0\underline{4}2$	C	$0\underline{4}2= 0\underline{2}1+0\underline{2}1$		C≈+	R1	C	+	+
4	$0\underline{2}1$	Not used (except: see #3)							
5	013		$013= 0\underline{1}1+022$	AB			B	B	-
6	$0\underline{2}2$	D	$0\underline{2}2= 0\underline{1}1+0\underline{1}1$		D≈+	R2	+	+	+
7	$0\underline{4}1$	Not used							
8	$14\underline{3}$	E					E	B	-
9	$13\underline{2}$		$13\underline{2}= 0\underline{1}1+14\underline{3}$	AE			E	B	-
10	$12\underline{3}$	F					+	+	+
11	$13\underline{4}$		$13\underline{4}=-0\underline{1}1+12\underline{3}$	AF			+	+	+
12	$1\underline{2}\underline{1}$		$1\underline{2}\underline{1}= 0\underline{4}2+12\underline{3}$	CF			C	+	+
13	$12\underline{1}$		$12\underline{1}= 0\underline{2}2+14\underline{3}$	DE			E	B	-
			$12\underline{1}= 0\underline{1}1+13\underline{2}$	A(AE)=E	D≈+	R3			
14	$10\underline{3}$		$10\underline{3}=-0\underline{2}2+1\underline{2}\underline{1}$	D(CF)			C	+	+
			$10\underline{3}=-022+12\underline{1}$	B(DE)	BCEF≈+	R4	BE		
15	150		$150= 013+14\underline{3}$	(AB)E			BE	+	+
			$150= 022+13\underline{2}$	B(AE)					
16	113	Not used							
17	$1\underline{5}1$	Not used							
18	$2\underline{3}1$	Not used							
19	$2\underline{2}1$	G					+	+	+
20	$3\underline{4}0$		$3\underline{4}0= 1\underline{2}\underline{1}+2\underline{2}1$	(CF)G			C	+	+
21	$30\underline{2}$		$30\underline{2}=-0\underline{4}2+3\underline{4}0$	C(CFG)=FG			+	+	+
			$30\underline{2}= 12\underline{3}+2\underline{2}1$	FG					
22	$3\underline{2}3$	H					H	H	-
23	$3\underline{1}2$		$3\underline{1}2=-0\underline{1}1+3\underline{2}3$	AH			H	H	-
24	$31\underline{3}$		$31\underline{3}=-0\underline{1}1+30\underline{2}$	A(FG)			+	+	+
			$31\underline{3}= 13\underline{4}+2\underline{2}1$	(AF)G					
25	301		$301=-0\underline{2}2+3\underline{2}3$	DH			H	H	-
			$301=-0\underline{1}1+3\underline{1}2$	A(AH)=H	D≈+	R5			

Note about the notation with symbols: AB is the sum of the phases A and B. Centrosymmetric AB is also a product of signs . AA=+ (0+0=180+180=0 mod 360).

- Reflection #3 is assigned symbol C. These top three reflection are important: they are used in several relationships listed in column 4, looking further down the table. But these three reflections are not attainable by any relationship using known phases.

- Column 4 lists the relationships which express the sign of the reflection in terms of 'known' signs (from higher up the table). These relationships now are actually being used. But: a special relationship is listed in column 4 for reflection #3. The triplet involves two identical reflections: the sign of the 0-42 is probably equal to the square of the sign of 0-21 , consequently, the 0-42 is probably positive. If that is true, then our letter-assigment gives: C≈+ . But is it really true? As a general strategy one should not rely too early on too few relationships. Therefor we continue with the letter-symbol C for the sign of reflection #3, but keep the symbol-relation in stock!

NOTE: such special sign or phase relationships are denoted *sigma*-1 relations; they may play an important role in the *a priori* determination of phases of *structure seminvariants*.

- Column 6 lists the symbol relations. For #3: [C ≈ +] to be referred to as 'R1'.
- Reflection #4 has been made use of in obtaining symbol relation R1. But otherwise this reflection is not used at all! It is not found anywhere down the table. We are never able to find its sign with the present direct method, so this reflection is rejected from now on.
- Reflection #5 does have the first useful sigma-2 relationship (up to this stage of the analysis): it is seen that the sign S(013) probably is equal to the product of signs of reflection #1 and #2.
- The symbolic result for the reflection is given in column 5: so S(013) ≈ AB.
- Reflection #6 again is a sigma-1 result: it is treated the same way as #3: a new symbol is assigned (D) and the probable symbol-relation [D ≈ + : R2] is given in column 6.
- Reflection #7 does not occur in any triplet: it is ignored. So are #16, #17, #18.
- For reflection #8 - #12: see what happens in the table. Two new symbols are assigned, and for three reflections probable signs are expressed in terms of the earlier symbols. Note that for reflection #11 reflection -(#1)=-(0-11) is used, meaning (01-1), which also has sign A. [Friedel'slaw: $\phi(\mathbf{h})=-\phi(-\mathbf{h})$, A=-A !]
- For reflection #13 two triplets are found: one leading straightforward to the sign DE . For the second triplet we must use the intermediate result for reflection #9: S(1-21) ≈ (EA)A=E.

The two results for one reflection indicate that the signs (DE) and E probably are the same. Otherwise we would have a contradiction (or a so-called inconsistency.) The probabilities for the triplets are decreasing while going down the table, but we are still using the most important triples, so that if all are correct so far, we have a new symbol relation: DE ≈ E or: D ≈ +. This is denoted symbol relation number R3: see columns 6 and 7.

What about the probabilities? Some qualitative comments may be in order. The symbol relations R1 and R2 come from sigma-1 relations: they have not the same but still comparable probabilities as the sigma-2 relations. But the symbol relation R3 can only be correct if three triplets simultaneously are correct: so the R3 relation certainly is less reliable and we have to be careful ....

- Reflection #14 shows similar features as #13. However: we have to make use of the sign of reflection #13, which has been determined as either ED or E. How to continue?? This and similar questions and the wide variety of answers constitute major differences in stategies of computer programs. In the present case ED is a straightforward result and using it we end up with a new symbol relation, R4, which relates four different symbols.

- Reflection #15 shows another important feature: two independent (?) triplets lead to the same sign: ABE . Each triplet uses symbols from forgoing determinations, which may not be very reliable, but the coincidence that the two results are equal increases the reliability of this sign. NOTE: this coincidence makes use of the sign AE for #9, thereby confirming the correctness of AE for #9, and thus increasing the probability of relation R3 .
- For reflections #19 - #25, see table: two more letters and one more symbol relation.

What to do now?

In order to fix the origin we select three linearly independent reflections which must be strong and used sufficiently often in the sign determination process. We select and set: #1, #10 and #19 equal to +. This implies that we have the following symbol relations (by definition!): A = F = G = + .

And let us look at the list of symbol-relations: it happens three times that D = + ! The result could be wrong: D could be negative, but it would imply too many inconsistencies! So we accept D = +.

- Column 8 gives the sign of each reflection after substitution of A=F=G=D=> +.

We are left with four letter symbols: each has two possible values, so we have obtained a 16-fold solution of the phase problem. It would be no problem to make 16 Fourier syntheses. But let us start with the most promising of the solutions: we still have symbol relations R1 [C ≈ +] and R4 [BCEF = BCE ≈ +]. The latter, with C = + becomes BE ≈ + . For a large structure, this looks like guessing, but for a small structure it is probably correct.

- Column 9 gives the signs after substitution of C => + and E => B .

We are now left with two unknown letter symbols, and no more relations between the symbols. This actually means that whatever we choose for the symbols (either + or -) the final consistency of the total assembly of sign relationships will not be affected. NOTE: in the present space group (as in all symmorphic space groups) there are no negative signs introduced by the symmetry relations among the contributing reflections. All triplets form a completely consistent set. Hence the importance of *negative quartets*! It is (almost) never possible that all signs are positive: we must introduce some minuses as well. In the evaluation of the present illustration it is just trivial to set the remaining symbols negative.

- The last column gives the final signs after defining B = H = - for statistical reasons.

So far the illustration of a (modified) symbolic addition method .

Concluding remark.

Many things in direct methods are simple. More precise 'elementary' definitions are given in the following lecture (H.Schenk). Later in this school you will come across complicated theories. It is hoped that the associated computer programs make life easier for macromolecular crystallographers.

AN ELEMENTARY INTRODUCTION TO DIRECT METHODS

H. Schenk
Laboratory for Crystallography, University of Amsterdam
Nieuwe Achtergracht 166, 1018 WV Amsterdam
The Netherlands

Introduction

In a diffraction experiment intensities $I_{hk\ell}$ are measured whereas $F_{hk\ell} = |F_{hk\ell}|\exp(i\varphi_{hk\ell})$ are necessary to image the electron density. Now $|F_{hk\ell}|$ can be calculated straightforwardly from $I_{hk\ell}$ but the relative phases $\varphi_{hk\ell}$ are lost in the experiment and cause the so-called phase problem. Direct methods try to evaluate phases $\varphi_{hk\ell}$ "directly" from the measured diffraction intensities $I_{hk\ell}$ by using relationships among the phases, relationships, whose values are based on the intensities only. Roughly it can be stated that, since the crystal structure can be described by a limited number of parameters (the positions of the atoms) and since many more intensities can be measured, relationships among the structure factors $F_{hk\ell}$, and thus among the phases $\varphi_{hk\ell}$, must exist. Direct Methods identify and use these phase relationships to solve the phase problem.

In a nutshell a Direct Method proceeds as follows: In the first step as many phase relationships as possible are collected, the origin is fixed by specifying the phases of a few suitable reflections numerically and then, using the phase relationships, new phases are calculated. In general, however, it will not be possible to phase all strong reflections and hence a few more starting reflections are selected, which act as unknowns (symbols, ambiguities) and from which new phases can be calculated using the phase relationships (the so-called phase extension). This process generally develops like a snow ball, provided a good choice of origin-defining reflections and unknowns has been made. Finally, when most of the strong reflections have got a phase, the numerical values of the unknowns are evaluated and using a Fourier summation (1) an image of the structure is produced.

In this introductory chapter this process will be explored in more detail and also some practical exercises will be introduced. Therefore, paragraphs in this chapter will successively deal with the origin of the phase problem, weak and strong structure factors, normalized structure factors, the physical meaning of the triplet relation, the origin definition in the centrosymmetric triclinic space group $P\bar{1}$ and the process of phase propagation.

The phase problem

The electron density $\rho(x,y,z)$, where x, y and z define a position in the unit cell in fractions of the cell edges, is given by the Fourier summation

$$\rho(x,y,z) = \sum_h \sum_k \sum_\ell |F_{hk\ell}|\cos[2\pi(hx+ky+\ell z)+\varphi_{hk\ell}] \qquad (1)$$

This expression shows that the electron density ρ is a superposition of planar density waves of the cosine form with amplitudes $|F_{hk\ell}|$, with a spatial direction and a wavelength defined by $hk\ell$, and with the position of the maxima of the planar waves with respect to the origin determined by the so-called phases $\varphi_{hk\ell}$. In the two upper rows of Fig. 1 a number of these planar density waves is depicted with their phases $\varphi_{hk\ell}$ all equal to 0, and their $|F_{hk\ell}|$'s all equal. In order to image the electron density ρ of a particular structure one needs to know the $|F_{hk\ell}|$-values and the corresponding phases

S. Fortier (ed.), Direct Methods for Solving Macromolecular Structures, 19–26.

$\varphi_{hk\ell}$ and is then able to calculate the Fourier summation (1). This process is depicted in the final row of Fig. 1, where the electron density of naphtalene is given and a few of the most important terms of (1) for that structure.

The central problem in a crystal structure determination follows from the fact that only intensities $I_{hk\ell} \approx |F_{hk\ell}|^2$ can be measured by means of X-ray diffraction, whereas for each term of expression 1 not only $|F_{hk\ell}|$ but also the phase $\varphi_{hk\ell}$ must be known in order to obtain the electron density $\rho(x,y,z)$. Except in very exceptional cases, these phases cannot be measured and therefore they have to be deducted in one way or another. This is generally referred to as the phase problem and solving the phase problem is the most important and sometimes most troublesome step of a crystal structure determination. To illustrate the phase problem it is recalled that in the final row of Fig. 1 the structure factor amplitudes were used in a Fourier summation (1) with their correct phases and that as a result the structure arises. However, if these phases were not known and taken arbitrarily to be equal to 0, then the Fourier summation (1) would show only one large maximum at the origin and indicate nothing of the structure at all. Another possibility would be to take phases at random, and then a completely grey image would result from expression 1, again not showing anything of the structure.

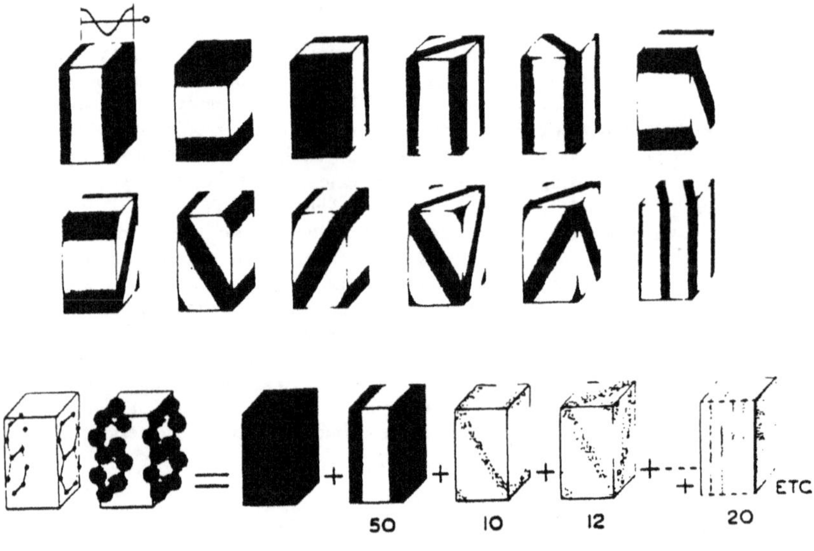

Fig. 1. Fourier components and the Fourier summation of naphtalene.

The phase problem thus is a serious problem; without solving it no image of the structure will be obtained. In contrast, even when the structure factor magnitudes are all set to 1 or taken from another crystal, the desired structure will show up in a Fourier summation, provided the correct phases have been used. Since wrong phases and correct amplitudes reveal no structure; the phase problem is a very serious problem indeed. Therefore methods to evaluate phases play a central role in crystal structure determinations. The so-called Direct Methods constitute one approach to solve the phase problem, Patterson Methods another. In protein crystallography two important methods to determine phases are based on isomorphous replacement and anomalous dispersion respectively.

Strong and weak structure factors

If in a crystal structure the atoms lie in the neighbourhood of the set of planes H (H is a compact notation for the indices $hk\ell$), as indicated in Fig. 2a, the reflection has a large intensity $I_{hk\ell}$. This intensity is small when the atoms are randomly distributed with respect to the same planes (Fig. 2b). This follows also directly from the structure factor expression

$$F_H = |F_H|\exp(i\varphi_H) = \sum_{j=1}^{N} f_j \exp[2\pi i(hx_j+ky_j+\ell z_j)] \quad (2)$$

because a large structure factor F_H will only be found if the atoms lie near positions for which $hx_j + ky_j + \ell z_j \simeq$ constant modulo 1 for all j.

Conversely, a strong intensity of H implies that the electron density will peak in planar regions which lie d_H apart and a weak intensity that the electron density is scattered with respect of the planes H. The choice of the origin with respect to these planar regions of electron density defines the phase φ_H in (2). Large and small structure factor amplitudes thus may be used to predict where in the unit cell electron density can approximately be expected.

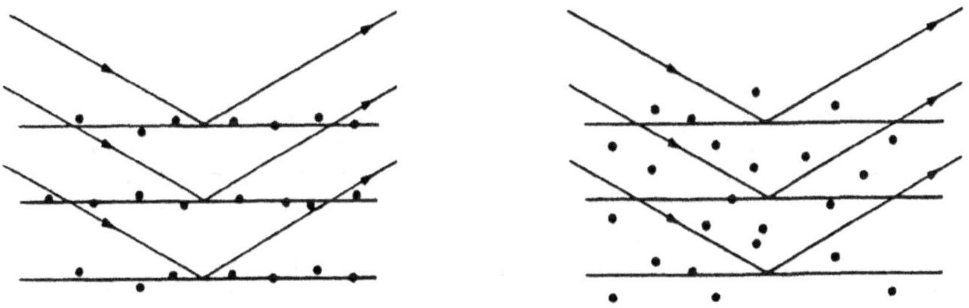

Fig. 2a Fig. 2b

Normalized structure factors

The amplitudes of different structure factors F_H cannot be compared directly since the form factor decreases with increasing reflection angle θ. Therefore the so-called normalized structure factors are used in direct methods instead. They are defined as:

$$|E_H|^2 = |F_H|^2 / (\sum_{j=1}^{N} f_j^2) \quad (3)$$

and have the property that $<|E_H|^2> = 1$ for all values of θ, because the expectation value of $|F_H|^2$ at given θ is expressed by:

$$<|F_H|^2> = \sum_{j=1}^{N} f_j^2(\theta) \quad (4)$$

The structure factor expression in terms of the normalized structure factor is then:

$$E_H = \frac{1}{\left(\sum_{j=1}^{N} f_j^2\right)^{1/2}} \sum_{j=1}^{N} f_j \exp 2\pi i(hx_j+ky_j+\ell z_j) \qquad (5)$$

If the form factor f_j has the same shape for all atoms ($f_j=Z_jf$), (5) can be written as

$$E_H = \frac{1}{\left(\sum_{j=1}^{N} Z_j^2\right)^{1/2}} \sum_{j=1}^{N} Z_j \exp 2\pi i(hx_j+ky_j+\ell z_j) \qquad (6)$$

This is clearly the structure factor formula of a point atom structure, because no θ-dependent factors are present any more. It follows easily from (6) that for a structure of N equal atoms the maximum possible value of $|E_H|$ is equal to $N^{1/2}$.

In the early beginning of Direct Methods also unitary structure factors U_H have been used extensively. For an equal atom structure the relation between U and E is given by $U_H = N^{-1/2}E_H$. Hence the maximum value of U_H is equal to unity.

The triplet relation

If two reflections H and K are both strong then the electron density is likely to be found in the neighbourhood of two sets of equidistant planes H and K at the same time, that is to say the electron density will be found near the lines of intersection of the planes H and K as indicated in projection in Fig. 3a. Geometrically the reflections H-K and -H-K fit on the grid formed by H and K. Now furthermore, a large $|E|$ for the reflection -H-K implies that the electron density will also peak in planes lying d_{-H-K} apart. It is therefore most likely that these planes run through the lines of intersection of the planes H and K, in other words that the three sets of planes have their lines of intersection in common (Fig. 3b).

The origin O has not been fixed yet and by choosing it at an arbitrary point in

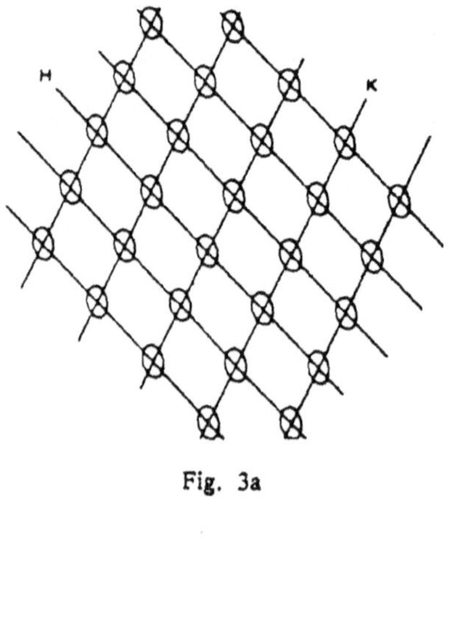

Fig. 3a

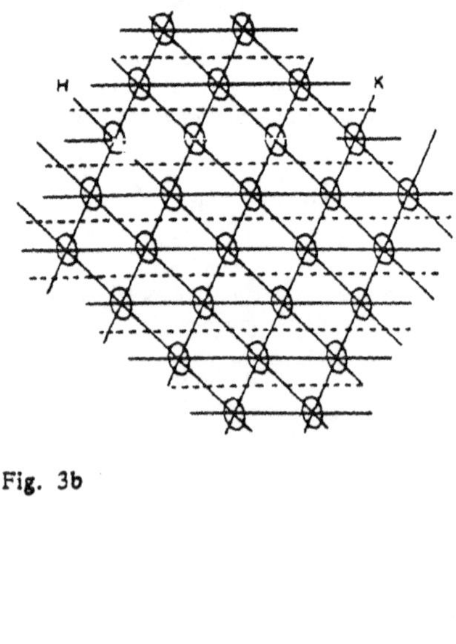

Fig. 3b

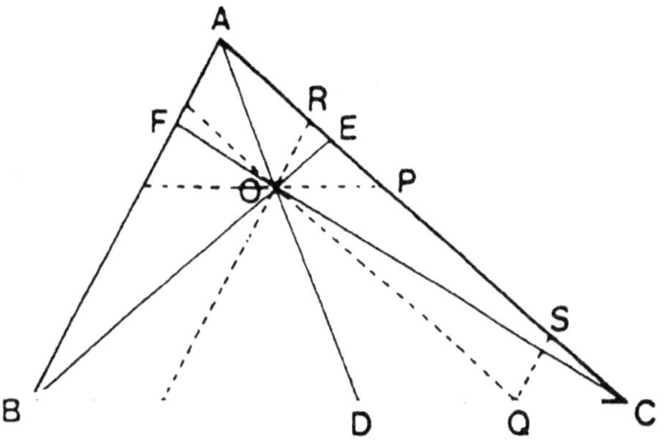

Fig. 4. In an arbitrary triangle ABC an origin O is chosen arbitrarily. Theorem: AO/AD + BO/BE + CO/CF = 2.
Proof: AO/AD = AP/AC; CO/CF = CR/AC; BO/BE = BQ/BC = AS/AC; because RP = SC, AP + CR + AS = 2AC

fig. 3b the triplet phase relationship

$$\varphi_H + \varphi_K + \varphi_{-H-K} = 0 \text{ (modulo } 2\pi) \tag{7}$$

follows easily from a planimetric theorem [2], as illustrated in Fig. 4. Since the origin has arbitrary been chosen, (7) is completely general; relations of this type are usually called structure invariants, although a better name would be origin invariants. Most probably the name structure invariant has been given by the early researchers in the field because these invariants carry structure information as is shown clearly in this paragraph.

In Fig. 3b the ideal situation is sketched; of course a small shift of the planes of largest density of −H−K does not affect the reasoning given above. However, the most unlikely position for these planes is the one indicated by the broken lines in Fig. 3b because then the planes −H−K of largest electron density just clear the lines of intersection of H and K. Therefore the triplet relation has a probability character and this is emphasized by writing it as

$$\varphi_H + \varphi_K + \varphi_{-H-K} \simeq 0 \tag{8}$$

in which ≃ means that the most probable value of (7) is 0. This relation holds more likely the larger the structure factors of H, K and −H−K or the larger the value of the product of the normalized structure factors $E_3 = N^{-1/2}|E_H E_K E_{-H-K}|$.

Origin definition in $P\bar{1}$

For a phase determination in a particular space group always the restrictive rules are employed as resulting from the conventional choice of the unit cell. In space group $P\bar{1}$ the only symmetry element is a centre of symmetry, and then the convention as given in the International Tables for X-ray Crystallography states that the unit cell has to be chosen with its origin at one of the available centres (see fig. 5). As a result all phases will have one out of two possible values only, 0 or π. In the unit cell eight different centres of symmetry are present; they all are differently surrounded by the electron density and lie half of the unit cell apart from each other. If one centre has been chosen as origin, the other centres then have fractional coordinates (1/2,0,0), (0,1/2,0), (0,0,1/2), (1/2,1/2,0), (1/2,0,1/2), (0,1/2,1/2) and (1/2,1/2,1/2). Of course for the description of the structure it does not matter which one of the eight centres has been fixed as origin and thus one is free to choose any of them. In Direct Methods

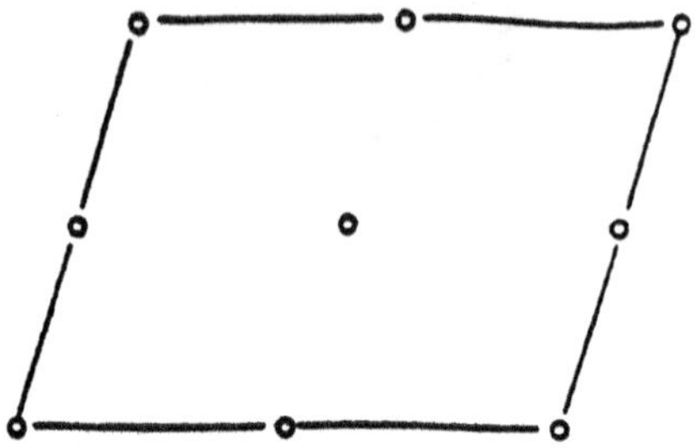

Fig. 5

generally such equivalent origin possibilities are referred to as permissible origins.

The problem now arises how to select one of the centres as the origin and firstly this process will be demonstrated in two dimensions. In fact all to be done is to ensure that the different centres of symmetry will be surrounded by different electron densities and that can be achieved by assigning numerical phases to a few terms in (1). In Fig. 6 the planes of reflection 210 are drawn. In case the phase of 210 is chosen to be 0 the maxima of the electron-density wave run through the centres 1 and 2 and the minima through 3 and 4. In case the sign of 210 is chosen to be π, the centres 3 and 4 are found at maxima and 1 and 2 at minima of the density wave. Thus the choice of either $\varphi_{210}=0$ or $\varphi_{210}=\pi$ breaks the group of four identically surrounded centres up into two groups: 1 and 2 have a different electron density than 3 and 4. By making a next choice, in which 1 and 3 form one group and 2 and 4 the other, we can ensure that the electron density at all 4 centres in the projection is different. For instance this can be done with reflection (120) as is shown in Fig. 7. Thus by choosing $\varphi_{210}=0$ and $\varphi_{120}=0$ the four centres of symmetry get a different electron density and hence the origin has been fixed in the two-dimensional unit cell.

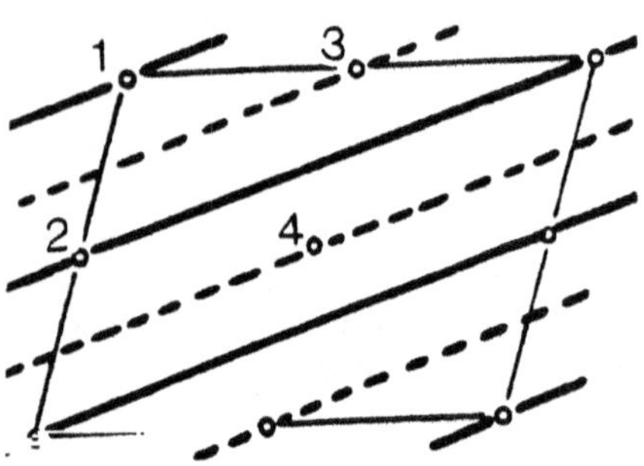

Fig. 6

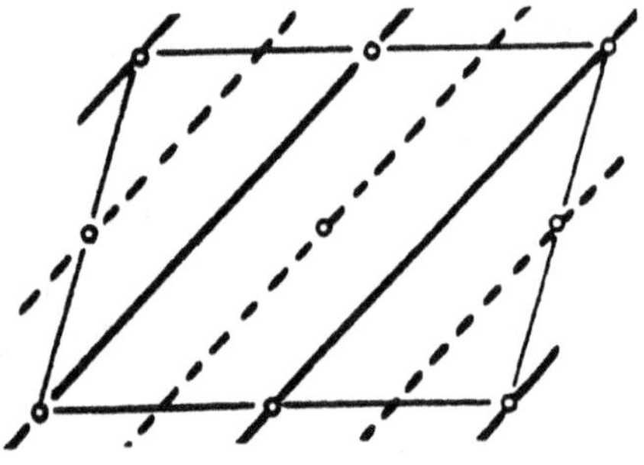

Fig. 7

It can also be seen easily that reflection 220 cannot be used for the origin definition at all. For this reflection all centres of symmetry lie either on the maximum of the density wave or in the minimum (see Fig. 8). The phases of these reflections are the same whatever origin is chosen and therefore unsuitable for origin selection.

Reflection 110 is, on the other hand, an alternative origin fixing reflexion, because it splits the centres again into two groups, now 1 and 4 in one and 2 and 3 in the other group (see Fig. 9). Thus the choice $\varphi_{110}=0$ and $\varphi_{210}=0$ equally well fixes the origin as $\varphi_{120}=0$ and $\varphi_{110}=0$ or as the above mentioned couple $\varphi_{120}=0$ and $\varphi_{210}=0$. Of course, in two dimensions not more than two reflections can be chosen to fix the origin; as soon as that has been done, all other reflections must be phased relative to the origin-fixing phases.

The above reasoning can be generalized, because all reflections with h even and k odd contribute similarly to the electron density as reflection 210, all reflections with h odd and k even similarly as 120, and all reflections with h odd and k odd similarly as 110. This implies that any reflection in the above three groups can be used to fix the origin and not just the example reflections.

Real crystals are three dimensional and therefore three reflections can be chosen following a similar reasoning as for two dimensions. There are now eight centres of symmetry with different electron density surroundings. Therefore the process of fixing the origin consists of choosing the phases of three reflections, whose contributions to the Fourier summation (1) have to be such that the eight centres will be surrounded by a different electron density. Without going through the details again, the final procedure is given here, using the eight parity groups eee, ooe, eoe, oee, eeo, oeo, eoo, and ooo where the e's and o's stand for even and odd respectively, and the position of the letter reflects the index h, k and ℓ respectively.

step 1: No origin fixing with eee reflections.

step 2: Choose origin defining reflection 1 from any of the 7 remaining parity groups. This parity group is used now and cannot be used again.

step 3: Choose reflection 2 from any of the 6 remaining parity groups. This parity group cannot be used again and also the parity group of the sum of the indices of reflection 1 and 2 cannot be chosen any longer. (e.g. reflection 1 is 324, reflection 2 is 410, then used are oee, eoe, but also ooe (334+410=734).

step 4: Choose reflection 3 from one of the remaining 4 parity groups.

step 5: Assign phases (e.g. φ=0) to the three reflections.

When the three reflections are chosen such that they have a large $|E|$-value and many phase relationships with large reliabilities, then the origin will be optimally fixed and the successive phase propagation will follow an optimal path.

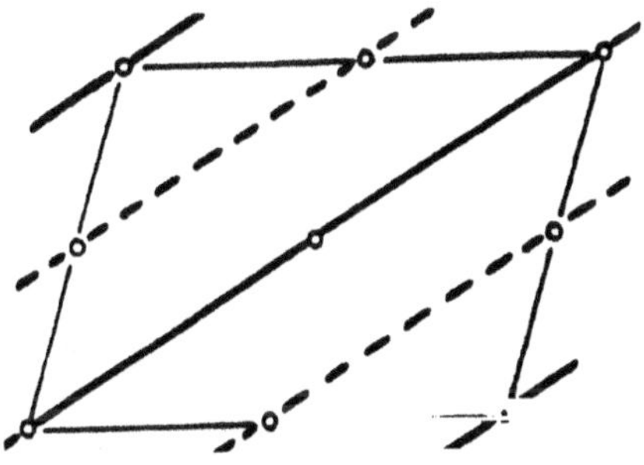

Fig. 8

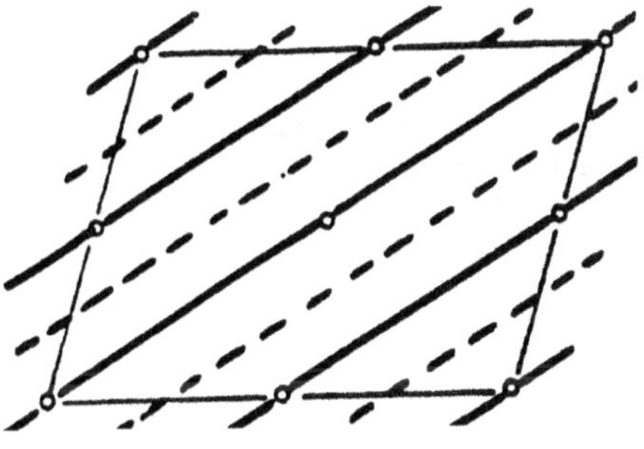

Fig. 9

Phase propagation

As stated before, in general it will not be possible to phase all reflections just from the origin defining phases. In space group P$\bar{1}$ it is even impossible, because then the final phases will all be equal to 0, and the expectation is that approximately half of them will be π and the other half 0. So at least a fourth reflection, whose phase has to be π, must be added to the three origin defining ones in order to solve the structure.

However, in most cases four phases in the starting set will again not be enough to phase all reflections and more starting reflections will be needed. How this can be done will be explained in later chapters. For now it suffices to know that it is possible to construct starting sets consisting of more reflections than the origin defining reflections alone which are capable to access the other phases through the use of phase relationships such as the triplet relation (8) by means of a process called phase propagation or phase extension. Of course, the phases of the not-origin defining reflections are unknown and they are referred to as unknown symbols or ambiguities.

The phase propagation is carried out either symbolically (symbolic addition) or numerically (multi solution). Therefore, the starting set gets first numerical values. This implies of course that ambiguities will get a number of different trial values, i.e. in centrosymmetric structures each ambiguity gets two different trial values, 0 or π. Thus, given a starting set with 3 origin defining reflections and 4 unknown extra reflections, there are $2^4 = 16$ starting points for the phase extension, and hence, in the multisolution approach, the phase extension has to be carried out 16 times. The actual phase extension uses in general just triplet relationships (8). When two of the reflections in 8 are known, the third one can be calculated and successively used in other triplets for calculating new phases again, this process being the actual phase propagation.

In order to carry out such a phase propagation for a structure, triplet relations for the strongest reflections and a pre-determined starting set are necessary. Different phase propagations will start from different starting points (different numerical values for the same starting phases), so simulating a multisolution approach. In a centrosymmetric structure it is easier to work with signs instead of phases. A phase of 0 is then represented by a sign of +1 and a phase of π by a sign of −1 and the triplet relation (8) is at the same time replaced by

$$s_H s_K s_{-H-K} \simeq +1 \qquad (9)$$

or $S_H = S_K S_{-H-K}$ because $S_H = S_{-H}$. In the propagation process new signs are simply calculated by using (9).

References

Schenk, H. (1979). Solving the Phase Problem in Crystal Structure Determination, a simple introduction to Direct Methods. J. Chem. Ed. 56, 383–384.

A SIMPLE INTRODUCTION TO PATTERSON METHODS

PAUL T. BEURSKENS
Crystallography Laboratory, University of Nijmegen,
Toernooiveld 1, 6525 ED Nijmegen, The Netherlands

Most crystallographers have been exposed to Patterson methods, in one way or another. Small-molecule crystallographers of course know the heavy atom method, although present-day experience will never match the Patterson-interpretation capabilities of the foregoing generation. Vector-search techniques generally are less known, although the term "molecular replacement" is well known. Nevertheless, wide differences exist in experience with Patterson methods among crystallographers, and because of that, various terms, definitions, ideas, methods and strategies related to the use of the Patterson function will be shortly discussed. It is hoped that this introductory lecture may fill existing gaps, and facilitates a better understanding of the lectures to come.

Why Patterson methods in a school on direct methods?
Because the majority of macromolecular structure determinations starts with the study of the Patterson function, in some way! Especially for the location of heavy atoms when using isomorphous replacement, or for molecular replacement when a molecular model is available. But some understanding of the principles underlying Patterson methods also is useful when studying the theory of various new methods. As long as *ab initio* direct methods are not perfect, Patterson methods remain of prime importance.

Formulae
$F_o(\mathbf{h})$ observed structure factor, $\mathbf{h}$ = hkl : diffraction vector *c.q.* reflection indices
$P_o(\mathbf{u})$ observed Patterson function, $\mathbf{u}$ is a vector in Patterson space
F and P are shorthand notations for $F(\mathbf{h})$ and $P(\mathbf{u})$, respectively.
$\Sigma_\mathbf{h}$ sum over all reflections
$\int_\mathbf{r}$ integral over the unit cell

The Patterson function is defined as: $P_o(\mathbf{u}) = V^{-1} \Sigma_\mathbf{h} |F_o|^2 \exp(2\pi i\, \mathbf{h.u})$.

It is seen that it is a Fourier synthesis with coefficients equal to $|F_o|^2$. It has the same periodicity as the crystal. Neglecting anomalous scattering, thus using $\phi(\mathbf{h}) = -\phi(-\mathbf{h})$, the two reflections forming a Friedel pair ($\mathbf{h}$ and $-\mathbf{h}$) add up to $2|F_o|^2 \cos(2\pi\, \mathbf{h.u})$, and $P_o(\mathbf{u})$ becomes a real function. Because absolute values $|F_o|^2$ are used, the Patterson function can

S. Fortier (ed.), Direct Methods for Solving Macromolecular Structures, 27–32.

be considered as a physical quantity: $P_o(\mathbf{u})$ is available by experiment.

From the theory of Fourier transformations it is known that the Fourier transform of a squared function [in reciprocal space] corresponds to an auto-convolution [in "direct" space, *i.e.* electron density and Patterson- = vector-space] [and *vice versa.*]. Consequently:

$$P(\mathbf{u}) = \rho(\mathbf{r}) * \rho(-\mathbf{r}) = \int_{\mathbf{r}} \rho(\mathbf{r})\ \rho(\mathbf{u}+\mathbf{r})\ d\mathbf{r}$$

which leads to the physical interpretation of the Patterson function: the value of P at point **u** tells us how important the correlation is between a point **r** in the electron density function and the point **u+r**. This is the physical background for the interpretation of a heavy-atom Patterson map.

Sharpening

The decrease of intensities with the scattering angle is related to the convergence of the Fourier series and the width of a Fourier peak. Sharpening functions are introduced to artificially enhance the high order reflections, which tends to sharpen up the Patterson peaks and especially effects the heavy-atom peaks. Too much sharpening, however, puts too much weight on weak and therefor less accurately measured high-order reflections, and it causes truncation effects.A good sharpening function is useful in Patterson-interpretation procedures. Replacing $|F|^2$ by $|E|^2$ usually is too much.

Heavy atom Patterson interpretation

If a structure contains heavy atoms (atoms which have a relatively large number of electrons) then the contributions of these atoms to the Patterson function are of relatively large importance and these contributions may be recognized as such.
If a structure contains one symmetry-independent heavy atom, then the Patterson function shows interatomic vectors $\mathbf{v}_{ij} = \mathbf{r}_j - \mathbf{r}_i$ where i and j refer to symmetry-related atoms.These vectors are called Harker vectors: they often (dependent on the space group) lay in restricted areas (Harker lines, Harker planes). Some examples:

Sp.gr. P1	$N_s = 1$	Harker point: 0,0,0 [$\mathbf{v}_{11} = \mathbf{r}_1 - \mathbf{r}_1 = \mathbf{0}$: the null vector = origin]
Sp.gr. P-1	$N_s = 2$	Harker area: the entire space [for $\mathbf{r}_2 = -\mathbf{r}_1$: $\mathbf{v}_{21} = 2\mathbf{r}_1$, $\mathbf{v}_{12} = -2\mathbf{r}$]
Sp.gr. $P2_1$	$N_s = 2$	Harker plane: *e.g.* for 2_1-axis // **b** through 000: {**v**}={x, 0.5, z}
Sp.gr. Pc	$N_s = 2$	Harker line: *e.g.* for c-plane ⊥ **b** through 000: {**v**} = { 0, y, 0.5}
Sp.gr. P222	$N_s = 4$	origin + 3 mutual perpendicular Harker planes

If a structure contains more than one symmetry-independent heavy atom, then the Patterson function shows all interatomic vectors $\mathbf{v}_{ij} = \mathbf{r}_j - \mathbf{r}_i$ where i and j refer to symmetry-independent atoms as well. The non-Harker vectors are called cross vectors.

The 'weight' of a peak in the Patterson representing vector $\mathbf{v}_{ij} = \mathbf{r}_j - \mathbf{r}_i$ is $Z_i Z_j$, where Z_j is the atomic number of atom j. When n Harker peaks coincide, the weight is $\Sigma_n Z_j^2$. Peak heights are approximately proportional to the peak weights.

Exercise. A platinum complex crystallizes in space group $P2_1/c$ with four platinum atoms in the unit cell. There are three outstanding peaks in the asymmetric unit of the Patterson: (u, v, w) with relative peak height:
(0, 0, 0.5) 100; (0.42, 0.5, 0.64) 50; (0.42, 0.5, 0.14) 50.
The symmetry relations are: x, y, z ; -x, -y, -z ; x , 0.5-y, 0.5+z ; -x , 0.5+y, 0.5-z.
Question: can you find the position of the platinum atom? [Just try !]

Origin and enantiomorph fixation
An axial system is defined in accordance with the translational symmetry of the crystal which can be known from experiment. For a given crystal, however, the positions of the atoms are not fixed relative to any axial system. In space group P1 at least one atom has to be fixed relative to the origin of the axial system. [We do not fix anything in the crystal: we only specify three parameters x_1,y_1,z_1 in order to be able to use an arbitrarily defined axial system to describe the structure.] Other atoms (in space group P1) then are fixed. Because the entire structure is a physical entity, the interatomic distances are physical quantities, thus the length of an interatomic vector $|\mathbf{v}_{ij}| = |\mathbf{r}_j - \mathbf{r}_i|$ is a physical quantity. The definition of the origin in an electron density map is a human activity or a mathematical concept, the origin in the Patterson map is defined by the underlying physics. All but the first atom (in space group P1) indeed are fixed in principle, but at the outset of a structure analysis we do have one ambiguity regarding the enantiomer of the crystal at hand: because of Friedel's law we do not know whether the trial structure or its inverted one is the correct structure. So the first asymmetric model introduced during the course of a structure analysis constitutes an arbitrary choice. [Of course, a crystallographic ambiguity often is resolved by the knowledge of the chemical - molecular structure.]

For symmetry higher than P1, we have the additional requirement that an origin shift must be compatible with the definition of the space group symmetry. These are *permissible* origin shifts. In P-1, for instance, there are eight different centers of symmetry in the unit cell; any vector between these centers is a permissible origin shift.

Answer to exercise-question: No, you cannot find the position of the platinum atom! There are exactly 32 'possible' positions in the unit cell which agree with the listed peaks. Here we list just 5 possibilities: (1) 0.21, 0.25, 0.32 (2) 0.71, 0.25, 0.32 (3) 0.79, 0.75, 0.68 (4) 0.21, 0.25, 0.82 (5) 0.21, 0.25, 0.07

We could use (1) to start solving the remainder of the structure. (2) is shifted from (1) by (0.5, 0, 0) which is a permissible origin shift: using (2) instead of (1) is an act of origin definition. (3) is obtained from (1) by a space-group symmetry operation (= inversion), so it is identical. (4) is complicated as it is related to (1) by the c-glide operation, but it also is a permissible origin shift (0, 0, 0.5). So ? Position (5), however, is a completely different position, not related to (1). The two heavy-atom structures associated with (1) and (5) are *homometric* Patterson solutions. We just have to try both, and only one will give the structure solution. Nevertheless: the structure solution will not be trivial: both (1) and (5) represent heavy-atom structures with extra translation symmetry (0, 0, 0.5). The choice (1) or (5) only partly fixes the origin: the structure at hand is a *superstructure*.

Vector search and molecular replacement

When we have *a priori* structural information, say our model is a sufficiently large known part of a molecule, we can evaluate the correlation between the observed Patterson function and the Patterson function calculated for the known model (real space), or the correlation between the observed structure factor magnitudes and those calculated from the partial structure (reciprocal space), in order to measure the goodness of fit of the model. A method to solve the crystal structure using the model is to evaluate the correlation as a function of the orientation and the position of the model. In general, the problem of finding the orientation and positon of the model involves a six-dimensional scanning procedure, but -as we shall see- the orientation and translation searches can be separated, so the problem is only a three-dimensional *orientation search* followed by an (up to)-three-dimensional *translation search.*

In the following we will discuss the basic principles of, and the similarity between Patterson strategies and reciprocal space correlation functions.

Notation

R rotation matrix to be applied to the known model or known fragment
t translation vecor (fractional) to be applied to the known oriented model
p as a subscript refers to the known model or partial structure
p^2 scattering fraction of the known model or partial structure: $p^2 = \Sigma_p Z^2 / \Sigma_j Z^2$
The summation Σ_p is over the known atoms (symmetry included),
and Σ_j is over all atoms in the unit cell. Z is the atomic number.
$p_1{}^2$ scattering fraction of the known fragment, symmetry not included

$P_p(R,\mathbf{t},\mathbf{u})$ Patterson function calculated for the partial structure (given R and **t**)
$F_p(R,\mathbf{t},\mathbf{h})$ structure factor calculated for the partial structure (including symmetry)
$E_o(\mathbf{h})$ and $E_p(R,\mathbf{t},\mathbf{h})$ observed and calculated normalized structure factors
$C(R,\mathbf{t})$ correlation function which measures in Patterson space the fit between $P_o(\mathbf{u})$ and $P_p(R,\mathbf{t},\mathbf{u})$ as a function of R and **t** of the model, or: in reciprocal space, it measures the fit between $|F_o(\mathbf{h})|^2$ and $|F_p(R,\mathbf{t},\mathbf{h})|^2$.
F and P are shorthand notations for F(**h**) and P(**u**), respectively.

Best known is the measure of fit between observed and calculated Patterson functions:

$$C_1(R,\mathbf{t}) = \int_{\mathbf{r}} P_o \,.\, P_p(R,\mathbf{t})\, d\mathbf{u}$$

This function is readily converted into the reciprocal space correlation functions:

$$C_2(R,\mathbf{t}) = \Sigma_{\mathbf{h}} |F_o|^2 \, |F_p(R,\mathbf{t})|^2 , \quad C_3(R,\mathbf{t}) = < |F_o|^2 \, |F_p(R,\mathbf{t})|^2 > , \quad C_4(R,\mathbf{t}) = < |E_o|^2 \, |E_p(R,\mathbf{t})|^2 >$$

where < > denote averaging over all reflections **h**. C_2 and C_3 just differ in scale, but C_4 differs in that sharpening is used. It should be noted that all correlation functions are dependent on the degree of sharpening. Both Patterson space and reciprocal space correlation functions can be calculated and can be used to find the orientation (R) of the model as well as the position (**t**) of the model relative to the symmetry elements.

The functions C_1 - C_4 can be modified in many ways. Sharpening has already been mentioned, which modifies the Patterson function and also the terms in C_2 - C_4.

The origin can be subtracted from the Patterson, which leads to a slight modification of expressions C_2 - C_4 . In the translation search (for given R) the intramolecular vectors can be removed. In reciprocal space one can modify the data set, for instance using the strong reflections only.

A more sophisticated approach closely related to C_1 is the so called 'sum function': instead of evaluating the integral in C_1 the Patterson function values are summed at positions given by interatomic vectors of the model. Define interatomic vectors as: $\mathbf{v}_{ij}(R,\mathbf{t})$ where ij denotes a pair of atoms, $\mathbf{v}_{ij}=\mathbf{r}_j-\mathbf{r}_i$. Corresponding Patterson values are $P_o[\mathbf{v}_{ij}(R,\mathbf{t})]$. The sum function (summing over all $\mathbf{v}_{ij}$) is:

$$C_5(R,\mathbf{t}) = \Sigma_{ij} P_o[\mathbf{v}_{ij}(R,\mathbf{t})]$$

When all interatomic vectors are used in this summation, the results for R and **t** are identical to the results obtained by the forgoing formulae. But considerable improvement is obtained by carefully selecting the vectors to be used in the summation for C_5 and also by applying a well designed weighting scheme to the individual terms in C_5 .

In macromolecular crystallography, where the known model usually is the almost complete molecule (available from another structure), the application of C_1 - C_4 is called *molecular replacement.* In small molecule crystallography, where the known part usually is a rather small fragment of the molecule (fragment obtained from the literature or by molecular modelling), the application of C_2 - C_5 is called *vector search* method. Equations C_2 - C_4 are calculated in reciprocal space but because the derivation started from the Patterson function, these reciprocal space correlation functions usually are also referred to as Patterson methods.

Orientation search

For all space groups but P1 the electron density can be described as the sum of symmetry-related molecules: $\rho = \Sigma_s \rho_s$ for s = 1, 2 ... N_s symmetry elements. As a consequence, the Patterson can be described as a double sum over all intermolecular convolutions: $P = \Sigma_s \Sigma_{s'} P_{ss'}$ where $P_{ss'}$ represents all interatomic vectors between molecule s and molecule s'.

P_{ss} is the *self*-Patterson for molecule s: it represents intramolecular vectors. It rotates when the molecule is rotated, but when the molecule is shifted relative to the symmetry elements, the P_{ss} vectors remain clustered around the origin.

$P_{ss'}$ ($s \neq s'$) is a *cross*-Patterson, representing intermolecular vectors, which are a nuisance for the orientation search, but later it gives us the position of the molecule.

P_{11} now is the *self*-Patterson for the given orientation of the known model.

$P_{11}(R)$ is P_{11} rotated by orientation matrix R (usually defined in terms of Eulerian angles A,B,C). $P_{11}(R)$ is a three-dimensional function in angular space.

The correlation function $C_1(R,t)$ is used on P_{11} the result of which is independent of **t** :

$$C(R) = \int_{r} P_o P_{11}(R)\, du$$

The highest peak in the function C(R) is expected to belong to the matrix R which gives the orientation of the model. Similar formulae can be based upon C_2 - C_4 . The calculations are straight forward for small molecules, but they may cost too much computing time for macromolecules. Complicated spherical-function transforms in reciprocal space, however, have been used to make fast procedures for macromolecules.

Translation search

In contrast to orientation functions, translation functions are fast, reliable, and easily programmed. When the orientation of the model is known, all symmetry elements are used simultaneously to find the shift vector **t** , using C_1 - C_4 . Here we show the application of $C_2(\mathbf{t})$ in space group P-1 .

For the well oriented model at starting position $\mathbf{t} = \mathbf{0}$, the structure factors F_1 are calcultated for all reflections **h** . For position **t** they will become: $F_1 \exp(2\pi i\, \mathbf{h.t})$. The symmetry related model is inverted; the structure factors for the inverted model are $F_2 = F_1^*$ (* = complex conjugate). When the original model is shifted over **t**, the inverted model will shift over **-t** , and its partial structure factor changes to $F_2 \exp(2\pi i\, \mathbf{h.-t}) = F_1^* \exp(2\pi i\, \mathbf{h.-t})$. The expression for the total structure factor is:

$$F_{\mathbf{h}}(\mathbf{t}) = F_1 \exp(2\pi i\, \mathbf{h.t}) + F_1^* \exp(-2\pi i\, \mathbf{h.t})$$ from which follows:

$$|F_{\mathbf{h}}(\mathbf{t})|^2 = 2|F_1|^2 + (F_1)^2 \exp(2\pi i\, \mathbf{2h.t}) + (F_1{}^*)^2 \exp(-2\pi i\, \mathbf{2h.t}) .$$

Substitution in C_2 leads to a Fourier transform. Leaving out the constant term $2|F_1|^2$ gives:

$$C_2(\mathbf{t}) = \Sigma_{\mathbf{h}} |F_o|^2 (F_1)^2 \exp(2\pi i\, \mathbf{2h.t}) + \text{terms for } \mathbf{-h} .$$

Substitution of $\mathbf{q} = -\,\mathbf{2t}$ gives a Fourier synthesis with coefficients $|F_o|^2 (F_1)^2$. The maximum in the resulting function C_2 is at position **q** which gives $\mathbf{t} = -\,\mathbf{q/2}$. For space group P-1 all calculations are simple. For higher symmetry the final result for $F_{\mathbf{h}}(\mathbf{t})$ is expressed as a summation over all symmetry elements.

Concluding remark. Classical as well as modern Patterson methods are powerful, but good computer programs are necessary to obtain results that solve 'difficult' or 'large' structures. Fortunately, good programs *are* available: references will be given in the following lecture notes of this school.

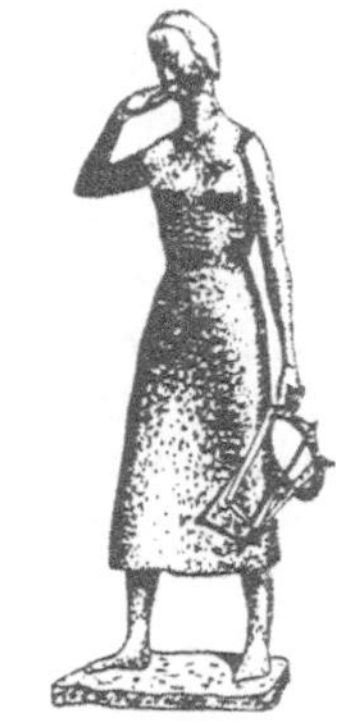

Mariken van Nieumeghen

AN INTRODUCTION TO PROBABILITY THEORY

C.J.Gilmore
Department of Chemistry
University of Glasgow
Glasgow G12 8QQ, Scotland, UK.

1. Introduction

In many ways a course on the phasing of macromolecules from diffraction data can be seen as a study of applied statistics: statistical theory pervades the whole subject, so it is important that the basics of the subject are described and understood at the outset. Statistics can be divided into three related disciplines:
(a) Probability theory: the fundamentals of the subject and its associated mathematics
(b) Data analysis: the collection, display and summary of the raw experimental data.
(c) Statistical Inference: drawing statistical conclusions from data using the rules of probability theory.
We will concentrate here on the fundamentals (a), although the course will deal with all these aspects. It is impossible to do more than sketch the subject here, and for further information the reader is referred to the following:

For a light-hearted but accurate account of statistics in action see Gonick and Smith [1]. In a crystallographic context Shmueli and Weiss is excellent with not a wasted word [2]. For Bayesian statistics in a non-crystallographic environment see Lee [3], and for a comprehensive study of circular data Fischer [4] is a good introduction.

2. Some Definitions

Let us start with some basic definitions:
(a) A *random experiment* is the process of observing or measuring the outcome of a chance event. It corresponds to carrying out a scientific experiment e.g. collecting a diffraction data set.
(b) The *elementary outcomes* are the possible results of a random experiment.
(c) The *sample space* is the set of all the possible elementary outcomes. It is often given the symbol Ω.
(d) An *elementary event* is a set of elementary outcomes. Sets are described using curly brackets: {......}.
The usual example to clarify these terms concerns the throwing of a die with faces numbered 1-6. The throwing of the die and recording the result is a random experiment;

S. Fortier (ed.), Direct Methods for Solving Macromolecular Structures, 33–44.

the elementary outcomes are the numbers 1-6, and the sample space is the set $\Omega=\{1,2,3,4,5,6\}$. A possible elementary event is that the number six appears or that the number is even. Now let us imagine a random experiment with *n* elementary outcomes labelled $O_{j,\ j=n}$; we want to assign a probability to each outcome which measures the likelihood with which it will occur. We can write that the probability of outcome *j* is $P(O_j)$. Thus for the case of throwing the die the probability of throwing a six is $P(6)=1/6$.

Probability deals with random variables. Rigorous definitions of probability and random variables can be very difficult, but in the crystallographic context it can take us too far afield for very little in the way of results if we are too pedantic. Let us therefore define a random variable as a variable whose value cannot be predicted without carrying out an experiment: the experiment gives us information concerning this variable. The definition of probability is equally complicated. (We have already referred to it as a form of likelihood in the previous paragraph.) In classical statistics a frequentist definition is used in which, if the experiment can be repeated, the probability of an elementary event is the proportion of the number of times the event occurs in a long run of random experiments. Bayesian statistics often uses the concept of personal or subjective probability - a belief in the chances of an event taking place.

What can we say about individual probabilities? It seems intuitively obvious that:

$$P(O_j)\geq 0; \qquad P(O_j)\leq 1 \tag{1}$$

i.e. probabilities cannot be negative or greater than 1. Similarly, the total probability of all the elementary outcomes must be unity:

$$\sum_{j=1}^{n} P(O_j) = 1 \tag{2}$$

i.e. probabilities are normalised. Elementary events are useful because we can combine them to get other events; to do this we use the operators *and*, *or* and *not*. For example, if we have two elementary events labelled E and F we can ask what is the probability that both E and F take place:

$$P(E\,and\,F) \quad or \quad P(E,F) \tag{3}$$

e.g. for a single die, what is the probability that the number thrown will be even and <5? This is an example of a *joint probability* and is usually written $P(E, F)$ i.e. we are looking jointly at the probabilities of events E and F. Some authors use the notation $P(E\wedge F)$ here. Similarly the probability that either event E or F takes place is written:

$$P(E\,or\,F) \tag{4}$$

e.g. in the case of the die what is the probability that the throw is either 3 or 5? We have to be careful with the exclusive or non-exclusive use of 'or' here. Finally, what is the probability that the event E does *not* take place:

$$P(not\,E) \tag{5}$$

e.g. what is the probability of not throwing a six? Note that $P(E) + P(not\ E) = 1$ i.e. an event is either true or false.

3. Conditional Probabilities and Joint Conditional Probabilities

The probabilities we have dealt with so far are all *unconditional.* Conditional probabilities extend the scope of the questions we can ask, and are of great importance. The conditional probability:

$$P(A|B) \tag{6}$$

gives the probability of event A given that event B has already occurred. The "|" symbol means *given.* We can see this in action by using two dice rather than one. With the throw of two dice, we can ask the question: what is the probability that the sum of the two faces equals 3? There are 36 elementary outcomes (6 on the first die x 6 on the second). Of these only two combinations (1 & 2; 2 & 1) add up to three so the probability that the faces will sum to three is 2/36 = 0.055. Now suppose we have thrown the first die and it shows a one, what is the probability that with the next throw the two faces will sum to three i.e. what is *P(3|1 in the first throw)*? We need to throw a two, the probability of this is 1/6 so that:

$$P(3|1\, in\, the\, first\, throw) = 1/6 \tag{7}$$

This idea is obviously extendible to joint conditional probabilities e.g.

$$P(A, B....|C, D....) \tag{8}$$

which is the joint conditional probability of events A, B... being true given that events C,D.. have occurred. We will use these a lot when looking at the fundamentals of probability theory in crystallography when we look at joint conditional probabilities of a set of structure factors given another set.

4. Independence

The concept of independence is going to be important. Two events E and F are said to be independent if the occurrence of one of them has no effect on the occurrence of the other i.e.

$$P(E\, and\, F) = P(E)\, P(F) \tag{9}$$

As an example when throwing two dice the probability of getting a six with the second is independent of whatever is thrown with the first.

An important example in crystallography involves the Sayre equation:

$$F_{\mathbf{h}} = \theta/V \sum_{\mathbf{k}} F_{\mathbf{k}} F_{\mathbf{h-k}} \tag{10}$$

where $F_{\mathbf{h}}$ is the structure factor for reflection ***h***, θ is a constant and V is the unit cell volume. The structure factors (both amplitudes and phases) are not independent of each other, but related by this equation.

5. Random Variables

Probability theory is grounded in the use of random variables. It is often useful, indeed essential, to define a partition of elementary events. For example we can toss a coin 5 times and define a partition that describes the occurrence of r tails in 5 tosses. The partition has 6 elements (r_0, r_1,....r_5) each of which is associated with a probability, and where r runs from 0 (no tails) to 5 (all tails). Rather than deal with the partition, we deal with the numbers themselves, and this is done through the concept of a random variable:

A random variable X assigns a different value x_i to each element E_i of a partition. The proposition E_i is identified with the proposition that $X=x_i$.

Note the use of the capital X for the random variable. As an example toss two dice. Let X represent the sum of the dots on both faces. X is then a random variable which must lie between 2 and 12.

In practice we want to look at the probabilities of the outcomes. For the probability that the random variable X takes the value x, we write:

$$\Pr(X=x)=p(x) \tag{11}$$

The function *p(x)* is a probability distribution which looks like this:

x	2	3	4	5	6	7	8	9	10	11	12
p(x)	1/36	2/36	3/36	4/36	5/36	6/36	5/36	4/36	3/36	2/36	1/36

It is very important to distinguish between X which can only be obtained from experiment and x which occurs as an argument to *p(x)* and is a specified parameter.

There are two sorts of random variable: discrete where X (and hence x) can have only certain values as in the table above, and continuous where it can take on a continuum of values. In general, it is easier to learn these concepts from the former case, but the latter is much more important, especially in crystallography.

6. Probability Density Functions

Let X be a random variable which is continuous. The probability density function or (usually abbreviated to pdf) of X is written *p(x)* and defined so that *p(x)dx* is the probability that X lies in the interval $(x, x+dx)$. We can define some very useful properties of the pdf:

$$p(x)\geq 0 \tag{12}$$

The probability that X lies between a and b is given by:

$$\Pr(a<X\leq b)=\int_a^b p(x)dx \tag{13}$$

Since X must lie somewhere in the range -∞ to ∞:

$$\int_{-\infty}^{\infty} p(x)dx = 1 \tag{14}$$

i.e. the pdf must be normalised. The cumulative distribution, *P(x),* is the probability that $X \leq x$ an is expressed as:

$$P(x) = \int_{-\infty}^{x} p(u)du \tag{15}$$

Note the use of the capital P. The *characteristic function*, often denoted $C(\omega)$ is defined as the Fourier transform of the pdf. As in crystallography itself, it is often much easier to work with the FT of a function than the function itself. The log of $C(\omega)$ is called the *cumulant generating function* $K(\omega)$.

6.1 MEAN AND VARIANCE OF RANDOM VARIABLES

It is important to distinguish between the properties of data sets and probability distributions. Properties of the data are called *sample* properties whereas properties of probability density functions are called *population* (or *model*) properties. We are concerned with the mean (or average) and its associated standard deviation. For the model we use the symbols μ and σ; for the data we use $\bar{x}$ and s.

μ and σ are examples of moments. The mean or expectation value is the first moment, and the standard deviation is related to the second moment. In general the *nth* moment $\langle X^n \rangle$ is defined as:

$$\langle X^n \rangle = \int_{-\infty}^{\infty} x^n p(x)dx \tag{16}$$

The standard deviation is given by:

$$\sigma^2 = \langle X^2 \rangle - \langle X \rangle^2 \tag{17}$$

This quantity can be loosely thought of as a measure of the width or dispersion of the associated pdf. σ^2 is called the *variance.* Higher moments measure, for example, the skewness (asymmetry of the distribution) and kurtosis (difference from the normal distribution) of the pdf.

6.2 MARGINAL DISTRIBUTIONS

Consider a situation in which we have a joint distribution of n variables $p(x_1, x_2, \ldots x_n)$. Often we want the pdf of a subset of m of these variables. We can obtain this by integrating out the n-m variables that we do not want to get a marginal distribution p_{marg}. There is one integral per redundant variable, and the limits of the integration are chosen to be the physical limits of that variable. In general these limits are $\pm\infty$, but for a phase angle we will use 0-2π. In the former case:

$$p_{\text{marg}}(x_1, x_2, \ldots x_m) = \int_{-\infty}^{\infty}\int_{-\infty}^{\infty}\cdots\int_{-\infty}^{\infty} p(x_1, x_2, \ldots x_n)\,dx_{m+1}dx_{m+2}\ldots dx_n \tag{18}$$

It can be regarded as the projection of an n-dimensional space on to an m-dimensional one.

7. Some Important Probability Density Functions

By far the most important pdf is the normal or Gaussian distribution:

$$p(x|\,\mu,\sigma) = \frac{1}{\sigma\sqrt{2\pi}}\exp\left(-\frac{1}{2}\left(\frac{x-\mu}{\sigma}\right)^2\right) \tag{19}$$

Its importance in part lies with the central limit theorem which will be discussed in the next section. Figure 1 below gives examples of two normal distributions.

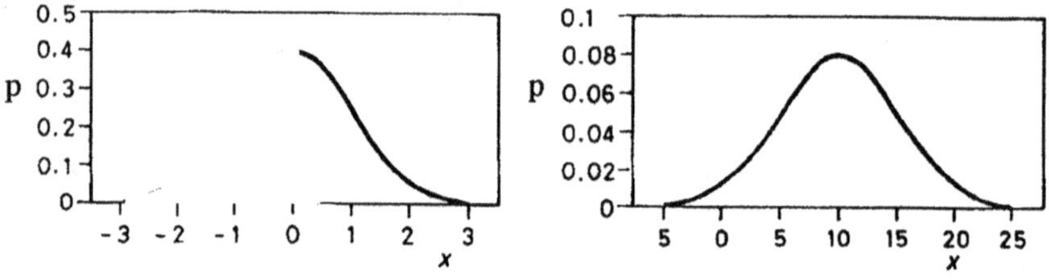

(a) (b)

Figure 1. *Two normal distributions; (a) has a mean of zero and a standard deviation of 1 (b) has a mean of 10 with a standard deviation of 5 (variance of 25).*

What happens when the normal distribution has more than one variable? Now we define **X** as a column vector of m variables $x_1, x_2, \ldots x_m$, and **m** is the corresponding column vector of means $\mu_1, \mu_2, \ldots \mu_m$. The distribution now takes the form:

$$p(\mathbf{X}|\mathbf{x},\mathbf{m}) = \frac{(\det\mathbf{M})^{1/2}}{(2\pi)^{m/2}}\exp\left\{-\frac{1}{2}(\mathbf{X}-\mathbf{m})^T\,\mathbf{M}^{-1}\,(\mathbf{X}-\mathbf{m})\right\} \tag{20}$$

where **M** is the variance-covariance matrix, its ii^{th} element is the variance of x_i and its ij^{th} element is the *covariance* of x_i and x_j which is defined via: $cov(x_i, x_j) = E\{(x_i - E(x_i)(x_j - E(x_j)\}$ where E denotes the first moment or expectation value.

A pdf that appears often in direct methods is the von Mises distribution:

$$p(\theta\,|\kappa,\mu) = [2\pi I_0(\kappa)]\exp[\kappa\cos(\theta-\mu)] \quad 0 \le \theta \le 2\pi, \quad 0 \le \kappa \le \infty \tag{21}$$

$I_0(\kappa)$ is the modified Bessel function of order zero. This pdf is an example of a circular distribution because of the exponential term in $\cos\theta$. Much of the statistics you will meet in this course is circular because of the use of phase angles as variables. The von Mises distribution first appeared in 1918 in a statistical study of the distribution of the deviation of atomic weights from integers. [5]. In direct methods it appears in the distribution of the sum of three phase angles (the triplet) given the associated normalised structure factors.

Consider the sum of three phase angles whose indices sum to zero:

$$\Phi_3 = \phi_{\mathbf{h}} + \varphi_{\mathbf{k}} + \varphi_{-\mathbf{h}-\mathbf{k}} \tag{22}$$

with associated normalised structure factor amplitudes |E(**h**)|, |E(**k**)|, |E(**-h-k**)| if there are N atoms, assumed equal, in the unit cell then:

$$p(\Phi_3|\kappa) = \frac{1}{2\pi I_0(\kappa)} \exp(\cos\Phi_3) \tag{23}$$

where κ is called the concentration parameter:

$$\kappa = \frac{2|E_{\mathbf{h}} E_{\mathbf{k}} E_{-\mathbf{h}-\mathbf{k}}|}{\sqrt{N}} \tag{24}$$

This is usually called the Cochran distribution. Graphs of this function for different values of κ are shown in Figure 2.

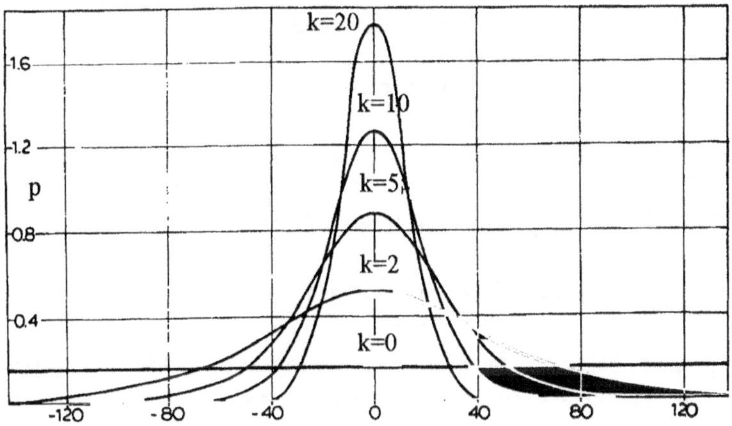

Figure 2 The Cochran distribution for varying values of the concentration parameter κ (From Giacovazzo, *Fundamentals of Crystallography*)

8. The Central Limit Theorem

Given a set of n independent random variables $x_1, x_2, \ldots x_n$ each with its own quite arbitrary pdf with mean μ_i and variance σ_i^2 consider the sum:

$$y = \sum_{i=1}^{n} a_i x_i \tag{25}$$

where the a_i are arbitrary constants. The central limit theorem states that as n increases, the distribution of y approaches the normal distribution with a mean μ given by:

$$\mu = \sum_{i=1}^{n} a_i \mu_i \tag{26}$$

and variance σ^2 given by:

$$\sigma^2 = \sum_{i=1}^{n} a_i^2 \sigma_i^2 \tag{27}$$

This is a quite remarkable theorem and does much to explain the near ubiquity of the normal distribution in statistics. There is also another form often called the "fuzzy central limit theorem": *data that are influenced by many small and unrelated random effects are approximately normally distributed.* Again, this explains the prevalence of this distribution. It is common, though sometimes incorrect, to assume that any unknown pdf is normal at least approximately.

9. Bayes' Theorem and Bayesian Statistics

Bayes' theorem, on which Bayesian statistics is based, appears in many guises. It was first derived by the 18th century English clergyman and talented mathematician Thomas Bayes, and found in his papers on his death. It was rediscovered by Laplace who used it to solve a wide variety of outstanding problems in astronomy. At its simplest, the theorem is uncontroversial. Let us suppose that we are carrying out an experiment to gain information on a parameter θ . Before commencing the experiment we have prior belief of the possible values of θ which we express as an unconditional probability distribution $p(\theta)$. If the experiment yields the result x, then Bayes' theorem states:

$$p(\theta \mid x) \propto p(x \mid \theta) p(\theta) \tag{28}$$

The constant of proportionality may be determined by normalisation. The term $p(\theta|x)$ is called the *posterior* distribution (or simply posterior); the function $p(x|\theta)$ is the *likelihood* which consults the data from the experiment, and $p(\theta)$ is the *prior* probability. Likelihood will prove to be very important throughout this course.

As written, this is an uncontroversial equation simply derived from the axioms of probability theory, and used by all statisticians to invert conditional probabilities i.e. to obtain $p(x|\theta)$ from $p(\theta|x)$ and *vice versa.* However, Bayesian statisticians take the topic a whole step further by using it as a model of scientific inference. In addition, probabilities are defined not in the narrow frequentist way but in the broader, and much more controversial, sense of *degree of belief,* which is ideally suited to science where measurements can be very difficult to relate to frequencies without the use of very convoluted argument, and a counter-intuitive use of language. Furthermore, the equation is used iteratively: the posterior becomes an updated prior, and the whole cyclic process repeated until convergence is reached. From these departures from classical statistical methods stem a great deal of sometimes very acrimonious dissent. It is not possible or desirable to pursue the intricacies of the Bayesian method; for a detailed discussion O'Hagen [6] is recommended. The classic text on likelihood is by Edwards [7], although this is quite strongly critical of entropy techniques. In the crystallographic context, there is a very clear exposition by French [8] who describes a three stage model in crystallography from a Bayesian viewpoint, and elegant papers concerning the crystallographic phase problem in a Bayesian context by Bricogne [9,10]

Here is a classic example of Bayesian statistics at work [1]: there is a rare disease which infects 1 out of every 1000 of the population. Suppose there is a good (but not perfect) test for this disease such that it gives a positive result 99% of the time. However it also produces some false indications and 2% of uninfected people also test positive. You have been tested positive: what is the probability that you truly have the disease?

We have two events:

A: Patient has the disease. $P(A)=0.001$ *i.e.* $1/1000$ people have the disease.
B: Patient tests positive. $P(B|A)=0.99$ i.e. probability of a positive test if you are infected = 0.99

We want: P(A|B).

From Bayes' theorem:

$$P(A|B) \propto P(B|A).p(A) \propto 0.99x0.001 \propto 9.9x10^{-4}$$

The problem here is that we only have a proportion, and we need to get the proportionality constant. This is quite easy. Some thought should convince you that:

$$P(A|not\ B) + P(A|B) = 1.0$$

(either you have the disease or you don't!)

$$P(A|not\ B) \propto P(not\ B|A).P(A) \propto 0.02x0.999 \propto 19.98x10^{-3}$$

This means that the required constant = $1.0/(9,9x10^{-4} + 19.98x10^{-3}) = 0.02097$, so that

$$P(A|B) = 0.0472$$

i.e. only 4.72% of the people who test positive actually have the disease! At first sight this is an extraordinary result, it arises from contributions from the very large group of non-infected people.

Bayesian statistics leads on quite naturally to the maximum entropy (ME) formalism. There are a number of ways of looking at this, but a common complaint concerning Bayesian methods is that a bias in the prior will lead to a bias in the posterior. The use of maximum entropy generates pdfs that are optimally unbiased. For a general review of the ME method in crystallography see Gilmore [11].

10. The Maximum Entropy Principle

Consider a discrete probability distribution for which we know a set of N normalised probabilities $P=\{p_1, p_2 \ldots\ldots p_N\}$. Such a distribution has an entropy, S, or information content I given by:

$$S = -I = -\sum_{J=1}^{N} p_j \log p_j \tag{29}$$

The logarithm can be to any base; often used bases are 2 (in which case the entropy is measured in *bits*), and e (natural units). Equation (29) was first defined and used in a non-thermodynamic context by Shannon [12] when working with problems of the capacity of communication channels, and the transmission of signals down noisy lines. Suppose we have m constraints, expressed as expectation values, $\langle A_r \rangle$, which represent our knowledge derived from experiment. These constraints can be written:

$$\sum_{j=1}^{N} A_{ri} p_i = \langle A_r \rangle \qquad r = 1, 2 \ldots\ldots\ldots m \tag{30}$$

and from this we wish to derive the set of probabilities {P}. Often the situation under study is a mathematically indeterminate one in which $N >> m$. The ME approach consists of maximising the entropy (29), or minimising the information content, subject to the constraints (30) to determine {P}.

10.1 THE BRANDEIS DICE PROBLEM

This classical problem may help to illuminate this formalism Consider a die. When it is thrown there are six possible results *i* where $1 \leq i \leq 6$. If it is thrown N times, an "honest" dice will give all $p_i = 1/6$, and an average throw of 3.5 when thrown repeatedly. Suppose, however, we have a dishonest or biased one for which the mean is 4.5 i.e.

$$\sum_{i=1}^{6} i p_i = 4.5 \tag{31}$$

Given this information *and nothing else* what probability, $p_{i,i=1,6}$ should we assign to the i spots on the next throw?

Clearly the problem is indeterminate, and it is possible to devise an infinity of solutions. (For example, the die only gives i=5 or i=4 with equal probabilities). The ME principle, however, generates a unique solution which, in the jargon of the subject, is *maximally unbiased*, has a *minimum information content*, and is declared *maximally non-committal with respect to missing information:*

There are two constraints: equation (7) and the normalisation condition:

$$\sum_{i=1}^{6} p_i = 1 \tag{32}$$

Assuming uniform prior probabilities, we maximise S subject to the constraints (31) and (32). This is a standard variation problem solved by the use of Lagrangian multipliers. A numerical solution using standard variation methods gives $\{p_1 \ldots p_6\}$={0.05435, 0.07877, 0.11416, 0.16545, 0.23977, 0.34749}, with an entropy of 1.61358 natural units.

Now what does this result mean? To quote Jaynes [13] " *..our result as it stands is only a means of describing a state of knowledge about the outcome of a single trial. It represents a state of knowledge in which one has only (1) the enumeration of the* *six possibilities, and (2) the mean value constraint...and no other information. The distribution is maximally noncommittal with respect to all other matters; it is as uniform ... as it can get without violating the given constraint."*

11. What is the Relevance of all this to Crystallography?

All this may seem a long way from the problems of solving crystal structures, but this is not so. There is a very good discussion of how these ideas can be used in a crystallographic context in [2] (pages 8-11) some of which is followed here. Let us look at the formula for the crystallographic structure factor:

$$F(h,k,l)=\sum_{j=1}^{N} f_j \exp 2\pi i\left(hx_j + ky_j + lz_j\right) \tag{33}$$

where there are N atoms in the unit cell with scattering factors f_j,, h,k,l are the Miller indices, and x_j, y_j, z_j the coordinates of the *j-th* atom in the cell. Let us fix the integers h, k, l and let each of the coordinates x_j, y_j, z_j be unknown (which of course they are in the initial stages of solving a crystal structure). Each of the fractional coordinates can be regarded as a random variable, and if we do not know where the atoms are then they can be described as uniformly distributed on (0,1) i.e. they take on any value between 0 and 1 i.e. they are somewhere in the unit cell. This assumption of a random distribution will be challenged many times during the remainder of this course, since it is patently untrue for vary small structures, and for the very large (i.e. proteins) where there are disordered solvent regions and ordered protein molecules. Symmetry can also give rise to excluded regions in the unit cell.

It is worth noting that there is an alternative description of the problem that leaves the coordinates fixed (but unknown) and treats the h,k,l indices as random variables. In general, it leads to similar if not identical results.

Once we use the ideas of random variables for the coordinates, the whole apparatus of probability theory and statistics in general becomes available to us. Wilson [14] was the first to apply these ideas to crystallography with studies in probability distributions for a single structure factor, and this work was greatly extended by Hauptman and Karle [15] starting in the early 1950's. It has proved to be a very fruitful way of dealing with problems in crystallography.

References

1. Gonick, L., and Smith, W. (1993) *The Cartoon Guide to Statistics* HarperPerennial Publishers, New York.
2. Shmueli, U., and Weiss, G.H. (1995) *Introduction to Crystallographic Statistics* International Union of Crystallography an Oxford University Press, Oxford.
3. Lee, P.M. (1989) *Bayesian Statistics* Oxford University Press, Oxford.

4. Fischer, N.I. (1993) *Statistical Analysis of Circular Data* Cambridge University Press, Cambridge.
5. von Mises, R (1918) Über die "Grazzahligkeit" der Atomgewichte und Verwandte Fragen *Physikal. Z.* **19**, 490-500.
6. O'Hagen, A. (1988) *Probability: Methods and Measurement* Chapman and Hall, London.
7. Edwards, A.W.F. (1972) *Likelihood* Cambridge University Press, Cambridge.
8. French, S. (1978) A Bayesian three stage model in crystallography *Acta Cryst.* **A34**, 728-738.
9. Bricogne, G. (1984) Maximum entropy and the foundations of direct methods, *Acta Cryst.* **A40**, 410-445.
10. Bricogne, G. (1988) A Bayesian statistical theory of the phase problem. I. A multichannel maximum entropy formalism for constructing generalized joint probability distributions of structure factors *Acta Cryst.* **A44** (1988). 517-545.
11. Gilmore, C.J. (1996) Maximum entropy and Bayesian statistics in crystallography: a review of practical applications *Acta Cryst.* **A52** (1996). 561-589.
12. Shannon, C.E. & Weaver, W. (1949). *The Mathematical Theory of Communication.* University of Illinois Press, Urbana, USA.
13. Jaynes, E.T. (1986). Where do we stand on maximum entropy? in J.H.Justice (ed.), *Maximum Entropy and Bayesian Methods in AppliedStatistics,* Cambridge University Press, Cambridge, pp 26-58.
14. Wilson, A.J.C. (1949) The probability distribution of X-ray intensities *Acta Cryst.* **2** 318-321.
15. Hauptman, H. and Karle, J. (1953) *The Solution of the Phase Problem. I. The Centrosymmetric Crystal ACA* Monograph no. 3. Polycrystal Book Service, New York.

HALF TITLE PAGE

2. TOOLS FOR SOLVING THE PHASE PROBLEM

2.1 Mathematical methods

2.2 Computational methods

2.3 Experimental methods

INTENSITY STATISTICS AND NORMALIZATION

ROBERT H. BLESSING, D.Y. GUO, and DAVID A. LANGS
Hauptman-Woodward Institute
73 High Street
Buffalo, New York 14203-1196, USA
E-mail: blessing@hwi.buffalo.edu

1. Introduction

The "fundamental theorem" of X-ray crystallography is (see, e.g., Coppens, 1996; Bricogne, 1996) the dual-space Fourier transform relationship

$$F(\mathbf{h}) \underset{\mathscr{F}^{-1}}{\overset{\mathscr{F}}{\rightleftharpoons}} \rho(\mathbf{r}) \quad (1)$$

between crystal structure factors in reciprocal space,

$$F(\mathbf{h}) = \mathscr{F}[\rho(\mathbf{r})] = \int_V d^3\mathbf{r}\ \rho(\mathbf{r}) \exp(+2\pi i\mathbf{h}\cdot\mathbf{r}) , \quad (2)$$

and the unit-cell electron density distribution in crystal space,

$$\rho(\mathbf{r}) = \mathscr{F}^{-1}[F(\mathbf{h})] = V^{-1} \Sigma_{\mathbf{h}} F(\mathbf{h}) \exp(-2\pi i\mathbf{h}\cdot\mathbf{r}) . \quad (3)$$

In crystal space $\mathbf{r} = x\mathbf{a} + y\mathbf{b} + z\mathbf{c} = \Sigma_i r^i \mathbf{a}_i$, and in reciprocal space $\mathbf{h} = h\mathbf{a}^* + k\mathbf{b}^* + \ell\mathbf{c}^* = \Sigma_i h_i \mathbf{a}^{*i}$, where $\mathbf{a}^{*i} = \mathbf{a}_j \times \mathbf{a}_k / V$, $V = (\mathbf{a}_i \times \mathbf{a}_j)\cdot\mathbf{a}_k$, $\mathbf{a}^{*i}\cdot\mathbf{a}_j = \delta^i_j$, $\delta^i_j = 1$ if $i = j$, $\delta^i_j = 0$ if $i \neq j$, and $i, j, k = 1, 2, 3$ (see, e.g., Shmueli, 1996; Sands, 1996). Due to the dual-space reciprocity, $\mathbf{h}\cdot\mathbf{r}$ reduces to $\mathbf{h}\cdot\mathbf{r} = hx + ky + \ell z$.

The structure factor is given by an atomic summation,

$$F(\mathbf{h}) = \sum_{a=1}^{N} f_a(\mathbf{h})\, W_a(\mathbf{h}) \exp(2\pi i\mathbf{h}\cdot\mathbf{r}_a) , \quad (4)$$

where, for atom *a* among the *N* atoms of the unit cell, $\mathbf{r}_a$ is the atomic position vector, $W_a(\mathbf{h})$ is the atomic Debye-Waller disorder and/or thermal vibration factor, and $f_a(\mathbf{h})$ is the atomic X-ray scattering factor. The Debye-Waller factor is the Fourier transform of the probability density of atomic displacements from the mean atomic position,

S. Fortier (ed.), Direct Methods for Solving Macromolecular Structures, 47–71.

$W_a(\mathbf{h}) = \mathscr{F}[p_a(\mathbf{r}_a - \langle\mathbf{r}_a\rangle)]$, and, at X-ray wavelengths sufficiently far removed from the crystal's resonant electronic absorption edges, the atomic scattering factor is the Fourier transform of the atomic electron density, $f_a(\mathbf{h}) = \mathscr{F}[\rho_a(\mathbf{r} - \mathbf{r}_a)] = \mathscr{F}[|\psi_a(\mathbf{r} - \mathbf{r}_a)|^2]$. For spherically symmetric or spherically averaged atomic densities, $f_a(\mathbf{h}) = f_a(|\mathbf{h}|) = f_a(h) = \mathscr{F}[4\pi r^2 |R_a(r)|^2]$, where $R_a(r)$ is the radial part of the atomic wavefunction $\psi_a(\mathbf{r})$. The Fourier transform product $(f_a W_a)$ in reciprocal space is the Fourier transform of the convolution product $(\rho_a * p_a)$ in crystal space, i.e., $f_a(\mathbf{h}) W_a(\mathbf{h}) = \mathscr{F}[\rho_a(\mathbf{r} - \mathbf{r}_a) * p_a(\mathbf{r}_a - \langle\mathbf{r}_a\rangle)]$. As the reciprocal space radius $|\mathbf{h}|$ increases through $0 \le |\mathbf{h}| < \infty$, the Debye-Waller factor decreases through $1 \ge W_a > 0$, and the atomic scattering factor decreases through $Z_a \ge f_a > 0$.

The electron density (3) is a real-valued, non-negative function, but the structure factor (2) or (4) is in general a complex-valued function,

$$\begin{aligned} F &= \Sigma_a f_a W_a \exp(i\phi_a) = \Sigma_a f_a W_a (\cos\phi_a + i \sin\phi_a) \\ &= \Sigma_a f_a W_a \cos\phi_a + i\Sigma_a f_a W_a \sin\phi_a = \Sigma_a A_a + i\Sigma_a B_a = A + iB , \\ F(\mathbf{h}) &= A(\mathbf{h}) + iB(\mathbf{h}) = |F(\mathbf{h})|[\cos\phi(\mathbf{h}) + i \sin\phi(\mathbf{h})] = |F(\mathbf{h})| \exp[i\phi(\mathbf{h})] . \quad (5) \end{aligned}$$

Structure factor magnitudes $|F(\mathbf{h})|$ can be obtained from kinematical diffraction measurements of Bragg reflection intensities,

$$I(\mathbf{h}) \propto |F(\mathbf{h})|^2 = F^*(\mathbf{h})\, F(\mathbf{h}) , \quad (6)$$

where $F^* = |F|e^{-i\phi} = A - iB$ is the complex conjugate of F, with the imaginary unit i everywhere replaced by $-i$. The corresponding structure factor phases $\phi(\mathbf{h})$, however, cannot be measured experimentally [although, in specialized experiments designed to measure dynamical diffraction effects (see, e.g., Weckert, Schwegle, and Hümmer, 1993; Weckert and Hümmer, 1997) certain three-phase sums, $\phi(\mathbf{h},\mathbf{k}) = \phi(\mathbf{h}) + \phi(\mathbf{k}) + \phi(-\mathbf{h}-\mathbf{k})$, can be measured]. Possible values for the unmeasurable phases are constrained by the requirement that, in conjunction with a sufficient subset that contains the largest elements of the in-principle infinite set of measurable magnitudes, the phases should produce via (3) an electron density distribution that is real, non-negative, and atomic, i.e., a unit-cell density of the form $\rho(\mathbf{r}) = \Sigma_a \rho_a(\mathbf{r} - \mathbf{r}_a)$, which exhibits distinct local maxima corresponding to a superposition of resolved atomic densities.

For hard, small-unit-cell crystals containing Z asymmetric crystal chemical units per unit cell and $N/Z \lesssim 100$ independent non-hydrogen atoms per crystal chemical unit, with the atoms packed more-or-less tightly in metallic, covalent, ionic, or molecular structures, the usual experimental situation is that the number n of Bragg reflection intensities measurable as significant above background greatly exceeds the number $3\,N/Z$ of unknown atomic coordinates; commonly, $n \gtrsim 100\,N/Z$. The unmeasurable phases are then, in principle, over-determined by systems of simultaneous equations based on (4). The equations are, however, transcendental, so straightforward analytical solution is precluded, and we are left with the crystallographic phase problem solvable in principle, but in practice difficult to solve.

For soft, large-unit-cell crystals of biological macromolecules containing $N/Z \gtrsim 1000$ independent non-hydrogen atoms in loosely packed biomolecules and solvate water molecules, the usual experimental situation is much less favorable, because the atomic Debye-Waller factors in large-molecule crystals are generally smaller than those in small-molecule crystals by factors of about $e^{-2} \approx 0.1$ to $e^{-10} \approx 5\times10^{-5}$. This results in much steeper fall-off of the atomic $[f_a(h)\, W_a(\mathbf{h})]$ values with increasing $|\mathbf{h}|$, so that $n \lesssim 10\,N/Z$ and experimental resolution limits $d_{min} = 1/|\mathbf{h}|_{max} \gtrsim 1.5$ Å are commonplace. In practice, $|\mathbf{h}|_{max}$ is defined operationally as the $|\mathbf{h}|$ at which the local average measurement precision, $\langle\sigma(|F(\mathbf{h})|^2)\rangle/\langle|F(\mathbf{h})|^2\rangle$, falls to ~50% due to the fall-off of $[f_a(h)\, W_a(\mathbf{h})]$ values with increasing $|\mathbf{h}|$. In consequence, biomolecular crystal structure analyses are usually carried out, not at atomic resolution, but at resolutions corresponding to small groups of atoms. Atomic structures within the groups are assigned from a database of known structures of chemical functional groups determined in atomic-resolution analyses of small-unit-cell crystals. Biochemical functional groups typically contain four to ten non-hydrogen atoms and have group diameters of ~3.5 Å. The functional groups in proteins include, of course, $C^{\alpha}{}_{1/2}$–CO–NH–$C^{\alpha}{}_{1/2}$ main-chain peptide groups and side-chain groups of the 20 common amino acids. For macromolecular crystals that do not diffract to atomic resolution, the "atomicity" constraint translates to a requirement for distinct high-density volumes enveloped by isodensity surfaces at $\rho \propto \sigma(\rho) = \langle(\rho - \langle\rho\rangle)^2\rangle^{1/2}$ surrounding resolved groups or chains of atoms separated by distances that exceed the experimental resolution limit.

With due allowance for experimental resolution limits, the over-determinacy, non-negativity, and atomicity constraints on the unknown phases imply magnitude-conditioned relationships among them. Based on the existence of such relationships, and using hypothetical random-atom structures as starting models, the probabilistic phase estimation methods that are the subject of this book have been developed. The probabilistic treatments of the crystallographic phase problem may be broadly classified as either frequentist or Bayesian. Frequentist approaches are formulated in terms of normalized structure factors,

$$E(\mathbf{h}) = F(\mathbf{h})/\{\sum_{a=1}^{N} [f_a(h)\, W_a(\mathbf{h})]^2\}^{1/2}\,, \tag{7}$$

and Bayesian approaches are formulated in terms of unitary structure factors,

$$U(\mathbf{h}) = F(\mathbf{h})/\sum_{a=1}^{N} f_a(h)\, W_a(\mathbf{h})\,. \tag{8}$$

As will be shown below, for structures in space group $P1$ or $P\bar{1}$, the square of the denominator on the right-hand side of (7) is equal to the probabilistic or statistical expectation value of the squared structure factor magnitude, $\langle|F(\mathbf{h})|^2\rangle = \Sigma_a [f_a(h)\, W_a(\mathbf{h})]^2$, so the E normalization (Hauptman and Karle, 1953) is such that $\langle|E|^2\rangle = 1$. From (4) and (5), the denominator on the right-hand side of (8) represents the maximum possible structure factor magnitude, with all atoms scattering in phase with $\phi_a = 0$, so the U normalization (Harker and Kasper, 1948) is such that $|U| \leq 1$.

For structures that can be fairly approximated as being composed of equal atoms with equal mean-square atomic displacements, substituting (4) into (7) and (8) reduces them to

$$E(\mathbf{h}) \approx N^{-1/2} \sum_{a=1}^{N} \exp(2\pi i \mathbf{h}\bullet\mathbf{r}_a), \quad U(\mathbf{h}) \approx N^{-1} \sum_{a=1}^{N} \exp(2\pi i \mathbf{h}\bullet\mathbf{r}_a), \quad \text{and} \quad U(\mathbf{h}) \approx E(\mathbf{h})/\sqrt{N}.$$

Thus, in terms of normalized or unitary structure factors, structures of equal atoms with equal mean-square displacements are equivalent to structures of equal point-atoms at rest. Structure factor normalization thus represents a simplifying idealization of crystal structure that has been widely employed in developing probabilistic theory for the crystallographic phase problem. The crystallographic literature on intensity statistics and normalization is sizeable, and excellent summary reviews of it (Shmueli and Wilson, 1996; Giacovazzo, 1996) are presented in the new, generally excellent, Volume B of the *International Tables for Crystallography.*

2. Debye-Waller Factors

The atomic Debye-Waller factor appearing in equations (4) through (8) is (see, e.g., Johnson and Levy, 1974) given by

$$W_a(\mathbf{h}) = \exp(-2\pi^2 \langle u_a^2\rangle / d_h^2)\,, \tag{9}$$

where $\langle u_a^2\rangle$ is the mean-square value of the displacement of atom a from its mean position, $\mathbf{u}_a = \mathbf{r}_a - \langle\mathbf{r}_a\rangle$, due to disorder and/or thermal vibration perpendicular to the Bragg reflecting crystal planes with Miller indices $(h, k, \ell) = (\mathbf{h}\bullet\mathbf{a}, \mathbf{h}\bullet\mathbf{b}, \mathbf{h}\bullet\mathbf{c})$ and interplanar spacing $d_h = 1/|\mathbf{h}|$, where $|\mathbf{h}| = (\Sigma_i \Sigma_j\, h_i h_j\, \mathbf{a}^{*i}\bullet\mathbf{a}^{*j})^{1/2} = 2(\sin\theta_h)/\lambda$. In general, atomic displacements are anisotropic and, for trivariate Gaussian distributions of rectilinear displacements about the mean atomic position, (9) becomes

$$W_a(\mathbf{h}) = \exp(-\mathbf{h}^T \mathbf{b}_a \mathbf{h}) = \exp(-2\pi^2 \mathbf{H}^T \mathbf{U}_a \mathbf{H})\,, \tag{10}$$

where we employ matrix instead of vector notation in the exponential arguments in which $\mathbf{h}^T = [h\ k\ \ell]$ and $\mathbf{H}^T = [ha^*\ k^*\ \ell c^*]$ are row matrices, $\mathbf{h}$ and $\mathbf{H}$ are the corresponding column matrices, and $\mathbf{b}$ and $\mathbf{U}$ are square matrices that represent positive-definite, symmetric, second-rank tensors with six ($i \leq j = 1, 2, 3$) independent components for an atom with site symmetry 1. The $\mathbf{b}$ and $\mathbf{U}$ matrix elements are related by

$$b^{ij} = 2\pi^2 a^{*i} a^{*j} U^{ij}\,, \quad U^{ji} = U^{ij}\,, \quad U^{ii} = \langle (u^i)^2 \rangle\,, \tag{11}$$

and necessary and sufficient conditions for positive definiteness of the mean-square displacement tensors are given by the determinantal inequalities

$$U^{11} > 0\,, \qquad \begin{vmatrix} U^{11} & U^{12} \\ U^{21} & U^{22} \end{vmatrix} > 0\,, \quad \text{and} \quad \begin{vmatrix} U^{11} & U^{12} & U^{13} \\ U^{21} & U^{22} & U^{23} \\ U^{31} & U^{32} & U^{33} \end{vmatrix} > 0\,. \tag{12}$$

In an isotropic approximation (9) reduces to

$$W_a(\mathbf{h}) = W_a(|\mathbf{h}|) = \exp\,[-B_{iso,\,a}(\sin\theta_h)^2/\lambda^2]\,, \tag{13}$$

where $B_{iso} = 8\pi^2 \langle u^2 \rangle$. Equivalent isotropic (scalar) values can be obtained by contraction of anisotropic (tensor) values,

$$B_{iso,\,eq} = (8\pi^2/3)\,\Sigma_i\,\Sigma_j\,(\mathbf{a}_i\bullet\mathbf{a}_j)\,a^{*i} a^{*j} U^{ij} = (4/3)\,\Sigma_i\,\Sigma_j\,(\mathbf{a}_i\bullet\mathbf{a}_j)\,b^{ij}\,, \tag{14}$$

and, conversely, equivalent anisotropic values can be generated by expansion of isotropic values,

$$U_{eq}^{\ ij} = [B_{iso}/(8\pi^2)](\mathbf{a}^{*i}\bullet\mathbf{a}^{*j})/(a^{*i} a^{*j})\,. \tag{15}$$

3. Statistical Expectation Values and Probability Distributions

In (7), the normalized structure factor E(**h**) is defined with respect to the statistical expectation value of the squared structure factor magnitude $\langle|F(\mathbf{h})|^2\rangle$. For any function *y* of a random variable *x* distributed according to a distribution density function p(x), the statistical expectation value for *y* is given by

$$\langle y \rangle = \int_{-\infty}^{+\infty} y(x)\,p(x)\,dx \,\Big/ \int_{-\infty}^{+\infty} p(x)\,dx\,. \tag{16}$$

If p(x) is a normalized probability density function, the denominator in (16) is unity, and

$$P = \int_a^b p(x)\,dx \tag{17}$$

gives the probability that $a \le x \le b$. For a representative finite sample of *n* values of the quantity *y*, (16) can be approximated by a weighted average,

$$\langle y\rangle = \sum_{i=1}^{n} w_i y_i / \sum_{i=1}^{n} w_i \,, \tag{18}$$

where the weight w_i for each $y_i = y(x_i)$ is proportional to the relative distribution density $p(x_i)$. For distributions $p(x)$ with finite variance,

$$\mathrm{var}(x) = \sigma^2(x) = \langle (x - \langle x\rangle)^2\rangle \,, \tag{19}$$

appropriate weights are the reciprocal variances,

$$w_i = 1/\sigma^2(y_i) \,, \tag{20}$$

where, to approximation by a first-order Taylor expansion,

$$\mathrm{var}(y) = \sigma^2(y) = (dy/dx)^2\, \sigma^2(x) \,. \tag{21}$$

If the $\sigma(y_i)$ values are constant, unit weights $w_i = 1$ are appropriate, and (18) reduces to the arithmetic average $\langle y\rangle = (\Sigma_i y_i)/n$.

If $y = y(\mathbf{x})$ is a multivariate function of n random variables, $\mathbf{x} = (x_1, x_2, \ldots, x_n)$, the joint probability density function of $\mathbf{x}$ is the function $p_J(\mathbf{x})$ such that

$$P = \int_{a_1}^{b_1} \ldots \int_{a_n}^{b_n} p_J(\mathbf{x})\, d^n\mathbf{x} \,. \tag{22}$$

gives the probability that $a_1 \le x_1 \le b_1, \ldots,$ and $a_n \le x_n \le b_n$. The marginal probability density function $p_M(x_i)$ of an element x_i of $\mathbf{x}$, irrespective of the values of the other elements, is

$$p_M(x_i) = \int_{-\infty}^{+\infty} \ldots \int_{-\infty}^{+\infty} p_J(\mathbf{x})\, d^{n-1}\mathbf{x} \,, \tag{23}$$

where the $(n - 1)$-fold integration is performed over all the elements of $\mathbf{x}$ except x_i. If $p_J(\mathbf{x}, \mathbf{y})$ is the combined joint probability density function for two sets of random variables $\mathbf{x}$ and $\mathbf{y}$, then the conditional probability density function for $\mathbf{x}$ given fixed, particular values for the elements of $\mathbf{y}$ is

$$p_C(\mathbf{x}|\mathbf{y}) = p_J(\mathbf{x}, \mathbf{y})/p_M(\mathbf{y}) \,. \tag{24}$$

Therefore,

$$p_J(\mathbf{x}, \mathbf{y}) = p_C(\mathbf{x}|\mathbf{y})\, p_M(\mathbf{y}) = p_C(\mathbf{y}|\mathbf{x})\, p_M(\mathbf{x}) \tag{25}$$

from which follows Bayes's Theorem,

$$p_C(\mathbf{x}|\mathbf{y}) = p_C(\mathbf{y}|\mathbf{x})\, p_M(\mathbf{x})/p_M(\mathbf{y}) \,. \tag{26}$$

The multivariate analog of the univariate relationship (21) is

$$\sigma^2(y) = \sum_{i=1}^{n} \sum_{j=1}^{n} (\partial y/\partial x_i)(\partial y/\partial x_j)\, \rho_{ij}\, \sigma(x_i)\, \sigma(x_j) \,, \tag{27}$$

where ρ_{ij} is the correlation coefficient,

$$\rho_{ij} = \frac{\mathrm{cov}(x_i\,,\, x_j)}{[\mathrm{var}(x_i)\ \mathrm{var}(x_j)]^{1/2}} = \frac{\langle (x_i - \langle x_i\rangle)\, (x_j - \langle x_j\rangle)\rangle}{[\langle (x_i - \langle x_i\rangle)^2\rangle\, \langle (x_j - \langle x_j\rangle)^2\rangle]^{1/2}} \,, \tag{28}$$

which has the property $-1 \leq \rho_{ij} \leq +1$. For the joint variation of two or more functions of the same set of random variables,

$$\mathrm{cov}(y_i\,,\, y_j) = \sum_{k=1}^{n} \sum_{\ell=1}^{n} (\partial y_i/\partial x_k)(\partial y_j/\partial x_\ell)\, \mathrm{cov}(x_k\,,\, x_\ell) \,. \tag{29}$$

If **x** represents a set of *n* parameters fitted by least-squares minimization to a set of $m > n$ data **y**, then

$$\mathrm{var}(x_i) = \mathrm{cov}(x_i\,,\, x_i) \qquad \text{and} \qquad \mathrm{cov}(x_i\,,\, x_j) = a^{ij}\, \chi^2/(m-n) \,, \tag{30}$$

where a^{ij} is an element of the matrix $[a^{ij}]$ inverse to the matrix $[a_{ij}]$ of coefficients of the least-squares normal equations, and $\chi^2 = \Sigma_i\, w_i\, [y_i - y(\mathbf{x})]^2$ is the minimized least-squares residual. Equations (21) and (27) through (30) form the basis of propagation-of-error calculations of the effects of experimental measurement uncertainties on the uncertainties of measurement-derived quantities.

4. Wilson Expectation Values $\langle |F(\mathbf{h})|^2\rangle$ and the Wilson Plot

Historically, the literature on intensity statistics and normalization dates back to an exchange of letters to *Nature* between S.H. Yü (1942) and A.J.C. Wilson (1942). Wilson considered the squared structure factor magnitude from (4) for a structure in space group *P*1,

$$\begin{aligned} |F(\mathbf{h})|^2 &= F(\mathbf{h})\, F^*(\mathbf{h}) \\ &= \Sigma_a \Sigma_b\, f_a(h)\, f_b(h)\, W_a(\mathbf{h})\, W_b(\mathbf{h})\, \exp\,[2\pi i \mathbf{h}\bullet(\mathbf{r}_a - \mathbf{r}_b)] \\ &= \Sigma_a\, [f_a(h)\, W_a(\mathbf{h})]^2 + \Sigma_a \Sigma_{b\neq a}\, f_a(h)\, f_b(h)\, W_a(\mathbf{h})\, W_b(\mathbf{h})\, \exp\,[2\pi i \mathbf{h}\bullet(\mathbf{r}_a - \mathbf{r}_b)] \,, \end{aligned} \tag{31}$$

and pointed out that if the $|F(\mathbf{h})|^2$ are locally averaged in spherical shells of $|\mathbf{h}| = 1/d_h = 2(\sin\theta_h)/\lambda$ then, in shells in which d_h does not greatly exceed the near-neighbor values of the interatomic distances $|\mathbf{r}_a - \mathbf{r}_b|$, the arguments $\phi = 2\pi\mathbf{h}\cdot(\mathbf{r}_a - \mathbf{r}_b)$ of the functions $\exp(i\phi) = \cos\phi + i\sin\phi$ will sample the range $0 \leq \phi \pmod{2\pi} < 2\pi$ more-or-less uniformly, so that $\cos\phi$ and $\sin\phi$ will oscillate between positive and negative values and average to practically zero. The statistical expectation value for (31) will then be

$$\langle|F(\mathbf{h})|^2\rangle = \Sigma_a\,[f_a(h)\;W_a(\mathbf{h})]^2\,. \tag{32}$$

Usually Bragg intensities (6) are measured, not on the absolute scale, but on a relative experimental scale, so that

$$|F|_{abs} = k\,|F|_{rel}\,, \tag{33}$$

and, under the approximation of isotropic mean-square atomic displacements (13) that are approximately the same for all atoms of the unit cell, (32) and (33) yield

$$\langle|F(\mathbf{h})|_{rel}^2\rangle = k^{-2}\exp\,[-2B_{iso}(\sin\theta_h)^2/\lambda^2]\;\Sigma_a\,f_a^2(h)\,, \tag{34}$$

from which the absolute scaling factor k and the overall mean-square atomic displacement parameter B_{iso} can be estimated by means of a least-squares straight line fitted to a plot of $\ell n\,\langle|F(\mathbf{h})|_{meas}^2/\Sigma_a f_a^2(h)\rangle_{|\mathbf{h}|}$ vs. $\langle(\sin\theta_h)^2/\lambda^2\rangle_{|\mathbf{h}|}$, where the notation $\langle x\rangle_{|\mathbf{h}|}$ denotes a local spherical $|\mathbf{h}|$-shell average. From empirical estimates for k and B_{iso}, experimental estimates of normalized (7) or unitary (8) structure factor magnitudes can be obtained as

$$|E(\mathbf{h})|_{meas} = |F(\mathbf{h})|_{meas}\,k\exp\,[+B_{iso}(\sin\theta_h)^2/\lambda^2]\;[\Sigma_a\,f_a^2(h)]^{-1/2}\,, \tag{35}$$

$$|U(\mathbf{h})|_{meas} = |F(\mathbf{h})|_{meas}\,k\exp\,[+B_{iso}(\sin\theta_h)^2/\lambda^2]\;[\Sigma_a\,f_a(h)]^{-1}\,. \tag{36}$$

5. The Wilson Distributions

Considering (31) and (32) further, Wilson (1949) derived the marginal probability density functions for structure factor magnitudes and intensities to be expected from uniform random unit-cell distributions of atomic positions in the space groups $P1$ and $P\bar{1}$. Assuming a uniform random distribution of the atomic phase components $0 \leq \phi_a = 2\pi\mathbf{h}\cdot\mathbf{r}_a \pmod{2\pi} < 2\pi$, the $P1$ derivation applies the central limit theorem separately to the real part, $A = \Sigma_a f_a W_a \cos\phi_a = |F|\cos\phi$, and the imaginary part, $B = \Sigma_a f_a W_a \sin\phi_a = |F|\sin\phi$, of the structure factor (4) and (5). The $P\bar{1}$ derivation follows the real part of the $P1$ derivation, allowing for halving of the number of independent atoms since the atoms occur in pairs at positions $+\mathbf{r}_a$ and $-\mathbf{r}_a$, which restricts the structure factor phases to $\phi(\mathbf{h}) = 0$ or π.

The resulting Wilson probability density functions for crystal structure factor magnitudes $|F| = (A^2 + B^2)^{1/2}$ are then the Gaussian forms:

Acentric *P*1 distribution

$$p_M(|F|) = (2|F|/\langle|F|^2\rangle) \exp(-|F|^2/\langle|F|^2\rangle) , \qquad \mathrm{var}(|F|) = \langle|F|^2\rangle \qquad (37)$$

Centric $P\overline{1}$ distribution

$$p_M(|F|) = [2/(\pi\langle|F|^2\rangle)]^{1/2} \exp[-|F|^2/(2\langle|F|^2\rangle)] , \qquad \mathrm{var}(|F|) = \langle|F|^2\rangle \qquad (38)$$

where $\langle|F|^2\rangle$ is the Wilson expectation value given by (32). For normalized structure factor magnitudes, $|E| = |F|/\langle|F|^2\rangle^{1/2}$, the probability density functions are:

Acentric
$$p_M(|E|) = 2|E| \exp(-|E|^2) , \qquad (39)$$
$$\lim_{|E|\to 0} p_M(|E|) = 0 , \qquad \mathrm{var}(|E|) = \langle|E|^2\rangle = 1 ;$$

Centric
$$p_M(|E|) = (2/\pi)^{1/2} \exp(-|E|^2/2) , \qquad (40)$$
$$\lim_{|E|\to 0} p_M(|E|) = (2/\pi)^{1/2}, \qquad \mathrm{var}(|E|) = 2\langle|E|^2\rangle = 2 .$$

And for intensities, $I = |F|^2$, the probability density functions are:

Acentric
$$p_M(I) = \langle I\rangle^{-1} \exp(-I/\langle I\rangle) , \qquad (41)$$
$$\lim_{I\to 0} p_M(I) = \langle I\rangle^{-1} , \qquad \mathrm{var}(I) = \langle I\rangle ;$$

Centric
$$p_M(I) = (2\pi I\langle I\rangle)^{-1/2} \exp[-I/(2\langle I\rangle)] , \qquad (42)$$
$$\lim_{I\to 0} p_M(I) = \infty , \qquad \mathrm{var}(I) = 2\langle I\rangle .$$

The probability density functions (39) through (42), and the corresponding cumulative distribution functions

$$N(x) = \int_0^x p_M(x)\, dx , \qquad (43)$$

are illustrated in Figure 1. The acentric *P*1 distributions are narrower than the centric $P\overline{1}$ distributions, with lower probabilities of very small or very large values of $|E|$, $|F|$, or I.

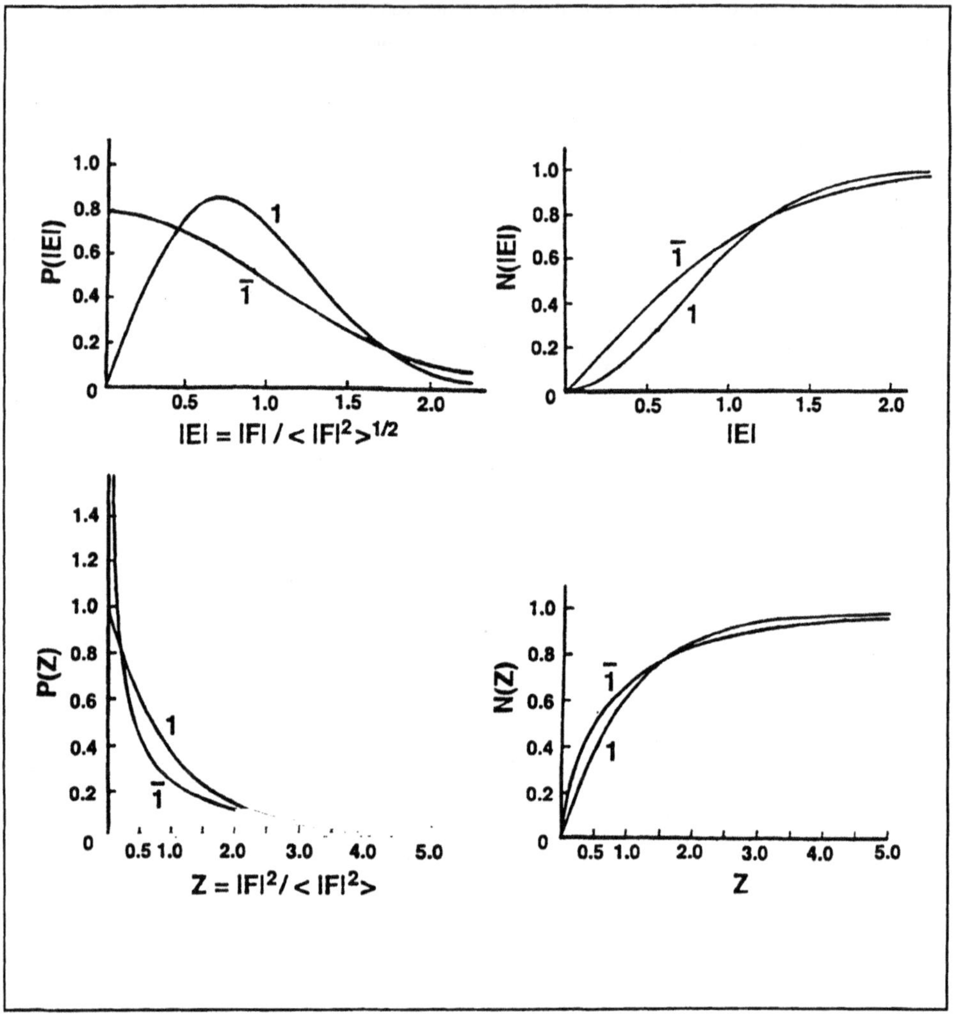

Figure 1. The Wilson distribution probability density functions, $p(|E|)$ and $p(|F|^2)$, and cumulative distribution functions, $N(|E|) = \int_0^{|E|} p(|E|)\ d|E|$ and $N(Z) = \int_0^Z p(Z)\ dZ$, where $Z = |F|^2/\langle|F|^2\rangle$.

Acentric $P1$	Centric $P\bar{1}$
$p(\lvert E\rvert) = 2\lvert E\rvert \exp(-\lvert E\rvert^2)$	$p(\lvert E\rvert) = (2/\pi)^{1/2} \exp(-\lvert E\rvert^2/2)$
$p(\lvert F\rvert^2) = \langle\lvert F\rvert^2\rangle^{-1} \exp(-\lvert F\rvert^2/\langle\lvert F\rvert^2\rangle)$	$p(\lvert F\rvert^2) = [2\pi\lvert F\rvert\langle\lvert F\rvert^2\rangle]^{-1/2} \exp[-\lvert F\rvert^2/(2\langle\lvert F\rvert^2\rangle)]$
$N(\lvert E\rvert) = 2\int_0^{\lvert E\rvert} x \exp(-x^2)\, dx = 1 - \exp(-\lvert E\rvert^2)$	$N(\lvert E\rvert) = (2/\pi)^{1/2}\int_0^{\lvert E\rvert} \exp(-x^2/2)\, dx = \mathrm{erf}(\lvert E\rvert/\surd 2)$
$N(Z) = \int_0^Z \exp(-x)\, dx = 1 - \exp(-Z)$	$N(Z) = (2\pi)^{-1/2}\int_0^Z x^{-1/2} \exp(-x/2)\, dx = \mathrm{erf}[(Z/2)^{1/2}]$

The $P1$ and $P\bar{1}$ Wilson distributions are archetypal. Their forms hold even for space groups of higher symmetry, if the asymmetric crystal chemical unit of the unit cell locally approximates a random-atom $P1$ or $P\bar{1}$ structure. The adaptation to higher symmetry is an integral factor $\epsilon \geq 1$ multiplying the distribution parameter $\langle|F|^2\rangle$ so that (32), (34) and (35) become

$$\langle|F(\mathbf{h})|^2\rangle = \epsilon(\mathbf{h}) \, \Sigma_a \, [f_a(h) \, W_a(\mathbf{h})]^2 \, , \tag{44}$$

$$\langle|F(\mathbf{h})|_{rel}{}^2\rangle = k^{-2} \exp [-2B_{iso}(\sin \theta_h)^2/\lambda^2] \, \epsilon(\mathbf{h}) \, \Sigma_a \, f_a{}^2(h) \, , \tag{45}$$

$$|E(\mathbf{h})|_{meas} = |F(\mathbf{h})|_{meas} \, k \exp [+B_{iso}(\sin \theta_h)^2/\lambda^2] \, [\epsilon(\mathbf{h}) \, \Sigma_a \, f_a{}^2(h)]^{-1/2} \, . \tag{46}$$

We call the factor $\epsilon(\mathbf{h})$ the *degeneracy* of the reciprocal lattice point **h** because it accounts for symmetry-dependent multiple enhancements of $P1$ or $P\bar{1}$ $|F(\mathbf{h})|^2$ expectation values. The degeneracy factor is given by

$$\epsilon(\mathbf{h}) = m_L \, \epsilon'(\mathbf{h}) \, , \tag{47}$$

where m_L is the lattice multiplicity,

$$m_L = \begin{cases} 1 & \text{for primitive P-lattices,} \\ 2 & \text{for C-, B-, A-, or I-centered lattices,} \\ 4 & \text{for F-centered lattices, or} \\ 3 & \text{for R-centered lattices on hexagonal axes,} \end{cases} \tag{48}$$

and

$$\epsilon'(\mathbf{h}) = 1, 2, 4, 8, 3, 6, \text{ or } 12 \tag{49}$$

is a projection symmetry multiplier for certain classes of zonal or axial reflections in particular reciprocal lattice point groups (Rogers, 1965, 1980). The m_L enhancements arise from the systematic extinction of a fraction $[1 - (1/m_L)]$ of the reflections due to lattice centering and the consequent concentration of the total scattering in the allowed fraction $1/m_L$ of the reflections. The ϵ' enhancements arise from superposition of symmetrically equivalent atoms in projection onto mirror planes or rotation axes. In the triclinic point groups 1 and $\bar{1}$, $\epsilon' = 1$ for all reflections. In all point groups, $\epsilon'(hk\ell) = 1$ for all non-axial, non-zonal, general reflections; but in the monoclinic point group 2/m (b-axis unique), for example, the zonal $h0\ell$ and axial $0k0$ reflections are special, and have $\epsilon'(h0\ell) = 2$ and $\epsilon'(0k0) = 2$, due to superposition in projection of mirror-equivalent and rotation-equivalent atoms, respectively. A useful complete table of ϵ' values has been given by Iwasaki and Ito (1977).

6. Wilson Normalization with a Statistical Expectation Value of the Debye-Waller Factor

Underlying the normalization equations (44) through (46) and (36) is the assumption that the atomic Debye-Waller factors vary little from atom to atom in the unit cell. For many crystals this is hardly the case. In small-unit-cell molecular crystals, mean-square atomic displacements due to thermal vibration are usually larger for atoms at the periphery of a molecule than for atoms near the molecular center of mass, and larger for conformationally flexible than for conformationally rigid functional groups. In crystals of biological macromolecules, displacements due to disorder and/or thermal vibration are generally larger for atoms at the biomolecular surface than for atoms in the biomolecular core, larger for side-chain than for main-chain atoms, and larger for solvate water molecules than for atoms of the biomolecule.

The crude approximation of constant atomic Debye-Waller factors can be replaced by a less crude approximation if, instead of factoring a squared-Debye-Waller factor W^2 out of the atomic sum (32), we factor-out the statistical expectation value $\langle W^2\rangle = \langle \exp(-2Bs^2)\rangle$, where $s = (\sin\theta)/\lambda$. Assuming that the unit-cell distribution of atomic B values can be fairly approximated by a normal distribution

$$p(B) = [(2\pi)^{1/2}\sigma_B]^{-1} \exp[-(B-\mu_B)^2/(2\sigma_B^2)] , \qquad (50)$$

with mean $\mu_B = \langle B\rangle$ and variance $\sigma_B^2 = \langle(B - \langle B\rangle)^2\rangle$, it has been shown (Blessing, Guo, and Langs, 1996) that (16) yields the expectation value

$$\langle W^2\rangle = \langle \exp(-2Bs^2)\rangle = \exp[-2(\mu_B - \sigma_B^2 s^2)s^2] . \qquad (51)$$

This indicates that, due to the spread of the unit-cell distribution of atomic mean-square displacements, the expectation value of the Debye-Waller factor corresponds to an effective overall B value, $B_{eff} = \langle B\rangle - \langle(B - \langle B\rangle)^2\rangle s^2$, that is smaller than the mean B and that decreases with increasing $(\sin\theta)/\lambda$.

Normalization effects of the spread of the distribution of mean-square atomic displacements can be sizeable because, especially in macromolecular crystals, $\langle(B - \langle B\rangle)^2\rangle^{1/2} \simeq \langle B\rangle$ is not uncommon for averages of structure-refined B values. The latter generally exhibit distributions that are positively skewed (since, of physical necessity, $B_{min} > 0$) and more sharply peaked than normal distributions. This suggests that (51) might be improved by employing an expansion about (50) to derive an expression for $\langle W^2\rangle$ that includes, in addition to the dispersion term in $(s^2)^2$, a skewness term in $(s^2)^3$ and a kurtosis term in $(s^2)^4$. It has, however, been shown that in practice, for data sets that extend to $d_{min} \lesssim 2.5$ Å resolution, such refinements are not necessary. Normalization via (51) of data from several well-determined protein crystal structures was shown to produce $|E_{obs}|$ values that agree with $|E_{calc}|$ values calculated from the refined $\mathbf{r}_a$ and B_a parameters as well as the un-normalized $|F_{obs}|$ values agree with the corresponding $|F_{calc}|$ values (Blessing, Guo, and Langs, 1996).

In terms of (51), with $s_h = (\sin \theta_h)/\lambda$, the empirical normalization equations (45), (46), and (36) become

$$\langle |F(\mathbf{h})|_{rel}^2 \rangle = k^{-2} \exp\,[-2(\mu_B - \sigma_B^2 s_h^2)s_h^2]\; \epsilon(\mathbf{h})\, \Sigma_a\, f_a^2(h)\ , \tag{52}$$

$$|E(\mathbf{h})|_{meas} = |F(\mathbf{h})|_{meas}\; k \exp\,[(\mu_B - \sigma_B^2 s_h^2)s_h^2]\; [\epsilon(\mathbf{h})\, \Sigma_a\, f_a^2(h)]^{-1/2}\ , \tag{53}$$

$$|U(\mathbf{h})|_{meas} = |F(\mathbf{h})|_{meas}\; k \exp\,[(\mu_B - \sigma_B^2 s_h^2)s_h^2]\; [\Sigma_a\, f_a(h)]^{-1}\ . \tag{54}$$

Via the Debye-Waller-factor relationships (10) and (13) through (15), these are readily recast in terms of overall anisotropic mean-square displacement parameters as

$$\langle |F(\mathbf{h})|_{rel}^2 \rangle = k^{-2} \exp\,\{-2[\mathbf{h}^T \boldsymbol{\mu}_b\, \mathbf{h} - (\mathbf{h}^T \sigma_b\, \mathbf{h})^2]\}\; \epsilon(\mathbf{h})\, \Sigma_a\, f_a^2(h)\ , \tag{55}$$

$$|E(\mathbf{h})|_{meas} = |F(\mathbf{h})|_{meas}\; k \exp\,[\mathbf{h}^T \boldsymbol{\mu}_b\, \mathbf{h} - (\mathbf{h}^T \sigma_b\, \mathbf{h})^2]\; [\epsilon(\mathbf{h})\, \Sigma_a\, f_a^2(h)]^{-1/2}\ , \tag{56}$$

$$|U(\mathbf{h})|_{meas} = |F(\mathbf{h})|_{meas}\; k \exp\,[\mathbf{h}^T \boldsymbol{\mu}_b\, \mathbf{h} - (\mathbf{h}^T \sigma_b\, \mathbf{h})^2]\; [\Sigma_a\, f_a(h)]^{-1}\ , \tag{57}$$

where here, as in (10), we employ matrix instead of vector notation in the exponential arguments in which $\mathbf{h}^T = [h\ k\ \ell]$ is a row matrix, $\mathbf{h}$ is the corresponding column matrix, and $\boldsymbol{\mu}_b$ and σ_b are symmetric matrices in which, to a first approximation from fitted isotropic, scalar values μ_B and σ_B,

$$\mu_{b,\,eq}^{\ ij} = \mu_B\, \mathbf{a}^{*i}\bullet\mathbf{a}^{*j}/4 \qquad \text{and} \qquad \sigma_{b,\,eq}^{\ ij} = \sigma_B\, \mathbf{a}^{*i}\bullet\mathbf{a}^{*j}/4\ . \tag{58}$$

7. Non-Wilson Characteristics of Intensity Distributions from Protein Crystals

The Wilson distributions have very wide ranges of effective applicability, but significant departures from the Wilson distributions do occur when the asymmetric crystal chemical unit does not locally approximate a *P*1 or $P\bar{1}$ structure of uniformly randomly distributed equal atoms. Deviant cases include: small, highly symmetric, highly heteroatomic structures; pseudosymmetric structures in which the crystallographic asymmetric unit has noncrystallographic symmetry or quasi-symmetry; and heavy-atom structures in which a small subset of the atoms of the asymmetric unit scatters much more strongly than the other atoms. Structure factor probability density functions for such cases have been derived as Edgeworth, Gram-Charlier, or Fourier-Bessel series expansions about the *P*1 or $P\bar{1}$ Wilson probability density functions, but discussion of these analyses would be beyond the scope of this chapter (see, e.g., Shmueli and Wilson, 1996; Castleden and Fortier, 1994).

Intensity and structure factor distributions from protein crystals depart from the Wilson distributions in several characteristic ways, because unit-cell distributions of atomic positions in protein crystals are characteristically nonuniform. Some 25 to 65% of the unit cell volume in protein crystals is occupied by solvent, mainly liquid-like water, filling the space between the large protein molecules (Matthews, 1968). Since $H_2O \cdots H_2O$ hydrogen bond distances $O \cdots O$ are ~2.8 Å, while protein C–O, C–N, C–C covalent bond lengths are ~1.2 to ~1.5 Å, average electron densities are lower in the solvent regions in protein crystals than in the protein regions. Simple empirical calculations (Blessing, Guo, and Langs, 1996) show that $\langle \rho_{protein} \rangle / \langle \rho_{solvent} \rangle \approx 4/3$. In addition, protein molecules have intricately folded polymeric $-C^{\alpha}-CO-NH-C^{\alpha}-$ structures with the fundamental repeat distances summarized in Table 1. These

Table 1. Fundamental repeat distances in protein crystals from standard bond lengths, valence angles, and conformation angles in peptides and water-water hydrogen bond geometry.

Repeat Unit	Repeat Distance
$C(\alpha_i) \cdots C(\alpha_{i+1})$	3.82 Å
$C(\alpha_{i-1}) \cdots C(\alpha_{i+1})$	5.42 Å in α-helices 6.92 Å in β-sheets
$H_2O \cdots H_2O$	2.75 Å $O \cdots O$ in ice
H_2O (central, H-bonded to two H_2O)	109.5° $O \cdots O \cdots O$
H_2O ······ H_2O	4.49 Å $O \cdots\cdots\cdots O$

ubiquitous molecular repeat distances in the range ~6 Å > $|\mathbf{r}_a - \mathbf{r}_b|$ > ~3 Å, along with $\cdots$protein$\cdots(H_2O)_x\cdots$protein$\cdots(H_2O)_x\cdots$ lattice or sub-lattice repeat distances in the range $|\mathbf{r}_a - \mathbf{r}_b|$ > ~30 Å, cause reflections with $d_h \gtrsim 3$ Å to violate the condition $d_h \lesssim |\mathbf{r}_a - \mathbf{r}_b|$ that underlies the deduction of the Wilson expectation values (32) and (44) (see also Harker, 1953). As a result, plots of $\ell n \langle |F(\mathbf{h})|_{meas}^2 / [\epsilon(\mathbf{h}) \Sigma_a f_a^2(h)] \rangle_{|\mathbf{h}|}$ vs. $\langle (\sin \theta_h)^2 / \lambda^2 \rangle_{|\mathbf{h}|}$ characteristically show pronounced nonlinear oscillations for $d = \lambda / (2 \sin \theta) \gtrsim 3$ Å, the most prominent deviations being a local minimum at $d \approx 6$ Å attributable to destructive interference of beams Bragg reflected from interleaved crystal planes corresponding to $C(\alpha_{i-1}) \cdots C(\alpha_i) \cdots C(\alpha_{i+1})$ repeats, and a local maximum at $d \approx 4$ Å attributable to constructive interference of beams reflected by adjacent planes corresponding to $C(\alpha_i) \cdots C(\alpha_{i+1})$ repeats. Typical examples of these effects are discussed in more detail and illustrated in Figures 1, 2, and 3 of Blessing, Guo, and Langs (1996); Figure 5 of Bricogne (1984); Figure 2 of French and Wilson (1978); and Figures 4 and 6 of Luzatti (1955).

8. Data Reduction and Error Analysis Procedures

Probabilistic phasing methods depend critically on normalized structure factor data sets that are as accurate and complete as possible. Foremost considerations include specimen crystal quality, instrument performance and calibration, and measurement strategy and technique. No less important are data processing procedures to reduce the in general multiply redundant set of raw intensity measurements to a unique set of structure factor magnitudes and, in the process, assess and preserve experimental accuracy and precision.

8.1. BACKGROUND SUBTRACTION, PEAK INTEGRATION, AND NET INTENSITY ESTIMATION

The procedures we employ for diffraction data from small-unit-cell crystals measured with four-circle diffractometers and single-reflection or point detectors have been described in some detail elsewhere (Blessing, 1987, 1989). The scheme of the procedures we employ for diffraction data from protein crystals measured using the oscillation method and area detectors is diagramed in Figure 2. Typically we process the oscillation frame images using the *Denzo* program (Otwinowski, 1993; Gewirth, Otwinowski, and Minor, 1995) to determine the crystal orientation and reflection indexing, the fully or partially recorded status of each reflection spot image, and the Lorentz- and polarization-corrected full- and partial-reflection net intensities $|F|^2 = (Lp)^{-1}(I_{peak} - I_{background})$ and their statistical experimental uncertainties $\sigma(|F|^2)$. In place of the *Scalepack* program that is part of the *Denzo* program package, we employ our programs *denzox* and *sortav* to evaluate interframe scale factors (Hamilton, Rollett and Sparks, 1968), scale the full and partial reflections, and sum the scaled partial reflections.

8.2. DATA MERGING AND EXPERIMENTAL UNCERTAINTY ESTIMATION

We also use the *sortav* program to evaluate and apply, when necessary, an empirical correction (Blessing, 1995) for residual anisotropic absorption-like errors not corrected by the interframe scaling; to average equivalent measurements using robust/resistant averaging weights to down-weight measurements that are outliers from multiple-measurement sample *medians* (Blessing, 1997a) [a procedure we have found to be superior to our earlier practice of normal-probability down-weighting of outliers from unit-weighted sample *means* (Blessing and Langs, 1987)]; and to perform a bivariate analysis of variance against $|F|^2$ and $(\sin\theta)/\lambda$ in order to improve the experimental uncertainty estimates obtained by propagation-of-error calculations applying (21) and (27) through (30) at each stage of the data processing.

```
                              Denzo
                    Peak-Background Integration
        |F(hkℓ)|² = (Lp)⁻¹(Ipk - Ibg),   σ(|F(hkℓ)|²)
                 Full/Partial Flag, Frame Number
                                ↓
                           "frames.x"
           (concatenated set of "denzo.x" files)
                                ↓
                        denzox/get fulls
                                ↓
                          "data.fulls"
                                ↓
"sortav.dat" (1)  →   sortav/interframe scale
                                ↓
                    "scalek.dat" + "frames.x"
                                ↓
                       denzox/sum partials
                                ↓
                          "data.total"
                                ↓
"sortav.dat" (2)  →     sortav/absorb/merge
                                ↓
                         "data.merged"
                                ↓
     "bayes.dat"  →             bayes
                                ↓
                          "data.bayes"
                                ↓
     "levy.dat"   →         levy or rogers      ←     "rogers.dat"
                                ↓
                    "eval.dat" + "data.bayes"
                                ↓
                               eval
                                ↓
                          "data.eval"
SIR or SAS cases                ↓
    "locscl.dat"  →            locscl
                                ↓
                   "data.locscl" + "eval.dat"
                                ↓
     "diffe.dat"  →            diffe
                                ↓
                          "data.diffe"
                                ↓
     DM, SIR, SAS, MIR, MAD, MR... Phasing Trials
```

Figure 2. Flow chart for processing diffraction data from protein crystals. Program names are indicated as program, and file names are indicated as "file". Files named "program.dat" are control data files, and files named "data.program" are reflection data files.

8.3. BAYESIAN POST-PROCESSING

The Laue-group or point-group unique data set resulting from the *sortav* processing is then post-processed with our program *bayes*, which applies a Bayesian statistical analysis (French and Wilson, 1978) to improve weak-reflection data with $|F|^2 \leq 3\sigma(|F|^2)$ and to derive appropriate values for the corresponding $|F|$ and $\sigma(|F|)$ data. Especially for protein crystals, the Bayesian post-processing can be important, because improving weak-reflection data is tantamount to increasing experimental resolution.

The post-processing applies Bayes's theorem (26) in the form

$$
\begin{aligned}
p_C(J|I) &\propto p_C(I|J)\, p_M(J) , \qquad (59)\\
\text{Posterior} &\propto \text{Likelihood} \times \text{Prior} ,\\
\text{Bayesian} &\propto \text{Normal} \times \text{Wilson} ,
\end{aligned}
$$

where $J = |F_o(\mathbf{h})|^2$ represents the "true" intensity, and $I = |F(\mathbf{h})|^2_{meas}$ represents the measured intensity, for a given reflection **h**. The *a priori* expectation $p_M(J)$ is the Wilson distribution (41) if **h** is an acentric reflection, or (42) if **h** is a centric reflection. From the central limit theorem, the likelihood is a normal distribution of measurement errors,

$$p_C(I|J) = [(2\pi)^{1/2}\sigma]^{-1} \exp[-(I - J)^2/(2\sigma^2)] , \qquad (60)$$

which, due to the statistical experimental uncertainty $\sigma = \sigma(I) = \sigma(|F(\mathbf{h})|^2_{meas})$, can yield $I < 0$ when $J \approx 0$ even though $J \geq 0$ is a physical necessity. The *a posteriori* Bayesian distributions are then

$$\text{Acentric} \quad p_C(J|I) \propto \exp[-(I - J)^2/(2\sigma^2)] \exp(-J/\langle J\rangle) , \qquad (61)$$

$$\text{Centric} \quad p_C(J|I) \propto J^{-1/2} \exp[-(I - J)^2/(2\sigma^2)] \exp[-J/(2\langle J\rangle)] , \qquad (62)$$

which, after "completing the square" and rearranging and collecting terms in the exponential arguments, become

$$
\text{Acentric} \quad p_C(J|I) \propto \exp\left[-\{J - [I - (\sigma^2/\langle J\rangle)]\}^2/(2\sigma^2)\right] ,
$$

$$
\begin{cases}
p_C(J|I) \propto \exp\left[-\{(J/\sigma) - [(I/\sigma) - (\sigma/\langle J\rangle)]\}^2/2\right] & \text{for } J \geq 0 , \\
p_C(J|I) = 0 & \text{for } J < 0 ;
\end{cases} \qquad (63)
$$

$$
\text{Centric} \quad p_C(J|I) \propto J^{-1/2} \exp\left\{-\left[J - \{I - [\sigma^2/(2\langle J\rangle)]\}\right]^2/(2\sigma^2)\right\} ,
$$

$$
\begin{cases}
p_C(J|I) \propto J^{-1/2} \exp\left\{-\left[(J/\sigma) - \{(I/\sigma) - [\sigma/(2\langle J\rangle)]\}\right]^2/2\right\} & \text{for } J \geq 0 , \\
p_C(J|I) = 0 & \text{for } J < 0 ;
\end{cases} \qquad (64)
$$

where the physical requirement $J \geq 0$, even if $I < 0$ due to peak-minus-background statistical fluctuations when $J \simeq 0$, is noted explicitly.

The probability density functions (63) and (64) are then used in (16) to improve the measured intensity data to Wilson-conditional Bayesian expectation values,

$$|F_o|^2 = \int_0^\infty J \;\; p_C(J|I)\, dJ \qquad \text{and} \qquad \sigma^2(|F_o|^2) = \int_0^\infty (J - |F_o|^2)^2 \;\; p_C(J|I)\, dJ\,,$$

and (65)

$$|F_o| = \int_0^\infty J^{1/2}\, p_C(J|I)\, dJ \qquad \text{and} \qquad \sigma^2(|F_o|) = \int_0^\infty (J^{1/2} - |F_o|)^2\, p_C(J|I)\, dJ\,,$$

all of which, including those for weak $J \simeq 0$ reflections that yield $I < 0$, are non-negative.

The experimental variables in (63) and (64) all appear in the dimensionless arguments $\{(I/\sigma) - [\sigma/(q\langle J\rangle)]\}$, where $I = |F(\mathbf{h})|^2_{meas}$, $\sigma = \sigma(|F(\mathbf{h})|^2_{meas})$, $q = 1$ for acentric $\mathbf{h}$, $q = 2$ for centric $\mathbf{h}$, and the Wilson distribution parameter $\langle J\rangle = \langle J(\mathbf{h})\rangle$ is estimated empirically from the local spherical $|\mathbf{h}|$-shell average of the measured intensities as $\langle J(\mathbf{h})\rangle \approx \epsilon(\mathbf{h})\langle|F|^2_{meas}/\epsilon\rangle_{|\mathbf{h}|}$. The integrals (65) have been evaluated numerically and tabulated against arguments, $-4 \leq (I/\sigma) - [\sigma/(q\langle J\rangle)] \leq +4$, which cover the range in which effects of the Bayesian treatment are significant (French and Wilson, 1978). The magnitudes of the changes $||F_o|^2 - |F|^2_{meas}|$ and $|\sigma(|F_o|^2) - \sigma(|F|^2_{meas})|$ depend primarily on I/σ and secondarily on $\sigma/(q\langle J\rangle)$; the magnitudes increase with decreasing I/σ and increasing $\sigma/(q\langle J\rangle)$. In general, the effects of the Bayesian treatment are: all $|F|^2_{meas} < 0$ are replaced by $|F_o|^2 \geq 0$, and most $|F|^2_{meas}$ with $0 < |F|^2_{meas} \lesssim 3\sigma(|F|^2_{meas})$ are replaced by $|F_o|^2 \lesssim |F|^2_{meas}$; $\sigma(|F_o|^2) \lesssim \sigma(|F|^2_{meas})$ since the Bayesian treatment tends to reduce measurement uncertainties by imposing the Wilson distribution requirements; and $|F_o| \leq (|F_o|^2)^{1/2}$ since, even if negative and positive errors in $|F|^2_{meas}$ are equally likely, negative errors are less likely than positive errors in the necessarily non-negative $|F_o|$ and $|F_o|^2$. If $I/\sigma >> \sigma/(q\langle J\rangle)$, the distributions (63) and (64) reduce to zero-mean, unit-variance normal distributions of $(I - J)/\sigma$. In practice, if $I \geq 3\sigma(I)$ the commonly used relationships $|F_o| = (|F_o|^2)^{1/2} = I^{1/2}$ and $\sigma(|F_o|) = \sigma(|F_o|^2)/(2|F_o|) = \sigma(I)/(2\,I^{1/2})$ are valid, and the Bayesian modifications are negligible.

8.4. STRUCTURE FACTOR NORMALIZATION

Since the Bayesian post-processing requires the evaluation of the local spherical $|\mathbf{h}|$-shell averaged intensities, our *bayes* program also produces a set of locally normalized data,

$$|E_o(\mathbf{h})| = |F_o(\mathbf{h})|/[\epsilon(\mathbf{h})\langle|F_o|^2/\epsilon\rangle_{|\mathbf{h}|}]^{1/2}. \tag{66}$$

To derive globally normalized data (53) or (56), the unique $|F_o|$ data set from the *bayes* program is analyzed with our program *levy* (Blessing, Guo, and Langs, 1996) to evaluate by least-squares fit the parameters k, μ_B, and σ_B of (52), or k, μ_b^{ij} and σ_b^{ij} of (55). In turn, the fitted parameters are used in our program *eval* to obtain the normalized structure factor magnitudes (53) or (56). For both the locally and globally normalized $|E|$ values, $\sigma(|E|)$ values are evaluated by propagation-of-error calculations based on (21) and (27) through (30) to include the effects of error-of-fit uncertainties of the normalization parameters. An important feature of the *levy* program is that it uses a logarithmically linearized least-squares fit based on (45) only to obtain a first approximation to the scale and mean-square displacement parameters, which the program then refines by properly weighted, iterative non-linear least-squares fit to the individual-reflection data, $|F_o(\mathbf{h})|^2/[\epsilon(\mathbf{h}) \Sigma_a f_a^2(h)]$, rather than logarithms of local spherical $|\mathbf{h}|$-shell data averages (Levy, Thiessen, and Brown, 1970). The individual-reflection fitting provides a direct evaluation of anisotropy of the mean-square-displacements distribution parameters μ_b^{ij} and σ_b^{ij}, and it allows the relatively many Wilson-distributed high-resolution data to overcome, or at least counteract, effects of non-Wilson distributions of the relatively few low-resolution data. [We have also found that the procedure in our program *levy* is often superior to a corresponding procedure in our earlier program *rogers* (Blessing and Langs, 1988), which estimates the parameters k and μ_b^{ij} through an analysis of the Patterson origin peak (Rogers, 1965; Nielsen, 1975).] For low-resolution data sets with $d_{min} > 2.5$ Å, global normalization via the *levy-eval* or *rogers-eval* programs might be unreliable, and it might be better to resort to the local normalization (66) provided by the *bayes* program.

9. Treatment of SIR and SAS Data

The cases of single isomorphous replacement (SIR) and single-wavelength anomalous scattering (SAS), and their extensions to the multiple isomorphous replacement (MIR) and multi-wavelength anomalous dispersion (MAD) cases, coupled with Patterson and molecular replacement (MR) analyses provide the classical tools of protein crystallography for dealing with the phase problem (see, e.g., Rossman and Arnold, 1996; Vijayan and Ramaseshan, 1996). Much of the current research on so-called direct methods (DM) probabilistic phasing is directed toward integrating DM with SIR, MIR, SAS, MAD, and MR techniques.

9.1. LOCAL SCALING

Given an SIR pair of data sets, $|F_{Nat}(\mathbf{h})|$ from a native protein crystal and $|F_{Der}(\mathbf{h})|$ from an isomorphous heavy-atom derivative crystal, or an SAS data set of Bijvoet or Friedel pairs, $|F(+\mathbf{h})|$ and $|F(-\mathbf{h})|$ from a crystal measured at an X-ray wavelength at which the crystal exhibits significant anomalous dispersion due to damped resonant scattering, classical SIR or SAS methods seek to determine the substructure of heavy atoms, or of atoms that dominate the anomalous scattering, from Patterson syntheses computed with squared difference coefficients $(|F_{Der}| - |F_{Nat}|)^2$ or $(|F_{+h}| - |F_{-h}|)^2$. These, like any analyses based on difference data, are highly susceptible to effects of experimental errors, since difference data $\Delta x = x_2 - x_1$ have uncertainties $\sigma(\Delta x) = [\sigma^2(x_1) + \sigma^2(x_2)]^{1/2}$ that are necessarily larger than either of the individual x_1 and x_2 data uncertainties.

To treat such cases, we employ our program *locscl* (Blessing, 1997b) to apply the local scaling procedure introduced by Matthews and Czerwinski (1975). The procedure assumes that errors that obscure the real differences between $|F_1|$ and $|F_2|$ pairs of data sets can be in large part empirically corrected by locally variable scale factors $q = q(\mathbf{h})$ defined by

$$\Delta|F| = |F_1| - q|F_2| \tag{67}$$

and estimated by least-squares fit minimizing

$$\chi^2 = \sum_{\mathbf{h}-\Delta\mathbf{h}}^{\mathbf{h}+\Delta\mathbf{h}} w[(|F_1|/|F_2|) - q]^2 , \tag{68}$$

where $w = w(\mathbf{h}) = \sigma^{-2}(|F_1|/|F_2|)$, and the notation

$$\sum_{\mathbf{h}-\Delta\mathbf{h}}^{\mathbf{h}+\Delta\mathbf{h}} x_{\mathbf{h}} = \sum_{\eta=h-\Delta h}^{h+\Delta h} \sum_{\kappa=k-\Delta k}^{k+\Delta k} \sum_{\lambda=\ell-\Delta\ell}^{\ell+\Delta\ell} x_{\eta\kappa\lambda} \tag{69}$$

denotes summation over a local block of reciprocal lattice points surrounding, but not including, the point of interest, $hk\ell$. For example, $\Delta h = \Delta k = \Delta\ell = 1$ defines a raster or three-dimensional moving window of (3×3×3) − 1 points for the local scale factor fit. The raster semidimensions need not, however, be equal. For a crystal with unit cell dimensions $c >> a > b$, and therefore reciprocal cell dimensions $c^* << a^* < b^*$, one might choose semidimensions $\Delta h = 2$, $\Delta k = 1$, $\Delta\ell = 4$ and use a raster of (5×3×9) − 1 points to sample local blocks of the reciprocal lattice that have edges of roughly equal length along a*, b*, and c*. Our program *locscl* chooses statistically optimum raster semidimensions by analyzing the global variation of the locally fitted scale factors and their error-of-fit uncertainties as the semidimensions Δh, Δk, and $\Delta\ell$ are iteratively varied in proportion to the unit cell dimensions *a*, *b*, and *c*, so that the raster retains the shape of an approximately rhombic parallelepiped with edges parallel to the *a**, *b**, and *c** axes of the reciprocal lattice. The *locscl* program also applies (21) and (27) through (30) to propagate the error-of-fit uncertainties of the local scale factors into the $\sigma(|F|)$ and $\sigma(|E|)$ values corresponding to the locally scaled $|F|$ and $|E|$ values.

9.2. DIFFERENCE STRUCTURE FACTOR NORMALIZATION

Among early efforts to exploit probabilistic phasing methods in protein crystallography were applications of the MULTAN program (see, e.g., Main, 1976, 1985) employing SIR (Wilson, 1978) or SAS (Mukherjee, Helliwell, and Main, 1989) difference-magnitude data. In connection with recent further work to develop stronger probabilistic methods for phasing difference magnitudes (Langs, Guo, and Hauptman, 1995; Smith, Nagar, Rini, Hauptman, and Blessing, 1997), we have developed a program *diffe* (Blessing, 1997c) that implements the following difference normalization procedures.

9.2.1. SIR Differences

In the SIR case the magnitude differences of interest are

$$\Delta = |F_{\text{Der}}| - |F_{\text{Nat}}| \,, \tag{70}$$

which, given the corresponding locally scaled $|E|$ magnitudes, can be calculated as

$$\Delta = (\epsilon_{\mathbf{h}} \Sigma_{\text{a,Der}} |f_a|^2)^{1/2} |E_{\text{Der}}| - (\epsilon_{\mathbf{h}} \Sigma_{\text{a,Nat}} |f_a|^2)^{1/2} |E_{\text{Nat}}| \,. \tag{71}$$

We recall that for structure factors $F = |F| \exp(i\phi)$ with $|F| \leq \Sigma_a f_a$, the Wilson distributions give the intensity expectation value $\langle |F|^2 \rangle = \epsilon_{\mathbf{h}} \Sigma_a f_a^2$. Therefore, for difference structure factors $F_\Delta = |\Delta| \exp(i\phi_{\text{Heavy}})$ with $|\Delta| \leq |F_{\text{Heavy}}| \leq \Sigma_{\text{a,Heavy}} |f_a|$, we expect squared SIR difference magnitudes with

$$\langle |\Delta|^2 \rangle \leq \langle |F_{\text{Heavy}}|^2 \rangle = \epsilon_{\mathbf{h}} \Sigma_{\text{a,Heavy}} |f_a|^2 = \epsilon_{\mathbf{h}} [(\Sigma_{\text{a,Der}} |f_a|^2) - (\Sigma_{\text{a,Nat}} |f_a|^2)] \,. \tag{72}$$

Thus, greatest-lower-bound estimates for SIR difference-E magnitudes can be calculated as

$$|E_\Delta| = \frac{|(\Sigma_{\text{a,Der}} |f_a|^2)^{1/2} |E_{\text{Der}}| - (\Sigma_{\text{a,Nat}} |f_a|^2)^{1/2} |E_{\text{Nat}}||}{q[(\Sigma_{\text{a,Der}} |f_a|^2) - (\Sigma_{\text{a,Nat}} |f_a|^2)]^{1/2}} \,, \tag{73}$$

where, following (51),

$$q = q_0 \exp(q_1 s^2 + q_2 s^4) \,, \quad \text{in which} \quad s = (\sin \theta)/\lambda, \tag{74}$$

is a least-squares-fitted normalization scaling function that imposes the requirement $\langle |E_\Delta|^2 \rangle = 1$ and is intended to empirically adjust for effects of: imperfect isomorphism of the derivative and native crystals; inaccurately known heavy-atom content due to multiple derivative sites and/or disordered partial site occupancies; and differences between the unit-cell distributions of mean-square atomic displacements in the heavy-atom substructure, the derivative crystal, and the native protein crystal.

The *diffe* program accepts user-supplied input cutoff values x_{min}, y_{min}, and s_{max} to limit the processing to data pairs with $\min[|E_{Nat}|/\sigma(|E_{Nat}|), |E_{Der}|/\sigma(|E_{Der}|)] \geq x_{min}$, $||E_{Der}| - |E_{Nat}||/[\sigma^2(|E_{Der}|) + \sigma^2(|E_{Nat}|)]^{1/2} \geq y_{min}$, and $s_h = (\sin\theta_h)/\lambda \leq s_{max}$, where typically $x_{min} = 3$, $y_{min} = 1$, and s_{max} is determined by inspection of a plot of the spherical shell averages $\langle|E_\Delta|^2\rangle_s$ vs. $\langle s\rangle$ from a preliminary run of the *diffe* program with unlimited s_{max}. The purpose of the s_{max} input cutoff is to prevent generating spuriously large $|E_\Delta|$ values for high-resolution data pairs measured with large uncertainties due to the general fall-off of scattering intensity with increasing scattering angle. The program propagates the error-of-fit uncertainties of the difference normalization scaling parameters into $\sigma(|E_\Delta|)$ values, and accepts a user-supplied output cutoff value z_{min} so that $|E_\Delta|$ values for which $|E_\Delta|/\sigma(|E_\Delta|) < z_{min}$, where typically $z_{min} = 3$, are rejected as too unreliable to be used in subsequent phasing calculations.

9.2.2. SAS differences

In the SAS case the magnitude differences of interest are

$$\Delta = |F_{+h}| - |F_{-h}| \,, \tag{75}$$

which, given the corresponding locally scaled $|E|$ magnitudes, can be calculated as

$$\Delta = (\epsilon_h \Sigma_a |f_a|^2)^{1/2} (|E_{+h}| - |E_{-h}|) \,,$$

$$\Delta = [\epsilon_h \Sigma_a (f_a^0 + f_a')^2 + (f_a'')^2]^{1/2} (|E_{+h}| - |E_{-h}|) \,. \tag{76}$$

Then, for structure factors $F_\Delta = |\Delta| \exp\{i[\phi_A + (\pi/2)]\}$ with $|\Delta| \leq 2|F''| \leq 2\Sigma_a f_a''$ and with ϕ_A representing the phase of the $F_A^0 + F'$ component of the total structure factor, $F = F^0 + F' + F'' = F_{non\text{-}A} + F_A^0 + F' + F''$, we expect squared SAS difference magnitudes with

$$\langle|\Delta|^2\rangle \leq 2\langle|F''|^2\rangle = 2\epsilon_h \Sigma_a (f_a'')^2 \,. \tag{77}$$

Thus, greatest-lower-bound estimates for SAS difference-E magnitudes can be calculated as

$$|E_\Delta| = \frac{(\Sigma_a |f_a|^2)^{1/2} \, ||E_{+h}| - |E_{-h}||}{2q[\Sigma_a (f_a'')^2]^{1/2}} \,, \tag{78}$$

where, again, $q = q_0 \exp(q_1 s^2 + q_2 s^4)$ with $s = (\sin\theta)/\lambda$ is a least-squares-fitted normalization scaling function that imposes the requirement $\langle|E_\Delta|^2\rangle = 1$ and, in the SAS case, is intended to empirically adjust for effects of: inaccurately known chemical composition of the unit cell; multiple sites and/or disordered partial site occupancies in the anomalously scattering substructure; inaccuracies in the values of the anomalous scattering corrections f′ and f″ and in the assumption that they are independent of both the magnitude and direction of **h**; and differences between the unit-cell distributions of mean-square atomic displacements in the strongly anomalously scattering substructure and the structure overall.

The *diffe* program again accepts user-supplied input cutoffs x_{min}, y_{min}, and s_{max}, propagates the error-of-fit uncertainties of the normalization parameters, and accepts an output cutoff z_{min}. The use of the data selection variables in the SAS case is directly analogous to their use in the SIR case described above.

10. Acknowledgements

Our own research related to the subjects of this chapter has been supported by grants from the US National Institutes of Health: at present grant no. GM46733, and in the past grants no. GM34073, HL32303, and DK19856. We are grateful for this support, and for helpful research discussions with our colleagues G. David Smith, Grant R. Moss, and Herbert A. Hauptman.

11. References

Blessing, R.H. (1987). Data Reduction and Error Analysis for Accurate Single-Crystal Diffraction Intensities. *Crystallogr. Rev.* **1**, 3-58

Blessing, R.H. (1989). DREADD - data reduction and error analysis for single crystal diffractometer data. *J. Appl. Cryst.*, **22**, 396-397.

Blessing, R.H. (1995). An Empirical Correction for Absorption Anisotropy. *Acta Cryst.*, A**51**, 33-38.

Blessing, R.H. (1997a). Outlier Treatment in Data Merging. *J. Appl. Cryst.* **30**, in press.

Blessing, R.H. (1997b). LOCSCL: a program to statistically optimize local scaling of SIR and SAS data. *J. Appl. Cryst.* **30**, 176-177.

Blessing, R.H. (1997c). Normalization of SIR and SAS Difference-Magnitudes for Phasing Heavy-Atom Substructures. *J. Appl. Cryst.* **30**, submitted.

Blessing, R.H., Guo, D.Y., and Langs, D.A. (1996). Statistical Expectation Value of the Debye-Waller Factor and |E(hkℓ)| Values for Macromolecular Crystals. *Acta Cryst.*, D**52**, 257-266.

Blessing, R.H., and Langs, D.A. (1987). Data averaging with normal down-weighting of outliers. *J. Appl. Cryst.* **20**, 427-428.

Blessing, R.H., and Langs, D.A. (1988). *A Priori* Estimation of Scale and Overall Anisotropic Temperature Factors From the Patterson Origin Peak. *Acta Cryst.*, A**44**, 729-735.

Bricogne, G. (1984). Maximum Entropy and the Foundations of Direct Methods. *Acta Cryst.* A**40**, 410-445.

Bricogne, G. (1996). Fourier transforms in crystallography: theory, algorithms, and applications. In *International Tables for Crystallography*, Vol. B, edited by U. Shmueli, pp. 23-106. Dordrecht, The Netherlands: Kluwer Academic Publishers.

Castleden, I.R., and Fortier, S. (1994). Intensity Statistics. I. *Acta Cryst.* A**50**, 9-17.

Coppens, P. (1996). The Structure Factor. In *International Tables for Crystallography*, Vol. B, edited by U. Shmueli, pp. 10-22. Dordrecht, The Netherlands: Kluwer Academic Publishers.

French, S., and Wilson, K.S. (1978). On the Treatment of Negative Intensity Observations. *Acta Cryst.* A**34**, 517-525.

Gewirth, D., Otwinowski, Z., and Minor, W. (1995). *The HKL Manual*, 4th ed. New Haven, Connecticut: Yale University, Department of Biophysics and Biochemistry.
Giacovazzo, C. (1996). Direct Methods. In *International Tables for Crystallography*, Vol. B, edited by U. Shmueli, pp. 201-229. Dordrecht, The Netherlands: Kluwer Academic Publishers.
Hamilton, W.C., Rollett, J.S., and Sparks, R.A. (1965). On the relative scaling of X-ray photographs. *Acta Cryst.* **18**, 129-130.
Harker, D. (1953). The Meaning of the Average of $|F|^2$ for Large Values of the Interplanar Spacing. *Acta Cryst.* **6**, 731-736.
Harker, D., and Kasper, J.S. (1948). Phases of Fourier Coefficients Directly from Crystal Diffraction Data. *Acta. Cryst.* **1**, 70-75.
Hauptman, H., and Karle, J. (1953). *Solution of the Phase Problem. I. The Centrosymmetric Crystal.* Am. Crystallogr. Assoc. Monograph No. 3. Dayton, Ohio: Polycrystal Book Service.
Iwasaki, H., and Ito, T. (1977). Values of ϵ for obtaining normalized structure factors. *Acta Cryst.* A**33**, 227-229.
Johnson, C.K., and Levy, H.A. (1974). Thermal Motion Analysis Using Bragg Diffraction Data. In *International Tables for X-Ray Crystallography*, Vol. IV, edited by J.A. Ibers and W.C. Hamilton, pp. 311-336. Birmingham, England: The Kynoch Press.
Langs, D.A., Guo, D.Y., and Hauptman, H.A. (1995). Use of 'Random-Atom' Phasing Methods to Determine Macromolecular Heavy-Atom Replacement Positions. *Acta Cryst.* D**51**, 1020-1024.
Levy, H.A., Thiessen, W.E., and (in part) Brown, G.M. (1970). A Least-Squares Method for Converting Observed Intensities into Normalized Structure Factors. Am. Crystallogr. Assoc. Meeting, Tulane Univ., New Orleans, Louisiana, March 1970. Abstract No. B6.
Luzatti, V. (1955). Sur l'Emploi des Méthodes Statistiques dans l'Étude de la Structure Cristalline des Protéines. *Acta Cryst.* **8**, 795-806.
Main, P. (1976). Recent Developments in the MULTAN System. The Use of Molecular Structure. In *Crystallographic Computing Techniques*, edited by F.R. Ahmed with K. Huml and B. Sedláček, pp. 97-105. Copenhagen: Munksgaard Publishers.
Main, P. (1985). MULTAN - a program for the determination of crystal structures, In *Crystallographic Computing 3: Data Collection, Structure Determination, Proteins, and Databases*, edited by G.M. Sheldrick, C. Krüger, and R. Goddard, pp. 206-215. Oxford, England: Oxford University Press.
Matthews, B.W. (1968). Solvent Content of Protein Crystals. *J. Mol. Biol.* **33**, 491-497.
Matthews, B.W., and Czerwinski, E.W. (1975). Local Scaling: A Method to Reduce Systematic Errors in Isomorphous Replacement and Anomalous Scattering Measurements. *Acta Cryst.* A**31**, 480-497.
Mukherjee, A.K., Helliwell, J.R., and Main, P. (1989). The Use of MULTAN to Locate Positions of Anomalous Scatterers. *Acta Cryst.* A**45**, 715-718.
Nielsen, K. (1975). Calculation of *E* Values by Means of the Origin Peak in the Patterson Function. *Acta Cryst.* A**31**, 762-763.
Otwinowski, Z. (1993). Oscillation Data Reduction Program. In *Proceedings of the CCP4 Study Weekend: Data Collection and Processing*, 29-30 January 1993, edited by L. Sawyer, N. Isaacs, and S. Bailey, pp. 56-62. Daresbury, England: SERC Daresbury Laboratory.
Rogers, D. (1965). Statistical Properties of Reciprocal Space. The Scaling of Intensities. In *Computing Methods in Crystallography*, edited by J.S. Rollett, pp. 117-132 and 133-148. Oxford, England: Pergamon Press.
Rogers, D. (1980). Calculation of $|E|$ Values. In *Theory and Practice of Direct Methods in Crystallography*, edited by M.F.C. Ladd and R.A. Palmer, pp. 82-92. New York: Plenum Press.
Rossman, M.G., and Arnold, E. (1996). Patterson and Molecular Replacement Techniques. In *International Tables for Crystallography*, Vol. B, edited by U. Shmueli, pp. 230-263. Dordrecht, The Netherlands: Kluwer Academic Publishers.
Sands, D.E. (1996). Dual Bases in Crystallographic Computing. In *International Tables for Crystallography*, Vol. B, edited by U. Shmueli, pp. 331-344. Dordrecht, The Netherlands: Kluwer Academic Publishers.

Shmueli, U. (1996). Reciprocal Space in Crystallography. In *International Tables for Crystallography*, Vol. B, edited by U. Shmueli, pp. 2-9. Dordrecht, The Netherlands: Kluwer Academic Publishers.

Shmueli, U., and Wilson, A.J.C. (1996). Statistical Properties of the Weighted Reciprocal Lattice. In *International Tables for Crystallography*, Vol. B, edited by U. Shmueli, pp. 184-200. Dordrecht, The Netherlands: Kluwer Academic Publishers.

Smith, G.D., Nagar, B., Rini, J.M., Hauptman, H.A., and Blessing, R.H. (1997). The Use of SnB to Determine an Anomalous Scattering Substructure. *Acta Cryst.* D**53**, submitted.

Vijayan, M., and Ramaseshan, S. (1996). Isomorphous Replacement and Anomalous Scattering. In *International Tables for Crystallography*, Vol. B, edited by U. Shmueli, pp. 264-279. Dordrecht, The Netherlands: Kluwer Academic Publishers.

Weckert, E., and Hümmer, K. (1997). Multiple-Beam X-ray Diffraction for Physical Determination of Reflection Phases and Its Applications. *Acta Cryst.* A**53**, 108-143.

Weckert, E., Schwegle, W., and Hümmer, K. (1993). Direct phasing of macromolecular structures by three-beam diffraction. *Proc. Roy. Soc. London,* A**442**, 33-46.

Wilson, A.J.C. (1942). Determination of Absolute from Relative X-Ray Intensity Data. *Nature*, **150**, 152.

Wilson, A.J.C. (1949). The Probability Distribution of X-Ray Intensities, *Acta Cryst.* **2**, 318-320. See also the discussion in Shmueli and Wilson (1996).

Wilson, K.S. (1978). The Application of MULTAN to the Analysis of Isomorphous Derivatives in Protein Crystallography. *Acta Cryst.* B**34**, 1599-1608.

Yü, S.H. (1942). A New Synthesis of X-Ray Data for Crystal Analysis. *Nature*, **149**, 638-639, 729. Determination of Absolute from Relative X-Ray Intensity Data. *Nature*, **150**, 151-152.

INVARIANTS, SEMINVARIANTS AND PHASE RELATIONSHIPS

H. Schenk
Laboratory for Crystallography, University of Amsterdam
Nieuwe Achtergracht 166, 1018 WV Amsterdam
The Netherlands.

1. Introduction

In direct methods quantities which are independent of the choice of origin are generally called structure invariants. Under certain conditions for some structure invariants numerical values can be estimated. Then they are called phase relationships, or sometimes also invariants. However, it is sensible to make a distinction between all invariants and those, for which it is possible to estimate their phase sums. For direct methods the latter are of course the only important ones, because reliably estimated phase relationships are the basis of a reliable phasing method.

The purpose of this chapter is to explore in a systematic way the nature of structure invariants and their space group dependent counterparts, the seminvariants, and also under what conditions they may prove to be useful phase relationships.

2. Structure invariants

The definition of a structure invariant states that it is a quantity the value of which remains unchanged when the origin of the unit cell is shifted over an arbitrary distance.

The most simple example of a structure invariant is of course the intensity of a reflection or $|F|^2$. Because in a diffraction experiment the intensity can be measured, and of course in that experiment no origin of the unit cell has been defined, it must be an invariant. On the other hand the structure factor F itself cannot be a structure invariant, otherwise the phase problem would not exist in crystallography. Both statements can be easily verified theoretically in terms of the structure factor equation:

$$F_H = |F_H| \exp i\varphi_H = \sum_{j=1}^{N} f_j \exp(2\pi i H.r_j) \tag{1}$$

With a shift Δr of the origin, (1) changes to

$$F_H' = \sum_{j=1}^{N} f_j \exp 2\pi i H.(r_j + \Delta r) = |F_H| \exp(2\pi i H.\Delta r).\exp i\varphi_H \tag{2}$$

So by shifting the origin by Δr the phase of a reflection is changed by $2\pi i H.\Delta r$ radians and $F \neq F'$. Thus F_H is not a structure invariant. In a similar way it is easily shown that the intensity $I \approx |F|^2 = F_H F_{-H}$ is a structure invariant:

$$|F_H|^2 = F_H' F_{-H}' = \exp(2\pi i H.\Delta r).F_H.\exp(-2\pi i H.\Delta r).F_{-H} = F_H F_{-H} \tag{3}$$

Less trivial and in the practice of Direct Methods very important is the fact that $F_H F_K F_{-H-K}$ is also an invariant:

$$F_H' F_K' F_{-H-K}' = F_H.\exp(2\pi i H.\Delta r).F_K.\exp(2\pi i K.\Delta r).F_{-H-K}\exp(2\pi i(-H-K).\Delta r) =$$

$$= F_H F_K F_{-H-K} \tag{4}$$

Since the moduli of the structure factors are invariants themselves the angular part of

This article was previously published in NATO ASI Series B: Vol. 274, 1991.

S. Fortier (ed.), Direct Methods for Solving Macromolecular Structures, 73–78.

$F_H F_K F_{-H-K}$ i.e. $\exp i\varphi_H . \exp i\varphi_K . \exp i\varphi_{-H-K}$, is also invariant:

$$\varphi_3 = \varphi_H + \varphi_K + \varphi_{-H-K} \tag{5}$$

In general, writing expressions similar as (4), it can be shown that if the sum of the indices equals zero the corresponding product of the structure factors is a structure invariant. Another example of a structure invariant is the quartet phase sum:

$$\varphi_4 = \varphi_H + \varphi_K + \varphi_L + \varphi_{-H-K-L} \tag{6}$$

Note that when a quantity is a structure invariant this does not imply that its value is known. However, in principle the values of structure invariants can only be functions of the values of other structure invariants, c.q. the intensities. So there is a fair chance that relations among structure invariants lead to useful values of some of them, such as (5).

3. Structure seminvariants

The structure seminvariants originate from space group symmetry, or more precisely, from the fact that the choice of origin in all but one space group is conventionally being limited to a set of special points in the unit cell. If the origin is shifted within this set of points, some of the non-invariants are now not changing their phases; they are called structure seminvariants.

Consider an example in P1: in this space group the origin is conventionally chosen at one of the eight centres of symmetry in the unit cell differing in coordinates x, y and z by fractions 0.5 or 0.0. In the structure factor expression (2) Δr is restricted to (0, 0, 0.5), (0, 0.5, 0) etc. and it can easily be seen that if H has ggg parity, $F'_H = F_H$ for all origin shifts permitted by this conventional origin choice. Therefore F_H with ggg-parity is called a structure seminvariant.

In the same way it can easily be seen that in the same space group also the product $F_H F_K$ is a structure seminvariant, provided that H+K is in the ggg-parity class:

$$F_H' F_K' = \exp(2\pi i H.\Delta r).F_H.\exp(2\pi i K.\Delta r).F_K = F_H F_K.\exp(2\pi i (H+K).\Delta r) \tag{7}$$

And as a result also the sum of the two phases involved ($\varphi_H + \varphi_K$) is a structure seminvariant in case the sum H+K is in the ggg-parity. Similarly the products of three and more structure factors and their respective phase sums are structure seminvariants provided that the sums of the indices involved in such a product have the ggg-parity.

Since the seminvariants depend on space group symmetry only, they are numerous and have to be derived separately for each space group. This has been done and the results are tabulated in [1].

4. Estimating phase sums of structure invariants and seminvariants

Estimating the phase sums of structure invariants and seminvariants is a problem which has intrigued workers in the field of direct methods for many years. The reason for this lies in the fact that, for example, knowledge of the actual values of φ_3 (5) reduces the solution of the phase problem to a simple numerical calculation. Therefore, reliable estimates of φ_3 or $|\varphi_3|$ would strengthen direct method procedures enormously. Instead, direct methods mostly use $|\varphi_3|=0$ for large values of E_3, and this may lead to failures when solving the phase problem.

Reliable estimates of phase sums have become available for invariants and seminvariants other than φ_3. The first results in this field were based on two observations, one that the estimation of the quartet invariant (6) is not only dependent on E_4 but also on the cross terms [2] and another that for special two-dimensional quartets the phase sum is equal to π [3], [4]. For both types of invariants applications were worked out at the same time.

This work was the starting point for a more systematic approach to the search for those structure factors upon which the structure (sem)– invariants most sensitively depend, ultimately resulting in the Neighborhood Concept of Hauptman [5], [6] and the Representation Theory of Giacovazzo [7]. These theories are based on three principles: a) the cross–term principle, b) the multipole or identity principle, and c) the extension principle. In the representation theory the space group symmetry is included in a systematic way, whereas this is done in a more implicit way in the neighborhood concept.

4.1. CROSS TERMS

Cross terms were used for the first time by Schenk [2], [4] and Schenk & De Jong [3] in order to estimate phase sums $|\varphi_4|$ of quartets (6). The example of quartets shows that cross terms originate from the fact that higher order invariants can be obtained by summing suitable lower order invariants. In general it can be stated that for any invariant

$$\varphi_n = \varphi_H + \varphi_K + \varphi_L + \varphi_M + \ldots\ldots \tag{8}$$

its value φ_n is controlled by the magnitudes of the main reflections H, K, L, M.... and those of the cross reflections H+K, H+L, ... K+L, ... H+K+L, ... Examples for quintets, sextets, etc., can be easily worked out.

4.2. IDENTITIES

The identity principle is most easily explained on the basis of quadrupoles of triplets, implicitly introduced by Sayre [8], and further explored by e.g. Vaughan [9], Karle and Hauptman [10], De Vries [11] and Viterbo [12]. A quadrupole is an identity among four triplet phase sums:

$$\left.\begin{aligned} &\varphi_H + \varphi_K + \varphi_{-H-K} = \varphi_3 \\ &\varphi_P + \varphi_{-H} + \varphi_{H-P} = b \\ &\varphi_{-P} + \varphi_{-K} + \varphi_{K+P} = c \\ &\varphi_{H+K} + \varphi_{P-H} + \varphi_{-K-P} = d \end{aligned}\right\} \quad \text{with } \varphi_3+b+c+d=0 \tag{9}$$

For φ_3 it follows that its value is also dependent on b, c, and d. Thus in case all reflections involved, H,K,–H–K,P,H–P and K+P, are strong and b,c, and d are likely to be near zero, the identity will support $|\varphi_3|=0$. The vector P is not defined and in order to get full support for $|\varphi_3|$ P must range over all reciprocal vectors. Thus where a triplet phase sum is primarily controlled by the $|E|$–values of H, K and –H–K, it is in second approximation dependent on the $|E|$–values of P, H–P and K+P where P ranges over all reciprocal space.

In the same way identities (or multipoles) can be set up for any invariant. Another example is the triple of quartets:

$$\left.\begin{aligned} &\varphi_H + \varphi_K + \varphi_L + \varphi_{-H-K-L} = \varphi_4 \\ &\varphi_{-H} + \varphi_{-K} + \varphi_P + \varphi_{H+K+P} = b \\ &\varphi_{-L} + \varphi_{H+K+L} + \varphi_{-P} + \varphi_{-H-K+P} = c \end{aligned}\right\} \quad \text{with } \varphi_4+b+c = 0 \tag{10}$$

From the cross–term principle we know already that $|\varphi_4|$ is controlled in first approximation by the $|E|$–values of the main reflections H, K, L and –H–K–L and in second approximation by those of the cross reflections H+K, H+L, and K+L. Now in third approximation the reflections P, H+K–P and the cross terms P–H, P–K, –P–L, and –P+H+K+L, where P ranges over all reciprocal space, can be added; this is the third neighbourhood introduced by Hauptman [5].

Many other identities are possible and the results have in common that the $|E|$'s which are identified to be important for the calculation of the phase sum of an invariant contain a free reciprocal lattice vector P. This implies that P always ranges over all reciprocal vectors and that calculations based on the multipole or identity principle are demanding as regards to computer-time.

4.3. REPRESENTATIONS

The third principle to be dealt with, the representation principle, is used primarily to identify the $|E|$-values controlling the phase sums of structure seminvariants. In general this principle works as follows: a structure (sem)invariant is summed with a second (sem)invariant to form an invariant, then this invariant is estimated and in case the second (sem)invariant has a trivial value the first (sem)invariant is estimated at the same time. All structure factors used in this process are then identified as being phase-sum controlling reflections. Although this principle was used implicitly from as early as 1948 [13], [14], it is used explicitly by Giacovazzo [7] and Hauptman [6] under the names representation and embedding (extension) respectively.

The most simple example of the principle is the Σ_1-relation or the one phase seminvariant φ_{2H} in $P\bar{1}$. By making the special triplet phase sum

$$\varphi_3' = \varphi_{2H} + \varphi_{-H} + \varphi_{-H} \tag{11}$$

it is seen that for large values of $|E_H|$ and $|E_{2H}|$ the magnitude of $|\varphi_{2H}|$ is known because φ_3 is most probably zero and in $P\bar{1}$ $\varphi_{-H}+\varphi_{-H}$ is trivially zero.

A next example are the 0g0 reflections in monoclinic space groups. These can be determined by means of a large number of invariants; for instance reflection 020 can be embedded in three phase invariants of the type

$$\varphi_3'' = \varphi_{020} + \varphi_{h-1\ell} + \varphi_{-h-1-\ell} \tag{12}$$

The seminvariant $\varphi_{h-1\ell} + \varphi_{-h-1-\ell}$ again is determined by symmetry only and thus for large $|E|$-values the invariant φ_3'' is known and thus φ_{020}.
Giacovazzo [7], [15] has tackled this principle very systematically, including also space-group symmetry.

4.4. EXAMPLES

The rules to identify all structure factor magnitudes which control the phase sum of a structure invariant or seminvariant are the following:

- For invariants first the cross terms are identified if present and if a next group of (less important) structure factors is needed the invariant is used in an identity or an extension.
- Seminvariants are first extended to invariants and then, by the cross-term principle, and if necessary by further embedding in identities, the important reflections are identified.

The following example seems rather trivial. Starting with the triplet relation the invariant $\varphi_P+\varphi_{-P}$, trivially equal to 0, is added and the resulting invariant is a quintet

$$\varphi_3 = \varphi_H + \varphi_K + \varphi_{-H-K} + \varphi_P + \varphi_{-P}.$$

This quintet may then be estimated using the $|E|$-values of the main terms H, K, -H-K, P, and -P and those of the cross terms H+K, H, K, O, ±P+H, ±P+K and ±P-H-K. The first four cross terms obviously do not add information, but the other six are new. In fact -P+H and P+K can be recognized as terms also occurring in the quadrupoles and by taking other triplet sequences it can be shown that the six cross terms involving P give rise to six quadrupoles. Thus the rather trivial trick of adding $\varphi_P+\varphi_{-P}$ is closely related to the quadrupole or identity principle.

In the literature [7] many more examples can be found and it can easily be seen that they all are based on the three principles treated here. Of course one has to take care of the expected effectiveness of the respective reflections. This can be worked out in an example for quartets

$$\varphi_4 = \varphi_H + \varphi_K + \varphi_L + \varphi_{-H-K-L}$$

The most important reflections, of course, are the four main terms H,K,L and -H-K-L. Then the next effective reflections are the cross terms H+K, H+L and K+L. The reflections P, H+K-P, P-H, P-K, -P-L and -P+H+K+L which follow from the multipole principle are third in order of importance [5]. An alternative third group may consist of those reflections obtained by taking $\varphi_4 + \varphi_P + \varphi_{-P}$, namely reflection P and the sextet cross terms, H±P, K±P, L±P, -H-K-L±P, H+K±P and K+L±P [7].

By using the three principles it is also possible to predict roughly the effect of the $|E|$'s of the reflections on the value of the invariant. An example again the quartet: we have seen that in case $|E_{H+K}|$, $|E_{H+L}|$ and $|E_{K+L}|$ are small, φ_4 is likely to be π. Now the third neighbourhood follows from the identity (10) and thus either b or c have to be π and the other 0. In case $b \simeq \pi$ it is expected that the cross terms P-H and P-K are small and the other terms -P-L and -P+H+K+L corresponding to c=0 are strong. Thus using the third neighbourhood of the quartet it can be stated that $|E_H|$, $|E_K|$, $|E_L|$ and $|E_{-H-K-L}|$ large, $|E_{H+K}|$, $|E_{H+L}|$ and $|E_{K+L}|$ small, $|E_P|$, $|E_{H+K-P}|$ large $|E_{P-H}|$ small and $|E_{-P-L}|$, $|E_{-P+H+K+L}|$ large, indicate that φ_4 is likely to be π.

5. Quartet relations

One example will be dealt with in some more detail, the quartet relation. Whereas the triplet relation is a two-dimensional phase relation and very successful in solving three-dimensional crystal structures, real three-dimensional phase relationships, the so-called quartet relations

$$\varphi_4 = \varphi_H + \varphi_K + \varphi_L + \varphi_{-H-K-L} = q \tag{13}$$

in which three linearly independent reflections H, K and L are involved. Quartets with $q \approx 0$ have been known since 1953 when they were published as Σ_5-relations in the monograph of Hauptman and Karle. Independently Simerska [16] derived quartets from a generalisation of the Sayre-Hughes equation for products of three reflections. In both cases it was shown that the probability that q = 0 was larger for larger values of

$$E_4 = N^{-1}|E_H E_K E_L E_{-H-K-L}| \tag{14}$$

or with other words the probability distribution of φ_4 is sharper for larger values of E_4. Simerska also tested the reliability of the quartet relations in a sample structure and suggested that they could be used for phasing in the same way as the triplet relation.

Quartets were not employed until 1973 when it was shown that, although quartets (13) with a probability distribution based on E_4 are weaker and thus less useful for actual phase determinations than triplets, by using the extra information of the so-called cross-term magnitudes $|E_{H+K}|$, $|E_{H+L}|$ and $|E_{K+L}|$, the quartets change into very useful relationships.

A simple explanation of the seven-magnitude positive ($q \approx 0$) and negative ($q \approx \pi$) quartet relations can be based on electron density considerations, following a reasoning as used in Schenk [17].

The positive quartet relation, expression (13) with $q \approx 0$, holds for large quartet products E_4. However, the reliability of this relation can be enhanced by combining the original quartet with an identical one constructed from two triplets:

$$\left.\begin{array}{l}\varphi_H + \varphi_K + \varphi_{-H-K} \simeq 0 \\ \varphi_L + \varphi_{-H-K-L} + \varphi_{H+K} \simeq 0\end{array}\right\} \quad \varphi_H + \varphi_K + \varphi_L + \varphi_{-H-K-L} \simeq 0 \tag{15}$$

which will hold for large E_4 and large $|E_{H+K}|$ [2]. Thus a large $|E_{H+K}|$ is an additional indication that the quartet holds. Similarly this applies to the other two cross terms K+L and H+L. In conclusion, a large value of E_4 and large $|E_{H+K}|$, $|E_{H+L}|$ and $|E_{K+L}|$ are four different indications that the positive quartet relation (13) with $q \approx 0$ is likely to be true, leading to improved reliabilities. As a result positive quartets are controlled by the magnitude of 7 structure factors and the number of useful quartets has been improved considerably. This implies that in actual phase determinations triplets and quartets can be used simultaneously and this will result in a more easy phase propagation and structure solution.

Quartet relations for which the sum $\varphi_H + \varphi_K + \varphi_L + \varphi_{-H-K-L} \approx \pi$ are referred to as negative quartets and such relations exist also for reasonably strong intensities for the reflections H, K, L and -H-K-L. In this case one expects that the triplets (15) do not exist and this is true only when the cross term reflection H+K has a neglectable E-value. In terms of electron densities it can even more explicitly explained [17]. As the same holds for the other two cross terms it can be stated that the negative quartet relation is likely to be true for reasonably large values of E_4 and small $|E_{H+K}|$, $|E_{H+L}|$ and $|E_{K+L}|$.

Negative quartets are not so numerous as positive quartets and triplets. Usually they comprise a small fraction of all relationships available, and as such they are in general not very important for the process of phase propagation. However, in a vast number of space groups they may play a crucial role when used in the Negative Quartet Criterion, a figure of merit to select the correct solution in multisolution methods or to determine the numerical values of the symbols in symbolic addition methods.

References

1. Karle, J. (1974). Direct Methods for Structure Determination: Origin Specification, Normalized Structure Factors, Formulas, and the Symbolic Addition Procedure for Phase Determination. In *International Tables for X-Ray Crystallography* IV. Edited by J.A. Ibers and W.C. Hamilton.The Kynoch Press, Birmingham, 337-358.
2. Schenk, H. (1973a). The use of Phase Relationships between Quartets of Reflexions. *Acta Cryst.* **A29**, 77-82.
3. Schenk, H. and De Jong, J.G.H. (1973). A Method for Direct Structure Determinations in P1bar and related Space Groups. *Acta Cryst.* **A29**, 31-34.
4. Schenk, H. (1973b). Direct Structure Determination in P1 and other Non-centrosymmetric Symmorphic Space Groups. *Acta Cryst.* **A29**, 480-481.
5. Hauptman, H. (1977). A Sequence of Nested Neighborhoods of the Structure Invariant $\varphi_h+\varphi_k+\varphi_l+\varphi_m$. *Acta Cryst.* **A33**, 553-555.
6. Hauptman, H. (1978). The Extension Concept and its Role in the Probabilistic Theory of the Structure Seminvariants. *Acta Cryst.* **A34**, 525-528.
7. Giacovazzo, C. (1977). A General Approach to Phase Relationships: The Method of Representations. *Acta Cryst.* **A33**, 933-944.
8. Sayre, D. (1952). The Squaring Method: a New Method for Phase Determination. *Acta Cryst.* 5, 60-65.
9. Vaughan, P.A. (1956). Am. Cryst. Ass. Ann. Meeting, Frenchlick, Ind.
10. Karle, J. and Hauptman, H. (1957). A Unified Algebraic Approach to the Phase Problem II. Space Group P1. *Acta Cryst.* 10, 515-524.
11. De Vries, A (1963), Thesis Utrecht.
12. Viterbo, D. (1976). In *Direct Methods in Crystallography*. Edited by H.A. Hauptman, Buffalo, p. 117.
13. Harker, D. and Kasper, J.S., (1948). Phase of Fourier Coefficients directly from Crystal Diffraction Data. *Acta Cryst.* 1, 70-75.
14. Hauptman, H. and Karle, J. (1953). A.C.A. Monograph No. 3. Pittsburgh, Polycrystal.
15. Giacovazzo, C. (1980). The Method of Representations of Structure Seminvariants II. New Theoretical and Practical Aspects. *Acta Cryst.* **A36**, 362-372.
16. Simerska, M. (1956). A Statistical Method for the Determination of the Signs of Structure Factors. Czech. J. Phys. **6**, 1-7.
17. Schenk, H. (1991). Quartets and other Phase Relationships. In *Direct Methods of Solving Crystal Structures*. Edited by H. Schenk, p103-110.

SAYRE EQUATION, TANGENT FORMULA AND SAYTAN

FAN HAI-FU
Institute of Physics, Chinese Academy of Sciences
Beijing 100080, P. R. China

1. Notation

V the volume of a unit cell
$\mathbf{r}$ real-space positional vector defined within a unit cell
$\rho(\mathbf{r})$ electron density function with $\mathbf{r}$ as argument
$\mathbf{h}$ reciprocal-lattice vector, which corresponds to the diffraction index hkl
$\mathbf{F_h}$ the structure factor with $\mathbf{h}$ as argument; the Fourier transform of $\rho(\mathbf{r})$
$\mathbf{F}^{sq}_{\mathbf{h}}$ the Fourier transform of $\rho^2(\mathbf{r})$
f_j the scattering factor of the jth atom in the unit cell
f^{sq}_j the scattering factor of the jth 'squared atom' in the unit cell
$F_\mathbf{h}$ the magnitude of $\mathbf{F_h}$
$\varphi_\mathbf{h}$ the phase of $\mathbf{F_h}$
$\Phi_3 = \varphi_{-\mathbf{h}}+\varphi_{\mathbf{h}'}+\varphi_{\mathbf{h}-\mathbf{h}'}$, the three-phase structure invariant
$\mathbf{E_h}$ the normalised structure factor corresponding to $\mathbf{F_h}$
$E_\mathbf{h}$ the magnitude of $\mathbf{E_h}$
$\sigma_n = \Sigma_j (Z_j)^n$, Z_j is the atomic number of the jth atom in the unit cell, n is an integer

2. Sayre's equation

The Sayre equation [1] is an exact equation linking structure factors. It holds under the following conditions:

i) positivity;
ii) atomicity;
iii) equal-atom structure.

Given a crystal structure represented by $\rho(\mathbf{r})$, we can construct a 'squared structure' expressed as

S. Fortier (ed.), Direct Methods for Solving Macromolecular Structures, 79–85.

$$\rho^2(\mathbf{r}) = \rho(\mathbf{r}) \times \rho(\mathbf{r}) \quad . \tag{1}$$

According to the convolution theorem, the Fourier transform of (1) yields

$$\mathbf{F}_{\mathbf{h}}^{sq} = \frac{1}{V}\sum_{\mathbf{h}'} \mathbf{F}_{\mathbf{h}'}\mathbf{F}_{\mathbf{h}-\mathbf{h}'} \quad , \tag{2}$$

where

$$\mathbf{F}_{\mathbf{h}} = \sum_{j=1}^{N} f_j \exp(i2\pi\,\mathbf{h}\cdot\mathbf{r}_j) \quad . \tag{3}$$

Since $\mathbf{F}^{sq}{}_{\mathbf{h}}$ is the Fourier transform of $\rho^2(\mathbf{r})$, according to (2) the 'squared structure' can be determined through the convolution of structure factors $\mathbf{F}_{\mathbf{h}}$. Now if we can find the relationship between $\rho^2(\mathbf{r})$ and $\rho(\mathbf{r})$, then the $\mathbf{F}^{sq}{}_{\mathbf{h}}$ in (2) can be converted to $\mathbf{F}_{\mathbf{h}}$ leading to an equation linking structure factors. If the first two conditions mentioned above are satisfied, i.e. we have i) $\rho(\mathbf{r}) \geq 0$, ii) the electron densities of different atoms do not overlap, then $\rho^2(\mathbf{r})$ and $\rho(\mathbf{r})$ will have the same number of maxima (atoms) situated at the same positions. We can write

$$\mathbf{F}_{\mathbf{h}}^{sq} = \sum_{j=1}^{N} f_j^{sq} \exp(i2\pi\,\mathbf{h}\cdot\mathbf{r}_j) \quad . \tag{4}$$

If the third condition is also satisfied, we have $\mathbf{F}_{\mathbf{h}} / \mathbf{F}^{sq}{}_{\mathbf{h}} = f/f^{sq}$ and hence

$$\mathbf{F}_{\mathbf{h}} = \frac{f}{f^{sq}V}\sum_{\mathbf{h}'} \mathbf{F}_{\mathbf{h}'}\mathbf{F}_{\mathbf{h}-\mathbf{h}'} \quad . \tag{5}$$

Equation (5) is Sayre's equation.

An important outcome of Sayre's paper, and two other papers published alongside that of Sayre by Cochran [2] and Zachariasen [3], was the relationship between the signs of structure factors in centrosymmetric case:

$$S_{\mathbf{h}} \sim S_{\mathbf{h}'} S_{\mathbf{h}-\mathbf{h}'} \quad , \tag{6}$$

where ~ means 'probably equals'. This can be seen from (5); if $\mathbf{F}_{\mathbf{h}}$ is a large structure factor then it is more likely than not that a large product on the right hand side will have the same sign (phase). The probability for (6) to be true was given by Woolfson [4] and more generally by Cochran and Woolfson [5]

$$P(s_{\mathbf{h},\mathbf{h}'}) = \frac{1}{2} + \frac{1}{2}\tanh\left(\frac{\sigma_3}{\sigma_2^{3/2}}\left|E_{\mathbf{h}}E_{\mathbf{h}'}E_{\mathbf{h}-\mathbf{h}'}\right|\right) \quad . \tag{7}$$

During the early days of direct methods, the sign relationships were used in various ways to solve centrosymmetric structures or centrosymmetric projections of non-centrosymmetric structures [3, 6, 7]. A very successful technique of applying Sayre's equation in *ab initio* phasing is the *SAYTAN* method, which will be described later in this paper. On the other hand, Sayre's equation and its variations have also been successfully used in phase extension and refinement for a wide variety of structures, from proteins to aperiodic crystals [8].

It should be noticed that, while the three conditions mentioned above are necessary for deriving the Sayre equation, they are not satisfied exactly in practice. It is useful to know what would happen when one or more conditions does not hold. In theory, any violation of the three conditions would lead to the collapse of equation (5). However even in this case equation (2) is still valid. Consequently results of applying Sayre's equation would tend to $\rho^2(\mathbf{r})$ rather than $\rho(\mathbf{r})$. The problem is: to what extent will $\rho^2(\mathbf{r})$ and $\rho(\mathbf{r})$ resemble each other? For example, in neutron diffraction the scattering factor of some elements is negative leading to 'negative atoms' in the density function $\rho(\mathbf{r})$. When Sayre's equation is used with neutron diffraction data, the result, approximately $\rho^2(\mathbf{r})$, will differ from $\rho(\mathbf{r})$ mainly in that the negative atoms are changed to positive ones. In another case, if the crystal contains heavy atoms together with light atoms, the map resulting from Sayre's equation will have heavy atoms heavier and light atoms lighter than those in $\rho(\mathbf{r})$. Hence in many cases Sayre's equation may still be applicable even when the three conditions are not completely fulfilled.

3. Cochran's distribution

The three-phase relationship

$$\varphi_{-\mathbf{h}} + \varphi_{\mathbf{h}'} + \varphi_{\mathbf{h}-\mathbf{h}'} \sim 0 \quad (\text{modulo } 2\pi) \quad . \tag{8}$$

and the probability distribution of

$$\Phi_3 = \varphi_{-\mathbf{h}} + \varphi_{\mathbf{h}'} + \varphi_{\mathbf{h}-\mathbf{h}'} \quad , \tag{9}$$

were given by Cochran [9]

$$P(\Phi_3) = \left[2\pi I_0(\kappa_{\mathbf{h},\mathbf{h}'})\right]^{-1} \exp\left(\kappa_{\mathbf{h},\mathbf{h}'}\cos\Phi_3\right) \quad , \tag{10}$$

where

$$\kappa_{\mathbf{h},\mathbf{h}'} = 2\sigma_3\sigma_2^{-3/2}E_{\mathbf{h}}E_{\mathbf{h}'}E_{\mathbf{h}-\mathbf{h}'} \tag{11}$$

for non-equal-atom structures and

$$\kappa_{\mathbf{h},\mathbf{h}'} = 2N^{-1/2}E_{\mathbf{h}}E_{\mathbf{h}'}E_{\mathbf{h}-\mathbf{h}'} \tag{12}$$

for equal-atom structures.

The derivation of (10) is based on the *central limit theorem*: Given a set of n independent random variables, x_i, with means $< x_i >$ and variances σ_i^2, the function

$$y = \sum_{i=1}^{n} a_i x_i \tag{13}$$

has a probability distribution that tends, as n becomes large, to a normal (Gaussian) distribution

$$P(y) = \frac{1}{\sigma_y (2\pi)^{1/2}} \exp\left[-\frac{(y - < y >)^2}{2\sigma_y^2} \right] \tag{14}$$

with mean $< y > = \sum_{i=1}^{n} a_i < x_i >$ and variance $\sigma_y^2 = \sum_{i=1}^{n} a_i \sigma_i^2$.

It has been shown by Kitaigorodskii [10] that, taking $\cos(2\pi\mathbf{h}.\mathbf{r}_j)$ as x_i and taking $\mathbf{F_h}$ as y, for a crystal in space group $P\bar{1}$ containing more than 10 atoms in the unit cell, the probability distribution of $\mathbf{F_h}$ tends, to a good approximation, to a Gaussian distribution. Now let the trigonometric factors

$$\begin{smallmatrix}\sin\\ \cos\end{smallmatrix}[2\pi(-\mathbf{h})\cdot\mathbf{r}_j]\begin{smallmatrix}\sin\\ \cos\end{smallmatrix}(2\pi\mathbf{h}'\cdot\mathbf{r}_j)\begin{smallmatrix}\sin\\ \cos\end{smallmatrix}[2\pi(\mathbf{h}-\mathbf{h}')\cdot\mathbf{r}_j] \quad ,$$

where $\begin{smallmatrix}\sin\\ \cos\end{smallmatrix}$ means 'sin' or 'cos', be the independent random variables x_i and the product $\mathbf{E}_{-\mathbf{h}}\,\mathbf{E}_{\mathbf{h}'}\,\mathbf{E}_{\mathbf{h}-\mathbf{h}'}$ be the function y. Assuming the probability distribution of both the real and the imaginary component of y tend to a Gaussian distribution, and assuming the amplitudes $E_{-\mathbf{h}}$, $E_{\mathbf{h}'}$ and $E_{\mathbf{h}-\mathbf{h}'}$ are known, we would finally obtain Cochran's distribution (10).

4. The tangent formula

From equations (9) and (10), if there are more than one pair of known phases, $\varphi_{\mathbf{h}'}$ and $\varphi_{\mathbf{h}-\mathbf{h}'}$, associated with the same $\varphi_{\mathbf{h}}$, then the total probability distribution for $\varphi_{\mathbf{h}}$ will be

$$P(\varphi_{\mathbf{h}}) = \prod_{\mathbf{h}'} P_{\mathbf{h}'}(\Phi_3) = N \exp\left(\sum_{\mathbf{h}'} \kappa_{\mathbf{h},\mathbf{h}'} \cos \Phi_3 \right) \quad , \tag{15}$$

where N is a normalising factor.

Let

$$\alpha \sin\beta = \sum_{\mathbf{h}'} \kappa_{\mathbf{h},\mathbf{h}'} \sin(\varphi_{\mathbf{h}'} + \varphi_{\mathbf{h}-\mathbf{h}'}) \tag{16}$$

and

$$\alpha \cos\beta = \sum_{\mathbf{h}'} \kappa_{\mathbf{h},\mathbf{h}'} \cos(\varphi_{\mathbf{h}'} + \varphi_{\mathbf{h}-\mathbf{h}'}) \tag{17}$$

(15) becomes

$$P(\varphi_{\mathbf{h}}) = \left[2\pi I_0(\alpha)\right]^{-1} \exp[\alpha \cos(\varphi_{\mathbf{h}} - \beta)] \quad . \tag{18}$$

By maximising $P(\varphi_{\mathbf{h}})$ in (18) we have $\varphi_{\mathbf{h}} = \beta$. Then from (16) and (17) we obtain

$$\tan \varphi_{\mathbf{h}} = \frac{\sum_{\mathbf{h}'} \kappa_{\mathbf{h},\mathbf{h}'} \sin(\varphi_{\mathbf{h}'} + \varphi_{\mathbf{h}-\mathbf{h}'})}{\sum_{\mathbf{h}'} \kappa_{\mathbf{h},\mathbf{h}'} \cos(\varphi_{\mathbf{h}'} + \varphi_{\mathbf{h}-\mathbf{h}'})} \quad . \tag{19}$$

with

$$\alpha = \left[\left(\sum_{\mathbf{h}'} \kappa_{\mathbf{h},\mathbf{h}'} \sin(\varphi_{\mathbf{h}'} + \varphi_{\mathbf{h}-\mathbf{h}'}) \right)^2 + \left(\sum_{\mathbf{h}'} \kappa_{\mathbf{h},\mathbf{h}'} \cos(\varphi_{\mathbf{h}'} + \varphi_{\mathbf{h}-\mathbf{h}'}) \right)^2 \right]^{1/2} \tag{20}$$

indicating the reliability of the estimation of $\varphi_{\mathbf{h}}$. This is the tangent formula introduced by Karle and Hauptman [11], which is the most widely used formula in direct methods. The form of the tangent formula given here differs a little from, but is equivalent to, that of the original one.

It is easily obtained from Sayre's equation a formula similar to (19) but with quite different meaning. By splitting equation (5) into the real and the imaginary parts and by dividing the imaginary part with the real part, it follows

$$\tan\varphi_{\mathbf{h}} = \frac{\sum_{\mathbf{h}'} F_{\mathbf{h}'}F_{\mathbf{h}-\mathbf{h}'}\sin(\varphi_{\mathbf{h}'}+\varphi_{\mathbf{h}-\mathbf{h}'})}{\sum_{\mathbf{h}'} F_{\mathbf{h}'}F_{\mathbf{h}-\mathbf{h}'}\cos(\varphi_{\mathbf{h}'}+\varphi_{\mathbf{h}-\mathbf{h}'})} \quad . \qquad (21)$$

Equation (21) may be regarded as the angular portion of Sayre's equation. It differs from (19) in that the summation in (21) should include all available **h**' terms, while that in (19) may just include a few or even only one term of **h**'. Besides, (21) is an exact equation, while (19) gives the most probable value of $\varphi_{\mathbf{h}}$. The tangent formula is much easier to use for *ab initio* phasing. A historical breakthrough on the application of direct methods was made by Karle and Karle [12] when they solved the non-centrosymmetric crystal structure of L-arginine dihydrate by the symbolic-addition procedure using the tangent formula. A few years later a systematic procedure to use the tangent formula and a computer program *MULTAN* (MULtisolution TANgent-formula method) [13] were introduced by Woolfson and his colleagues. The development and application of the *MULTAN* and related procedures led to the domination of direct methods in solving small molecular structures.

5. *SAYTAN*

The basic idea behind *SAYTAN* is to use not only the relationship among phases but also that among amplitudes implied in Sayre's equation. The philosophy is that a good set of phases should satisfy a system of Sayre equations.

$$\mathbf{E}_{\mathbf{h}} = \frac{K}{g_{\mathbf{h}}}\sum_{\mathbf{h}'}\mathbf{E}_{\mathbf{h}'}\mathbf{E}_{\mathbf{h}-\mathbf{h}'} \quad , \qquad (22)$$

where $g_{\mathbf{h}}$ is the scattering factor for 'squared' atoms and K is an overall scaling constant, which allows for the fact that only structure factors with large magnitude are included on the right-hand side. The derivation started with the following residual for a system of Sayre equations:

$$R = \sum_{\mathbf{h}}\left| g_{\mathbf{h}}\mathbf{E}_{\mathbf{h}} - K\sum_{\mathbf{h}'}\mathbf{E}_{\mathbf{h}'}\mathbf{E}_{\mathbf{h}-\mathbf{h}'}\right|^2 \quad . \qquad (23)$$

As a condition that R should be a minimum it is necessary that

$$\frac{\partial R}{\partial \varphi_{\mathbf{h}}} = 0 \quad \text{for all } \mathbf{h}$$

and this leads to the Sayre-equation tangent formula [14]

$$\varphi_{\mathbf{h}} = \text{phase of}\left[\sum_{\mathbf{h}''}\left(g_{\mathbf{h}} + g_{\mathbf{h}''} + g_{\mathbf{h}-\mathbf{h}''}\right)\mathbf{E}_{\mathbf{h}''}\mathbf{E}_{\mathbf{h}-\mathbf{h}''} - 2K\sum_{\mathbf{h}''}\mathbf{E}_{\mathbf{h}-\mathbf{h}''}\sum_{\mathbf{h}'}\mathbf{E}_{\mathbf{h}'}\mathbf{E}_{\mathbf{h}''-\mathbf{h}'}\right] . \tag{24}$$

As is seen the Sayre-equation tangent formula includes both triplet and quartet terms. A distinctive feature of the Sayre-equation tangent formula is that it can use the information from Sayre equations for which the values of $E_{\mathbf{h}}$ are small, ideally zero. As is well known in powder-method crystal-structure analysis, a good structure model should satisfy the weakest reflections as well as the strongest. The Sayre-equation tangent formula tends to develop phase sets, which satisfy the smallest magnitudes as well as the largest. Since it uses extra information *SAYTAN* is more effective than *MULTAN*, either giving a solution in fewer trials or giving a solution where *MULTAN* would not.

Reference

1. Sayre, D. (1952) The squaring method: a new method for phase determination, *Acta Cryst.* **5**, 60-65.
2. Cochran, W. (1952) A relation between the signs of structure factors, *Acta Cryst.* **5**, 65-67.
3. Zachariasen, W. H. (1952) A new analytical method for solving complex crystal structures, *Acta Cryst.* **5**, 68-73.
4. Woolfson, M. M. (1954) The statistical theory of sign relationships, *Acta Cryst.* **7**, 61-64.
5. Cochran, W. and Woolfson, M. M. (1955) The theory of sign relations between structure factors, *Acta Cryst.* **8**, 1-12.
6. Woolfson, M. M. (1957) An efficient process for solving crystal structures by sign relationships, *Acta Cryst.* **10**, 116-120.
7. Grant, D. F., Howells, R. G. and Rogers, D. (1957) A method for the systematic application of sign relations, *Acta Cryst.* **10**, 489-497.
8. Woolfson, M. M. and Fan, H. F. (1995) *Physical and Non-physical Methods of Solving Crystal Structures*, Cambridge University Press, Cambridge, pp. 208-233; 235-238.
9. Cochran, W. (1955) Relations between the phase of structure factors, *Acta Cryst.* **8**, 473-478.
10. Kitaigorodskii, A. I. (1957) *Theory of structure analysis*, Academic Press, USSR, Moscow. (in Russian)
11. Karle, J. and Hauptman, H. (1956) A theory of phase determination for the four types of non-centrosymmetric space groups 1P222, 2P22, $3P_12$, $3P_22$, *Acta Cryst.* **9**, 635-651.
12. Karle, I. L. and Karle, J. (1964) An application of the symbolic addition method to the structure of L-arginine dihydrate, *Acta Cryst.* **17**, 835-841.
13. Germain, G. and Woolfson, M. M. (1968) On the application of phase relationships to complex structures, *Acta Cryst.* **B24**, 91-96.
14. Debaerdemaeker, T., Tate, C. and Woolfson, M. M. (1988) On the application of phase relationships to complex structures. XXVI. Developments of the Sayre-equation tangent formula, *Acta Cryst.* **A44**, 353-357.

PROBABILISTIC PHASING METHODS VERSUS MACROMOLECULAR CRYSTALLOGRAPHY

C. GIACOVAZZO
Istituto di Ricerca per lo Sviluppo di Metodologie Cristallografiche CNR, c/o Dipartimento Geomineralogico, Campus Universitario, Via Orabona 4, 70125 Bari, Italy.

1. Introduction

Direct methods solved the phase problem for small molecules, and are on the way of solving the structures of medium size molecules [1] (e.g. 300-600 atoms in the asymmetric unit). The question is then: how useful can they be for normal size or large size macromolecules? We will show that:
a) direct methods cannot solve *ab initio* normal size or large size macromolecules;
b) If some prior information is available, they can be effectively useful for: phase refinement and extension, as an isomorphous replacement technique, as a molecular replacement technique.
Some preliminary discussions about the various probabilistic approaches are necessary to state the points a) , b) on a solid ground.

2. About the machinery of the probabilistic approaches

The method of the joint probability distribution functions of structure factors [2] may be so described:
1) fix the set of reflexions to involve in the probabilistic problem: say $E_1, E_2, ..., E_m$.
2) calculate the characteristic function

$$C(\vartheta_1, \vartheta_2, ..., \vartheta_m, \rho_1, \rho_2, ..., \rho_m), \qquad (1)$$

where ϑ_i, ρ_i are the carrying variables respectively associated with the phases ϕ_i and with the moduli R_i of the normalized structure factors chosen at the point 1).
3) calculate the joint probability distribution

$$P(\phi_1, ..., \phi_m, R_1, ..., R_m) \qquad (2)$$

via the Fourier transform of (1). From (2) the conditional probabilities

$$P(\phi_i | \{\phi, R\}) \qquad (3)$$

may be obtained, where ϕ_i is the ith phase and $\{\phi, R\}$ is a set of phases and moduli which are supposed to be known. Alternatively, the conditional probability

$$P(\Phi_n | R_1, ..., R_m) \qquad (4)$$

S. Fortier (ed.), Direct Methods for Solving Macromolecular Structures, 87–97.

may be obtained, where

$$\Phi_n = \phi_1 + \phi_2 + ... + \phi_n$$

is a suitable structure invariant or seminvariant. Usually $n \le m$.
The first problem to solve is to define the set of the m magnitudes on which Φ_n primarily depends. Such a point will be discussed in the section 4.
If $n = m = 3$ then (4) reduces to the Cochran relationship [3]

$$P(\Phi_3) = [2\pi I_0(G_3)]^{-1} \exp(G_3 \cos \Phi_3) \qquad (5)$$

where

$$\Phi_3 = \phi_{\mathbf{h}_1} + \phi_{\mathbf{h}_2} + \phi_{\mathbf{h}_3},$$
$$\mathbf{h}_1 + \mathbf{h}_2 + \mathbf{h}_3 = 0,$$
$$G_3 = 2|E_{\mathbf{h}_1} E_{\mathbf{h}_2} E_{\mathbf{h}_3}| / \sqrt{N}. \qquad (6)$$

N is the number of atoms in the unit cell. If some phases are known (i.e., the r pairs $(\phi_{\mathbf{k}_j}, \phi_{\mathbf{h}-\mathbf{k}_j})$) then (3) may be used to estimate $\phi_\mathbf{h}$ [4] [5]:

$$P(\phi_\mathbf{h} | ...) = [2\pi I_0(\alpha_\mathbf{h})]^{-1} \exp[\alpha_\mathbf{h} \cos(\phi_\mathbf{h} - \vartheta_\mathbf{h})] \qquad (7)$$

where $\vartheta_\mathbf{h}$, the most probable value of $\phi_\mathbf{h}$, is defined by

$$\tan \vartheta_\mathbf{h} = \frac{\sum_{j=1}^{r} G_j \sin(\phi_{\mathbf{k}_j} + \phi_{\mathbf{h}-\mathbf{k}_j})}{\sum_{j=1}^{r} G_j \cos(\phi_{\mathbf{k}_j} + \phi_{\mathbf{h}-\mathbf{k}_j})} = \frac{T_\mathbf{h}}{B_\mathbf{h}} \qquad (8)$$

and

$$\alpha_\mathbf{h} = [T_\mathbf{h}^2 + B_\mathbf{h}^2]^{1/2}. \qquad (9)$$

is the reliability parameter. Small values of α suggests that the relation $\phi_\mathbf{h} \approx \vartheta_\mathbf{h}$ is unreliable.

3. The *z* test

In absence of any information on the phases the $\langle\alpha\rangle$ value could be used as a first criterion for predicting or excluding the success of direct methods. Since each α, and therefore each $\langle\alpha\rangle$, is calculated from a limited sample of triplets extracted from an ideally infinite population, also the variance of α is a useful supplementary parameter. The distribution $P(\alpha)$ has been derived by Cascarano, Giacovazzo, Burla, Nunzi, Polidori [6]: it was found that

$$P(\alpha_\mathbf{h}) \approx \mathcal{N}(\alpha_\mathbf{h}; \langle\alpha_\mathbf{h}\rangle, \sigma_{\alpha_\mathbf{h}}^2) \qquad (10)$$

where $\mathcal{N}$ denotes the normal distribution, and

$$\langle\alpha_\mathbf{h}\rangle = \sum_{j=1}^{r} G_j D_1(G_j), \qquad (11)$$

$$\sigma^2_{\alpha_h} \approx \frac{1}{2}\sum_{j=1}^{r} G_j^2\left[1+D_2\left(G_j\right)-2D_1^2\left(G_j\right)\right]. \qquad (12)$$

While α_h cannot be calculated in absence of phase information, $\langle\alpha_h\rangle$ and σ_{α_h} are completely defined in terms of the experimental data (i.e., the structural complexity and the measured diffraction data). Both $\langle\alpha_h\rangle$ and σ_{α_h} increase if r increases: thus the information on $\langle\alpha_h\rangle$ has always to be coupled with the complementary information on σ_{α_h} and *vice versa.* In this respect of basic importance is the parameter $z_h = \langle\alpha_h\rangle/\sigma_{\alpha_h}$; it may be considered as a "signal-to-noise ratio " [7]. Large values of z correspond to large values of $\langle\alpha\rangle$ divided by small values of the variance. A simplified numerical study of z may be performed by assuming constant values of G_j , for j=1, ..., r. In Table 1 the values of $\langle\alpha\rangle$, σ_α and z are shown for specific values of r and G. For example, for G_j =1.2, j=1, ..., 15, we obtain $\langle\alpha\rangle$=9.18, σ_α =2.61 and z = 3.52. It may be noted that z may assume quite different values for equal values of $\langle\alpha\rangle$: e.g., for G=2.5 and r = 5 it is $\langle\alpha\rangle$ = 9.59 and z = 5.29 ; for G =0.2 and r = 500 it is $\langle\alpha\rangle$ = 10.1 but z = 3.3.

Table 1. The values of $\langle\alpha\rangle/\sigma_\alpha = z$ are given for selected values of r and G.

G	r=5	r=15	r=50
0.1	0.02/0.16=0.12	0.07/0.28=0.26	0.24/0.51=0.48
0.2	0.10/0.31=0.33	0.30/0.54=0.56	1.01/0.98=1.03
0.4	0.40/0.61=0.66	1.20/1.05=1.15	4.01/1.92=2.09
0.8	1.49/1.13=1.32	4.46/1.95=2.28	14.87/3.57=4.1
1.2	3.06/1.51=2.03	9.18/2.61=3.52	30.59/4.76=6.4
1.8	5.95/1.78=3.35	17.86/3.08=5.81	59.55/.62=10.6
2.5	9.59/1.81=5.29	28.76/3.14=9.16	95.86/5.73=16.
4.0	17.25/1.79=9.64	51.76/3.10=16.70	172.54/5.66=30

G	r=100	r=200	r=500
0.1	0.50/0.70= 0.71	1.00/1.00=1.00	2.40/1.60=1.50
0.2	2.00/1.40=1.50	4.10/2.00=2.10	10.10/3.10=3.3
0.4	8.02/2.71=2.96	16.03/3.83=4.18	40.08/6.06=6.6
0.8	29.74/5.04=5.90	59.48/7.13=8.34	148.69/11.27=1
1.2	61.18/6.74=9.08	122.37/9.53=12.8	305.92/15.06=2
1.8	119.09/7.94=14.	238.18/11.23=21.	595.46/17.76=3
2.5	191.72/8.11=23.	383.44/11.46=33.	958.60/18.13=5
4.0	345.09/8.00=43.	690.18/11.32=60.	1725.40/17.90=

Diffraction data for which high values of z may be calculated for most of the strong reflections constitute a set suitable for the successful application of Direct Methods. Indeed : a) owing to the large values of α (implicit in large z values) phases can be accurately defined ; b) the α's calculated from correct sets of phases should be closely distributed about the corresponding $\langle\alpha\rangle$'s. Then the Sayre's equation is approximately satisfied in terms of moduli and of phases and suitable figures of merit can in principle discriminate the correct from the false solutions. If small values of z are calculated for most of the strong reflexions then $\alpha_{\mathbf{h}} \exp(i\vartheta_{\mathbf{h}})$ is distributed about $\langle\alpha_{\mathbf{h}}\rangle \exp(i\phi_{\mathbf{h}})$ according to a rather flat distribution. Then the tangent formula will hardly work and true solutions will hardly be discriminated from the false's.

The situation for proteins may be described in simple terms on assuming that the G_j 's have very small values. Accordingly $D_1(G_j) \approx G_j/2$ and

$$\langle\alpha_{\mathbf{h}}\rangle \approx \sum_{j=1}^{r} G_j^2/2, \tag{13}$$

$$\sigma_{\alpha_{\mathbf{h}}}^2 = \frac{1}{2}\sum_{j=1}^{r} G_j^2 \left\{1 + \left[1 - \frac{2}{G_j} D_1(G_j)\right] - 2D_1^2(G_j)\right\}$$

$$\approx \sum_{j=1}^{r} G_j^2/2. \tag{14}$$

In conclusion for proteins

$$\langle\alpha_{\mathbf{h}}\rangle \approx \sigma_{\alpha_{\mathbf{h}}}^2,$$

and therefore $z_{\mathbf{h}} \approx \langle\alpha_{\mathbf{h}}\rangle^{1/2}$. If for most of the strong reflexions $\langle\alpha_{\mathbf{h}}\rangle < 1$ then the signal is generally smaller than the noise : the tangent formula should not efficiently work and correct solutions should not be recognized among the various trials. The signal is generally stronger than the noise if $\langle\alpha_{\mathbf{h}}\rangle > 1$ for most of the strong reflections. This situation can only occur for high resolution data : for example, it may be seen from Table 1 that $z = 0.48$ for $G_j = 0.1$ and r = 50, while $z = 1.5$ for $G = 0.1$ and $r = 500$. It may be concluded that signal-to-noise ratios larger than unity are only produced by high resolution data : in absence of them ab initio direct solution of proteins structures is impossible.

The results described in the preceding section suggest [7] that in principle direct methods can successfully be applied to a given set of data if, for a sufficiently high percentage of large normalized structure factors,

$$z \geq T, \tag{15}$$

where T is a threshold to be fixed. T represents the acceptable inferior limit for the "signal-to-noise" ratio : a reasonable choice may be $T \approx 2,3$. For crystallographers such a choice is not surprising : the practical rule $F_{obs} > T\sigma$ is often adopted for crystal structure analysis. Of course (15) is a necessary but not a sufficient condition : it only suggests that both the structure solution and the identification of the "correct" set of phases among numerous trials are theoretically possible. Conversely, if $z \leq T$ for the large majority of the reflexions, remarkable deviations from Sayre relationships are expected : the correct set of phases is hardly obtained by application of the tangent formula and the correct trials cannot be recognized among the others.

We checked our z tests on some medium size structures recently solved by *Shake and Bake* [1]. The corresponding $P(z)$ distributions are shown in Fig. 1.

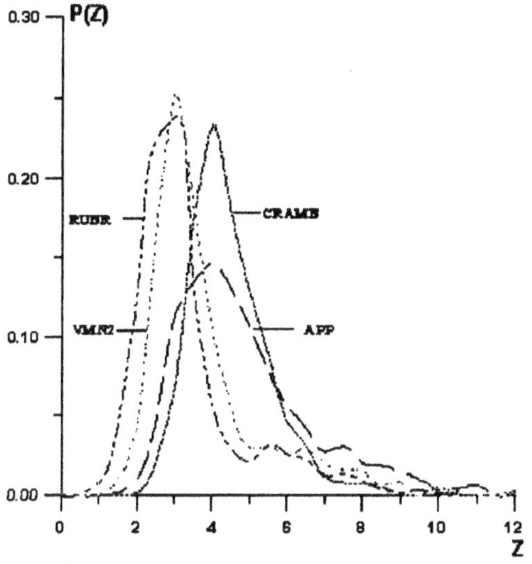

Fig.1 - P(z) test for some structures solved by *Shake and Bake* (see [1] for references)

4. The idea of representation

How to fix the set of magnitudes on which a structure invariant primarily depends ? The problem may be solved by means of the theory of representations [8] [9] (for a related theory see [10]) : for any structure invariant Φ_n the reflexions may be arranged in a sequence of sets in order of their expected effectiveness for the estimation of Φ_n (see Fig. 2a).

$\mathbf{h}_1, \mathbf{h}_2, \ldots, \mathbf{h}_n$ are said to be the *basis vectors* of Φ_n. If the crystal symmetry is higher that triclinic then in favourable cases a number of symmetry operators may be found such that one or more structure invariants

$$\Psi_n = \phi_{\mathbf{h}_1\mathbf{R}_s} + \phi_{\mathbf{h}_2\mathbf{R}_t} + \ldots + \phi_{\mathbf{h}_n\mathbf{R}_m} \tag{16}$$

arise (i.e., for which $\mathbf{h}_1\mathbf{R}_s + \mathbf{h}_2\mathbf{R}_t + \ldots + \mathbf{h}_n\mathbf{R}_m = 0$) when $\mathbf{R}_s$, $\mathbf{R}_t$, $\mathbf{R}_m$ vary over the set of rotation matrices.

The *first representation* of Φ_n is the set of the invariants Ψ_n ; the distinct vectors

$$m_1\mathbf{h}_1\mathbf{R}_s + \ldots + m_n\mathbf{h}_n\mathbf{R}_m = 0 \qquad m_i = 0, 1$$

are called *cross-vectors.*

The moduli of the basis vectors, joined with the moduli of the cross vectors for any of the invariants Ψ_n, constitute the *first phasing shell* of Φ_n. We give only few simple examples (no explicit use of the crystallographic symmetry will be made) :

i) Let us consider the triplet

$$\Phi_3 = \phi_{\mathbf{h}_1} + \phi_{\mathbf{h}_2} + \phi_{\mathbf{h}_3} \tag{17}$$

In P1 the first representation contains only the triplet (17) and $\left\{R_{\mathbf{h}_1}, R_{\mathbf{h}_2}, R_{\mathbf{h}_3}\right\}$ constitutes the first phasing shell.

ii) Let us consider the quartet

$$\Phi_4 = \phi_{\mathbf{h}_1} + \phi_{\mathbf{h}_2} + \phi_{\mathbf{h}_3} + \phi_{\mathbf{h}_4}\ , \qquad (\mathbf{h}_1 + \mathbf{h}_2 + \mathbf{h}_3 + \mathbf{h}_4 = 0).$$

$\{\mathbf{h}_1, \mathbf{h}_2, \mathbf{h}_3, \mathbf{h}_4\}$ is the set of basis vectors, $\{\mathbf{h}_1 + \mathbf{h}_2,\ \mathbf{h}_1 + \mathbf{h}_3,\ \mathbf{h}_2 + \mathbf{h}_3\}$ is the set of cross vectors, $\{R_{\mathbf{h}_1}, \ldots, R_{\mathbf{h}_4}, R_{\mathbf{h}_1+\mathbf{h}_2}, R_{\mathbf{h}_1+\mathbf{h}_3}, R_{\mathbf{h}_2+\mathbf{h}_3}\}$ is the first phasing shell of Φ_4. (see Fig. 2b). Thus Φ_4 primarily depends on 7 cross-magnitudes.

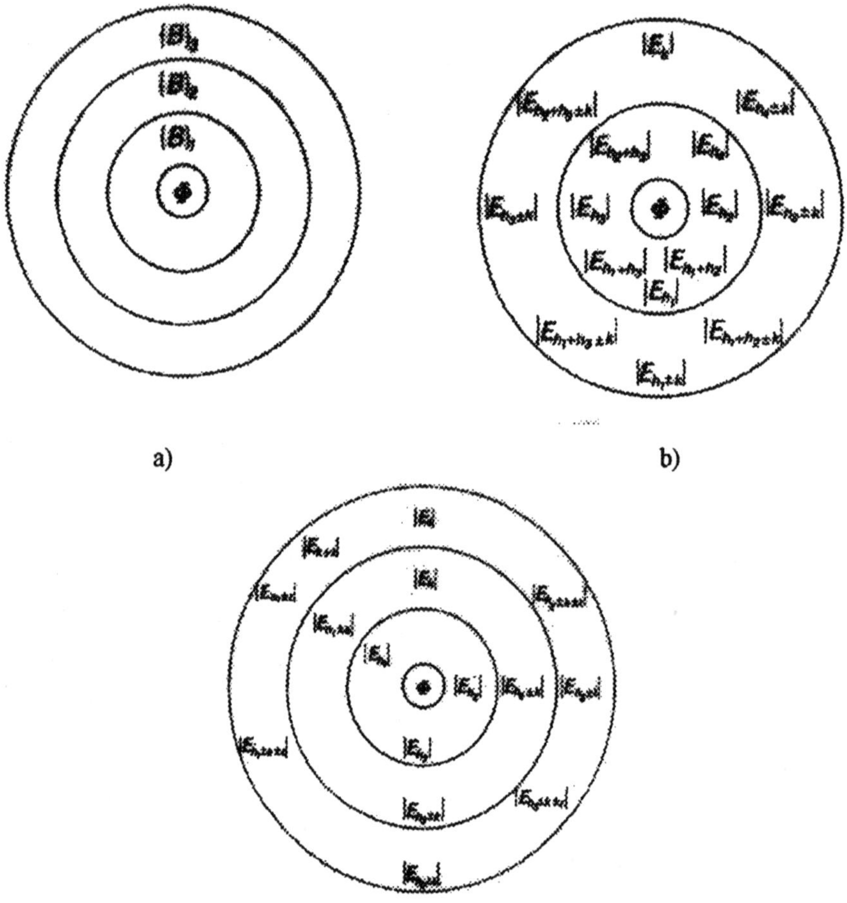

a) b) c)

Fig.2 a) General scheme arranging the set of reflections according to the representation theory; b) scheme for quartet invariants; c) scheme for triplet invariants

The *second representation* of Φ_n is the collection of invariants

$$\Psi_{n+2} = \Phi_n + \phi_{\mathbf{k}} - \phi_{\mathbf{k}}$$

where $\mathbf{k}$ is a free vector. Now

$$\mathbf{h}_1,\ \mathbf{h}_2, \ldots, \mathbf{h}_n,\ \mathbf{k}$$

are the basis vectors of Ψ_n, and

$$m_1\mathbf{h}_1+...+m_n\mathbf{h}_n+m_{n+1}\mathbf{k}-m_{n+2}\mathbf{k}, \text{ with } m_i=0,1$$

will create the cross-vectors. The second phasing shell of Φ_n is constituted by the collection of magnitudes whose indices are basis or cross vectors of any of the invariants Ψ_{n+2}. As an example, if $n = 3$

$$\Phi_3 = \phi_{\mathbf{h}_1} + \phi_{\mathbf{h}_2} + \phi_{\mathbf{h}_3}$$

$$\Psi_3 = \Phi_3$$

$$\Psi_5 = \phi_{\mathbf{h}_1} + \phi_{\mathbf{h}_2} + \phi_{\mathbf{h}_3} + \phi_{\mathbf{k}} - \phi_{\mathbf{k}}$$

The second phasing shell is then the set of 10 magnitudes (see Fig. 2c)

$$\mathbf{h}_1,\ \mathbf{h}_2,\ \mathbf{h}_3,\ \mathbf{k},\ \mathbf{h}_1 \pm \mathbf{k},\ \mathbf{h}_2 \pm \mathbf{k},\ \mathbf{h}_3 \pm \mathbf{k}\ .$$

A formula estimating triplets via their second representation (the P_{10} formula) has been described by [11]. P_{10} is the default formula in the SIR92 package [12] and is responsible of its many successes (and failures).

5. The representation theory extended to isomorphous structures

The representation theory may be usefully applied to macromolecules if it is integrated with the concept of isomorphous data [13]. Isomorphous data (generated by one or more heavy atom derivatives or by anomalous dispersion effects) play a central role in the procedures for the crystal structure solution of macromolecules. An additional type of isomorphism occurs when prior information on part of the structure is available. In all the above cases, for a given index $\mathbf{h}$, a collection of moduli $R_{j,\mathbf{h}}$

$$R_{j,\mathbf{h}} \qquad\qquad j=1,....,b$$

might be available. For example:

1) if diffraction data from two heavy atom derivatives are available (besides native protein data), then b=3: $R_{1,\mathbf{h}}$ may correspond to the native protein while $R_{2,\mathbf{h}}$and $R_{3,\mathbf{h}}$ refer to the isomorphous derivatives.

2) If measurements are made at a specific wavelength generating anomalous dispersion effects then b=1. The isomorphous pairs are then the Friedel related reflections:

$$R_{1,\mathbf{h}} = R_{\mathbf{h}} \qquad , \qquad R_{1,-\mathbf{h}} = R_{-\mathbf{h}}\,.$$

If two wavelengths are used, then b=2 and

$$R_{1,\mathbf{h}} = R_{\lambda_1,\mathbf{h}} \qquad\qquad R_{1,-\mathbf{h}} = R_{\lambda_1,-\mathbf{h}}$$

$$R_{2,\mathbf{h}} = R_{\lambda_2,\mathbf{h}} \qquad\qquad R_{2,-\mathbf{h}} = R_{\lambda_2,-\mathbf{h}}$$

3) If a molecular fragment is correctly oriented and positioned the measured $R_{\mathbf{h}}$ value and the calculated $R_{p\mathbf{h}}$ value, corresponding to the known partial structure, are simultaneously available. Then b=2 and

$$R_{1,\mathbf{h}} = R_{\mathbf{h}} \qquad , \qquad R_{2,\mathbf{h}} = R_{p\mathbf{h}}\,.$$

We state here the necessary definitions and properties: all of them will hold provided a common origin for the isomorphous structures has been fixed.

Invariance condition. The phase

$$\Phi = \phi_{i,\mathbf{h}_1} + \phi_{j,\mathbf{h}_2} + ... + \phi_{r,\mathbf{h}_n} \qquad (18)$$

is a structure invariant if

$$\mathbf{h}_1 + \mathbf{h}_2 +,...+\mathbf{h}_n = 0$$

no matter if $i \neq j \neq ... \neq r$.

The first representation of a structure invariant. For a given structure invariant

$$\Phi = \phi_{i,\mathbf{h}_1} + \phi_{j,\mathbf{h}_2} + ... + \phi_{r,\mathbf{h}_n}$$

the collection of the distinct structure invariants

$$\Psi_1 = \phi_{i,\mathbf{h}_1\mathbf{R}_s} + \phi_{j,\mathbf{h}_2\mathbf{R}_t} + ... + \phi_{r,\mathbf{h}_n\mathbf{R}_v}\ ,\ \left(\mathbf{h}_1\mathbf{R}_s + \mathbf{h}_2\mathbf{R}_t + ... + \mathbf{h}_n\mathbf{R}_v = 0\right)$$

obtained when $\mathbf{R}_s, \mathbf{R}_t, ..., \mathbf{R}_v$ vary over the set of the rotation matrices of the space group and $i,j,...,r$ over the different isomorphous data sets, is called the first representation of Φ. The first phasing shell of Φ is given by the collection $\{B\}_1$ of the magnitudes which are basis or cross-magnitudes of at least one structure invariant Ψ_1. We make some practical examples:

1) The phase

$$\Phi = \phi_{i,\mathbf{h}} + \phi_{j,-\mathbf{h}} = \phi_{i,\mathbf{h}} - \phi_{j,\mathbf{h}}$$

is a structure invariant even when $i \neq j$, when the set i corresponds to the native protein and the set j to one of the derivatives (or i corresponds to one derivative and j to another).

2)
$$\Phi = \phi_{i,\mathbf{h}} + \phi_{i,-\mathbf{h}}$$

is a non-trivial structure invariant if the set i corresponds to data collected at a wavelength exciting anomalous dispersion. More generally,

$$\Phi = \phi_{i,\mathbf{h}} + \phi_{j,-\mathbf{h}}$$

is a non-trivial two-phase invariant when at least one of i and j corresponds to wavelengths exciting anomalous dispersion.

3) In the case of a native protein and one heavy atom derivative the first representation of the triplet invariant

$$\Phi = \phi_{1,\mathbf{h}_1} + \phi_{1,\mathbf{h}_2} + \phi_{1,\mathbf{h}_3}$$

is the collection of the eight triplets:

$$\begin{array}{ll}
\phi_{1,\mathbf{h}_1} + \phi_{1,\mathbf{h}_2} + \phi_{1,\mathbf{h}_3}\ , & \phi_{1,\mathbf{h}_1} + \phi_{1,\mathbf{h}_2} + \phi_{2,\mathbf{h}_3}\ , \\
\phi_{1,\mathbf{h}_1} + \phi_{2,\mathbf{h}_2} + \phi_{1,\mathbf{h}_3}\ , & \phi_{2,\mathbf{h}_1} + \phi_{2,\mathbf{h}_2} + \phi_{1,\mathbf{h}_3}\ , \\
\phi_{1,\mathbf{h}_1} + \phi_{2,\mathbf{h}_2} + \phi_{2,\mathbf{h}_3}\ , & \phi_{2,\mathbf{h}_1} + \phi_{1,\mathbf{h}_2} + \phi_{2,\mathbf{h}_3}\ , \\
\phi_{2,\mathbf{h}_1} + \phi_{2,\mathbf{h}_2} + \phi_{1,\mathbf{h}_3}\ , & \phi_{2,\mathbf{h}_1} + \phi_{2,\mathbf{h}_2} + \phi_{2,\mathbf{h}_3}\ .
\end{array}$$

Accordingly

$$\{B_1\} = \left\{R_{1,\mathbf{h}_1}, R_{1,\mathbf{h}_2}, R_{1,\mathbf{h}_3}, R_{2,\mathbf{h}_2}, R_{2,\mathbf{h}_2}, R_{2,\mathbf{h}_3}\right\}.$$

4) In the case of native protein data with anomalous dispersion effects (one wavelength) the first representation of the triplet

$$\Phi_3 = \phi_{\mathbf{h}_1} + \phi_{\mathbf{h}_2} + \phi_{\mathbf{h}_3}$$

is the collection of the eight triplets

$$\phi_{\mathbf{h}_1}+\phi_{\mathbf{h}_2}+\phi_{\mathbf{h}_3}\quad, \qquad \phi_{\mathbf{h}_1}+\phi_{\mathbf{h}_2}-\phi_{-\mathbf{h}_3},$$
$$\phi_{\mathbf{h}_1}-\phi_{-\mathbf{h}_2}+\phi_{\mathbf{h}_3}\quad, \qquad -\phi_{-\mathbf{h}_1}+\phi_{\mathbf{h}_2}+\phi_{\mathbf{h}_3},$$
$$\phi_{\mathbf{h}_1}-\phi_{-\mathbf{h}_2}-\phi_{-\mathbf{h}_3}\quad, \qquad -\phi_{-\mathbf{h}_1}+\phi_{\mathbf{h}_2}-\phi_{-\mathbf{h}_3},$$
$$-\phi_{-\mathbf{h}_1}-\phi_{-\mathbf{h}_2}+\phi_{\mathbf{h}_3}, \qquad -\phi_{\mathbf{h}_1}-\phi_{-\mathbf{h}_2}-\phi_{-\mathbf{h}_3},$$

Accordingly the first phasing shell is the collection

$$\{B_1\}=\left\{R_{\mathbf{h}_1}, R_{\mathbf{h}_2}, R_{\mathbf{h}_3}, R_{-\mathbf{h}_1}, R_{-\mathbf{h}_2}, R_{-\mathbf{h}_3}\right\}.$$

5) Let us consider the case of a native protein and one heavy atom derivative (b=2). Then the first representation of the quartet invariant

$$\Phi=\phi_{1,\mathbf{h}_1}+\phi_{1,\mathbf{h}_2}+\phi_{1,\mathbf{h}_3}+\phi_{1,\mathbf{h}_4} \qquad (\mathbf{h}_1+\mathbf{h}_2+\mathbf{h}_3+\mathbf{h}_4=0)$$

is the collection of the 16 quartets

$$\phi_{1,\mathbf{h}_1}+\phi_{1,\mathbf{h}_2}+\phi_{1,\mathbf{h}_3}+\phi_{1,\mathbf{h}_4}, \qquad \phi_{1,\mathbf{h}_1}+\phi_{1,\mathbf{h}_2}+\phi_{1,\mathbf{h}_3}+\phi_{2,\mathbf{h}_4},$$
$$\phi_{1,\mathbf{h}_1}+\phi_{1,\mathbf{h}_2}+\phi_{2,\mathbf{h}_3}+\phi_{1,\mathbf{h}_4}, \qquad \phi_{1,\mathbf{h}_1}+\phi_{2,\mathbf{h}_2}+\phi_{1,\mathbf{h}_3}+\phi_{1,\mathbf{h}_4},$$

....

$$\phi_{2,\mathbf{h}_1}+\phi_{2,\mathbf{h}_2}+\phi_{2,\mathbf{h}_3}+\phi_{1,\mathbf{h}_4}, \qquad \phi_{2,\mathbf{h}_1}+\phi_{2,\mathbf{h}_2}+\phi_{2,\mathbf{h}_3}+\phi_{2,\mathbf{h}_4}.$$

The first phasing is then the collection of 14 magnitudes:

$$\{B_1\}=\{R_{1,\mathbf{h}_1}, R_{1,\mathbf{h}_2}, R_{1,\mathbf{h}_3}, R_{1,\mathbf{h}_4}, R_{2,\mathbf{h}_1}, R_{2,\mathbf{h}_2}, R_{2,\mathbf{h}_3}, R_{2,\mathbf{h}_4},$$
$$R_{1,\mathbf{h}_1+\mathbf{h}_2}, R_{1,\mathbf{h}_1+\mathbf{h}_3}, R_{1,\mathbf{h}_2+\mathbf{h}_3}, R_{2,\mathbf{h}_1+\mathbf{h}_2}, R_{2,\mathbf{h}_1+\mathbf{h}_3}, R_{2,\mathbf{h}_2+\mathbf{h}_3}\}.$$

Upper representations of a structure invariant. The second representation of the structure invariant (18) is the set of quintet invariants

$$\{\Psi\}=\left\{\phi_{i,\mathbf{h}_1}+\phi_{j,\mathbf{h}_2}+\ldots+\phi_{r,\mathbf{h}_n}+\phi_{\alpha,\mathbf{k}}-\phi_{\beta,\mathbf{k}}\right\}$$

where **k** is a free vector and i, j, r, α, β can assume any value between 1 and b.

Let us give two examples:

1) In the case of a native protein and one heavy atom derivative the second representation of the two-phase invariant

$$\Phi=\phi_{1,\mathbf{h}}+\phi_{2,-\mathbf{h}}$$

is the collection of the special quartets

$$\phi_{1,\mathbf{h}}+\phi_{2,-\mathbf{h}}+\phi_{1,\mathbf{k}}+\phi_{1,-\mathbf{k}}, \qquad \phi_{1,\mathbf{h}}+\phi_{2,-\mathbf{h}}+\phi_{2,\mathbf{k}}+\phi_{2,-\mathbf{k}},$$
$$\phi_{1,\mathbf{h}}+\phi_{2,-\mathbf{h}}+\phi_{1,\mathbf{k}}+\phi_{2,-\mathbf{k}}, \qquad \phi_{1,\mathbf{h}}+\phi_{2,-\mathbf{h}}+\phi_{1,-\mathbf{k}}+\phi_{2,\mathbf{k}}.$$

Then

$$\{B\}_2=\left\{R_{1,\mathbf{h}}, R_{2,\mathbf{h}}, R_{1,\mathbf{k}}, R_{2,\mathbf{k}}, R_{1,\mathbf{h}+\mathbf{k}}, R_{1,\mathbf{h}-\mathbf{k}}\right\}$$

k is a free vector spanning all the reciprocal space.

2) In the case of a native protein and one-heavy atom derivative, the second representation of the triplet

$$\Phi=\phi_{1,\mathbf{h}_1}+\phi_{1,\mathbf{h}_2}+\phi_{1,\mathbf{h}_3}$$

is the collection of special quintets

$$\{\Psi\}_2=\left\{\phi_{i,\mathbf{h}_1}+\phi_{j,\mathbf{h}_2}+\phi_{r,\mathbf{h}_3}+\phi_{\alpha,\mathbf{k}}-\phi_{\beta,\mathbf{k}}\right\}$$

or, more explicitly, the collection of special quintets

$$\phi_{1,\mathbf{h}_1} + \phi_{2,\mathbf{h}_2} + \phi_{1,\mathbf{h}_3} + \phi_{1,\mathbf{k}} - \phi_{1,\mathbf{k}}$$

$$\phi_{1,\mathbf{h}_1} + \phi_{1,\mathbf{h}_2} + \phi_{1,\mathbf{h}_3} + \phi_{1,\mathbf{k}} - \phi_{2,\mathbf{k}}$$

$$\phi_{1,\mathbf{h}_1} + \phi_{1,\mathbf{h}_2} + \phi_{1,\mathbf{h}_3} + \phi_{2,\mathbf{k}} - \phi_{1,\mathbf{k}}$$

....

$$\phi_{2,\mathbf{h}_1} + \phi_{2,\mathbf{h}_2} + \phi_{2,\mathbf{h}_3} + \phi_{2,\mathbf{k}} - \phi_{2,\mathbf{k}}$$

The above definitions can be easily extended to higher order structure invariants.

6. The applications of direct methods to macromolecular crystallography

Let us suppose that isomorphous data are available. We will consider three cases:
a) protein data and one isomorphous derivative (SIR case). Then, according to the § 5, the distribution

$$P\left(E_{p\mathbf{h}_1}, E_{p\mathbf{h}_2}, E_{p\mathbf{h}_3}, E_{d\mathbf{h}_1}, E_{d\mathbf{h}_2}, E_{d\mathbf{h}_3}\right) \tag{19}$$

may be studied [14] [15] where p stands for "protein" and d for "derivative". A phasing procedure based on (19) has been proposed which proved to be competitive with respect to traditional SIR techniques [16] [17] [18] [19] [20].
b) protein data with anomalous dispersion effects (SAS case). Then according to the § 5 the distribution

$$P\left(E_{\mathbf{h}_1}, E_{\mathbf{h}_2}, E_{\mathbf{h}_3}, E_{-\mathbf{h}_1}, E_{-\mathbf{h}_2}, E_{-\mathbf{h}_3}\right) \tag{20}$$

may be studied [21] [22]. A phasing procedure based on (20) has been recently proposed which proved able to phase proteins ab initio [23].
c) part of the structure is correctly positioned. In this case (see § 5) the distribution

$$P\left(E_{\mathbf{h}_1}, E_{\mathbf{h}_2}, E_{\mathbf{h}_3} | E_{p\mathbf{h}_1}, E_{p\mathbf{h}_2}, E_{p\mathbf{h}_3}\right) \tag{21}$$

may be studied [24] where E_p now represents the normalized structure factor of the partial structure. Quite recently a feasibility study [25] has been described with particular emphasis for the application of (21) to macromolecules. It seems clear now [26] that the use of (21) is the counterpart in reciprocal space of the electron density techniques so widely used in macromolecular crystallography for phase extension and refinement.
A further case occurs when a protein is correctly oriented but has to be translated in the correct position. Then direct methods can profit by the available supplementary information for efficiently solving the translation problem [27] [28].

References

1 Weeks, C. & Miller, R. (1996). Proceedings of the Macromolecular Crystallography Computing School, Western Washington University.

2 Hauptman, H., & Karle, J. (1953). The solution of the Phase Problem I. The Centrosymmetric Crystal. ACA Monograph no 3, Polycristal Book Service, New York
3 Cochran, W. (1955). *Acta Cryst.* **A33**, 933-944
4 Karle, J. & Hauptman, H. (1956) *Acta Cryst.* **9**, 635-651
5 Karle, J. & Karle, I. (1966). *Acta Cryst.* **21**, 849-859
6 Cascarano, G., Giacovazzo, C., Burla, M. C., Nunzi, A. & Polidori, G. (1984) *Acta Cryst.* **A40**, 389-394
7 Giacovazzo, C., Guagliardi, A., Ravelli, R. & Siliqi, D. (1993). *Z. für Kristallogr.* **209**, 136-142
8 Giacovazzo, C. (1977). *Acta Cryst.* **A33**, 934-944
9 Giacovazzo, C. (1980). *Acta Cryst.* **A36**, 362-373
10 Hauptman, H. (1976). *Acta Cryst.* **A32**, 934-940
11 Cascarano, G., Giacovazzo, C., Camalli, M., Spagna, R., Burla, M. C., Nunzi, A. & Polidori, G. (1994). *Acta Cryst.* **A40**, 278-283
12 Altomare, A., Cascarano, G., Giacovazzo, C., Guagliardi, A., Burla, M. C., Polidori, G. & Camalli, M. (1994). *J. Appl. Cryst.* **27**, 343-350
13 Giacovazzo, C. (1984). Direct Methods of Solving Crystal Structures. International School of Crystallography, 10th course, Erice, Italy.
14 Hauptman, H. (1982). *Acta Cryst.* **A38**, 289-294
15 Giacovazzo, C., Cascarano, G. & Zheng, C. D. (1988). *Acta Cryst.* **A44**, 45-51
16 Giacovazzo, C., Siliqi, D. & Ralph, A. (1994). *Acta Cryst.* **A50**, 503-510
17 Giacovazzo, C., Siliqi, D. & Spagna, R. (1996). *Acta Cryst.* **A50**, 609-621
18 Giacovazzo, C., Siliqi, D. & Zanotti, G. (1995). *Acta Cryst.* **A51**, 177-188
19 Giacovazzo, C., Siliqi, D. & Gonzalez-Platas, J. (1995). *Acta Cryst.* **A51**, 811-820
20 Giacovazzo, C., Siliqi, D., Gonzalez-Platas, J., Hecht, H-J., Zanotti, G. & York, B. (1995). *Acta Cryst.* D**52**, 813-825
21 Hauptman, H. (1982). *Acta Cryst.* **A38**, 632-641
22 Giacovazzo, C. (1983). *Acta Cryst.* **A39**, 585-592
23 Hauptman, H. (1996). Proceedings of the Macromolecular Crystallography Computing School, Western Washington University.
24 Giacovazzo, C. (1983). *Acta Cryst.* **A39**, 685-592
25 Giacovazzo, C. & Gonzalez-Platas, J. (1995). *Acta Cryst.* **A51**, 398-404
26 Giacovazzo, C., Siliqi, D. & Gonzalez-Platas, J. (1997). Direct Methods of Solving Macromolecular Structures, School of Crystallography, 25th Course, Erice, Italy.
27 Main, P. (1976). In Crystallographic Computing Techniques, edited by F.R.Ahmed, pp 97-105. Copenhagen: Munksgaaard.
28 Giacovazzo, C., Manna, L., Siliqi, D., Bolognesi, M. & Rizzi, M. (1997) *Acta Cryst.* **A53** submitted.

INTRODUCTION TO THE USE OF NON-CRYSTALLOGRAPHIC SYMMETRY IN PHASING

MICHAEL S. CHAPMAN
Department of Chemistry & Institute of Molecular Biophysics, Florida State University, Tallahassee, FL 32306-3015, USA.

Abstract: The methodology is reviewed with particular attention to developments since prior review, and to aspects most important to the improvement and extension of crudely approximated initial phases as might be encountered in *ab initio* structure determinations. The fundamental similarity between symmetry and multiple crystal averaging is emphasized.

1. Introduction

Non-crystallographic symmetry (NCS) alone has not yet been used to determine *ab initio* a macromolecular structure. However, it provides one of the most powerful methods for improving approximate phases, and has become an important part of *ab initio*, experimental and molecular replacement phase determinations. Its use required the development of the methods of molecular replacement [1]. By the mid 1970's real-space averaging algorithms conceptually similar to those currently in use had been implemented [2]. Bricogne's paper [2] laid out some of the problems that have been addressed at least partially in the last 20 years. With increasing confidence in the methods, NCS became widely exploited in the 1980's. The methodology has matured rapidly and an excellent review has been published [3]. In the prior review and in other literature, "symmetry averaging" sometimes comes under the umbrella term, "molecular replacement", due to methodological similarities with the use of one structure to determine the phases of another. Here, the focus is "averaging", including the use of non-crystallographic symmetry (NCS) and multiple crystal forms. Molecular replacement is covered elsewhere in this volume.

Prior to the widespread use of NCS, multiple copies of a molecule in an asymmetric unit was considered a substantial disadvantage in structure determination. Likewise, the tendency of proteins to crystallize in different forms was at best an inconvenience. A

S. Fortier (ed.), Direct Methods for Solving Macromolecular Structures, 99–108.

testament to the newly realized power of averaging is the joy in at least the author's laboratory with which redundancy in the same crystal or different crystals is greeted.

Redundancy is the source of the phase information. A diffraction pattern is the continuous molecular transform sampled at discrete lattice points. A set of structure amplitudes may contain only ~ ½ the information required to reconstruct the molecular transform at a given resolution, because the phases are missing. Data from another crystal form samples the same molecular transform at different points. If the resolutions are the same, no detail is added. Rather, with the doubling of the number of samplings, there is, in principle "only one" set of phases to the molecular transform that can give both sets of amplitudes. Incorporation of multiple copies of a molecule within the same crystal has a similar effect. The unit cell size is increased, and more reflections can be measured at a given resolution, but the unique part of the structure remains the same.

For computational expediency, the redundancy is usually exploited with iterative real-space calculations:

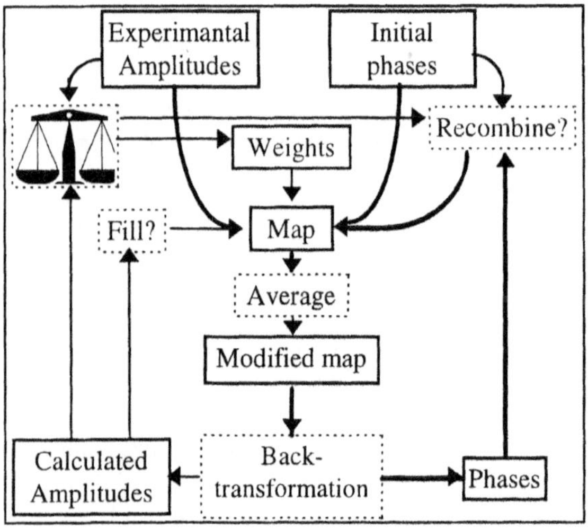

Figure 1: Phase refinement by iterative real-space averaging. The basic procedure follows thick arrows: (a) a map is calculated from the observed amplitudes and initial phases; (b) the map is replaced by one that has been averaged. (Other modifications can also be applied, such as solvent flattening.); (c) Back-transformation yields new phases which can be merged with the initial ones, or combined directly with the observed amplitudes for calculation of the improved map for the next iteration. Common embellishments to the basic mechanism are shown with thin arrows and are explained in the text.

2. Pre-requisites

2.1. ORENTATION AND LOCATION OF SYMMETRY ELEMENTS

Success of the method depends on use of the correct symmetry operators. It is only occasionally that the initial phases will be of such quality that this is best done by direct comparison of electron density values. Usually, the transformations are determined from the diffraction data using variants of the rotation and translation functions that are used for molecular replacement [1]. By searching in Patterson space, it is possible to split a computationally intractable 6-dimensional search into a 3-D orientation search [4-6] followed by a translational search [7]. This factoring results from an assumption that is only approximately true – that peaks close to the origin (short interatomic vectors) are only dependent on the molecular orientation, and that distant peaks depend only on the relative positions of molecules. The inexactness of this and other approximations, combined with the inherent noise of Patterson functions can lead to considerable ambiguity in the interpretation of rotation, and especially translation functions. Optimal strategies of overcoming these problems are covered in other sections of this volume. Here, it is noted only that the detection of spurious solutions is more difficult than with molecular replacement, because it may be many months before it is possible to build an atomic model – special care is worth the effort!

Since the last review [3], improvements in rotation and translation functions have made the process at least slightly easier. *AMoRe* is a particularly fast implementation of both rotation and translation functions [8] that facilitates systematic examination of the interdependence of rotation and translation functions. Advances targeted towards molecular replacement -- in the sensitivity and convergence radius of searches, in rigid refinement protocols and in accounting for domains that may have different orientations [9, 10] help in cases where averaging is to be used to improve molecular replacement phase determinations. In such cases, one's confidence can be increased by checking that cross-rotation functions between model and each molecule in the unknown crystal(s) are consistent with self(/cross)-rotation functions for the diffraction data set(s) that give the relative orientations of molecules in the unknown crystal(s). Within a single crystal form, NCS is often closed point group. "Locked" rotation and translation functions apply the point group as a constraint and lead to very significant gains in sensitivity and precision [11, 12].

2.2. MOLECULAR ENVELOPE

Unlike crystallographic symmetry, NCS operators are local and do not apply to the whole unit cell. It is therefore necessary to define which parts of the unit cell should be averaged with each other. With very poor initial phases, even the distinction between protein and solvent may be difficult. Distinguishing neighboring molecules that might be operated on by different local symmetry is even more challenging.

For a large number of NCS cases, in which molecules are related by a closed point group, the challenge is lesser. It is necessary only to define an envelope containing the whole assembly, because the same operations will apply to all points within the assembly. Requirements are only that an envelope include all of the protein, and is consistent with the crystallographic symmetry.

Recent progress is in two areas. One implementation of averaging, *ENVELOPE* [13], has demonstrated the feasibility of mask refinement during averaging. The values of points in the map are compared to other points to which they might be related with adjustment of the envelope. With high order NCS, biases due to prior mis-assignment can be overcome. Both *ENVELOPE* and *MAMA* [14] (for example) have iterative procedures for forcing masks to be consistent with crystallographic symmetry, smooth and unfragmented, while the latter interfaces with interactive graphics programs for manual adjustment, and has routines to help define masks, which with appropriate transformation are consistent with two crystal forms, a requirement of averaging between crystals.

2.3. STRUCTURE AMPLITUDES & STARTING PHASES

The power of averaging comes from over-sampling of the molecular transform. If the N x fraction of data observed (N = order of NCS) falls below 1, then little phase information can be obtained. The relative effects of partial data, error *etc.* have been estimated elsewhere [15]. Exact equivalence of related points in a map can only be expected with error-free complete data. The absence of reflections has all of the bad consequences of arbitrarily setting their amplitudes to zero. Partial data sets are a fact of life for virus crystallographers, and after much argument, it has become common practice to fill in missing observations with those calculated from the back-transform of the most recent averaged map. Viruses are atypical in fraction of data typically measured, and in their higher order NCS leading to good estimates for missing observations. The author is not aware of systematic tests of filling for more typical protein structure determinations.

3. Algorithms

3.1. NCS AS DENSITY MODIFICATION OR AS A RECIPROCAL-SPACE CONSTRAINT

It has long been thought possible that an equivalent procedure could be applied exclusively in reciprocal space [16, 17], and its feasibility has recently been demonstrated with the re-determination of bacteriophage ϕX174 phases [18]. Due to their computational complexity, reciprocal-space algorithms have not yet widely replaced their real-space counterpart. The reader is referred to Tong & Rossmann [18] for a more complete treatment, but here an (over-) simplified summary will be given, because it is helpful in understanding properties of the real-space algorithms.

The phased (vector) structure factor, $\vec{F}$ can be calculated from a sum of other structure factors. For the simple case of N equivalents related by non-crystallographic point group symmetry operators, T_n, about the origin in space group P1:

$$\vec{F}_{h1} \propto \sum_{h2} \vec{F}_{h2} \sum_{n}^{N} G\{h_1, h_2, T_n\} \qquad (1)$$

where h_1, h_2 are reflection indices, and G is a complicated integral over the molecular volume. Note that equation 1 is a formula for iteration: improved phases can be calculated providing that there are already approximations for at least some of the phases. However, it has been suggested [19] that it may be possible to extend from F_{000} by arbitrarily setting it to one of the centric possibilities[1]. G is a correlation function between the structure factors that is similar to the interference function of rotation functions [5]. Simplification to a G-function appropriate for a spherical molecular envelope was adequate for ϕX174 (but perhaps not for proteins with lower symmetry), and is conceptually instructive. Although not necessarily monotonically, the magnitude of G decreases with $\Delta h = |h_1 - h_2[T_n]|$ (Figure 2), i.e. the phase of a reflection depends mostly on the structure factors that are neighbors after symmetry transformation.

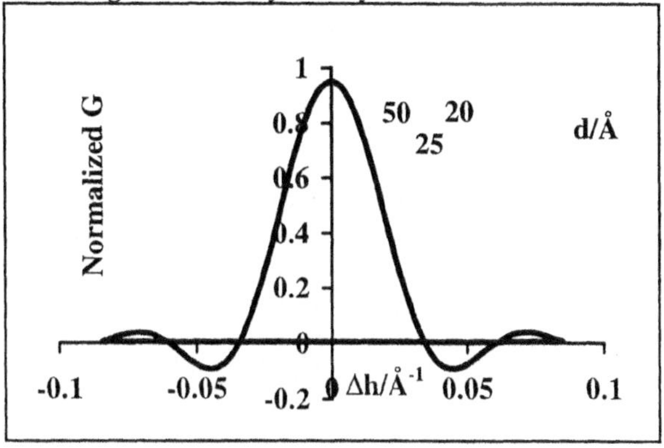

Figure 2: The normalized G correlation function for a spherical molecule of 20 Å radius is the Fourier transform of a sphere. The x-axis corresponds to the magnitude of a difference vector between two points in reciprocal space.

3.2. ALGORITHMS

3.2.1. Averaging & Incorporation of other constraints

With more powerful computers, averaging has ceased to be a computational challenge for all but the large virus unit cells (covered in [20]). A new generation of programs has come to the fore, including *RAVE* [21], *SQUASH* [22] and *DM* [23]. Improvements are a sign of a maturing field: user-friendliness, portability, robustness, generality, integration (combining several steps into a single program) and combination of averaging with other methods of phase improvement (*e.g.* Sayre's equation [24]) and density modification [25], such as solvent flattening [26] and histogram-matching [27]. Most recently, support has been added for multiple crystal forms and domains with different relative orientations [28, & Cowtan, K. personal communication].

In general, the NCS-equivalent of a grid point falls between other grid points, and its electron density needs to be interpolated. The precision of interpolation is increased with more finely spaced grid points. Although Shanon's theory [29] shows that no benefit results from the calculation of the *averaged* map much finer than ½ the resolution limit, Bricogne [2] showed that when linear interpolation was used, the *pre-*

---

[1] Such an extension might have equal chances of generating enantiomeric or Babinet solutions – see accompanying article [20]

averaged map needed to be sampled with a spacing finer than 1/5th resolution limit to avoid serious errors. The large unaveraged maps are now commonly avoided through the use of high-order (e.g. 11 point) interpolation through which accuracy can be achieved without fine grids. Aliasing problems are overcome with grids slightly finer than the Shannon limit, and it is now common to compute both maps (for convenience) with a grid 1/2.5 to 1/3.0 times the resolution limit.

3.2.2. *Phase recombination and weighting*

Phases calculated by back transformation have no figure of merit, and no width to their (unimodal) phase probability distributions. How does one calculate the map of least error? If initial phase information is worth keeping, how should the back-transformed phases be combined with it? Perhaps maximum likelihood methods will provide rigorous answers, but here let us consider methods that are currently widely used.

Most methods assume that inaccuracy of the back-transformed phases is likely to be correlated with inaccuracy of calculated structure amplitudes. Although this may be true only sometimes for individual reflections, overall, it seems reasonable. Focusing on individual relfections, aribitrary weights, w, have been used [30]:

$$w = \exp\left\{-\left|F_o - F_c\right| / \left\langle F_o \right\rangle\right\} \qquad (2)$$

Bricogne [2] adapted Sim-weighting [31, 32] to the approximation of back-transformed probability distributions:

$$P(\alpha) \propto \exp\left\{\left[2F_o F_c \cos(\alpha - \alpha_c)\right] / \left\langle F_u^2 \right\rangle\right\} \qquad (3)$$

where o, c and u subscripts refer to observed, calculated and unknown components, and where the unknown component is estimated from the average difference between F_0 and F_c. Combinations of these two weighting schemes have been particularly successful [33]. Filled reflections are currently usually given the mean weight of observed reflections at that resolutions, although this is likely over-weighting these reflections. Recent programs have been enhanced through incorporation of σ_A-weighting [34] that should lead to improved weights and map coefficients, especially when partial structure is part of the initial phasing, but also when not.

Finally, a distinction has recently been made between the majority of current phase refinement methods that are designed to progress smoothly toward convergence, and search algorithms that evaluate discretely different phase sets [35]. The latter may be an exception, as it may not be beneficial to down-weight the effects of reflections with biggest disagreements.

4. Quality Assessment

A common method of monitoring progress is correlation (or residuals) of the symmetry related electron density. Monitoring in reciprocal space is also common with calculation of correlation coefficients or R-factors between observed structure amplitudes, and those back-transformed from the current map. Unfortunately, it is easy to obtain a map with near-exact NCS, even when there are errors that should be corrected. Numerical values should appear much "better" than for comparisons to atomic model, and should be "better" for low-order NCS. Use of correlation coefficients as a diagnostic indicator is covered in the accompanying article [20].

Recent advances include methods of cross-validation [36] that may provide better indicators of quality. In real-space, a sphere (and its symmetry equivalents) may be omitted from any density modification or averaging and can be used for "free" statistic evaluation [37]. Similarly in reciprocal space, a portion of the data can be omitted prior to map calculation, and this data can be used for calculation of statistics (and improved weighting coefficients) [38]. To avoid the problems introduced by omission of data, the process is repeated so that one cycle is used for statistics, another for phase calculation.

5. Applications, Further Considerations and Proof of the pudding...

Early applications in the 1960's did little to convince the crystallographic community of the general worth of NCS-phasing. With the derivation and implementation of real-space methods [2, 39], it became possible in the 1970's to determine the atomic structures of two viruses [40, 41] by improving poor initial phases. This began exponential growth of the use of NCS in phase refinement.

Although phase improvement became accepted quickly, there was much skepticism about the use of NCS to extend phases beyond the resolution of the experimental phases. These doubts were only laid to rest when NCS was used to generate maps that could be interpreted with atomic models, having started at resolutions where this would not have been possible. Haemocyanin and human rhinovirus 14 [42, 43] were among the earliest examples. What is the basis for the generation of phases where none previously existed?

Considered in reciprocal space, for a reflection, $\vec{h}$, just beyond the current phasing limit, equation 1 has half of the terms required for a precise determination of its phase. Each of the unphased reflections will have weight of zero, so $\Delta\vec{h} = 0$ will not contribute, as will the ~ ½ of the immediately neighboring reflections that happen to be at higher resolution (Figure 1). The nearly ½ of the neighbors for which phases have previously been estimated will allow a very crude approximation of $\vec{F}_h$. Addition of even a crude initial approximation for α_h facilitates its refinement, because the summation of equation 1 is more complete. Following convergence, it can also be used in the phase estimation of $\vec{h}' \geq \vec{h}$, providing a basis for phase extension.

Extension, once shown feasible, has been extended to lower and lower resolution, culminating in cases, reviewed in [20], with high order NCS where it has been possible to refine and extend from low resolution *ab initio* phases in the absence of experimental phase information. The processes of extending over a large or modest resolution range are essentially the same. One additional challenge of the *ab initio* applications is knowing that the phase determination is correct at resolutions lower than those at which molecular features should be recognizable.

The reciprocal-space understanding demonstrates the importance of several experimental factors, especially in the case of phase extension. It is important to include as many of the terms in equation 1 as possible. By increasing the resolution in steps ≤ 1 reciprocal lattice unit, nearly half of the terms can be included. Although steps of 2 or 3 reciprocal lattice units have been used occasionally for computational expediency, usually this generates poorer results, because equation 1 is missing not only all of the higher resolution terms, but some of the lower resolution terms. Clearly, anything that further detracts from the initial approximation of equation 1 is deleterious. Extension may not be able to recover from incorrect starting phases. Figure 1 shows that it is highest resolution shell that is most critical, so it is necessary to start extension at a resolution for which there is confidence in the starting phases. Incomplete or imprecise amplitudes, can be seen to affect extension through equation 1, and it may be necessary to limit the resolution to which phases are extended. Finally it is noted, that because refinement of the extended phases depends on their first approximation, it is possible for phase refinement to head towards an incorrect solution and never recover.

The usefulness of NCS is no longer questioned. With high order NCS, excellent phases can be obtained. Considerable effort has recently gone into the determination of accurate experimental phases for proteins [44] using anomalous dispersion. For the virus crystallographers lucky enough to have 60-fold NCS, although usually at lower resolution, electron density maps are routinely just as accurate [45], in spite of being calculated with amplitudes that are > 2-fold less precise.

6. Acknowledgments and References

The author is grateful to have had the opportunity to learn about averaging through research in the laboratory of one of the most preeminent of many contributors to this field, Michael Rossmann. The author is also grateful to members of his laboratory: Eric Blanc, Genfa Zhou and Qing Xie, who keep him thinking about NCS and averaging through the interesting problems their research poses, and for the financial support of the National Science Foundation and Lucille P. Markey Charitable Trust.

1 Rossmann, M. G. (1972) *The Molecular Replacement Method,* Gordon and Breach, New York

2 Bricogne, G. (1976) Methods and Programs for Direct Space Exploitation of Geometric Redundancies, *Acta Crystallographica* **A32**, 832-47

3 Rossmann, M. G. (1990) The Molecular Replacement Method, *Acta Cryst.* **A46**, 73-82

4 Rossmann, M. G. (1972) The Rotation Problem, in Rossmann, M. G. (ed.), *The Molecular Replacement Method,*, Gordon and Breach, New York, pp. 16-44

5 Rossmann, M. G. and Blow, D. M. (1962) The Detection of Sub-Units Within the Crystallographic Asymmetric Unit, *Acta Crystallographica* **15**, 24-31

6 Rossmann, M. G. and Blow, D. M. (1964) The Relative Positions of Independent Molecules Within the Same Asymmetric Unit, *Acta Crystallographica* **17**, 338-42

7 Crowther, R. A. and Blow, D. M. (1967) A Method of Positioning a Known Molecule in an Unknown Crystal Structure, *Acta Crystallographica* **23**, 544-8

8 Navaza, J. (1994) *AMoRe*: an Automated Package for Molecular Replacement, *Acta Crystallographica* **A50**, 157-163

9 Brünger, A. T. (1997) Patterson Correlation Searches and Refinement, *Methods in Enzymology* ***in press***

10 Zhang, X.-J. and Matthews, B. W. (1994) Enhancement of the Method of Molecular Replacement by Incorporation of Known Structural Information, *Acta Crystallographica* **D50**, 675-686

11 Tong, L. and Rossmann, M. G. (1990) The Locked Rotation Function, *Acta Crystallographica* **A46**, 783-792

12 Tong, L. (1996) The Locked Translation Function and Other Applications of a Patterson Correlation Function, *Acta Crystallographica* **A52**, 476-79

13 Rossmann, M. G., McKenna, R., Tong, L., Xia, D., Dai, J.-B., Wu, H., Choi, H.-K. and Lynch, R. E. (1992) Molecular Replacement Real-Space Averaging, *J. Appl. Cryst.* **25**, 166-80

14 Kleywegt, G. J. and Jones, T. A. (1993) Masks Made Easy, in (ed.), *ESF/CCP4 Newsletter,*vol. 28, 56-59

15 Arnold, E. and Rossmann, M. G. (1986) Effect of errors, redundancy, and solvent content in the molecular replacement procedure for the structure determination of biological macromolecules , *Proc. Natl. Acad. Sci. USA* **83** , 5489-93

16 Main, P. and Rossmann, M. G. (1966) Relationships among Structure Factors due to Identical Molecules in Different Crystallographic Environments, *Acta Crystallographica* **21**, 67-72

17 Crowther, R. A. (1967) A Linear Analysis of the Non-Crystallographic Symmetry Problem, *Acta Crystallographica* **22**, 758-764

18 Tong, L. and Rossmann, M. G. (1995) Reciprocal-Space Molecular-Replacement Averaging, *Acta Crystallographica* **D51**, 347-353

19 Tong, L. and Rossmann, M. G. (1995) Reciprocal-space molecular-replacement averaging, *Acta Crystallographica* **D51**, 347-53

20 Chapman, M. S., Blanc, E., Johnson, J. E., McKenna, R., Munshi, S., Rossmann, M. G. and Tsao, J. (1997) Use of non-crystallographic symmetry for *ab initio* phasing of virus structures, in Fortier, S. (ed.), *NATO Advanced Study Institute on Direct Methods for Solving Macromolecular Structures,* Kluwer Acad. Publ., Dordrecht, The Netherlands

21 Jones, T. A. (1992) a, yaap, asap, @£? A set of averaging programs, in Dodson, E., Gover, S. and Wolf, W. (ed.), *Proceedings of the CCP4 Study Weekend,*, Science and Engineering Research Council, 91 - 105

22 Zhang, Y. J. (1993) SQUASH - combining constraints for macromolecular phase refinement and extension, *Acta Crystallographica* **D49**, 213-222

23 Cowtan, K. D. and Main, P. (1993) Improvement of macromolecular electron-density maps by the simultaneous application of real and reciprocal space constraints, *Acta Crystallographica* **D49**, 148-57

24 Sayre, D. (1952) The Squaring Method: a New Method for Phase Determination, *Acta Crystallographica* **5**, 60-65

25 Cannillo, E., Oberti, R. and Ungaretti, L. (1983) Phase Extension and Refinement by Density Modification in Protein Crystallogaphy , *Acta Cryst.* **A39** , 64-74

26 Wang, B. C. (1985) Resolution of the Phase Ambiguity in Macromolecular Crystallography , *Methods in Enzymology* **115** , 90-112

27 Lunin, V. Y. (1994) Electron-density histograms and the phase problem, *Acta Crystallographica* **D49**, 90-99

28 Schuller, D. J. (1996) MAGICSQUASH: More Versatile Non-crystallographic Averaging with Mulitple Constraints, *Acta Crystallographica* **D52**, 425-34

29 Shannon, C. E. (1949) Communication in the presence of noise, *Proc. Inst. Radio Engrs. New York* **37**, 10-21

30 Rayment, I., Baker, T. S., Caspar, D. L. D. and Murakami, W. T. (1982) Polyoma virus capsid structure at 22.5 Å resoution , *Nature* **295** , 110-15

31 Sim, G. A. (1959) The distribution of phase angles for structures containing heavy atoms., *Acta Crystallographica* **12**, 813-15

32 Sim, G. A. (1960) A note on the heavy atom method , *Acta Cryst.* **13** , 511-12

33 Arnold, E. and Rossmann, M. G. (1988) The Use of Molecular-Replacement Phases for the Refinement of the Human Rhinovirus 14 Structure, *Acta Crystallographica* **A44**, 270-282

34 Read, R. J. (1985) Improved Fourier Coefficients for Maps Using Phases from Partial Structures with Errors, *Acta Crystallographica* **A42**, 140-9

35 Miller, S. T., Hogle, J. M. and Filman, D. J. (1996) A Genetic Algorithm for the *Ab Initio* Phasing of Icosahedral Viruses, *Acta Crystallographica* **D52**, 235-251

36 Brünger, A. T. (1992) Free R value: a novel statistical quantity for assessing the accuracy of crystal structures , *Nature* **355** , 472-5

37 Abrahams, J. P. and Leslie, A. G. W. (1996) Methods Used in the Structure Determination of Bovine Mitochondrial F_1 ATPase, *Acta Crystallographica* **D52**, 30-42

38 Cowtan, K. D. and Main, P. (1996) Phase combination and cross validation in iterated density-modification calculations, *Acta Crystallographica* **D52**, 43-48

39 Bricogne, G. (1974) Geometric sources of redundancy in intensity data and their use for phase determination, *Acta Crystallographica* **A30**, 395-405

40 Bloomer, A. C., Champness, J. N., Bricogne, G., Staden, R. and Klug, A. (1978) Protein disk of tobacco mosaic virus at 2.8 Ångstrom resolution showing the interactions within and between subunits, *Nature* **276**, 362-8

41 Harrison, S. C., Olson, A., Schutt, C. E., Winkler, F. K. and Bricogne, G. (1978) Tomato bushy stunt virus at 2.9 Å resolution , *Nature* **276** , 368-373

42 Gaykema, W. P. J., Hol, W. G., Vereijken, J. M., Soeter, N. M., Bak, H. J. and Beintema, J. J. (1984) 3.2 Å structure of the copper-containing oxygen-carrying protein Panulirus interruptus haemocyanin, *Nature* **209**, 23-29

43 Rossmann, M. G., Arnold, E., Erickson, J. W., Frankenberger, E. A., Griffith, J. P., Hecht, H., Johnson, J. E., Kamer, G., Luo, M., Mosser, A., Rueckert, R., Sherry, B. and Vriend, G. (1985) Structure of human common cold virus and functional relationship to other picornaviruses. , *Nature* **317** , 145-153

44 Burling, F. T., Weis, W. I., Flaherty, K. M. and Brünger, A. (1996) Direct Observation of Protein Solvation and Discrete Disorder Using Experimental Crystallographic Phases, *Science* **271**, 72-77

45 Zhou, G., Wang, J., Blanc, E. and Chapman, M. S. (1997) Determination of the Relative Precision of Atoms in a Macromolecular Structure, *Acta Crystallographica* ***submitted***

AN INTRODUCTION TO MAXIMUM ENTROPY IN ACTION

C.J.Gilmore
Department of Chemistry
University of Glasgow
Glasgow G12 8QQ, Scotland, UK.

1. Introduction

1.1 SOME NECESSARY THEORY AND NOTATION

Let us start with a set of unitary structure factors $|U_{\mathbf{h}}|^{obs}$, which are derived from the intensity data by standard normalisation procedures. (For N point atoms of unit weight in space group P1 $|U_{\mathbf{h}}|^{obs} = 1/\sqrt{N}\,|E_{\mathbf{h}}|^{obs}$) The notation $\boldsymbol{U}$ will be used to define an arbitrary vector of *nU's*, $\mathbf{U} = (U_{\mathbf{h1}}, U_{\mathbf{h2}}, U_{\mathbf{h3}}, \ldots\ldots U_{\mathbf{hn}})$. We also have a set of phase angles $\varphi_{\mathbf{h}}$ most of which are either unknown (the *ab initio* case) or only approximately determined. For *ab initio* structure solution, because of rules of origin and enantiomorph definition, some phases can usually, but not always (it depends on the space group) be assigned subject to certain well known rules, whereas in the case of macromolecular crystallography, there will be a large set of such phases, albeit rather inaccurately known at times, derived from isomorphous replacement methods. In electron diffraction, the Fourier transform of the corresponding image may give useful phase information.

Those reflections which are phased comprise the basis set {H}; the disjoint set, the non-basis set, of unphased amplitudes is {K}; the phase problem is one of phasing $|U_{\mathbf{h}\in K}|^{obs}$ from $U_{\mathbf{h}\in H}$. In electron diffraction the unmeasured set {U} may also be important. This is shown diagramatically in Figure 1.

At first sight, solving a crystal structure seems straightforward: a set of reflections is chosen to form the basis set. These reflections are used as constraints in an entropy maximisation to generate a maximum entropy map $q^{ME}(\boldsymbol{x})$. Because it is a maximum entropy map $q^{ME}(\underline{x})$ satisfies the following conditions:

(1) It is optimally unbiased.

(2) Its Fourier transform reproduces the constraints to within experimental error.

S. Fortier (ed.), Direct Methods for Solving Macromolecular Structures, 109–117.

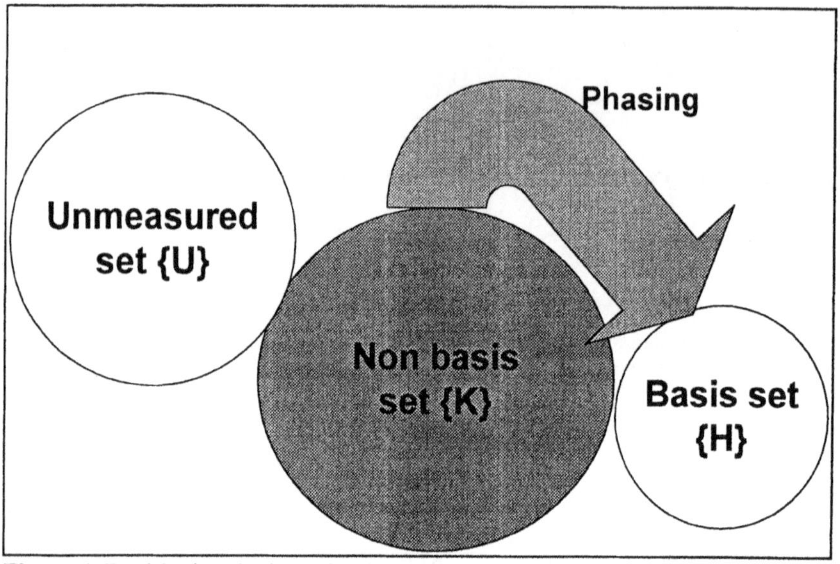

Figure 1. Partitioning the intensity data.

(3) The Fourier transform of $q^{ME}(\underline{x})$ generates estimates of amplitudes and phases for reflections in {K}. This process is called *extrapolation.* This is shown diagramatically in Figure 2 below.

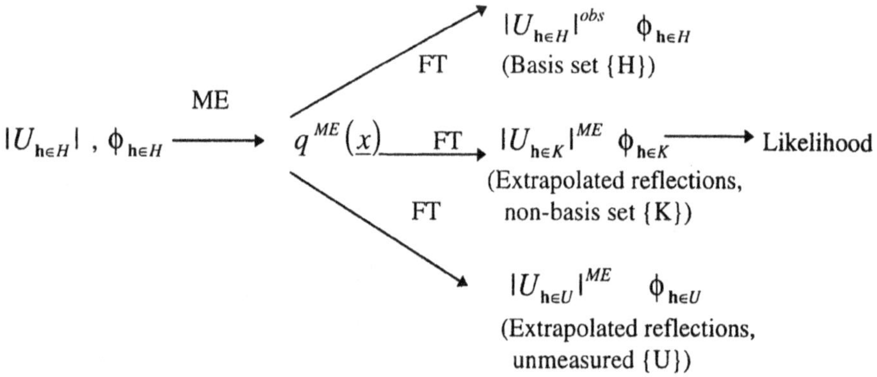

Figure 2. Phase extrapolation

At this stage, for a small basis set, the extrapolation is rather weak, and the structure is certainly not visible in any map one cares to generate. So what can be done? The most obvious answer is to examine the most strongly extrapolated reflections, and to use the phase information from them coupled with the observed U magnitudes, and to pass these into the basis set {H}. This usually leads to disaster: the ME solution gets trapped in a local optimum point in phase space, whole subsets of reflections become wrongly phased, and the structure remains unsolved. This is a manifestation of the *branching problem*: phases are being selected without exploring the relevant U-space in sufficient detail, so that what appears to be an unambiguous choice of new phase is, in reality, no such thing. So what can be done? Extra unphased reflections must be added to the starting set with permuted phases giving rise to a multisolution environment just as in conventional direct methods, and these reflections should be those which are very weakly extrapolated, so that their inclusion in a new basis set offers maximum surprise to the calculations. How then do you select the correct permutation set? Can you choose those which give maps of maximum entropy? The answer to this is *'no'*, and this is discussed in a later chapter, but the Bricogne formalism [1-7] has a way forward.

2. The Bricogne Formalism

Bricogne has outlined a methodology of solving crystal structures based on the ME (or saddlepoint) method coupled with likelihood evaluation, that directly addresses the branching problem [1-6].

What have we done when we have defined a basis set and performed the entropy maximisation? We have explored *P(U)* around the basis set phase sub-region in *U* space. Why do we need maximum entropy? We only require *P(U)* to reproduce those U-magnitudes which are members of the basis set. It makes no direct reference to non-basis set reflections, and so it is highly indeterminate. The ME method is an obvious way of overcoming this: for such a phase choice maximises the entropy of the distribution subject to the constraint that it must reproduce the basis set amplitudes and phases to generate a non-uniform distribution of atoms $q^{ME}(\underline{x})$. *P(U)* is thus multimodal, and this is the source of the branching problem.

There is a remarkable, and unexpected benefit of using the ME method here: the approximations to *P(U)* are now accurate for large U magnitudes, and we have re-centred the usual Edgeworth series expansion for *P(U)* away from $U = 0$. In real space this is equivalent to leaving behind the Wilson uniform distribution of atoms to a non-uniform one - this is one of the long-term goals of direct methods.

So far all ME phasing methods perform in this way whether we realise it or not, but what can we do next? How do we stop ourselves locking into an incorrect solution? Let us explore new *sub-regions P(U)* where we test various new phase combinations to overcome the multimodality or branching problem. In practice we take the current basis set and add to it a few reflections with large associated U magnitudes which optimally enlarge the second neighbourhood of the basis set. The second neighbourhood is defined by reflections $\mathbf{h}_1 \pm {}^t\underline{\mathbf{R}}_g.\mathbf{h}_2$ for $\mathbf{h}_1, \mathbf{h}_2 \in H$, where ${}^t\underline{\mathbf{R}}_g$ is the transpose of a rotation matrix obtained from the crystal space group. For acentric reflections phase choices can be $\pm\pi/4$, $\pm 3\pi/4$ i.e. quadrant permutation; for centrics both possible values are used e.g. 0, π or $\pm\pi/2$. We now have several possible basis sets and we now explore *P(U)* around each of these phase sub-regions by performing a constrained entropy maximisation for each phase permutation, and so derive associated phase and amplitude extrapolation. However, we have made no judgement as to which sub-region is the more likely to be the correct one. In the Bricogne formalism, this decision is made using likelihood. For every sub-region, *P(U)* is a multivariate Gaussian centred around the vector $\mathbf{U}^{ME}$ corresponding to $q^{ME}(\mathbf{x})$. It is possible to express this in an analytical form. Several approximations exist, but we will concentrate here on the simplest: the diagonal form in which the extrapolates in *U* are decoupled and treated as independent i.e. the covariances between them are zero. This gives rise to readily tractable likelihood functions.

For each acentric extrapolated, non-basis set reflection **k** the likelihood measure, in its diagonal approximation, can be written [1,7,8]

$$\Lambda_{\mathbf{k}} = \frac{|U_{\mathbf{k}}|^{obs}}{\varepsilon_{\mathbf{k}}\Sigma+\sigma_{\mathbf{k}}^{2}} \exp\left\{-\frac{1}{2}\frac{\left(|U_{\mathbf{k}}|^{obs}\right)^{2}+\left|U_{\mathbf{k}}^{ME}\right|^{2}}{\varepsilon_{\mathbf{k}}\Sigma+\sigma_{\mathbf{k}}^{2}}\right\} I_{0}\left(\frac{|U_{\mathbf{k}}|^{obs}\left|U_{\mathbf{k}}^{ME}\right|}{\varepsilon_{\mathbf{k}}\Sigma+\sigma_{\mathbf{k}}^{2}}\right) \tag{1}$$

where $\varepsilon_{\mathbf{k}}$ is the statistical weight of reflection **k**, $\sigma_{\mathbf{k}}^{2}$ the variance of $|U_{\mathbf{k}}|^{obs}$ and Σ a refinable measure of unit cell contents $\Sigma \approx 1/(2N)$ for N point atoms in the unit cell. The distribution (1) is a Rice distribution comprising a Gaussian (the exponential term) with an offset represented by the Bessel function term (I_0). Note also that this expression is a measure of agreement between $|U_{\mathbf{k}}|^{obs}$ and $\left|U_{\mathbf{k}}^{ME}\right|$, indeed it has a maximum where $|U_{\mathbf{k}}|^{obs} = \left|U_{\mathbf{k}}^{ME}\right|$. For the centric case the Bessel function is replaced by a cosh term:

$$\Lambda_{\mathbf{k}} = \frac{2|U_{\mathbf{k}}|^{obs}}{\pi\left(2\varepsilon_{\mathbf{k}}\Sigma+\sigma_{\mathbf{k}}^{2}\right)} \exp\left\{-\frac{1}{2}\frac{\left(|U_{\mathbf{k}}|^{obs}\right)^{2}+\left|U_{\mathbf{k}}^{ME}\right|^{2}}{2\varepsilon_{\mathbf{k}}\Sigma+\sigma_{\mathbf{k}}^{2}}\right\} \cosh\left(\frac{|U_{\mathbf{k}}|^{obs}\left|U_{\mathbf{k}}^{ME}\right|}{2\varepsilon_{\mathbf{k}}\Sigma+\sigma_{\mathbf{k}}^{2}}\right) \tag{2}$$

In the spirit of traditional likelihood analysis, we define a corresponding null hypothesis for the situation of null extrapolation, $\left|U_{\mathbf{k}}^{ME}\right| = 0$, which gives the Gaussian distribution of Wilson statistics. For acentric reflections:

$$\Lambda_{\mathbf{k}}^{0} = \frac{|U_{\mathbf{k}}|^{obs}}{\varepsilon_{\mathbf{k}}\Sigma+\sigma_{\mathbf{k}}^{2}} \exp\left\{-\frac{1}{2}\frac{\left(|U_{\mathbf{k}}|^{obs}\right)^{2}}{\varepsilon_{\mathbf{k}}\Sigma+\sigma_{\mathbf{k}}^{2}}\right\} \tag{3}$$

The extension to centric reflections is obvious. Define:

$$L_{\mathbf{k}} = \log\frac{\Lambda_{\mathbf{k}}}{\Lambda_{\mathbf{k}}^{0}} \tag{4}$$

Then the global log-likelihood gain (LLG) is:

$$LLG = \sum_{\mathbf{k}} L_{\mathbf{k}} \tag{5}$$

The LLG will be largest when the phase assumptions for the basis set lead to predictions of deviations from Wilson statistics for the unphased reflections, and in this context it is used as a powerful figure of merit. However, rather than just choose those phase sets with high associated LLG, which is a somewhat subjective process, the Student t-test is used [9]. The LLGs are analysed for phase indications using the t-test. The simplest example involves the detection of the main effect associated with the sign of a single centric phase. The LLG average, μ^{+}, and its associated variance V^{+} is computed for those sets in which the sign of this permuted phase under test is +. The calculation is then repeated for those sets in which the same sign is - to give the corresponding μ^{-}, and variance V^{-}. The t-statistic is then:

$$t = \frac{\left|\mu^{+} - \mu^{-}\right|}{\sqrt{V^{+} + V^{-}}} \tag{6}$$

The use of the t-test enables a sign choice to be derived with an associated significance level. This calculation is repeated for all the single phase indications, and is then extended to combinations of two and three phases. An extension to acentric phases is straightforward by employing two signs to define the phase quadrant both in permutation and in the subsequent analysis. In general, only relationships with associated significance levels <2% are used, but this is sometimes relaxed. This sort of calculation is repeated for all the single phase indications, and is then extended to combinations of two and three phases. Each of the *m* phase relationships, *i*, so generated is given an associated weight w_i

$$w_i = \left(1 - \frac{I_1(s_i)}{I_0(s_i)}\right) \tag{7}$$

where I_1 and I_0 are the appropriate Bessel functions and s_i is the significance level of the *i-th* relationship from the t-test. This weighting function reflects the need for a scheme in which the absolute values of the significance levels are not given undue emphasis since they are themselves subject to errors arising from the nature of the likelihood function used and the lack of error estimates for the LLGs themselves. Each node n is now given a score, s_n :

$$s_n = LLG_n \sum_{j=1}^{m} w_j \tag{8}$$

where the summation spans only those phase relationships where there is agreement between the basis set phases and the t-test derived phase relationships [10]. The scores are sorted and only the top 8-16 nodes are kept; the rest are discarded. New reflections are then permuted and a corresponding set of ME solutions is generated. In this way we build a *phasing tree* in which each phase choice is represented as a *node*, and has a score, or figure of merit, based on its log likelihood gain (LLG). The root node of the tree is defined by the origin defining reflections. The first set of phase permutations defines the second level. Those which do not pass the analysis of likelihood are discarded, then further phase permutations are used to generate the third level, and this continues until a recognisable structure or structural fragment appears. Figure 3 shows a simple outline of a three-level phasing tree.

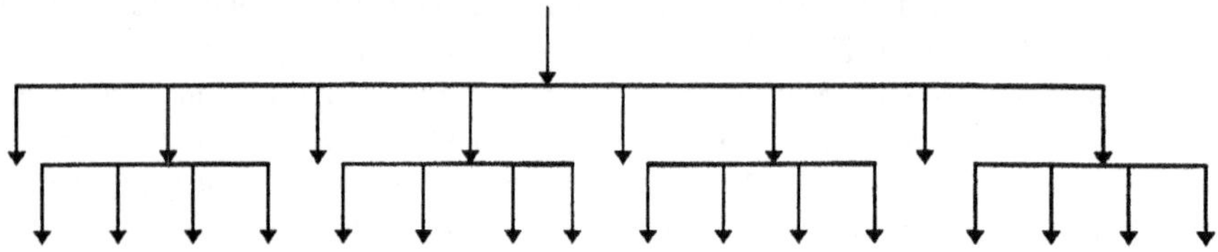

Figure 2. A Three level, 25 node phasing tree.

There is an interesting example of the ME method in action using idealised, one-dimensional data by Carter [11]. Note that likelihood is not used in this approach in the Bayesian way of multiplying the prior, instead it becomes a very powerful figure of merit which controls the branching problem by selecting the correct nodes. However, there is a Bayesian connection in the may in which the prior, $q^{ME}(x)$, is continually updated from level to level in the phasing tree by adding new phase information. We shall see later that the power of likelihood is such that it can be used to make other decisions concerning envelopes for macromolecules, parameters for non-crystallographic symmetry and to measure the effective number of atoms in the unit cell at low resolution.

Now $q^{ME}(\underline{x})$ is a probability distribution, and not a map in the traditional sense (although its peaks do correspond to atom positions), and thus it needs conversion to a more conventional one. The trial electron density maps generated in this approach are called *centroid maps*, and can be visualised as Sim filtered maximum entropy maps. For **k** acentric, the Fourier coefficients are:

$$|U_{\mathbf{k}}|^{obs}\left[I_1(X_{\mathbf{k}})/I_0(X_{\mathbf{k}})\right]\exp\left(i\varphi_{\mathbf{k}}^{ME}\right) \tag{9}$$

where:

$$X_{\mathbf{k}} = (2N/\varepsilon_{\mathbf{k}})|U_{\mathbf{k}}|^{obs}\left|U_{\mathbf{k}}^{ME}\right| \tag{10}$$

For **k** centric, these coefficients become:

$$|U_{\mathbf{k}}|^{obs}\tanh(X_{\mathbf{k}})\exp\left(i\varphi_{\mathbf{k}}^{ME}\right) \tag{11}$$

with:

$$X_{\mathbf{k}} = (N/\varepsilon_{\mathbf{k}})|U_{\mathbf{k}}|^{obs}\left|U_{\mathbf{k}}^{ME}\right| \tag{12}$$

The practical differences between centroid and maximum entropy maps can be seen with reference to membrane data in Gilmore, Shankland & Fryer [12].

One final practical point needs to be made here concerning the tightness with which one fits the constraints in the entropy maximisation which can be very important in a phasing environment. If the fit between $|U_{\mathbf{h}}|^{obs}$ and $\left|U_{\mathbf{h}}^{ME}\right|$ is very slack the ME extrapolation is weaker than it can be, and the phasing power and the discriminating power of likelihood is correspondingly reduced. If however there is overfitting, spurious details (often looking like small stones) appear in $q^{ME}(\boldsymbol{x})$ which give false phase indications. The latter situation can often be detected by the use of likelihood: if the LLG is monitored through the iterative cycles of entropy maximisation, a maximum is often reached and this can be used as a place to stop. Alternatively, the χ^2 statistic can be used with a default choice of unity as a place to stop.

$$\chi^2 = \frac{1}{n}\sum_{h \in H}\frac{1}{\sigma_h^2}(|U_{\mathbf{h}}|^{obs} - |U_{\mathbf{h}}^{ME}|)^2 \tag{13}$$

where n is the total number of degrees of freedom (2 for each acentric reflection in the basis set, and one for each centric). In practice, for small basis sets either method works well, but for large basis sets maximum LLG is preferred.

The MICE computer program is a practical implementation of a part of the ME formalism [7,8] as is BUSTER [6]. Surveys of the method and its practical applications can be found in [13,14].

Other developments of great importance which will be described elsewhere include:

- The use of error correcting codes to generate amazingly efficient phase permutations [6].
- The use of envelopes [15] including the resolution of envelope ambiguities using likelihood [16].
- Non-crystallographic symmetry [17].

The ME method can incorporate these concepts in a natural way.

3. Where has the method been used in structure solution?

The method has been used successfully to solve:

- Small molecule structures (but it cannot compete with modern direct methods) [8]
- Powder data sets [18-20].
- Membrane proteins using two-dimensional electron diffraction data [12,21].
- Problems with isomorphism and envelope definition, and as an alternative to regular envelope flattening in protein crystallography [15,16].
- Heavy atom positions in protein crystallography [17,22].
- Surface structures [10].
- Electron diffraction data sets on a number of small molecules [14,23,24].

References

1. Bricogne, G. (1984) Maximum entropy and the foundations of direct methods, *Acta Cryst.* **A40**, 410-445.

2. Bricogne, G. (1988) A Bayesian statistical theory of the phase problem. I. A multichannel maximum entropy formalism for constructing generalized joint probability distributions of structure factors, *Acta Cryst.* **A44** (1988). 517-545.
3. Bricogne, G. (1988) Maximum entropy methods in the crystallographic phase problem, in N.W. Isaacs and M.R. Taylor (eds.), *Crystallographic Computing 4,* Clarendon Press, Oxford, pp. 60-79.
4. Bricogne, G, (1991) The X-ray crystallographic phase problem, in B.Buck and V.A.Macaulay (eds.), *Maximum entropy in action*, Oxford University Press, Oxford pp. 187-216.
5. Bricogne, G. (1991) Maximum entropy as a common statistical basis for all phase determination methods, in D.Moras, A.D. Podjarny and J.C. Thierry (eds.), *Crystallographic Computing 5: From chemistry to biology* Oxford University Press, Oxford, pp. 257-297.
6. Bricogne, G. (1993) Direct phase determination by entropy maximisation and likelihood ranking: status report and perspectives. *Acta Cryst.* **D49**, 37-60.
7. Bricogne, G. and Gilmore, C.J. (1990) A multisolution method of phase determination by combined maximisation of entropy and likelihood. I. Theory, algorithms and strategy, *Acta Cryst.* **A46,** 284-297.
8. Gilmore, C.J., Bricogne, G. and Bannister, C. (1990) A multisolution method of phase determination by combined maximisation of entropy and likelihood. II Application to small molecules, *Acta Cryst.* **A46,** 297-308.
9. Shankland, K., Gilmore, C.J., Bricogne G. and Hashizume, H. (1993) A multisolution method of phase determination by combined maximisation of entropy and likelihood. V Automatic Likelihood analysis *via* the student t-test, with an application to the powder structure of magnesium boron nitride, Mg_3BN_3. *Acta Crystallogr.* ***A49,*** 493-501.
10. Gilmore, C.J., Marks, L.D., Grozea, D., Collazo, C., Landree, E. and Twesten R.D, Direct Solutions of the Si (111) 7x7 Structure, *Surface Science*, in press.
11. Carter, C.W.Jr. (1994) Entropy maximisation permutation, and likelihood scoring methods for improving macromolecular electron density maps, in W.Wolf, E.J.Dodson and S.Glover (eds.), *From first Map to Final Model, Proceedings of the Study Weekend held at Daresbury* Daresbury Laboratory, Warrington, pp. 41-58.
12. Gilmore, C.J., Shankland, K. and Fryer, J.R. (1993), Phase extension in electron crystallography using the maximum entropy method and its application to two-dimensional purple membrane data from *Halobacterium halobium. Ultramicroscopy,* **49**, 147-178.
13. Gilmore, C.J. (1996) Maximum entropy and Bayesian statistics in crystallography: a review of practical applications. *Acta Cryst.* **A52** 561-589.
14. Gilmore, C.J., Shankland, K. and Bricogne, G. (1993) Applications of the maximum entropy method to powder diffraction and electron crystallography, *Proc.Roy.Soc.Ser. A.* **442,** 97-111.
15. Xiang, S., Carter, C.W.C.Jr, Bricogne,G., and Gilmore, C.J., Entropy maximisation constrained by solvent flatness: a new method for macromolecular phase extension and map improvement, *Acta Cryst.* (1993), **D49,** 193-212.

16. Doublié,S., Xiang, S., Gilmore,C.J., Bricogne,G. and Carter C.W.C. Jnr. Overcoming Non-Isomorphism by Phase Permutation an Likelihood Scoring: Solution of the TrpRS Crystal Structure, *Acta Cryst.* (1994), **A50**, 164-182.
17. Gilmore, C.J. & Bricogne, G. (1997) The MICE Computer Program in C.W.C. Carter Jnr *Methods in Enzymology* Academic Press, New York In press.
18. Gilmore, C.J., Henderson, K. and Bricogne G. (1991) A multisolution method of phase determination by combined maximisation of entropy and likelihood. IV The *ab initio* solution of crystal structures from their X-ray powder data, *Acta Cryst.* **A47,** 830-841.
19. Tremayne, M., Lightfoot, P., Mehta, M.A., Bruce, P.G., Harris, K.D.M., Shankland, K., Gilmore, C.J. and Bricogne G. (1992) An *ab initio* determination of $LiCF_3SO_3$ from X-ray powder diffraction data using entropy maximisation and likelihood ranking. *J. Solid State Chem.* **100,** 191-196.
20. Tremayne, M., Lightfoot, Glidewell, C., Harris K.D.M., Shankland, K., Gilmore, C.J., Bricogne G. and Bruce, P.G. (1992) Application of the combined entropy and likelihood method to the *ab initio* determination of an organic crystal structure from X-ray powder diffraction data. *J. Mater. Chem.* **2,** 1301-1302.
21. Gilmore, C.J., Nicholson, W.V. and Dorset, D.L. (1996) Direct methods in protein electron crystallography: the ab initio structure determination of two membrane protein structures in projection using maximum entropy and likelihood, *Acta Cryst.* **A52,** 937-946.
22. Gilmore, C.J., Henderson, A.N. and Bricogne G. (1991) A multisolution method of phase determination by combined maximisation of entropy and likelihood. V The use of likelihood as a discriminator of phase sets produced by the SAYTAN program for a small protein. *Acta Cryst.* **A47,** 832-846.
23. Voigt-Martin, I.G.,Han, D.H., Gilmore,C.J., Shankland, K. and Bricogne, G (1994) The Use of Maximum Entropy and Likelihood Ranking to Determine the Crystal Structure of 4-(4'-(N,N-dimethyl)aminobenzylidene)-pyrazolidine-3,5-dione at 1.4Å Resolution from Electron Diffraction and High Resolution Electron Microscopy Image Data, *Ultramicroscopy,* **56**, 271-288.
24. I.G.Voigt Martin, D.H.Yan, C.J.Gilmore and G.Bricogne (1995) Structure Determination by Electron Crystallography Using both Maximum Entropy and Simulation Approaches, *Acta Cryst,* **A51,** 849-868.

CRYSTAL STRUCTURE REFINEMENT INCORPORATING CHEMICAL INFORMATION

GEORGE M. SHELDRICK
University of Göttingen
Institut für Anorganische Chemie, Tammannstraße 4,
D-37077 Göttingen, Germany

Abstract

Crystal structure refinement using the new release of the SHELX program is discussed with emphasis on the incorporation of chemical information via constraints and restraints. It would be desirable to use refinement techniques in the structure solution as well, so as to incorporate chemical information into the phasing process. One such approach that looks very promising is ARP.

1. Introduction

A common feature of macromolecular refinements and electron crystallography is the comparatively poor typical data to parameter ratio (compared to small-molecule X-ray structure analysis). In both cases it is desirable to supplement the diffraction data with chamical information in the form of constraints or restraints. A new version of the SHELX refinement program (SHELXL-97) is presented here, with the emphasis on the use of chemical information.

Recent advances in cryogenic techniques, area detectors, and the use of synchrotron radiation enable macromolecular data to be collected to higher resolution than was previously possible. In practice this tends to complicate the refinement because it is possible to resolve finer details of the structure; it is often necessary to model alternative conformations, and in a few cases even anisotropic refinement is justified. Although SHELXL [1] provides a number of other features not found in many macromolecular refinement programs (e.g. refinement against data from twinned crystals), it is probably the restrained anisotropic refinement, the flexible treatment of disorder and the estimation of esds that are most likely to be of interest to macromolecular crystallographers.

SHELXL is designed to be easy to use and general for all space groups, but uses a conventional structure-factor calculation rather than a FFT summation; the latter would be faster, but in practice involves some small approximations and is not very suitable for the treatment of dispersion or anisotropic thermal motion. The price to pay

S. Fortier (ed.), Direct Methods for Solving Macromolecular Structures, 119–130.

for the extra generality and precision is that SHELXL is much slower than programs written specifically for macromolecules, but this is to some extent compensated for by the good convergence properties, reducing the amount of manual intervention required (and also the *R*-factor). Two input files are required to run SHELXL: an intensity data file name.hkl and a file name.ins that contains crystal data, atoms, restraints, instructions, etc. All the information in the .ins file is given by four-letter keywords followed by numerical data and references to atoms etc. in free format.

An auxiliary program SHELXPRO is provided as an interface to other protein programs. SHELXPRO is able to generate an .ins file for input to SHELXL from a file in PDB format, including the appropriate restraints etc. SHELXPRO can also generate map files for O and other map interpretation programs, and can display the refinement results in the form of Postscript plots, as well as including the updated coordinates in the .ins file for the next refinement. SHELXL produces PDB and CIF format files that can be read by SHELXPRO and used for archiving.

Macromolecular structures are conventionally divided up into residues, for example individual amino-acids. In SHELXL residues may be referenced either individually, by '_' followed by the appropriate residue number, or as all residues of a particular class, by '_' followed by the class. For example 'DFIX 2.031 SG_9 SG_31' could be used to restrain a disulfide distance between two cystine residues, whereas 'FLAT_PHE CB > CZ' would apply planarity restraints to all atoms between CB and CZ inclusive in all PHE (phenylalanine) residues. Plus and minus signs refer to the next and previous residue numbers respectively, so 'DFIX_* 1.329 C_- N' applies a bond length restraint to all peptide bonds ('_*' after the command name applies it to all residues). This way of referring to atoms and residues is not restricted to proteins; it is equally suitable for oligonucleotides, polysaccharides, or to structures containing a mixture of all three. It enables the necessary restraints and other instructions to be input in a concise and relatively self-explanatory manner. These instructions are checked by the program for consistency and appropriate warnings are printed.

Although the unique features of SHELXL are primarily useful for refinement against very high resolution data, tests on SHELXL-93 indicated that only small changes would be required to extend its range of applicability to medium resolution data (say 2.5Å or better). The most important of these changes, implemented in SHELXL-97, were improved diagnostics and more sophisticated anti-bumping restraints, and the addition of non-crystallographic symmetry (NCS) restraints.

2. Constrained and Restrained Refinement

A *constraint* is an exact mathematical condition that leads to the elimination of one or more least-squares variables; a *restraint* is an additional piece of information that is not exact but is associated with an esd. Restraints are normally applied as additional observational equations, i.e. they improve the data to parameter ratio by increasing the number of data but leaving the number of parameters unchanged. The function to be minimized then becomes:

$$M = \sum w_x(F_o^2 - F_c^2)^2 + \sum w_r(t - t_c)^2$$

where the first term is summed over all the X-ray data and the second over all the restraints. t is a target value for a restraint and t_c is the current value of the corresponding quantity. In SHELXL the restraint weights w_r are treated as absolute, and are given the values $1/\sigma^2$, where σ is the esd assigned to the restraint, and the X-ray weights w_x are then brought onto an approximately absolute scale by dividing the values set by the X-ray weighting scheme (e.g. $w=[\sigma^2(F_o^2)+(0.1P)^2]^{-1}$ where $P=(F_o^2+2F_c^2)/3$) by the mean value of $w(F_o^2-F_c^2)^2$. This has the desirable side-effect of weighting down the X-ray data releative to the restraints in the early stages of refinement when the agreement between the obseved and calculated intensities is poor, and increasing the contribution of the X-ray data as the agreement improves. In practice the restraint esds are independent of the resolution.

2.1 CONSTRAINTS

Typical constraints are the fixing of parameters or the imposition of s strict linear relation between parameters, e.g. coordinates or displacement parameters of atoms on special positions, and the refinement of a single parameter p to describe the occupancies of the two components of a disordered side-chain as p and $1-p$. Rigid group constraints enable a structure to be refined with very few parameters, especially when the (thermal) displacement parameters are held fixed (using the BLOC instruction in SHELXL). After a structure has been solved by molecular replacement using a rather approximate model for the whole protein or oligonucleotide, it may well be advisable to divide the structure up into relatively rigid domains (using a few AFIX 6 and AFIX 0 instructions) and to refine these as rigid groups, initially for a limited resolution shell (e.g. SHEL 8 3), then stepwise extending the resolution (using the new STIR instruction). Restraints may still be required to define flexible hinges and prevent the units from flying apart. In view of the small number of parameters and the high correlations introduced by rigid group refinement, L.S. (full-matrix refinement) should be used for this stage (but CGLS will be necessary for the subsequent refinement). After this initial step, which exploits the large convergence radius of rigid-group refinement, in general the more flexible restraints will be used in preference to constraints for the rest of the refinement.

2.2 GEOMETRICAL SIMILARITY AND NCS RESTRAINTS

Very often there are chemically equivalent units in the structure, but it is not easy to establish reliable target values for their geometrical parameters. A good approach is to restrain chemically but not crystallographically equivalent 1,2- and 1,3-distances to be equal, but without the use of target values. For example, chemically equivalent bonds in a heme unit can be restrained to be the same; with 4- or 8-fold redundancy this is a powerful and sensible restraint. The SADI instruction in SHELXL sets individual distances to be equal, and the SAME instruction may be used to generate SADI instructions to restrain complete molecules to have the same 1,2- and 1,3-distances etc.

We can take this idea one stage further to generate local NCS (non-crystallographic symmetry) restraints. The use of *NCS restraints* considerably improves the effective data to parameter ratio, and the resulting Fourier maps often look as though they were calculated with higher resolution data than were actually used (because the phases are more accurate). Two types of NCS restraint may be generated automatically with the help of the NCSY instruction. The first type uses the connectivity table to define equivalent 1,4-distances, which are then restrained to be equal. The second restrains the isotropic *U*-values of equivalent atoms to be equal. It is not normally necessary to restrain equivalent 1,2- and 1,3-distances to be equal because the DFIX and DANG restraints will have this effect anyway; but SAME may be used to add such restraints in the absence of DFIX and DANG. The use of *restraints* rather than applying NCS as an exact *constraint* (e.g. in the structure factor calculation) is more flexible (but slower) and does not require the specification of transformation matrices and real-space masks. Experience indicates that NCS restraints should be used wherever possible; it is not difficult to relax them later (e.g. for specific side-chains involved in interactions with other non-NCS related molecules) should this prove to be necessary. Note that restraining equivalent 1,4-distances to be equal is not quite as restrictive as restraining equivalent torsion angles: the 1,4-distances are equal for gauche$^+$ and gauche$^-$ conformations. On the other hand such conformational differences are chemically plausible for exposed side-chains.

2.3 OTHER GEOMETRICAL RESTRAINTS

SHELXL provides distance, planarity and chiral volume restraints, but not torsion angle restraints or specific hydrogen bond restraints. For oligonucleotides, distance restraints [2] may be used, but for reasonably high resolution data it is probably better to assume that for the sugars and phosphates the chemically equivalent 1,2- and 1,3-distances are equal (using the SAME and SADI restraints) without the need to specify target values. In this way the effect of the pH on the protonation state of the phosphates and hence the P-O distances does not need to be predicted, but it is assumed the whole crystal is at the same pH. For proteins, since some amino-acid residues occur only a small number of times in a given protein, it is probably better to use 1,2- and 1,3-target distances based on the study of Engh and Huber [3]; these are employed in the restraints added by SHELXPRO to the .ins file.

The three bonds to a carbonyl carbon atom may be restrained to lie in the same plane by means of a *chiral volume restraint* [4] with a target volume of zero (e.g. 'CHIV_GLU 0 C CD'). Chiral volume restraints are also useful to prevent the inversion of α-carbon atoms and the β-carbons of Ile and Thr, e.g. 'CHIV_ILE 2.5 CA CB' Chiral volume restraints may even be applied to non-chiral atoms such as CB of valine and CG of leucine in order to ensure conformity with conventional atom-labeling schemes (from the point of view of the atom names, these atoms could be considered to be chiral!). The FLAT instruction restrains planar groups by restraining the (chiral) volumes of a sufficient number of atomic tetrahedra to be zero; this works well but is somewhat unconventional, and so the program also prints the r.m.s. deviations from the planes.

2.4 ANTI-BUMPING RESTRAINTS

Anti-bumping restraints are distance restraints that are only applied if the two atoms are closer to each other than the target distance. They can be generated automatically by SHELXL, taking symmetry equivalent atoms and disorder into account. Since this step is relatively time-consuming, in the 1993 release it was performed only before the first refinement cycle, and the anti-bumping restraints were generated automatically only for the solvent (water) atoms (however they could be inserted by hand for any pairs of atoms). In practice this proved to be too limited, so in SHELXL-97 the automatic generation of anti-bumping restraints was extended to all C, N, O and S atoms (with an option to include H...H interactions) and was performed each refinement cycle. Anti-bumping restraints are not generated automatically for (a) atoms connected by a chain of three bonds or less in the connectivity array (unless separated by more than a specified number of residues), (b) atoms with different non-zero PART numbers, and (c) pairs of atoms for which the sum of occupancies is less than 1.1. The target distances for the O...O and N...O distances are less than for the other atom pairs to allow for possible hydrogen bonds. The H...H anti-bumping restraints are applied to all pairs of hydrogen atoms not bonded to the same atom; they help to ensure chemically reasonable conformations for flexible side-chains.

2.5 RESTRAINED ANISOTROPIC REFINEMENT

There is no doubt that macromolecules are better described in terms of anisotropic displacements, but the data to parameter ratio is very rarely adequate for a free anisotropic refinement. Such a refinement often results in 'non-positive definite' (NPD) displacement tensors, and at the best will give probability ellipsoids that do not conform to the expected dynamical behavior of the molecule. Clearly constraints or restraints must be applied to obtain a chemically sensible model. It is possible to divide a macromolecule up into relatively rigid domains, and to refine the 20 TLS parameters of rigid body motion for each domain [5]. This may be a good model for the bases in oligonucleotides and for the four aromatic side-chains in proteins, but otherwise macromolecules are probably not sufficiently rigid for the application of TLS constraints, or they would have to be divided up into such small units that too many parameters would be required. As with the refinement of atomic positions, restraints offer a more flexible approach.

The *rigid-bond restraint* (DELU) [6] assumes that the components of the anisotropic displacement parameters (ADPs) along bonded (1,2-) or 1,3-directions are zero within a given esd. This restraint should be applied with a low esd, i.e. as a 'hard' restraint. For many non-planar groups of atoms, rigid-bond restraints effectively impose TLS conditions of rigid body motion [7]. Although rigid-bond restraints involving 1,2- and 1,3-distances reduce the effective number of free ADPs per atom from 6 to less than 4 for typical organic structures, further restraints are often required for the successful anisotropic refinement of macromolecules.

The *similar ADP restraint* (SIMU) restrains the corresponding U_{ij}-components to be approximately equal for atoms which are spatially close (but not necessarily bonded because they may be in different components of a disordered group). The isotropic version of this restraint has been employed frequently in protein refinements. This restraint is consistent with the characteristic patterns of thermal ellipsoids in many organic molecules; on moving out along side-chains, the ellipsoids become more extended and also change direction gradually.

Neither of these restraints are suitable for isolated solvent (water) molecules. A linear restraint (ISOR) restrains the ADP's to be *approximately isotropic*, but without specifying the magnitude of the corresponding equivalent isotropic displacement parameter. Both SIMU and ISOR restraints are clearly only approximations to the truth, and so should be applied as 'soft' restraints with high esds. When all three restraints are applied, structures may be refined anisotropically with a much smaller data to parameter ratio, and still produce chemically sensible ADP's. Even when more data are available, these restraints are invaluable for handling disordered regions of the structure.

An ensemble distribution created by molecular dynamics is an alternative to the harmonic description of anisotropic motion [8,9], and may be more appropriate for structures with severe conformational disorder that do not diffract to high resolution.

2.6 ADVANTAGES OF RESTRAINED REFINEMENT

Constraints and restraints greatly increase the radius and rate of convergence of crystallographic refinements, so they should be employed in the early stages of refinement wherever feasible. The difference electron density syntheses calculated after such restrained refinements are often more revealing than those from free refinements. In large small-molecule structures with poor data to parameter ratios, the last few atoms can often not be located in a difference map until an anisotropic refinement has been performed with geometrical and ADP restraints. Atoms with low displacement parameters that are well determined by the X-ray data will be relatively little affected by the restraints, but the latter may well be essential for the successful refinement of poorly defined regions of the structure. Premature removal or softening of the restraints (to improve the *R*-value !) often impedes further progress.

3. The Free *R*-factor

The question of whether the restraints can be removed in the final refinement, or what the best values are for the corresponding esds, can be resolved elegantly by the use of R_{free} [10]. To apply this test, the data are flagged (using SHELXPRO) as a working set (90 to 95% of the reflections) and a reference set (the remaining 5 to 10%). The reference set is only used for the purpose of calculating a conventional *R*-factor that is called R_{free}. It is very important that the structural model is not in any way based on the reference set of reflections, so these are left out of *all* refinement and Fourier map calculations. If the original model was in any way derived from the same data, then many refinement cycles are required to eliminate memory effects. This ensures that the *R*-factor for the reference set provides an objective guide as to whether the introduction of additional parameters or the weakening of restraints has actually improved the model, and not just reduced the *R*-factor for the data employed in the refinement ('*R*-factor cosmetics').

R_{free} is invaluable in deciding whether a restrained anisotropic refinement is significantly better than an isotropic refinement. Experience indicates that both the resolution and the quality of the data are important factors, but that restrained anisotropic refinement is unlikely to be justified for crystals that do not diffract to better than 1.5 Å

Despite the overwhelming arguments for using R_{free} to monitor macromolecular refinements, it is only a single number, and is itself subject to statistical uncertainty because it is based on a limited number of reflections. Thus R_{free} may be insensitive to small structural changes, and small differences in R_{free} should not be taken as the last word; one should always consider whether the resulting geometrical and displacement parameters are *chemically reasonable*. The final refinement and maps should always be calculated with the *full* data, but without introducing additional parameters or changing the weights of the restraints. R_{free} is most useful for establishing refinement protocols; for a series of closely similar refinements (e.g. for mutants to similar resolution) the R_{free} tests only need to be applied to the first.

4. Disorder Made Simple

To obtain a chemically sensible refinement of a disordered group, we will probably need to constrain or restrain a sum of occupation factors to be unity, to restrain equivalent interatomic distances to be equal to each other or to standard values (or alternatively apply rigid group constraints), and to restrain the displacement parameters of overlapping atoms. In the case of a tight unimodal distribution of conformations, restrained anisotropic refinement may provide as good a description as a detailed manual interpretation of the disorder in terms of two or more components, and is much simpler to perform. With high-resolution data it is advisable to make the atoms anisotropic *before* attempting to interpret borderline cases of side-chain disorder; it may well be found that no further interpretation is needed, and in any case the improved phases from the anisotropic refinement will enable higher quality difference maps to be examined.

Typical warning signs for disorder are large (and pronounced anisotropic) apparent thermal motion (in such cases the program may suggest that an atom should be split and estimate the coordinates for the two new atoms), residual features in the difference electron density and violations of the restraints. This information is summarized by the program on a residue by residue basis, separately for main-chain, side-chain and solvent atoms. In the case of two or more discrete conformations, it is usually necessary to model the disorder at least one atom further back than the maps indicate, in order that the restraints on the interatomic distances are fulfilled. The different conformations should be assigned different PART numbers so that the connectivity array is set up correctly by the program; this enables the correct rigid bond restraints on the anisotropic displacement parameters and idealized hydrogen atoms to be generated automatically even for disordered regions (it is advisable to model the disorder before adding the hydrogens).

Several strategies are possible for modeling disorder with SHELXL, but for macromolecules the simplest is to include all components of the disorder in the same residues and use the same atom names, the atoms belonging to different components being distinguished only by their different PART numbers. This procedure enables the standard restraints etc. to be used unchanged, because the same atom and residue names are used. No special action is needed to add the disordered hydrogen atoms, provided that the disorder is traced back one atom further than it is visible (so that the hydrogen atoms on the PART 0 atoms bonded to the disordered components are also correct). The following extract from an .ins file illustrates the action necessary:

```
RESI 38 SER
N  3 0.7714 0.9267 0.0062  11.0 0.1093
CA 1 0.7887 0.9740 0.0744  11.0 0.1370
PART  1
CB 1 0.8386 1.0427 0.0551  41.0 0.1188
OG 4 0.8994 1.0027 0.0230  41.0 0.1820
PART  2
CB 1 0.8414 1.0366 0.0653 -41.0 0.1493
OG 4 0.8368 1.1036 0.0102 -41.0 0.1732
PART  0
C  1 0.7414 1.0167 0.1038  11.0 0.0840
O  4 0.7072 1.0231 0.0690  11.0 0.1018
```

Atoms in PART 1 may bond to other atoms in PART 1 and also to those in PART 0, but not to those in PART 2 etc. Component (PART) 1 has been assigned an occupancy equal to free variable number 4, and component 2 has been assigned an occupancy equal to one minus free variable 4 (this is specified by the codes 41 and –41), so that a single occupancy parameter is refined, and the occupancies of the disordered atoms sum to unity. All other instructions are the same as for non-disordered residues. The last column contains the (isotropic) U-values ($B = 8\pi^2 U$). The program works out itself how to apply the restraints, add H-atoms etc. Note that this very simple and effective treatment of disorder was not available in SHELXL-93.

5. The Solvent Model

It is relatively common practice in the refinement of macromolecular structures to insert water molecules with partial occupancies at the positions of difference electron density map peaks in order to reduce the R-factor (another example of 'R-factor cosmetics'). Usually when two different determinations of the same protein structure are compared, only the most tightly bound waters, which usually have full occupancies and smaller displacement parameters, are the same in each structure. The refinement of partial occupancy factors for the solvent atoms (in addition to their displacement parameters) is rarely justified by R_{free}, but sometimes the best R_{free} value is obtained for a model involving some water occupancies fixed at 1.0 and some at 0.5.

Regions of diffuse solvent may be modeled using *Babinet's principle [11]*; the same formula is employed in the program TNT [12], but the implementation is somewhat different. In SHELXL it is implemented as the SWAT instruction and usually produces a significant but not dramatic improvement in the agreement of the very low angle data. Anti-bumping restraints may be input by hand or generated automatically by the program, taking symmetry equivalents into account. After each refinement job, the displacement parameters of the water molecules should be examined, and waters with very high values (say U greater than 0.8 Å^2, corresponding to a B of 63) eliminated. The F_o-F_c map is then analyzed automatically to find the highest peaks that involve no bad contacts and make at least one geometrically plausible hydrogen bond to an electronegative atom. These peaks are then included with full occupancies and oxygen scattering factors in the next refinement job. This procedure is repeated several times; in general R_{free} rapidly reaches its minimum value, although the conventional R-index continues to fall as further waters are added. It should be noted that the automatic generation of anti-bumping restraints is less effective when the water occupancies are allowed to have values other than 1.0 or 0.5. This approach provides an efficient way of building up a chemically reasonable (but not necessarily unique) network of waters that are prevented from diffusing into the protein, thus facilitating remodeling of disordered side-chains etc. The occupancies of specific waters may also be tied (using free variables) to the occupancies of particular components of disordered side-chains where this makes chemical sense. This procedure may be facilitated by using SHELXPRO to convert the .res output file from one refinement job to the .ins file for the next, or fully automated using the program SHELXWAT that calls SHELXL repeatedly. A related but much more sophisticated approach (ARP) described by Lamzin and Wilson [13] may also be used for automatic irrigation (ARP will also be discussed later in a more general context).

6. The Radius of Convergence

A crucial aspect of any macromolecular refinement program is the radius of convergence. A larger radius of convergence reduces the amount of time-consuming manual intervention using interactive graphics. There are probably a number of contributing factors to the good convergence typically observed for SHELXL, e.g. the refinement against properly weighted F^2 values for *all* data, the inclusion of important off-diagonal terms in the least-squares algebra [4], the ability to refine all parameters at once (i.e. coordinates and displacement parameters in the same cycle), and the restriction to unimodal restraint functions; multimodal restraint functions such as torsion angles or hydrogen bonds tend to increase the number of spurious local minima. It is much better to reserve the multimodal chemical information such as torsion angles for verifying the structure with an independent program such as PROCHECK [14], and to use the unimodal information as restraints. The errors in the FFT calculation of derivatives are larger that those in the structure factors (for the same grid intervals); this would also impede convergence.

Many claims that SHELXL gives *R*-factors one or two percent lower than other programs have been tracked down either to subtle differences in the model or to not getting trapped in local minima. The differences in the model include the treatment of diffuse solvent and hydrogen atoms, and the ability to refine common occupancies for disordered groups. The inclusion of dispersion terms and the use of a conventional rather than a FFT structure factor summation are also more precise; the approximations in the FFT summation may become significant for high resolution data and atoms with small displacement parameters.

7. Estimated Standard Deviations

No small molecule crystallographer would contemplate publishing a structure without estimated standard deviations, but they are rarely quoted for macromolecules, and then usually only in the form of a Luzzati plot (which is rather inappropriate and was never intended for the purpose [15]!). Provided that there are appreciably more data than parameters, it is in fact possible to invert the full least-squares normal matrix (or at least large blocks of it) from the refinement of a macromolecule, and so derive the esds in all parameters by small-molecule methods. SHELXL uses the full covariance matrix for the estimation of the esds in all dependent parameters such as bond lengths, torsion angles etc.

The structure should be refined to convergence by conjugate gradient least-squares (CGLS) so that the matrix needs to be inverted only once, at the end of the refinement. It turns out that the inversion produces sensible esds even when the calculated shifts would lead to instability. The esds take the restraints into account (in a Bayesian sense) so all restraints should be switched off for this final full-matrix cycle, which is performed with L.S. 1 and DAMP 0 0. This DAMP instruction specifies zero damping (which would otherwise artificially reduce the esds) and zero shift multipliers. All the reflection data should of course be used.

If the full-matrix cycle would take longer than a week or require the purchase of extra memory, an adequate compromise is to use BLOC 1 N_1 > LAST (or something similar) to set up a full-matrix block consisting of all positional but no thermal displacement parameters.

SHELXPRO can be used to plot the atomic positional and bond length esds (a BOND instruction is needed for SHELXL to generate the latter) against the B or B_{eq} values. Preliminary tests suggest that a formula recently proposed by Durward Cruickshank [15] models the dependence of the esds on B_{eq}, effective atomic number, the $R1$-value and the completeness of the data rather well (much better that the Luzzati method).

8. Structure Refinement Techniques in Structure Solution

Since current techniques for solving small-molecule and macromolecular structures by *ab initio* methods appear to require data to almost atomic resolution, yet - with the help of restraints incorporating chemical information - there is no problem in refining structures at much more modest resolution, it is worth asking if refinement algorithms provide a way of incorporating the necessary chemical information into the structure solution process itself. One promising approach is the ARP method [13,16]. In this method, alternate refinement cycles and Fourier maps are calculated; at each stage, a few atoms with high B-values and negative difference electron density are eliminated, and replaced by the most promising candidate atoms found by a sophisticated analysis of the resulting electron density maps. Recent improvements in this method [16] are the use of maximum-entropy as well as least-squares refinement, and sigma-A weighting for the maps. Although geometrical information is taken into account in selecting the new atoms from the maps, the refinement is performed without restraints. At the end of the procedure the final electron density maps have to be refitted; it appears to be best to run several ARP jobs in parallel with slightly different criteria for placing the new atoms, and to calculate a weighted average density from all of them.
The method works well for phase extension (sometimes much better than density modification methods) provided that the resolution is fairly high; for example it is possible to expand the structure of rubredoxin from the positions of the iron and sulfur atoms with data truncated to 1.5Å, and larger fragments (e.g. from molecular replacement with poor sequence identity) can be improved at resolutions as low as 2.0Å.

A possible area for future development of ARP would be to include some form of general restraints, e.g. based on a predicted radial probability distribution function, into the refinement part of the procedure.

I am very grateful to a large number of SHELX users, too many to name here, who made many helpful comments and suggestiongs that have found there way into the current SHELXL refinement program. We also thank the the Eropean Commission for support: contract number ERBCHBGCT940731.

9. References

1. Sheldrick, G.M. and Schneider, T.R. (in press) SHELXL: High-resolution refinement, in C.W. Carter and R.M. Sweet (eds.), *Methods in Enzymology,* Academic Press, New York.

2. Parkinson, G., Voitechovsky, J., Clowney, L., Brünger, A.T., and Berman, H.M. (1996) New parameters for the refinement of nucleic acid-containing structures, *Acta Cryst.* **D52**, 57–64.

3. Engh, R.A. and Huber, R. (1991) Accurate bond and angle parameters for X-ray protein structure refinement, *Acta Cryst.* **A47**, 392–400.

4. Hendrickson, W.A. (1985) Stereochemically restrained refinement of macromolecular structures, in H.W. Wyckoff, C.H.W. Hirs and S.G. Timasheff (eds.), *Methods in Enzymology,* Academic Press, New York, Vol. **115B**, pp. 252–270.

5. Driessen, H., Haneef, M.I.J., Harris, G.W., Howlin, B., Khan, G., and Moss, D.S. (1989) RESTRAIN: Restrained structure-factor least-squares refinement program for macromolecular structures, *J. Appl. Cryst.* **22**, 510–516.

6. Rollett, J.S. (1970) Least-squares procedures in crystal structure analysis, in F.R. Ahmed, S.R. Hall and C.P. Huber (eds.), *Computing in Crystallography*, Munksgaard, Copenhagen, pp. 167–181.

7. Didisheim, J.J. and Schwarzenbach D. (1987) Rigid-link constraints and rigid-body molecules, *Acta Cryst.* **A43**, 226–232.

8. Gros, P., van Gunsteren, W.F., and Hol, W.G. (1990) Inclusion of thermal motion in crystallographic Structures by restrained molecular dynamics, *Science*, **249**, 1149–1152.

9. Clarage, J.B. and Phillips, G.N. (1994) Cross-validation tests of time-averaged molecular dynamics refinements for determination of protein structures by X-ray crystallography, *Acta Cryst.* **D50**, 24–36.

10. Brünger, A.T. (1992) Free R value: a novel statistical quantity for assessing the accuracy of crystal structures, *Nature*, **355**, 472–475.

11. Moews, P.C. and Kretsinger, R.H. (1975) Refinement of the structure of carp muscle calcium-binding parvalbumin by model building and difference Fourier analysis, *J. Mol. Biol.* **91**, 201–228.

12. Tronrud, D.E., Ten Eyck, L.F., and Matthews, B.W. (1987) An efficient general-purpose least-squares refinement program for macromolecules, *Acta Cryst.* **A43**, 489–501.

13. Lamzin, V.S. and Wilson, K.S. (1993) Automated refinement of protein models, *Acta Cryst.* **D49**, 129–147.

14. Laskowski, R.A., MacArthur, M.W., Moss, D.S. and Thornton, J.M. (1993) PROCHECK: a program to check the stereochemical quality of protein structures, *J. Appl. Cryst.* **26**, 283–291.

15. Cruickshank, D.W.J. (1994) Protein precision re-examined: Luzzati plots do not estimate final errors, Poster Abstract, I.U.Cr. Meeting, Seattle.

16, Perrakis, A.,Sixma, T.A., Wilson, K.S., and Lamzin, V.S. (1997) The world according to wARP: improvement and extension of crystallographic phases, CCP4 Meeting, York, January 1997.

LOCATION OF HEAVY ATOMS BY AUTOMATED PATTERSON INTERPRETATION

GEORGE M. SHELDRICK
University of Göttingen
Institut für Anorganische Chemie, Tammannstraße 4,
D-37077 Göttingen, Germany

Abstract

Automated computer interpretation of the Patterson function using superposition techniques provides an effective way of locating heavy atoms in small moiety structures, and - normally from isomorphous or anomalous difference data - also in macromolecules. Another promising approach is to locate potential heavier atoms by random start real/reciprocal space iteration, and then to use the Patterson superposition function as a figure of merit to identify the correct solutions.. Under particularly favorable circumstances, the location of heavier atoms with the help of the Patterson function, combined with automated E-Fourier recycling, provides an ***ab initio*** method of solving the phase problem for small proteins from the native data alone, without the use of heavy-atom derivatives, anomalous dispersion, or chemical information in the form of molecular fragments of known geometry.

1. Introduction

Although overshadowed by the omnipotent direct methods, computer automated heavy atom location via the Patterson function provides an efficient approach to the solution of small-moiety structures. The Patterson function also serves as a good 'noise filter' for isomorphous or anomalous ΔF data, but the scaling of the data and suppression of aberrant reflections is essential for success and requires careful attention. If anomalous ΔF data are used, the missing centric reflections probably present a less serious impediment for Patterson methods than for classical direct methods; MAD experiments involving measurements at three or more different wavelengths may be able to circumvent this problem. The Patterson superposition algorithm described here is equally applicable to the analysis of native and difference data, and is included in the recent (1997) release of the SHELX program system.

S. Fortier (ed.), Direct Methods for Solving Macromolecular Structures, 131–141.

2. The Patterson Interpretation Algorithm

The Patterson vector superposition minimum function approach employed here was suggested in the early 1950's, and a review of the early literature may be found in Buerger's 1959 book [1], where it is referred to as the 'vector shift' method. The approach was revived by Richardson and Jacobson [2], who showed that a computer analysis of a single vector superposition could be used to solve quite complex problems. We have adopted a similar general strategy [3,4], but (unlike Richardson and Jacobson) all operations are carried out in real space.

'Super-sharp' Pattersons, using coefficients $(E^3F)^{1/2}$ instead of F^2, are employed throughout this algorithm. E is a structure factor F normalized by dividing by the smoothed $<F^2>^{1/2}$ for the resolution range in question (the statistical correction factor ε is not used in the Patterson calculations). For isomorphous ΔF data it may be better not to apply sharpening because the lack of isomorphism becomes more pronounced at higher resolution, but as we shall show, it is essential to weight up the high angle data to find and exploit the heavier atoms when using native data alone.

2.1. A SIMPLE DEMONSTRATION

The vector superposition approach can be demonstrated effectively in two dimensions by the use of overhead transparencies. Figure 1 shows a demonstration structure consisting of five atoms with one symmetry element (a horizontal mirror plane).

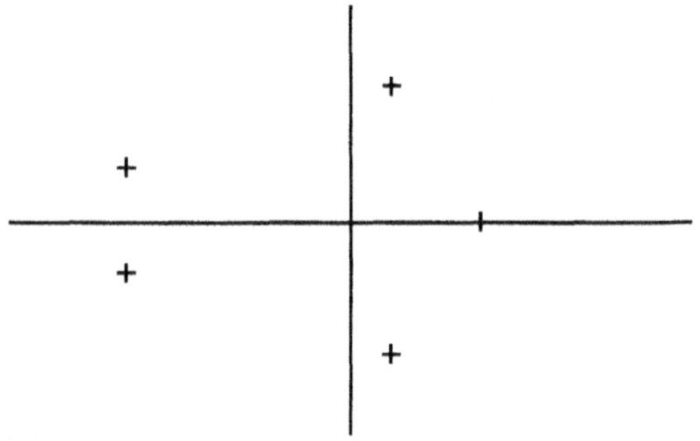

Figure 1. Two-dimensional demonstration structure with m symmetry.

The Patterson function may be regarded as a vector map, so we can generate it by placing each atom in turn at the origin, and marking the positions of the other atoms on a superimposed overhead transparency. This Patterson function (Figure 2) has retained the mirror plane and also possesses a centre of symmetry, since for each interatomic vector A→B there must be an inverse vector B→A. This way of generating the Patterson makes it clear that it contains *N* (here 5) images of the structure, plus *N* inverted images (because of the inversion center).

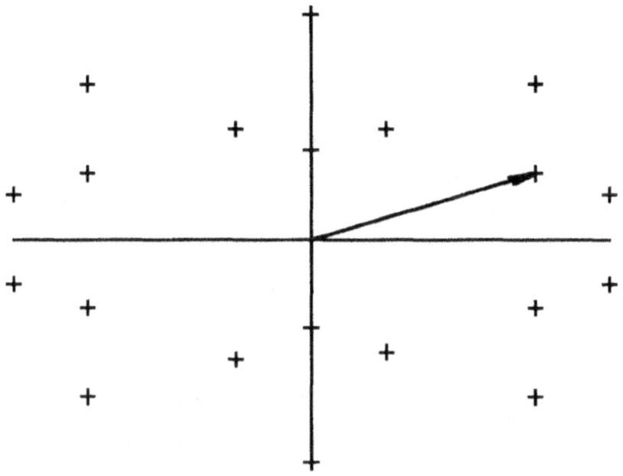

Figure 2. The Patterson function of the structure shown in Figure 1. The m symmetry has been retained and an inversion center added. A suitable superposition vector is indicated (and is used to generate Figure 3).

We can use the reverse procedure to deconvolute the Patterson. If we superimpose two copies of the Patterson, displaced from each other by one of the vectors in the Patterson (Figure 3), then two of the images will coincide exactly, as must two of the inverted images.

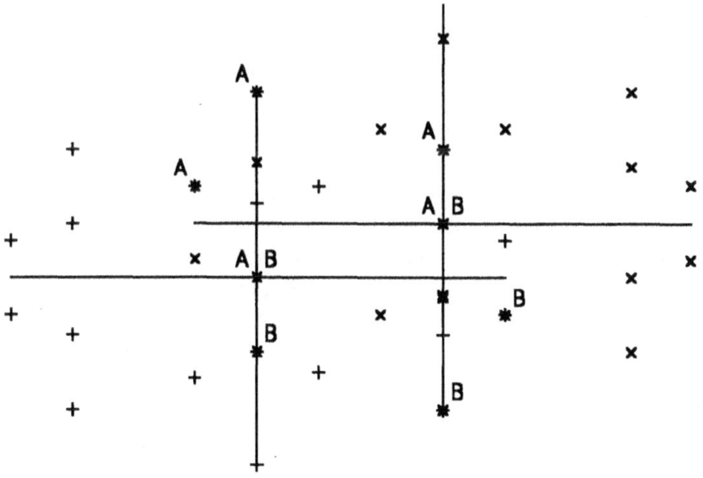

Figure 3. The superposition map (*) consists of a true image B and an inverted image A. The mirror has gone but we may locate either A or B by finding a group of atoms possessing m symmetry.

2.2 THE SUPERPOSITION MINIMUM FUNCTION

By forming the *superposition minimum function*, which possesses peaks only where peaks are present in both maps, we get a much cleaner map which only contains ca. $2N$ peaks rather than the ca. N^2 of the original Patterson. Note that this map has retained the centre of symmetry but lost the mirror plane; in three dimensions all space group symmetry is lost but an inversion centre is always present. Note also that the points corresponding to the origins of the two Pattersons must belong both to the image and to the inverse image. If a multiple vector had been used for the superposition (there are none in this simple demonstration) then in general a multiple image ($4N$ or more images) would result.

In the special case of a $2x,2y,2z$ vector in a centrosymmetric space group being used for the superposition, the image of the structure will coincide with the inverse image and the superposition map is (in theory) the true structure, with its origin midway between the origins of the two Patterson functions. In all other cases the image is shifted by an unknown vector from its true origin. The problem thus reduces to the (relatively tractable) one of finding an origin shift which gives as many atoms as possible which are related to one another by the symmetry operations of the space group. In the case of the demonstration structure, the image (or its inverse) can readily be located by moving the transparency vertically until a group of atoms are found with a horizontal mirror plane of symmetry. We can then isolate our structure of five atoms (one of which happens to lie on a special position on the plane) by discarding the remaining three atoms, for which there are no symmetry equivalents.

2.3 IMPLEMENTATION IN SHELXS-97

The procedure used in SHELXS-97 is as follows. First between 1 and 20 suitable peaks are selected as superposition vectors. It is useful to specify a minimum possible length for these vectors; for example, protein secondary structure may give pronounced features around the Patterson origin that would not be good candidate heavy-atom to heavy-atom vectors. For each of these vectors $\mathbf{r}$, a superposition minimum function is calculated by overlaying two copies of the sharpened Patterson which have been displaced from the origin by $+\mathbf{r}/2$ and $-\mathbf{r}/2$. The next stage is the analysis of the peak list to locate potential origin shifts which will move one of the two images so that its constituent atoms conform to the symmetry of the space group; this also enables it to be separated from the other image (which after the move will in general not obey the space group symmetry). Both this stage and the choice of the original superposition vector generate multiple solutions. Finally each solution is presented in the form of a *crossword table* that contains one row and one column per potential atom. For each row/column combination, two numbers are given: the minimum distance between two atoms, and the minimum Patterson density at all vectors between the two atoms, taking symmetry into account in both cases. The most effective figure of merit for comparing the different solutions appears to be the correlation coefficient (based on E_o and E_c) of Fujinaga and Read [5].

3. Examples of Structures Solved by Patterson Superposition

The Patterson superposition approach will be illustrated here by two structures, ureido-balhimycin [6] and a cytochrome c_{6}, [7], that were first solved by this method. The direct methods available at the time were not able to solve these structures, and limited attempts using molecular replacement had also failed; in neither case was a suitable search fragment then available. In both cases the key to the structure solution was the location of the heavier atoms by computer-aided interpretation of the super-sharp Patterson calculated from the single-wavelength native data alone. In both cases the full structures were developed from the heavier atoms (four chlorine atoms in balhimycin and an iron and three sulfurs in the cytochrome) by means of iterative real/reciprocal space recycling. Since this structure expansion procedure will be described in detail in later lectures in this School, it will not be discussed further at this point.

3.1. LOCATION OF THE CHLORINE ATOMS IN UREIDO-BALHIMYCIN

Since ureido-balhimycin [6] crystallizes in the space group $P1$, it is not possible to use the symmetry of the space group to recognise one of the two images of the structure. The result is a superposition of the image of the four chlorine atoms with its inverse, with two chlorines (those related by the superposition vector) common to both images. In the crossword table shown in Table 1, these two images are marked A and B. The higher number in the box linking a row and column is the shortest distance between the atoms representing the row and the column, and underneath it is given the value of the Patterson function at the vector linking the two atoms. For higher symmetry space groups, the minimum of the Patterson function values at all vectors linking one atom and the other or its symmetry equivalents would appear here. In order to resolve the two images, one needs to find two groups of four atoms, both containing the first two atoms, in which all Patterson values are high (say greater than 20 - they are scaled so that the Patterson origin peak has a height of 999), and none of the distances between chlorine atoms in the same image are less than about 4Å, the shortest inter- or intramolecular distance expected on chemical grounds. The last two 'atoms' in Table 1 are not chlorine atoms; they might be oxygen atoms from one or other image, or they might be noise. It is necessary to select one of the images by hand in order to complete the structure by E-Fourier recycling. The 'self-vectors' are absent in space group $P1$. In general, in space groups other than $P1$ this approach should yield a single image of the structure, although double (or multiple) images can appear if the original superposition vector chosen from the Patterson involved overlap of two or more interatomic vectors.

TABLE 1. Patterson interpretation for balhimycin

x	*y*	*z*	*Minimum distances / Patterson function value*							
0.296	0.454	0.214	A,B							
0.704	0.546	0.786	12.78	A,B						
			41.8							
0.584	0.289	0.336	6.82	8.15	A					
			35.5	38.3						
0.416	0.711	0.664	8.15	6.82	11.51	B				
			38.3	35.5	0.0					
0.576	0.234	0.361	7.59	8.12	1.14	11.91	B			
			40.5	32.2	3.5	23.9				
0.424	0.766	0.639	8.12	7.59	11.91	1.14	10.60	A		
			32.2	40.5	23.9	3.5	26.0			
0.178	0.397	0.362	4.12	14.61	8.49	6.01	8.83	6.40	?	
			32.4	21.5	0.0	6.5	9.4	0.0		
0.822	0.603	0.638	14.61	4.12	6.01	8.49	6.40	8.83	11.19	?
			21.5	32.4	6.5	0.0	0.0	9.4	3.4	

3,2. CYTOCHROME C_6: *AB INITIO* SOLUTION OF AN UNKNOWN PROTEIN

Cytochrome c_6 from *Monoraphidium braunii* is an 89 amino-acid protein containing one heme unit [7]. It contains one iron and three sulfurs in the asymmetric unit in the space group $R3$. Data were collected at EMBL c/o DESY on the synchrotron to 1.3Å at a wavelength of about 0.9Å, and later to 1.1Å.. The Patterson superposition approach was indeed able to locate plausible positions for the iron and three sulfurs from the original 1.3Å data, and the iron (but not the sulfur) positions could be confirmed from the anomalous ΔF Patterson. Using the 1.1Å data, it proved possible to build up the structure in stages, with manual intervention at each stage to add only those peaks that made chemical sense. About 90% of the protein atoms could be placed correctly by purely 'small-molecule' methods; however it then proved necessary to examine F_0-F_c and $2F_0$-F_c maps with interactive computer graphics to complete the structure. In particular two regions with high apparent thermal motion could not be interpreted correctly on the basis of the peak-lists alone: these were the first five residues at the N-terminus, and a loop between residues 40 and 45.

Table 2 illustrates the location of the iron and three sulfur atoms; the 'crossword table' shown here is based on the 1.1Å data, and is a little clearer than that based on the 1.3Å data used successfully for the original location of these atoms. The two cysteine sulfurs are attached to corners of the heme unit and must lie between 6.5 and 7Å from the iron atom, and about 9Å from each other; the methionine sulfur coordinates directly to the iron atom (Fe-S should be about 2.4Å) and lies on the other side of the heme unit to the cysteine sulfurs.

TABLE 2. Heavy-atom location for cytochrome c6 in space group *R*3

At.No.	x	y	z	Minimum distances / PATSMF (self first)								
27.8	.8260	.2120	.1728	19.25								
				27.7								Fe
15.1	.9073	.2178	.3050	18.94	*6.83*							
				6.8	*11.8*							S(Cys)
8.3	.8052	.2166	.2095	21.34	1.61	6.16						
				.4	9.4	.0						
8.2	.8175	.2305	.2456	22.02	3.13	4.62	1.83					
				1.8	10.3	1.7	.0					
7.9	.9744	.1297	.1151	7.77	*6.63*	*8.87*	7.86	8.51				
				4.8	*10.6*	*1.6*	.0	.0				S(Cys)
7.6	.9412	.1193	.1423	9.97	5.59	8.10	6.69	7.40	1.66			
				.0	7.3	.0	.0	.0	.0			
7.5	.0475	.1732	.0617	6.24	9.36	10.11	10.66	11.10	3.86	5.48		
				2.5	4.9	.1	.0	.0	.0	.0		
7.3	.9821	.2229	-.0255	12.80	9.36	10.45	10.75	11.07	6.28	6.03	4.76	
				.0	7.5	1.2	.0	.0	.0	.0	.0	
7.0	.8124	.2726	.1485	21.58	*2.50*	*7.80*	3.07	4.05	*8.09*	7.43	10.01	
				8.8	*12.9*	*3.5*	.0	.0	*1.6*	.0	.0	S(Met)

4. The Role of the Resolution

Traditionally, protein crystallographers have solved structures using *F*-values, but small-molecule direct methods always employ *E*-values. In the location of heavier atoms from the native Patterson it appears to be essential to use super-sharp Patterson based essentially on *E*-values. In Figure. 1 the relative scattering factors *f*(Fe)/*f*(C) and *f*(S)/*f*(C) are shown as a function of the resolution *d*. Although the ratio of the iron to carbon scattering factors increases from 4.3 to about 7 in the resolution range corresponding to the data as a result of the inner core electrons, it will be seen that if the difference in apparent thermal motion (*B*-values) is also taken into account, the ratio becomes very much larger at very high resolution. At 1.0Å resolution, if we assume that the *B*-value of the iron is five units less than that of most of the protein atoms, it behaves like uranium! On the other hand, both effects are relatively small at 'normal' protein resolution (say 2-3Å).

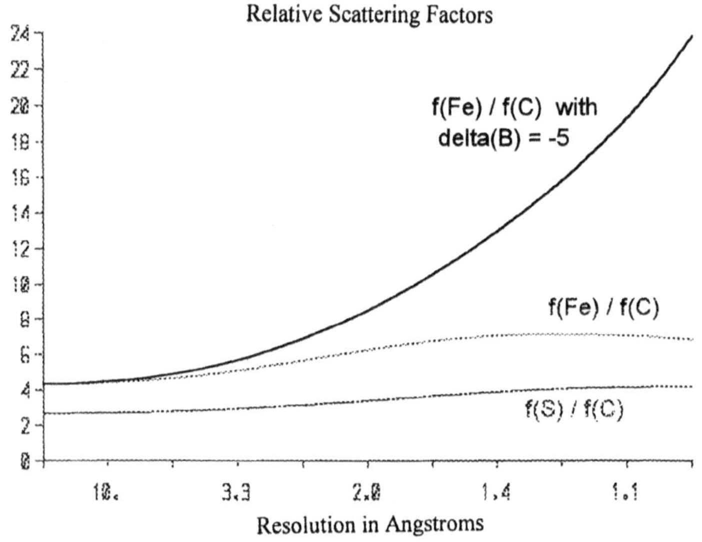

Figure 4. Relative scattering power as a function of resolution.

From the isotropic refinement of cytochrome c_6, the following mean *B*-values were obtained: Fe, 10.8; S, 12.6; heme, 12.4; mainchain, 19.5; side-chain, 25.3; water, 38.4. It is clear that the smaller *B*-values of the iron and sulfur atoms make a major contribution to the ease of location of these atoms by the Patterson superposition approach, and also to the effective phasing power of these atoms, given data to very high resolution.

5. Combined Patterson and Direct Methods

Although small-molecule direct-methods programs such as MULTAN and SHELXS have been frequently (mis)used to locate the positions of heavy atoms from isomorphous and anomalous ΔF-data, they were never designed for this purpose. These ΔF estimates represent lower bounds, so for example the tangent formula is still valid if only the large (pseudo-)E-values are employed, but phase relations such as negative quartets that depend on the weak reflections are of no use. In fact most of the advances of the last 20 years in conventional small-molecule direct methods are of little help in locating the heavy atoms!

A more appropriate direct methods approach, to be described in detail in later lectures at this School, involves real/reciprocal-space iteration. In the implementation used here, to find M heavy atoms one starts with say $2M$ random atoms, and eliminates M of them to maximize $\Sigma(E_o^2-1)E_c^2$ summed over the largest E-values only. These M atoms are then used to calculate phases that are expanded and refined using the tangent formula and then used to calculate an E-map. This map is searched for the strongest $2M$ peaks (possibly imposing a minimum distance constraint) and the whole process repeated several times. This method uses only the most reliable (i.e. largest) E-values and is relatively insensitive to missing data, but we still need to find a suitable figure of merit for identifying the correct solution; in classical direct methods the weak E-values are often used for this purpose, but here they would be inappropriate. The Patterson superposition function, in the form of the 'crossword table' illustrated above, provides an excellent figure of merit. This use of the Patterson function is more robust than the Patterson interpretation algorithm described in section 2, because a few missing or overlapping Patterson peaks do not cause the correct answer to be rejected prematurely.

In the implementation of the MAD method, a problem has arisen involving the location of the anomalous scatterers from anomalous ΔF or MAD F_A values. Patterson methods work well if there are only a few such atoms, but the complexity increases as the square of the number of atoms and becomes prohibitive, even for automated computer interpretation, when the number of anomalous scatterers is more than about 12. One would have expected that classical direct methods should be able to solve this problem, since they are capable of finding at least 100 equal atoms, and the anomalous scatterers are usually separated from one another by distances much greater than the limiting resolution of the reflection data, but in practice they invariably fail to locate say 20 independent selenium atoms. Preliminary tests on MAD and other anomalous difference data-sets indicate that real/reciprocal space recycling combined with the Patterson superposition minimum function as a figure of merit is indeed able to find such a large number of anomalous scatterers. This approach can take advantage of the fact that the number of anomalous scatterers (e.g. selenomethionines or iron atoms in a metalloprotein) is usually known, although conformationally disordered selenomethionines can complicate matters. An example of a crossword table obtained in this way for a structure containing 8 unique anomalous scatterers is shown in Table 3. These HiPIP test data were kindly provided by Hazel Holden and Gary Wesenberg;

they were measured to 1.5Å on a Cu-Kα rotating anode with multiwire proportional counter [8]. Hand solution of the 1.5Å anomalous Patterson was not successful. The data have been truncated to 2.0Å to make this test more demanding!

TABLE 3. Crossword table for the second best solution from the Cu-Kα anomalous ΔF values for a HiPIP protein with two Fe_4S_4 clusters in the asymmetric unit [8]. The upper row gives the minimum distance between the atom defining the row and the atom defining the column, the lower row gives the corresponding Patterson superposition minimum function.

```
Try 89, CC(HA)=35.74%, PATFOM=39.67

Peak   x     y     z    self  cross-vectors
99.9 0.389 0.736 0.176 29.2
                       41.0

98.4 0.432 0.746 0.249 30.1  2.6
                       51.0 66.6

90.7 0.399 0.696 0.194 29.4  2.2  3.3
                        0.0 47.5 33.0

89.9 0.914 0.187 0.126 27.9 14.0 16.6 14.4
                       53.2 34.6 49.1 74.7

88.1 0.354 0.742 0.255 31.4  2.6  2.9  3.4 14.6
                       45.7 69.3 73.2 56.5 57.4

82.3 0.960 0.160 0.043 26.6 14.6 17.0 14.8  3.2 14.7
                       67.6 42.5 37.9 54.7 27.5 37.8

71.1 0.901 0.125 0.082 27.7 14.0 16.5 13.8  3.5 14.5  3.0
                       22.2 27.9 32.6 34.9 25.3 32.8 47.8

67.4 0.973 0.342 0.132 27.4 16.6 18.8 18.0  8.4 16.8  9.9 11.8
                       41.7  0.9 49.0 20.0  0.5 31.5  0.0  0.0

-------------------------------------------------------------------

46.8 0.966 0.143 0.145 27.6 16.4 18.9 16.5  3.1 16.8  3.0  3.1 10.4
                       38.3 34.2 43.8 19.7 22.5 25.3 26.7 45.8  0.0

41.3 0.500 0.749 0.286 28.8  5.1  2.7  5.3 19.1  5.4 19.6 18.9 21.3
                        0.0  4.5 46.5  0.7 14.7 21.6  5.8  2.8 22.7
```

The second best solution is illustrated; the Patterson superposition minimum function values clearly show that the atoms 1-7 and 9 correspond to the eight expected iron atoms. They form two Fe_4 clusters with Fe•••Fe distances of about 3Å. The PATFOM figure of merit is simply the mean of the Patterson superposition minimum function values for the top N_H atoms. The solution with the best PATFOM, but the second best CC, gave atoms 1-8 as the correct iron atoms.

I am very grateful to T. Bes, M.A. Carrondo, Z. Dauter, C. Frazao, E. Parisini, H.M. Holden, H. Hope, K. Johnsen, P. Lubini, E. Paulus, E. Pohl, I. Rayment, M. Schäfer, T.R. Schneider, L.Sieker, I. Uson, L. Vertesy, G. Wesenberg and K.S. Wilson for their excellent data, help and advice in the course of this work. We thank the European Commission for support through contract ERBCHBGCT940731.

6. References

1. Buerger, M.J. (1959). *Vector Space*, Wiley, New York.

2. Richardson, J. W. and Jacobson, R. A. (1987) Computer-aided analysis of multi-solution Patterson superpositions, in J.P. Glusker, B. Patterson and M. Rossi (eds.), *Patterson & Pattersons*, I.U.Cr. and Oxford University Press, Oxford, pp. 311–317.

3. Sheldrick, G.M. (1991) Tutorial on automated Patterson interpretation to find heavy atoms, in D. Moras, A.D. Podjarny and J.C. Thierry (eds.), *Crystallographic Computing* **5**, I.U.Cr. and Oxford University Press, Oxford, pp. 145–157.

4. Sheldrick, G.M., Dauter, Z., Wilson, K.S., Hope, H., and Sieker, L.C. (1993) The application of direct methods and Patterson interpretation to high-resolution native protein data. *Acta Cryst.* **D49**, 18–23.

5. Fujinaga, M. and Read, R.J. (1987) Experiences with a new translation-function program. *J. Appl. Cryst.* **20**, 517–521.

6. Sheldrick, G.M., Paulus, E., Vertesy, L., and Hahn, F. (1994) Structure of ureido-balhimycin, *Acta Cryst.* **B51**, 89–98.

7. Frazão, C., Soares, C.M., Carrondo, M.A., Pohl, E., Dauter, Z., Wilson, K.S., Hervás, M., Navarro, J.A., De la Rosa, M., and Sheldrick, G.M. (1995) *Ab initio* determination of the crystal structure of cytochrome c_6 and comparison with plastocyanin, *Structure*, **3**, 1159–1169 .

8. Rayment, I., Wesenberg, G., Meyer, T.E., Cusanovich, M.A., and Holden, H.M. (1992) Three-dimensional structure of the high-potential iron-sulfur protein isolated from the purple phototropic bacterium *Rhodocyclus tenuis* determined and refined at 1.5Å resolution, *J. Mol. Biol.*, **228**, 672–686.

NEW APPLICATIONS OF SIMULATED ANNEALING IN CRYSTALLOGRAPHIC REFINEMENT

AXEL T. BRÜNGER
The Howard Hughes Medical Institute and
Department of Molecular Biophysics and Biochemistry,
Yale University, New Haven, CT 06520, USA.

PAUL D. ADAMS
Department of Molecular Biophysics and Biochemistry,
Yale University, New Haven, CT 06520, USA.

AND

LUKE M. RICE
Department of Molecular Biophysics and Biochemistry,
Yale University, New Haven, CT 06520, USA.

1. Introduction

Over the last decade, developments in molecular biology, X-ray diffraction instrumentation, and computational methods have allowed nearly exponential growth of macromolecular structural studies. The analysis of data from these studies generally requires sophisticated computational procedures culminating in refinement and structure validation. These procedures can be formulated as the chemically–constrained or restrained non–linear optimization of a target function, which usually measures the agreement between observed data and data computed from an atomic model. The ultimate goal is to optimize the simultaneous agreement of an atomic model with observed data and with *a priori* chemical information.

The target function used for this optimization normally depends on several atomic parameters, but most importantly on atomic coordinates. The large number of adjustable parameters (typically at least three times the number of atoms in the model) give rise to a very complicated target func-

S. Fortier (ed.), Direct Methods for Solving Macromolecular Structures, 143–157.

tion. This in turn produces what is known as the multiple minima problem: the target function contains many local minima in addition to the global minimum. These local minima tend to defeat gradient–descent optimization techniques such as conjugate gradient or least-squares methods [1]. These methods are simply not capable of sampling molecular conformations thoroughly enough to find the most optimal model if the starting one is far from the correct structure.

Simulated annealing [2] is an optimization technique particularly well suited to overcoming the multiple minima problem. Unlike gradient–descent methods, simulated annealing can cross barriers between minima and thus can explore a greater volume of the parameter space to find better models in deeper minima. Following its introduction to crystallographic refinement [3], there have been major improvements of the original method in four principal areas: the measure of model quality, the search of the parameter space, the target function, and the modelling of conformational variability.

For crystallographic refinement, the introduction of cross–validation (the "free" R-value) [4] has significantly reduced the danger of overfitting the diffraction data. The complexity of the conformational space has been reduced by the introduction of torsion–angle molecular dynamics [5], which decreases the number of adjustable parameters that describe a model approximately tenfold. The target function has been improved by incorporating the concept of maximum–likelihood which takes into account model error, model incompleteness, and errors in the experimental data [6, 7]. Finally, the sampling power of simulated annealing can be used for exploring the molecule's conformational space in cases where the molecule undergoes dynamic motion or static disorder through multi–conformer models for X–ray crystallography [8, 9, 10].

2. The Target Function

In essence, crystallographic refinement is a search for the global minimum of a target function

$$E = E_{\text{chem}} + w_{\text{data}} E_{\text{data}} \tag{1}$$

as a function of the parameters of an atomic model, in particular atomic coordinates. E_{chem} comprises empirical information about chemical interactions; it is a function of all atomic positions, describing covalent (bond lengths, bond angles, torsion angles, chiral centers, and planarity of aromatic rings) and non–bonded (intra-molecular as well as inter–molecular and symmetry–related) interactions [11]. E_{data} describes the difference between observed and calculated data, and w_{data} is a weight appropriately chosen to balance the gradients (with respect to atomic parameters) arising from the two terms.

2.1. *A PRIORI* CHEMICAL INFORMATION

The parameters for the covalent terms can be derived from average geometry and root-mean-square (r.m.s.) deviations observed in a small-molecule data base. Extensive statistical analyses were undertaken for the chemical moieties of proteins [12] and of polynucleotides [13] using the Cambridge crystallographic database [14]. Analysis of the ever increasing number of atomic resolution macromolecular crystal structures will no doubt cause some modifications of these parameters in the future [15, 16, 17].

In X-ray crystallography, it is common to use a purely repulsive quartic function ($E_{\text{repulsive}}$) for the nonbonded interactions which are included in $E_{\text{chem,}}$ [11],

$$E_{repulsive} = \sum_{ij} ((cR_{ij}^{min})^n - R_{ij}^{\,n})^m \tag{2}$$

where R_{ij} is the distance between two atoms i, j, $R_{ij}^{\min}$ is the van der Waals radius for a particular atom pair ij, $c \leq 1$ is a constant that is sometimes used to reduce the radii, and $n = 2$, $m = 2$ or $n = 1$, $m = 4$. Van der Waals attraction and electrostatic interactions are usually not included in crystallographic refinement. These simplifications are valid since the experimental data contains information that is able to produce atomic conformations consistent with actual nonbonded interactions. In fact, atomic resolution crystal structures can be used to derive parameters for electrostatic energies [18]. If the experimental information is insufficient to fully determine the macromolecular structure, use of electrostatic and simulated solvent interactions can bias the structure towards the theoretical non-bonded model.

Geometric energy functions are related to empirical energy functions that were developed for energy-minimization and molecular-dynamics studies of macromolecules (see [19] for an introduction). These empirical energy functions were not designed for structure determination, and therefore required some modification for use in macromolecular structure refinement [20, 21, 22, 23]. Recently, crystallographic simulated annealing refinement was implemented with a purely geometric energy function [7] which provides uniformity among different crystallographic refinement programs and simplifies the generation of parameters for new chemical compounds.

2.2. X-RAY DIFFRACTION DATA

The conventional form of E_{xray} consists of the crystallographic residual E^{LSQ}, defined as the sum over the squared differences between the observed ($|F_o|$) and calculated ($|F_c|$) structure factor amplitudes for a particular atomic model:

$$E_{xray} = E^{LSQ} = \sum_{hkl} (|F_o| - k|F_c|)^2 \tag{3}$$

where hkl are the indices of the reciprocal lattice points of the crystal, and k is a relative scale factor.

Reduction of E^{LSQ} can result from improvement in the atomic model, but also from an accumulation of systematic errors in the model or from fitting noise in the data [24]. The least–squares residual is therefore poorly justified when the model is far away from the correct one or incomplete [6]. An improved target for macromolecular refinement can be obtained using a maximum–likelihood formulation [25, 26, 27, 6, 7]. The goal of the maximum–likelihood method is to determine the probability of making a measurement, given the model, estimates of the model's errors and those of the measured intensities. The effects of model errors (incorrectly–placed and missing atoms) on the calculated structure factors are first quantified with σ_A values, which correspond roughly to the fraction of each structure factor that is expected to be correct. However, overfitting of the diffraction data causes the model bias to be underestimated and undercorrected in the σ_A values. The effect of this overfitting can be reduced by cross–validating σ_A values, i.e. by computing them from a randomly selected test set which is excluded from the summation on the right hand side of Eqs. 3 and 4 [4, 28]. The expected values of $\langle|F_o|\rangle$ and the the corresponding variance (σ^2_{ML}) are derived from σ_A, the observed $|F_o|$, and calculated $|F_c|$ [6]. These quantities can be readily incorporated into a maximum–likelihood target function

$$E_{\mathrm{xray}} = E^{ML} = \sum_{hkl \in \mathrm{working\ set}} \left(\frac{1}{\sigma^2_{ML}}\right)(|F_o| - \langle|F_o|\rangle)^2 \qquad (4)$$

In order to achieve an improvement over the least-squares residual (Eq. 3), cross–validation was found to be essential [7] for the computation of σ_A and its derived quantities in Eq. 4.

2.3. ADDITIONAL INFORMATION

Additional constraints or restraints may be used to improve the ratio of observables to parameters. For example, atoms can be grouped so that they move as rigid bodies during refinement, or bond lengths and bond angles can be kept fixed [30, 29, 5]. The existence of non–crystallographic symmetry can be used to average over equivalent molecules and thereby to reduce noise in the diffraction data [21].

Phase information can be incorporated into E_{xray} in order to improve the ratio of observables to parameters and, thus, to obtain a more accurate model especially when only low–resolution diffraction data is available [5]. Since the experimental phase information for macromolecules is usually not very accurate, the errors of the phase observations have to be taken into account. Ideally, this can be accomplished by computing a phase probability distribution for each reflection and to refine the model phases against these distributions (A.T.B., unpublished). Several approximations to this computationally expensive procedure have been developed. A simple representation of the phase probability distribution consists of a square–well potential around the phase centroid where the width of the square–well is determined by the individual figure of merit [45]. For well–defined phases, for example for those obtained from highly redundant virus crystal structures, structures solved by multiple wavelength anomalous dispersion (MAD) phasing [34], or good multiple isomorphous replacement (MIR) solutions, one can restrain the real and imaginary parts A, B of the structure factor simultaneously [33, 5]. When phase information is incomplete, as is often the case, the phase–restraint target can be combined with an amplitude–based target using figure–of–merit weighting [31, 32].

2.4. WEIGHTING

The weight w_{data} (Eq. 1) balances the forces arising from E_{data} and E_{chem}. The choice of w_{data} can be critical: if w_{data} is too large, the refined structure will show unphysical deviations from ideal geometry; if w_{data} is too small, the refined structure will not satisfy the observed data. Automated protocols to provide initial estimates for optimal weighting have been developed [20, 7]. However, independent information must be used (e.g. cross–validation) to objectively obtain the best possible weight for X–ray diffraction data [4].

3. Searching Conformational Space

Annealing denotes a physical process wherein a solid is heated until all particles randomly arrange themselves in a liquid phase, and then is cooled slowly so that all particles arrange themselves in the lowest energy state. By formally defining the target E (Eq. 1) to be the equivalent of the potential energy of the system, one can simulate the annealing process [2]. There is no guarantee that simulated annealing will find the global minimum (except in the case of an infinitely long search) [35]. Compared to conjugate–gradient minimization where search directions must follow the gradient, simulated annealing achieves more optimal solutions by allowing motion against the gradient [2]. The likelihood of *uphill* motion is determined by a control parameter referred to as *temperature*. The higher the temperature, the more likely it is that simulated annealing will overcome barriers. It should be noted that the simulated annealing temperature normally has no physical meaning and merely determines the likelihood of overcoming barriers of the target function.

The simulated annealing algorithm requires a generation mechanism to create a Boltzmann distribution at a given temperature T. Simulated annealing also requires an annealing schedule, that is, a sequence of temperatures $T_1 \geq T_2 \geq \cdots \geq T_l$ at which the Boltzmann distribution is computed. Implementations of the generation mechanism differ in the way they generate a transition, or move , from one set of parameters to another which is consistent with the Boltzmann distribution at given temperature. The two most widely used generation mechanisms are Metropolis Monte Carlo [36] and molecular dynamics [37] simulations. For X–ray crystallographic refinement, molecular dynamics proved extremely successful [3] whereas Monte Carlo methods have yet to be shown to be effective.

3.1. MOLECULAR DYNAMICS

A suitably chosen set of atomic parameters can be viewed as generalized coordinates that are propagated in time by the classical (Hamilton) equations of motion [38]. If the generalized coordinates represent the x, y, z positions of the atoms of a molecule, the Hamilton equations of motion reduce to the more familiar Newton's second law:

$$m_i \frac{\partial^2 \vec{r}_i}{\partial t^2} = -\nabla_i E. \quad (5)$$

The quantities m_i and $\vec{r}_i$ are respectively the mass and coordinates of atom i, and E is given by Eq. 1. The solution of the partial differential equations (Eq. 5) is achieved numerically using finite–difference methods [37]. This approach is referred to as molecular dynamics [37].

Initial velocities for the integration of Eq. 5 are usually assigned randomly from a Maxwell distribution at the appropriate temperature. Assignment of different initial velocities will produce a somewhat different structure after simulated annealing. By performing several refinements with different initial velocities one can therefore improve the chances of success of simulated–annealing refinement. Furthermore, this improved sampling can be used to determine discrete disorder and conformational variability (see below).

Although Cartesian (i.e., flexible bond lengths and bond angles) molecular dynamics places restraints on bond lengths and bond angles (through E_{chem}, Eq. 1), one might want to implement these restrictions as constraints, i.e., fixed bond lengths and bond angles [30]. This is supported by the observation that the deviations from ideal bond lengths and bond angles are usually small in X–ray crystal structures. Indeed, fixed–length constraints have been applied to crystallographic refinement by least–squares minimization [30]. It is only recently, however, that efficient and robust algorithms have become available for molecular dynamics in torsion–angle space [40, 41, 42, 5, 43]. We chose an approach that retains the Cartesian-coordinate formulation of the target function and its derivatives with respect to atomic coordinates so that calculation remains relatively straightforward and topology independent [5]. In this formulation, however, the expression for the acceleration becomes a function of positions and velocities. Iterative equations of motion for constrained dynamics in this formulation can be derived and solved by finite difference methods [39]. This method is numerically very robust and has a significantly increased radius of convergence in crystallographic refinement compared to Cartesian molecular dynamics [5].

3.2. TEMPERATURE CONTROL

Simulated annealing requires the control of the temperature during molecular dynamics. The current temperature of the simulation (T_{curr}) is computed from the kinetic energy

$$E_{\mathrm{kin}} = \sum_{i}^{n_{\mathrm{atoms}}} \frac{1}{2} m_i \left(\frac{\partial r_i}{\partial t}\right)^2 \tag{6}$$

of the molecular dynamics simulation,

$$T_{\mathrm{curr}} = \frac{2E_{\mathrm{kin}}}{3nk_b}. \tag{7}$$

Here n is the number of degrees of freedom, m_i is the mass of the atom, and k_b is Boltzmann's constant. One commonly used approach to control the temperature of the simulation consists of coupling the equations of motion to a heat bath. A "friction" (γ_i) term [44]

$$-m_i \gamma_i v_i \left(1 - \frac{T}{T_{\mathrm{curr}}}\right) \tag{8}$$

can be added to right hand side of Eq. 5 to lower the temperature where v_i are the velocities of the atoms. This method generalizes the concept of "friction" by allowing a negative "friction" coefficient and by determining the friction coefficient and its sign by the ratio of the current simulation temperature to the target temperature T_{curr}.

3.3. WHY DOES SIMULATED ANNEALING WORK?

The goal of any optimization problem is to find the global minimum of a target function. In the case of crystallographic refinement, one searches for the conformation or conformations of the molecule that best fit the experimental data and that simultaneously maintain reasonable covalent and non-covalent interactions. Simulated annealing refinement has a much larger radius of convergence than conjugate-gradient minimization (see below). It must therefore be able to find a lower minimum of the target E (Eq. 1) than the local minimum found by simply moving along the negative gradient of E. Paradoxically, the very reasons that make simulated annealing such a powerful refinement technique (the ability to overcome barriers in the target energy function) would seem to prevent it from working at all. If it crosses barriers so easily, what allows it to stay in the vicinity of the global minimum?

It is most easy to visualize this property of simulated annealing in the case of molecular dynamics. By specifying a fixed temperature, the system essentially gains a certain inertia which allows it to cross energy barriers of the corresponding target function (Eq.7). The target temperature must be large enough to overcome smaller barriers but low enough to ensure that the system will not "climb out" out of the global minimum if it manages to arrive there. While temperature itself is a global parameter of the system, temperature fluctuations arise principally from local conformational transitions - for example from an amino acid sidechain falling into the correct orientation. These local changes tend to lower the value of the target E, thus increasing the kinetic energy, and hence the temperature, of the system. Once the temperature coupling (Eq. 8) has removed this excess kinetic energy through "heat dissipation", the reverse transition is very unlikely, since it would require a localized increase in kinetic energy where the conformational change occurred in the first place (figure 1). Temperature coupling maintains a sufficient amount of kinetic energy to allow local conformational corrections, but does not supply enough to allow escape from the global minimum. This explains the observation that on average the agreement with the experimental data will improve rather than worsen with simulated annealing.

3.4 PRACTICAL CONSIDERATIONS

The simulated annealing temperature needs to be high enough to allow conformational transitions but not too high to avoid moving too far away from the initial structure. The optimum temperature for a given starting structure is a matter of trial and error. We empirically determined starting temperatures for a variety of simulated annealing protocols [45, 7] which should work for the average case. However, it might be worth trying different temperatures if a particularly difficult refinement problem is encountered. In particular, significantly higher temperatures are attainable using torsion-angle molecular dynamics. Note, that each simulated annealing refinement run is subject to "chance" by using a random number generator

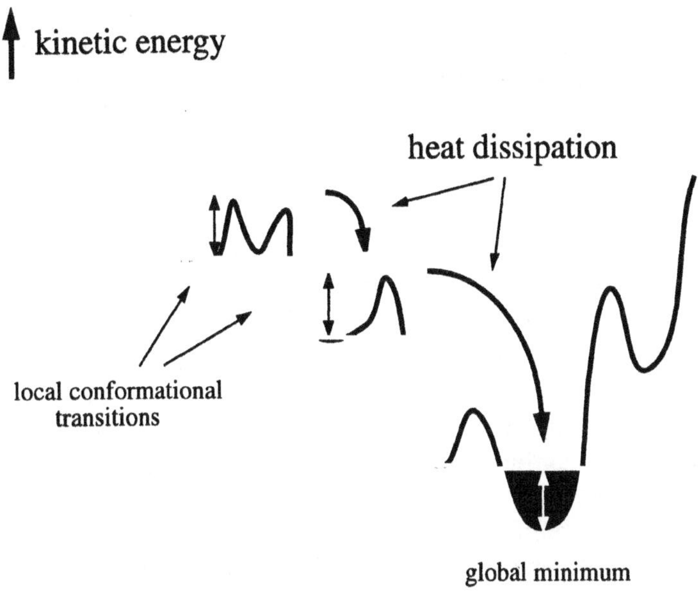

Figure 1. Why simulated annealing works. The kinetic energy of the system allows local conformational transitions with barriers smaller than the kinetic energy. If a larger drop in energy is encountered the excess kinetic energy is dissipated through the friction term (Eq. 8). It is thus unlikely that the system can climb out of the global minimum once it has reached it.

to generate the initial velocities. Thus, multiple refinements must be run if systematic trends resulting from changes of certain parameters of the annealing schedule are to be studied. The best structure(s) among a set of refinements using different initial velocities and/or temperatures should be taken for further refinement or averaging (see below).

The annealing schedule run can in principle be any function of the simulation step (or "time" domain). The two most commonly used protocols are linear slow–cooling or constant–temperature followed by quenching. A slight advantage is obtained with slow–cooling [22]. The duration of the annealing schedule is another parameter. Too short a protocol does not allow sufficient sampling of conformational space. Too long a protocol may waste computer time since it is more efficient to run multiple trials as opposed to one long refinement protocol (unpublished results).

4. Crystallographic Refinement

In the crystallographic case, the limited radius of convergence of refinement arises not only from the high dimensionality of the parameter space, but also from the crystallographic phase problem. For new crystal structures, initial electron density maps must be computed from a combination of observed diffraction amplitudes and experimental phases where the latter are typically of poorer quality and lower resolution than the former. A different problem arises when structures are solved by molecular replacement [46, 47] which uses a similar structure as a search model. In this case the resulting electron density maps can be severely "model–biased", that is, they seem to confirm the existence of the search model without providing clear evidence of actual differences between it and the true crystal structure. In either case, initial atomic models usually require extensive refinement.

Many examples have shown that simulated annealing refinement starting from initial models (obtained by standard crystallographic techniques) produces significantly better final models compared to those produced by least–squares or conjugate–gradient minimization. In recent tests [5, 7], arbitrarily "scrambled" models were generated from an initial model of α–amylase inhibitor built using experimental phase information from multiple isomorphous replacement. Scrambling of this initial model was obtained by increasingly long molecular dynamics simulations at 600K computed without reference to the X–ray diffraction data. These scrambled structures were energy minimized in order to provide chemically reasonable starting models. It was shown that these models represent realistic test cases for phasing by molecular replacement: the scrambled structures contained sufficient information to solve the crystal structure through an extensive molecular replacement search [7].

Both conjugate gradient minimization and simulated annealing were carried out in order to compare the performance of the maximum-likelihood target against the least-squares residual. The maximum–likelihood target produces models with lower phase error. For conjugate gradient minimization the average phase improvement, compared to the least–squares residual, is approximately 5°. It can also be seen that simulated annealing with the least–squares target has a larger radius of convergence than minimization with maximum–likelihood, and that this convergence is further improved by the use of the maximum–likelihood target. For the most scrambled model an average phase improvement, compared to the least–squares residual, of more than 15° is obtained. The resulting structures are very close to the published crystal structure. Similar results were obtained for a test case at lower (2.8 Å) resolution and for a real case involving a new crystal structure starting with an initial model built into a poor electron density map [7].

Cross–validation is essential in the calculation of the maximum likelihood target [28, 6, 7]. Maximum–likelihood refinement without cross–validation gives much poorer results, as indicated by higher free R-values, higher $R_{\text{free}} - R$ differences, and phase errors [7]. It should be noted that the normal R value in general increases upon using the cross–validated maximum likelihood formulation. This is a consequence of the reduction of overfitting by this method.

Simulated annealing refinement is most useful when the initial model is relatively crude. Given a well–refined model, it offers little advantage over conventional methods, with the exception of providing information about the accuracy and conformational variability of the refined structure.

5. Averaging of Independently Refined Structures

As mentioned above, multiple simulated annealing refinements will generally produce somewhat different structures, some of which may be better (as assessed, for example, in terms of the free R value). This approach offers several advantages. First, a more optimum structure can be obtained from multiple trials as opposed to a single simulated annealing calculation. Thus, by examining the ensemble during model–building, one may gain insights into possible local conformations of the molecule. Third, the structure factors of all structures of the family may be averaged in the X–ray crystallographic case. This averaging will reduce the effect of local errors (noise) that are presumably different in each member of the family [32].

6. Ensemble Models

In cases of conformational variability or discrete disorder, there is not a single correct solution to the optimization problem Eq. 1. Rather, the X–ray diffraction data represent a spatial and temporal average over all conformations that are assumed by the molecule. Ensembles of structures, which are simultaneously refined against the observed data, may thus be a more appropriate description of the data. This has been used for some time when alternate conformations are modelled locally. Alternate conformations can be generalized to global conformations [49, 8, 9], i.e., the model is duplicated n–fold, the corresponding calculated structure factors are added and refined simultaneously against the observed X–ray diffraction data, and each member of the family is chemically "invisible" to all other members. The optimal number n can be determined by cross–validation [9, 10].

An advantage of a multi–conformer model is that it directly incorporates many possible types of disorder and motion (global disorder, local sidechain disorder, local wagging and rocking motions). Furthermore, it can be used to automatically detect the most variable regions of the molecule by inspecting the atomic root–mean–square difference around the mean as a function of residue number. Thermal factors of single conformer models may sometimes be misleading by underestimating the degree of motion or disorder [50] and, thus, the multiple-conformer model is a more faithful representation of the diffraction data. A disadvantage of the multi–conformer model is that it introduces many more degrees of freedom.

Although there are some similarities between averaging individually refined structures and multi–conformer models there are also fundamental differences. For example, averaging seeks to improve the calculated electron density map by averaging out the noise present in the individual models, each of which is still a good representation of the diffraction data. This method is most useful at the early stages of refinement when the model still contains errors. In contrast, multi–conformer refinement seeks to create an ensemble of structures at the final stages of refinement which, taken together, best represent the data. It should be noted that each each individual conformer of the ensemble does not necessarily remain a good description of the data since the whole ensemble is refined against the data. Clearly, this method requires high–quality data and a high observable–to–parameter ratio.

7. Conclusions

Simulated annealing has improved the efficiency of crystallographic refinement. A case in point is the combination of torsion angle molecular dynamics with a cross–validated maximum–likelihood target for X–ray crystallography which interact synergistically to produce less model bias than any other method to date. The combined method also increases the radius of convergence allowing the refinement of poor initial models, e.g., those obtained by weak molecular replacement solutions [5, 7]. However, simulated annealing refinement alone is still insufficient to refine a structure automatically without human intervention. For example, crystallographic refinement using simulated annealing typically cannot correct chain tracing errors such as register shifts. Fully automatic structure determination remains a distant goal that will be likely to require significant new algorithmic developments.

Simulated annealing can also be used to provide new physical insights into molecular function which may depend on conformational variability. The sampling characteristics of simulated annealing allow the generation of multi–conformer models which can represent molecular motion and discrete disorder, especially when combined with the acquisition of high–quality data [10]. Simulated annealing is thus also a stepping stone towards development of improved models of macromolecules in solution and in the crystaline state.

8. Acknowledgements

The recent computational developments discussed in this review will become available on the internet (URL: http//atb.csb.yale.edu) in the near future. L.M.R. is an HHMI predoctoral fellow. This work was funded in part by grants from the National Science Foundation to A.T.B. (BIR 9514819 and ASC 93-181159).

References

1. Press, W.H., Flannery, B.P., Teukolosky, S. A. and Vetterling, W.T. (eds.) (1986). Numerical Recipes, Cambridge, Cambridge University Press, 498-546.
2. Kirkpatrick, S., Gelatt, C.D. and Vecchi, Jr., M.P., (1983). Optimization by simulated annealing. *Science* **220**, 671-680.
3. Brünger, A.T., Kuriyan, J. and Karplus, M. (1987). Crystallographic R factor refinement by molecular dynamics. *Science* **235** , 458-460.
4. Brünger, A.T. (1992). The Free R value: a Novel Statistical Quantity for Assessing the Accuracy of Crystal Structures. *Nature* **355** , 472-474.
5. Rice, L.M. and Brünger, A.T. (1994). Torsion Angle Dynamics: Reduced Variable Conformational Sampling Enhances Crystallographic Structure Refinement. *Proteins: Structure, Function, and Genetics* **19**, 277-290.
6. Pannu, N.S. and Read, R.J. (1996). Improved Structure Refinement Through Maximum Likelihood. *Acta Cryst. A* **52**, 659-668.
7. Adams, P.D., Pannu, N.S., Read, R.J. and Brünger, A.T. (1997). Cross-validated Maximum Likelihood Enhances Crystallographic Simulated Annealing Refinement. *Proc. Natl. Acad. Sci. USA*, submitted.
8. Kuriyan, J., Ösapay, K., Burley, S.K., Brünger, A.T., Hendrickson, W.A. and Karplus, M. (1991). Exploration of Disorder in Protein Structures by X-ray Restrained Molecular Dynamics. *Proteins* **10**, 340-358.
9. Burling, F.T. and Brünger, A.T. (1994). Thermal Motion and Conformational Disorder in Protein Crystal Structures: Comparison of Multi-Conformer and Time-Averaging Models. *Israel Journal of Chemistry* **34**, 165-175.
10. Burling, F.T., Weis, W.I., Flaherty, K.M. and Brünger, A.T. (1996). Direct Observation of Protein Solvation and Discrete Disorder With Experimental Crystallographic Phases, *Science* **271**, 72-77.
11. Hendrickson, W.A. (1985). Stereochemically restrained refinement of macromolecular structures. *Meth. Enzymol.* **115**, 252-270.
12. Engh, R.A. and Huber, R. (1991). Accurate bond and angle parameters for X-ray structure refinement. *Acta Cryst. A* **47**, 392-400.
13. Parkinson, G., Vojtechovsky, J., Clowney, L., Brünger, A.T. and Berman, H.M. (1996). New Parameters for the Refinement of Nucleic Acid Containing Structures. *Acta Cryst. D* **52**, 57-64.
14. Allen, F.H., Kennard, O. and Taylor, R. (1983). Systematic Analysis of Structural Data as a Research Technique in Organic Chemistry. *Acc. Chem. Res.* **16**, 146-153.
15. Dauter, Z., Lamzin, V.S. and Wilson, K.S. (1995). Proteins at atomic resolution. *Curr. Opin. Struct. Biol.* **5**, 784-790.
16. Sevcik, J., Dauter, Z., Lamzin, V.S. and Wilson, K.S. (1996). Ribonuclease from Streptomyces aureofaciens at atomic resolution, *Acta Cryst. D* **52**, 327-344.
17. Stec, B., Zhou, R. and Teeter, M.M. (1995). Full-matrix refinement of the protein crambin at 0.83 Å and 130 K. *Acta Cryst. D* **51**, 663-681.
18. Pearlman, D.A. and Kim, S.-H. (1990). Atomic charges for DNA constituents derived from single-crystal X-ray diffraction data. *J. Mol.Biol.* **211**, 171-187.

19. Karplus, M. and Petsko, G.A. (1990). Molecular dynamics simulations in biology. *Nature* **347**, 631-639.
20. Brünger, A.T., Karplus, M. and Petsko, G.A. (1989). Crystallographic Refinement by Simulated Annealing: Application to a 1.5 Å Resolution Structure of Crambin. *Acta Cryst. A* **45**, 50-61.
21. Weis, W.I., Brünger, A.T., Skehel, J.J. and Wiley, D.C. (1989). Refinement of the Influenza Virus Haemagglutinin by Simulated Annealing. *J. Mol. Biol.* **212**, 737-761.
22. Brünger, A.T., Krukowski, A. and Erickson, J. (1990). Slow-Cooling Protocols for Crystallographic Refinement by Simulated Annealing. *Acta Cryst. A* **46**, 585-593.
23. Fujinaga, M., Gros, P. and van Gunsteren, W.F. (1989). Testing the method of crystallographic refinement using molecular dynamics. *J. Appl. Cryst.* **22**, 1-8.
24. Silva, A.M. and Rossmann, M.G. (1985). The refinement of southern bean mosaic virus in reciprocal space. *Acta Cryst. B* **41**, 147-157.
25. Read, R.J. (1990). Structure-factor probabilities for related structures. *Acta Cryst. A* **46**, 900-912.
26. Bricogne, G. (1991). A multisolution method of phase determination by combined maximization of entropy and likelihood. III. Extension to powder diffraction data. *Acta Cryst. A* **47**, 803-829.
27. Bricogne, G. (1993). Direct phase determination by entropy maximization and likelihood ranking: status report and perspectives. *Acta Cryst. D* **49**, 37-60.
28. Kleywegt, G.J. and Brünger, A.T. (1996). Cross-validation in crystallography: practice and applications. *Structure* **4**, 897-904.
29. Sussman, J.L., Holbrook, S.R., Church, G.M. and Kim, S.-H. (1977). Structure-factor least-squares refinement procedure for macromolecular structure using constrained and restrained parameters. *Acta Cryst. A* **33**, 800-804.
30. Diamond, R. (1971). A real-space refinement procedure for proteins. *Acta Cryst. A* **27**, 436-452.
31. Fujinaga, M. (1993). Crystallographic Refinement. In: Computer Simulation of Biomolecular Systems: Theoretical and Experimental Applications (W.F. van Gunsteren, P.K. Weiner, A.J. Wilkinsin, Eds.) ESCOM Science Publishers B.V., Leiden, pp. 371-381.
32. Rice, L.M., Shamoo, Y., Brünger, A.T., manuscript in preparation.
33. Arnold, E. and Rossmann, M.G. (1988). The use of molecular-replacement phases for the refinement of the human rhinovirus 14 structure. *Acta Cryst. A* **44**, 270-282.
34. Hendrickson, W.A. (1991). Determination of macromolecular structures from anomalous diffraction of synchrotron radiation, *Science* **254**, 51-58.
35. Laarhoven, P.J.M. and Aarts, E.H.L.(eds.) (1987). Simulated Annealing: Theory and Applications. Dordrecht: D. Reidel Publishing Company, pp. 187.
36. Metropolis, N., Rosenbluth, M., Rosenbluth, A.Teller, A. and Teller, E. (1953). Equation of state calculations by fast computing machines. *J. Chem. Phys.* **21**, 1087-1092.
37. Verlet, L. (1967). Computer Experiments on classical fluids. I. Thermodynamical properties of Lennard-Jones molecules. *Phys. Rev.* **159**, 98-105.
38. Goldstein, H. (1980). Classical Mechanics 2nd ed., Addison-Wesley Pub. Co., Reading, Massachusetts.
39. Abramowitz, M. and Stegun, I. (1968). Handbook of Mathematical Functions, Applied Mathematics Series, vol. 55, Dover Publications, New York, pp. 896.
40. Bae, D.-S. and Haug, E.J. (1987). A recursive formulation for constrained mechanical system dynamics: Part I. Open loop systems. *Mech. Struct. and Mach.* **15**, 359-382.
41. Bae, D.-S. and Haug, E.J. (1988). A recursive formulation for constrained mechanical system dynamics: Part II. Closed loop systems. *Mech. Struct. and Mach.* **15**, 481-506.
42. Jain, A., Vaidehi, N. and Rodriguez, G. (1983). A Fast Recursive Algorithm for Molecular Dynamics Simulation. *J. Comp. Phys.* **106**, 258-68.

43. Mathiowetz, A.M., Jain, A., Karasawa, N. and Goddard, W.A. (1994). Protein Simulations Using Techniques Suitable for Very Large Systems: The Cell Mulipole Method for Nonbond Interactions and the Newton-Euler Inverse Mass Operator Method for Internal Coordinate Dynamics. *Proteins* **20**, 227-247.
44. Berendsen, H.J.C., Postma, J.P.M., van Gunsteren, W.F., DiNola, A. and Haak, J.R. (1984). Molecular dynamics with coupling to an external bath. *J. Chem. Phys.* **81**, 3684-3690.
45. Brünger, A.T. (1988). Crystallographic refinement by simulated annealing: application to a 2.8 Å resolution structure of aspartate aminotransferase. *J. Mol. Biol.* **203**, 803-816.
46. Hoppe, W. (1957). Die Faltmolek—lmethode - eine neue Methode zur Bestimmung der Kristallstruktur bei Ganz oder Teilweise bekannter Molek—lstruktur. *Acta Cryst.* **10**, 750-751.
47. Rossmann, M.G. and Blow, D.M. (1962). The detection of sub-units within the crystallographic asymmetric unit. *Acta Cryst. A* **15** , 24-51.
48. Pflugrath, J.W., Wiegand, G., Huber, R. and Vertesey, L., (1986). Crystal structure determination, refinement and the molecular model of the alpha-amylase inhibitor Hoe-467A. *J. Mol. Biol.* **189(2)**, 383-386.
49. Gros, P., van Gunsteren, W.F. and Hol, W.G.J. (1990). Inclusion of Thermal Motion in Crystallographic Structures by Restrained Molecular Dynamics. *Science* **249** , 1149-1152
50. Kuriyan, J., Petsko, G.A., Levy, R.M. and Karplus, M. (1986). Effect of Anisotropy and Anharmonicity on Protein Crystallographic Refinement. *J. Mol. Biol.* **190**, 227-254.

FIGURES OF MERIT FOR EVALUATING PHASES

C.J.Gilmore
Department of Chemistry
University of Glasgow
Glasgow G12 8QQ, Scotland, UK.

1. Introduction

Virtually all direct methods techniques are multisolutional: they produce more than one phase set, and, indeed, often produce thousands and even more. The problem then arises as to how to establish which phase set is correct in the sense of having the lowest phase error (assuming there is such a set). To do this, figures of merit (FOMs) are used in which the phases are tested against invariants and seminvariants to establish consistency criteria. It is the purpose of this chapter to examine such criteria, and to look critically at their efficacy in a macromolecular phasing environment.

2. Traditional Figures of Merit

There are four standard figures of merit used in traditional direct methods, based on triplet and quartet invariants.

2.1 ABSFOM

The 'absolute figure of merit' ABSFOM is defined by the following equations:

$$ABSFOM = \sum_{\mathbf{h}} (\alpha_{\mathbf{h}} - \alpha_{R\mathbf{h}}) \Big/ \sum_{\mathbf{h}} (\alpha_{E\mathbf{h}} - \alpha_{R\mathbf{h}}) \tag{1}$$

where:

$$\alpha_{E\mathbf{h}}^2 = \sum_{\mathbf{k}} A_{\mathbf{hk}}^2 + \sum_{\mathbf{k}} \sum_{\mathbf{k}'} A_{\mathbf{hk}} A_{\mathbf{hk}'} \frac{I_1(A_{\mathbf{hk}})}{I_1(A_{\mathbf{hk}})} \frac{I_1(A_{\mathbf{hk}'})}{I_1(A_{\mathbf{hk}'})} \tag{2}$$

$$\alpha_{R\mathbf{h}} = \left(\sum_{\mathbf{h}} A_{\mathbf{hk}}^2 \right)^{1/2} \tag{3}$$

and:

S. Fortier (ed.), Direct Methods for Solving Macromolecular Structures, 159–167.

$$A_{\mathbf{hk}} = 2\left|E_{\mathbf{h}} E_{\mathbf{k}} E_{-\mathbf{h}-\mathbf{k}}\right| / \sqrt{N} \tag{4}$$

The |E|'s are normalised structure factors for N atoms, assumed equal, in the unit cell and α_h is estimated from each phase set from the tangent formula:

$$\tan \phi_{\mathbf{h}} = \frac{\sum_{\mathbf{k}} w_{\mathbf{k}} w_{\mathbf{h}-\mathbf{k}} |E_{\mathbf{k}}| |E_{\mathbf{h}-\mathbf{k}}| \sin(\phi_{\mathbf{k}} + \phi_{\mathbf{h}-\mathbf{k}})}{\sum_{\mathbf{k}} w_{\mathbf{k}} w_{\mathbf{h}-\mathbf{k}} |E_{\mathbf{k}}| |E_{\mathbf{h}-\mathbf{k}}| \cos(\phi_{\mathbf{k}} + \phi_{\mathbf{h}-\mathbf{k}})} = \frac{T_{\mathbf{h}}}{B_{\mathbf{h}}} \tag{5}$$

$$\alpha_{\mathbf{h}}^2 = T_{\mathbf{h}}^2 + B_{\mathbf{h}}^2 \tag{6}$$

Usually good phase sets have values of ABSFOM between 1.0-1.3.

Comments: Only strong reflections are used in the tangent formula, and only triple phase relationships employed. The tangent formula will itself maximise ABSFOM which therefore becomes highly correlated with it - this is not an independent phase indicator. Since the tangent formula breaks down with macromolecular phase sets, so will this figure of merit.

2.2 RESID

RESID is an R-factor defined either on α or on E-magnitude. For the former case:

$$R_{\alpha} = 100 \times \sum_{\mathbf{h}} \left| \alpha_{\mathbf{h}} - \alpha_{E\mathbf{h}} \right| \Big/ \sum_{\mathbf{h}} \alpha_{E\mathbf{h}} \tag{7}$$

The relationship to the conventional crystallographic R factor is obvious. A value for R_{α} around 20-25% is a good indicator of phase consistency.

Comments: As for ABSFOM, this is unlikely to be a good phase indicator in the macromolecular situation.

2.3 PSI-ZERO

The Ψ_0 figure of merit is defined as:

$$\Psi_0 = \frac{\sum_{\mathbf{l}} \left(\left| \sum_{\mathbf{k}} E_{\mathbf{k}} E_{\mathbf{l}-\mathbf{k}} \right| \right)}{\sum_{\mathbf{l}} \left(\sum_{\mathbf{k}} |E_{\mathbf{k}} E_{\mathbf{l}-\mathbf{k}}|^2 \right)^{1/2}} \tag{8}$$

The reflections involved in these triplets have $|E_k|$ and $|E_{l-k}|$ large but $|E_l|$ *small.* This differs from the other figures of merit in that it establishes a consistency between the amplitudes and phases of the strong reflections with a small set (50-100) of weak reflections. Correct phase sets usually give a value of Ψ_0 close to unity.

Comments: A useful figure of merit, but the Ψ_0 triplets will be very unreliable in the macromolecular case. The figure of merit is, however, largely independent of the tangent formula, and this is of considerable benefit in the small molecule case.

The Ψ_0 triplets can be used actively [1]; this seems to be useful with small molecules, but there will be a loss of independence of figure of merit and phasing technique.

2.4 NQEST

$$NQEST = \frac{\sum_{h,k,l} w_{hkl} \cos(\phi_h + \phi_k + \phi_l + \phi_{-h-k-l})}{\sum_{h,k,l} w_{hkl}} \tag{9}$$

The summation is over the negative quartets (those with small or zero cross terms i.e. the 2nd neighbourhood) and the weights, *w*, are obtained from a variety of quartet formulae. The correct set of phases should display a negative value for NQEST. The minimum possible value when all the negative quartets are satisfied (an unlikely event!) is -1.0. It is especially useful for symorphic structures.

Comments: If negative quartets are not used actively in the tangent formula, then this figure of merit is independent of it. Unfortunately for macromolecules there is a 1/N dependence of quartet reliability (as opposed to a $1/N^{1/2}$ dependence for triplets) so that negative quartet reliability becomes a problem, although third neighbourhood formulae may help.

There is a good discussion of these FOMs by Cascarano, Giacovazzo and Guagliardi [2] who also introduce some extensions and modifications. In particular, they define an NALF figure of merit based on various α parameters as in 2.1, but also use the second moments of α_h, and an NPS FOM utilising α_h. and probability distributions for α. These are undoubtedly more efficient in many circumstances, but are still limited by triplet phase reliability for proteins.

It is customary to combine all the figures of merit into a single combined figure in which the individual contributors are suitably weighted. The weighting scheme used can be important in difficult cases.

3. Adapting the Standard Figures of Merit for Macromolecules

Giacovazzo and co-workers have extended the use of small molecule FOMs to macromolecules [3] where they incorporate derivative information. They found it difficult to generalise as to which figures of merit were the most efficient: it varied from structure to structure. This is not, however, an ab initio protein structure solution in the sense of working with native data only - derivatives play a vital role.

4. The Minimal Function.

Hauptman and co-workers have used the minimal function to spectacular effect for small proteins where the data are at atomic resolution i.e. around 1.1Å [4,5]. Phases are refined such as to minimise the function:

$$R(\phi)=\Big(\sum_{\mathbf{h},\mathbf{k}} A_{\mathbf{hk}}\left\{\cos(\phi_{\mathbf{h}}+\varphi_{\mathbf{k}}+\varphi_{-\mathbf{h}-\mathbf{k}})-\left[I_1(A_{\mathbf{hk}})/I_0(A_{\mathbf{hk}})\right]\right\}^2+$$

$$|B_{\mathbf{lmn}}|\left\{\cos(\varphi_{\mathbf{l}}+\varphi_{\mathbf{m}}+\varphi_{\mathbf{n}}+\varphi_{-\mathbf{l}-\mathbf{m}-\mathbf{n}})-\left[I_1(B_{\mathbf{lmn}})/I_0(B_{\mathbf{lmn}})\right]\right\}^2 \qquad (10)$$

$$\times\left(\sum_{\mathbf{h},\mathbf{k}} A_{\mathbf{hk}}+\sum|B_{\mathbf{lmn}}|\right)^{-1}\Big)$$

where:

$$B_{\mathbf{lmn}}=(2/N)|E_{\mathbf{l}}E_{\mathbf{m}}E_{\mathbf{n}}E_{-\mathbf{l}-\mathbf{m}-\mathbf{n}}|\left[\left(|E_{\mathbf{m}+\mathbf{n}}|^2+|E_{\mathbf{l}+\mathbf{m}}|^2+|E_{\mathbf{n}+\mathbf{l}}|^2\right)-2\right] \qquad (11)$$

This has been extended to SAS maximal principle when single-wavelength anomalous scattering diffraction data are available [6]. In this case the function is much simpler:

$$M(\phi)=\left(1\Big/\sum_{\mathbf{h},\mathbf{k}} A_{\mathbf{hk}}\right)\sum_{\mathbf{h},\mathbf{k}} A_{\mathbf{hk}}\left\{\cos(\phi_{\mathbf{h}}+\varphi_{\mathbf{k}}+\varphi_{-\mathbf{h}-\mathbf{k}}-\omega_{\mathbf{hk}})\left[I_1(A_{\mathbf{hk}})/I_0(A_{\mathbf{hk}})\right]\right\} \qquad (12)$$

This contains new terms, ω_{hk}, which are the non-zero estimates of the triplets arising from the SAS estimates. There are myriad local maxima in $R(\varphi)$ or $M(\varphi)$, of course, but the global maximum is expected to be very isolated and thus readily identified.

Comments: In this use of a minimal function the value of the function itself is used as a figure of merit. The global optimum is identified by its isolation in phase space. Only the strongest reflections contribute to the triplets and so only the strongest are included in the relevant function.

5. R-free

Brünger [7] has proposed the use of a free R-factor in which the intensity data are partitioned into two sets of unique reflections: a working set {W} and a much smaller test set {T}. When a structure is refined, only set {W} is used throughout. At the end of the refinement an R factor is calculated based only on set {T}:

$$R_T^{free}=\frac{\sum_{\mathbf{h}\in T}\left||F_{\mathbf{h}}^{\text{obs}}|-|F_{\mathbf{h}}^{calc}|\right|}{\sum_{\mathbf{h}\in T}\left|F_{\mathbf{h}}^{\text{obs}}\right|} \qquad (13)$$

This is claimed to offer an unbiased criterion of goodness of fit, but it is not completely so, because structure factors are not independent but interlinked *via* the Sayre equation.

$$F_{\mathbf{h}} = \theta/V \sum_{\mathbf{k}} F_{\mathbf{k}} F_{\mathbf{h-k}} \tag{14}$$

Obviously the free R-factor is not a figure of merit in the sense of choosing correct phase sets, which is the main thrust of this chapter, but it is related to problems in protein structure determination and so is worth discussing here.

6. Likelihood

Let us consider a crystal structure as a form of three-dimensional image (which it is). We can write Bayes' theorem in a new form:

$$p(image|data) \propto p(image) \times p(data|image) \tag{15}$$

The constant of proportionality is determined by normalisation. Remember that the term *p(data|image)* is called the likelihood, and we can now see that the likelihood consults the data once an image or structure or corresponding phase set is postulated. The use of likelihood as a figure of merit in a multisolution phasing environment has already been discussed in previous chapters, but we need to revise several features here.

We have generated a maximum entropy map $q^{ME}(x)$ whose Fourier transform gives the extrapolates $|U_{\mathbf{h}}^{ME}|$. The observed U-magnitudes are $|U_{\mathbf{h}}|^{obs}$. For each acentric extrapolated, reflection **k** the likelihood measure, in its diagonal approximation where the reflections are decoupled, can be written [8-11]:

$$\Lambda_{\underline{k}} = \frac{|U_{\mathbf{k}}|^{obs}}{\varepsilon_{\mathbf{k}}\Sigma + \sigma_{\mathbf{k}}^2} \exp\left\{ -\frac{1}{2} \frac{\left(|U_{\mathbf{k}}|^{obs}\right)^2 + |U_{\mathbf{k}}^{ME}|^2}{\varepsilon_{\mathbf{k}}\Sigma + \sigma_{\mathbf{k}}^2} \right\} I_0\left(\frac{|U_{\mathbf{k}}|^{obs}|U_{\mathbf{k}}^{ME}|}{\varepsilon_{\mathbf{k}}\Sigma + \sigma_{\mathbf{k}}^2} \right) \tag{16}$$

where $\varepsilon_{\mathbf{k}}$ is the statistical weight of reflection **k**, $\sigma_{\mathbf{k}}^2$ the variance of $|U_{\mathbf{k}}|^{obs}$ and Σ a refinable measure of unit cell contents $\Sigma \approx 1/(2N)$ for N point atoms in the unit cell. The distribution (1) is a Rice distribution comprising a Gaussian (the exponential term) with an offset represented by the Bessel function term (I_0). Note also that this expression is a measure of agreement between $|U_{\mathbf{k}}|^{obs}$ and $|U_{\mathbf{k}}^{ME}|$, indeed it has a maximum where $|U_{\mathbf{k}}|^{obs} = |U_{\mathbf{k}}^{ME}|$. For the centric case the Bessel function is replaced by a *cosh* term:

$$\Lambda_{\mathbf{k}} = \frac{2|U_{\mathbf{k}}|^{obs}}{\pi\left(2\varepsilon_{\mathbf{k}}\Sigma + \sigma_{\mathbf{k}}^2\right)} \exp\left\{ -\frac{1}{2} \frac{\left(|U_{\mathbf{k}}|^{obs}\right)^2 + |U_{\mathbf{k}}^{ME}|^2}{2\varepsilon_{\mathbf{k}}\Sigma + \sigma_{\mathbf{k}}^2} \right\} \cosh\left(\frac{|U_{\mathbf{k}}|^{obs}|U_{\mathbf{k}}^{ME}|}{2\varepsilon_{\mathbf{k}}\Sigma + \sigma_{\mathbf{k}}^2} \right)$$

(17)

We now define a corresponding null hypothesis for the situation of null extrapolation, $\left|U_{\mathbf{k}}^{ME}\right|=0$, which gives Wilson statistics. For acentric reflections:

$$\Lambda_{\mathbf{k}}^{0}=\frac{\left|U_{\mathbf{k}}\right|^{obs}}{\varepsilon_{\mathbf{k}}\Sigma+\sigma_{\mathbf{k}}^{2}}\exp\left\{-\frac{1}{2}\frac{\left(\left|U_{\mathbf{k}}\right|^{obs}\right)^{2}}{\varepsilon_{\mathbf{k}}\Sigma+\sigma_{\mathbf{k}}^{2}}\right\} \tag{18}$$

Define:

$$L_{\mathbf{k}}=\log\frac{\Lambda_{\mathbf{k}}}{\Lambda_{\mathbf{k}}^{0}} \tag{19}$$

Then the global log-likelihood gain (LLG) is:

$$LLG=\sum_{\mathbf{k}}L_{\mathbf{k}} \tag{20}$$

The LLG should be largest for correct phase sets.

Comments: LLG uses all the intensity data - it does not partition reflections into small and large magnitudes, and furthermore it employs their standard deviations in an easy way which can be helpful with poor quality data sets. In the macromolecular environment the diagonal approximation may be too extreme, although alternatives now exist [11], and as structures get larger U-magnitudes will get smaller, thus decreasing the dynamic range of the LLG. However, structures with up to 1500 carbon atoms in the unit cell have worked [12].

7. Is Entropy a Viable Indicator of Phase Correctness?

The short answer is *no*, but further amplification is obviously needed:

(i) Naively, there is no reason why entropy should be such an indicator. Entropy can be considered a measure of map flatness; there is no reason why the correct map among many possible maps each with different phases but the same amplitudes, should be the flattest.

(ii) Lemaréchal and Navaza [13] in a critique of Prince [14] give arguments concerning the phases assigned to the Lagrangian multipliers used in the entropy maximisation procedure itself to demonstrate that entropy cannot be used as a reliable figure of merit.

(iii) Gilmore, *et al.* [15], Shankland *et al.* [16] give examples where the use of entropy in this way is disastrous.

(iv) Sato [17] however, disagrees and claims that entropy is a valid indicator for data sets of resolution greater than 1.5Å., but in this case he is working with very large basis sets of accurate phases extracted from a refined small molecule structure, $C_{10}H_{13}N_5O_3$.

Is it a question of resolution: can entropy work as a figure of merit at low resolution where one expects maps free of large peaks? Entropy is a measure of map flatness, so should perhaps be able to select the 'best' one(s). The trouble here is that density builds quite quickly even at low resolution. To demonstrate this, Table 1 presents a simple, one level phasing tree for halorhodopsin using electron diffraction data from Havelka, Henderson, Heymann and Oesterhelt [18] At 15Å it can be seen that LLG is an excellent indicator of phase correctness whereas entropy is simply incorrect.

Node	LLG	Entropy	Phase Error
1	0.787	-0.027	18.8
2	-0.095	-0.023	69.3
3	0.000	-0.026	58.5
4	0.472	-0.023	127.3
5	-0.034	-0.025	103.9
6	0.050	-0.025	99.5
7	0.076	-0.025	88.7
8	-0.014	-0.024	139.2
9	-0.042	-0.026	43.6
10	0.061	-0.024	94.1
11	0.065	-0.025	83.2
12	-0.001	-0.024	133.8
13	0.775	-0.027	73.8
14	-0.087	-0.024	124.3
15	-0.012	-0.026	113.4
16	0.494	-0.023	164.0
17	0.810	-0.027	0.0
18	-0.035	-0.023	50.5
19	-0.066	-0.026	39.7
20	0.501	-0.023	90.2
21	0.058	-0.026	30.2
22	0.080	-0.024	80.7
23	0.043	-0.025	69.9
24	-0.063	-0.024	120.4
25	0.050	-0.025	24.8
26	0.092	-0.024	75.3
27	0.033	-0.025	64.5
28	0.033	-0.025	115.0
29	0.796	-0.027	55.0
30	-0.026	-0.023	105.5
31	-0.079	-0.026	94.7
32	0.513	-0.023	145.2

Table 1. A phasing tree for halorhodopsin at 15Å resolution.

Likelihood, in the author's experience, is the optimum figure of merit to use in certain environments. Entropy can be included by using $NS+LLG$ (S is the entropy), but there are problems here with the definition of N, which cannot be simply equated with the number of atoms in the unit cell. There is a summary of a discussion on this matter at a conference on Direct Methods of Phasing in Macromolecular Crystallography in Florida, USA in 1992 by Sayre [19] which defines the controversies on these points quite succinctly.

References

1. Cascarano, G. and Giacovazzo, C. (1995) The joint probability distribution of any set of phases given any set of diffraction magnitudes. IV The active use of psi-zero triplets *Acta Cryst.* **A51**, 820-825.
2. Cascarano, G., Giacovazzo, C., and Guagliardi, A. (1992) Improved figures of merit for direct methods *Acta Cryst.* **A48**, 859-865.
3. Giacovazzo, C., Siliqi, D. and Spagna, R. (1994) The ab initio structure solution of protein structures by direct methods II. The procedure and its first applications *Acta Cryst.* **A50**, 609-621.
4. DeTitta, G.T., Weeks, C.M., Thulman, P., Miller, R, and Hauptman, H.A. (1994) Structure solution by minimal-function phase refinement and Fourier filtering. I. Theoretical basis *Acta Cryst.* **A50,** 203-210.
5. Weeks, C.M., Hauptman, H.A., Smith, G.D., Blessing, R.H., Teeter, M.M. and Miller, R. (1995) Crambin: a direct solution for a 400 atom structure *Acta Cryst.* **D51**, 33-38.
6. Hauptman, H.A. (1996) The SAS maximal principle: a new approach to the phase problem *Acta Cryst.* **A52,** 490-496.
7. Brünger, A.T. (1993) Assessment of phase accuracy by cross validation: The free R value, methods and applications *Acta Cryst.* **D49**, 24-36.
8. Bricogne, G. (1984) Maximum entropy and the foundations of direct methods *Acta Cryst.* **A40**, 410-445.
9. Bricogne, G. and Gilmore, C.J. (1990) A multisolution method of phase determination by combined maximisation of entropy and likelihood. I. Theory, algorithms and strategy *Acta Cryst.* **A46,** 284-297.
10. Bricogne, G. (1988) A Bayesian statistical theory of the phase problem. I. A multichannel maximum entropy formalism for constructing generalized joint probability distributions of structure factors *Acta Cryst.* **A44** (1988). 517-545.
11. Bricogne, G. (1993) Direct phase determination by entropy maximisation and likelihood ranking: status report and perspectives *Acta Cryst.* **D49**, 37-60.
12. Gilmore, C.J., Nicholson, W.V., and Dorset, D.L., (1996) Direct methods in protein electron crystallography: the ab initio structure determination of two membrane protein structures in projection using maximum entropy and likelihood, *Acta Cryst.* **A52,** 937-946.
13. Lemaréchal, C. and Navaza, J. (1991) On the *ab initio* phase determination by maximum entropy *Acta Cryst.* **A47,** 631-632.
14. Prince, E. (1993) Construction of maximum-entropy density maps, and their use in phase determination and extension. *Acta Cryst.* **D49,** 61-65.

15. Gilmore, C.J., Shankland, K. and Bricogne, G. (1993) Applications of the maximum entropy method to powder diffraction and electron crystallography *Proc.Roy.Soc.Ser. A.* **442**, 97-111.
16. Shankland, K., Gilmore, C.J., Bricogne G. and Hashizume, H. (1993) A multisolution method of phase determination by combined maximisation of entropy and likelihood. V Automatic Likelihood analysis via the student t-test, with an application to the powder structure of magnesium boron nitride, Mg_3BN_3 *Acta Cryst.* **A49**, 493-501.
17. Sato, T. (1992) Maximum entropy method: phase refinement *Acta Cryst.* **A48**, 842-850.
18. Havelka, W.A., Henderson, R., Heymann, J.A.W. and Osterhelt, D. (1993) Projection structure of halorhodopsin from Halobacterium halobium at 6Å resolution obtained by electron cryo-microscopy *J.Mol.Biol.* **234**, 837-846.
19. Sayre, D. (1993) Discussion after talk by J.M.Stewart *Acta Cryst.* **D49**, 107

INTRODUCTION TO DENSITY MODIFICATION

K. D. COWTAN
Dept. of Chemistry, University of York
Heslington, York, YO1 5DD. United Kingdom

1. Aim

In practice it is fairly easy to tell by looking at it a poor map from a good one. The purpose of phase improvement by density modification is to find the nearest 'good looking' map to the initial bad map. If the bad map contains enough information, it should be possible to proceed to structure solution.

2. Method

The aim of a phase improvement calculation is to calculate a set of phases, and a corresponding set of weights, which combine all of the available information which is known about a structure.

Sources of information include the following:

1. The observed native magnitudes.
2. Experimental phase estimates based on isomorphous/anomalous differences.
3. Solvent content in the crystal. The solvent can usually be located from very weak phase information, and thus this part of the crystal structure can be solved.
4. Atomicity. This can be employed strongly as Sayre's equation at high resolution and as a known density histogram at lower resolutions.
5. Quaternary structure. The complexing of molecules into a multimer either in solution or on crystallisation provides redundancy in the map which is informative in phasing.
6. Tertiary structure can be exploited through skeletonisation, which allows isolation and enhancement of of protein-like features in the density.

Combining information from these sources is difficult, since the first two sources are in reciprocal space and include good estimates of error,

S. Fortier (ed.), Direct Methods for Solving Macromolecular Structures, 169–179.

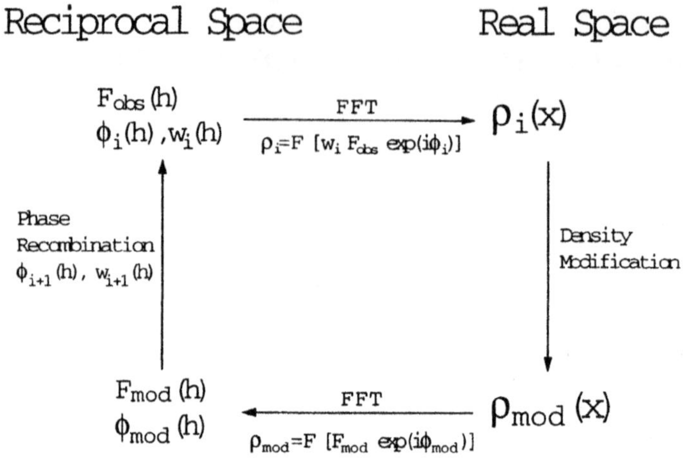

Figure 1. Density modification calculation, showing iterative application of real-space and reciprocal space fitting.

whereas the other sources are in real space, and in general represent our expectations about the structure, which are frequently hard to quantify.

As a result the method which has been adopted is cyclic and divided into real and reciprocal space procedures. A weighted map is calculated, and used as a basis for applying all the real-space modifications. The modified map is then back-transformed to produce a set of magnitudes and phases. The agreement between the map-magnitudes and the observed values is then used to estimate weights for the modified phases, which are used to combine the new phases experimental phases. This process is shown diagrammatically in Figure (1).

3. Density Modifications

A variety of density modification procedures have been applied to maps as part of phase improvement calculations. The most established, and most powerful modifications are solvent flattening and averaging, however histogram matching may also be routinely applied with good results, and other techniques may help in some cases.

3.1. SOLVENT FLATTENING

When protein molecules pack together to form a crystal, there are invariably voids between them. These are usually filled with solvent, which is typically disordered from cell to cell. Thus in the electron density maps it appears to be smooth and featureless. If we know where the solvent is in the unit

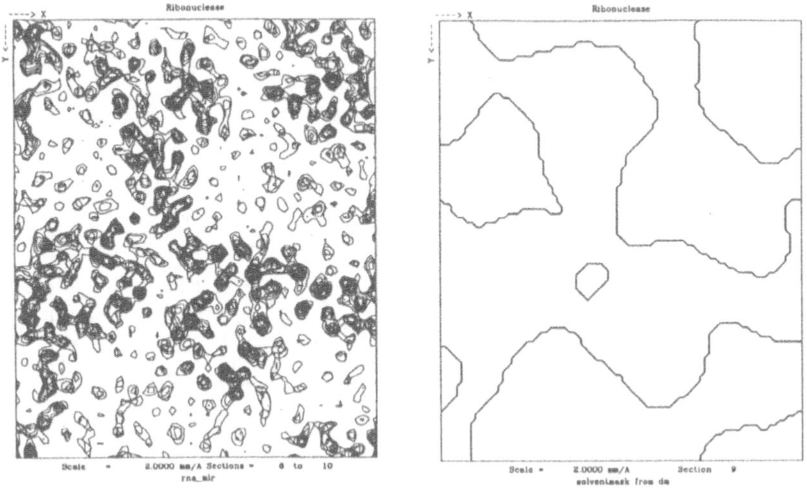

Figure 2. A map section and the corresponding automatically determined solvent mask.

cell, then this portion of the unit cell, and its contribution to the unknown phases, is essentially known.

In practice by simply flattening the density in the solvent region (setting the whole region to the expected mean value) a map is obtained which contains significant additional phase information. Combination of the improved phases with the observed magnitudes in turn improves the unknown density in the protein region.

The solvent region is automatically and routinely identified even from weak phase information by Wang's method [12], or its derivatives. The method is as follows: Electron density below the solvent level is truncated, then a local average or low resolution filter is applied to the map. The resulting map shows high values in the protein region and low values in the solvent. A cutoff level is chosen to give an appropriate solvent fraction (Figure 2).

A variant employed by J. P. Abrahams is solvent flipping. In this case the solvent density is inverted about the solvent level. He has shown [1] that this reduces the dependence between the initial and modified maps, thus reducing the overestimation of weights which otherwise occurs.

3.2. HISTOGRAM MATCHING

Histogram matching is a technique from image processing, in which the density values in the map are rescaled by a function which varies nonlinearly with density. The aim of this is to alter the map in such a way that the frequency distribution of density values (the histogram) matches the

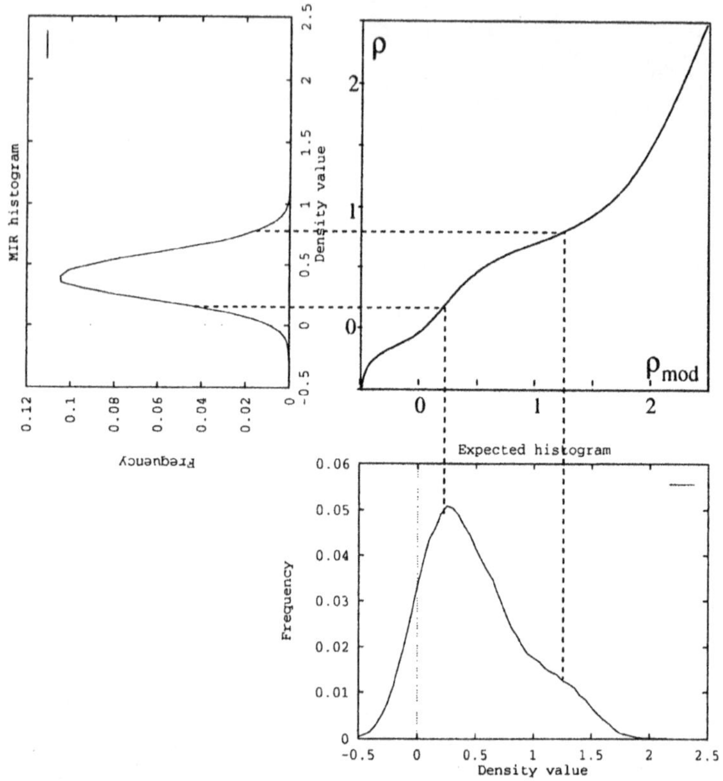

Figure 3. Histogram matching: a non-linear rescaling function is applied to the density in order to obtain the desired distribution. The rescaling function can calculated from the cumulative distribution functions.

frequency distribution for a perfect map [15].

Tests indicate that once the effects of solvent and temperature factor are removed, the expected histogram for a perfect map is largely independent of structure. The histogram of a typical MIR map will differ significantly from this ideal, and so it is generally possible to improve the electron density map of an unknown structure by the application of histogram matching.

Figure (3) shows a typical histogram for the protein region of MIR and perfect maps, along with the rescaling function to map one to the other. The rescaling function may be calculated by cross-referencing the cumulative distributions of the current and desired histograms.

In the case of small metalloproteins it is possible for the histogram to be significantly influenced by the presence of metal atoms. In this case two approaches are possible: either the histogram may be adjusted to include the resolution-dependent heavy atom contribution, or the unit cell must be divided into three volumes: solvent to be flattened, protein to be histogram matched, and density to be preserved. The latter approach is usually simpler.

Histogram matching has been tested on many structures and dramatically increases the effectiveness of phase improvement calculations. However some users remain sceptical, since it is such a 'weak' modification of the map. It is possible that its effectiveness is derived in part through interaction with the phase combination method, however even if this is the case it does not change the fact that the maps are better.

4. Averaging

Averaging is possible when there is additional symmetry in the arrangement of molecules in the crystal beyond those symmetry which map the lattice onto itself. The symmetries are 'local' in that they only map a limited volume of density onto a matching region, and are frequently inexact, since the related molecules find themselves in different chemical environments [3].

These non-crystallographic symmetries provide phase information by the following mechanisms: Firstly, the related regions of the map may be averaged together, increasing the ratio of signal to noise in the map. Secondly, the differently oriented copies of the molecule provide complementary information through the fact that they offer different samplings of the molecular transform. Thus averaging a map can lead to a significant improvement in the phases.

Averaging calculations are generally more complex than solvent flattening, but some progress is being made to the automation of the process. The self-rotation and translation function are now routinely solved using 'AMORE' [6]. Once the averaging operators are determined, the mask can be determined using the local density correlation function as implemented in the DEMON suite [11], or fully automatically using the 'dm' program. This is achieved by a systematic search for extended peaks in the local density correlation, which must be carried out over a volume several unit cells in order to guarantee finding the whole molecule.

In those cases where automatic determination of the averaging operators fails, it is sometimes possible to identify where in the cell the molecule lies, and cut out a representative volume of density to use as a molecular replacement search model.

Some thought is required when using automatic averaging mask determination methods. The local correlation function distinguishes those volumes of crystal space which map onto similar density under transformation by the averaging operator. Thus, in the case of improper NCS, the local correlation mask should cover only one monomer, and the volume masked

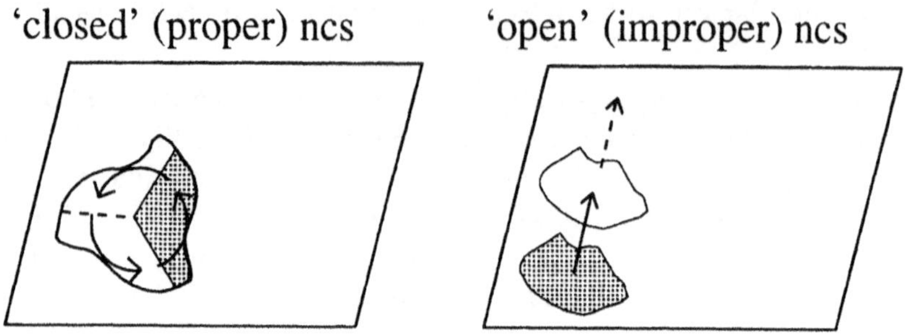

Figure 4. Proper and improper NCS. In the case of proper NCS the mask may cover monomer or multimer, in the case of improper NCS it must cover a monomer.

should reflect this. In the case of a proper symmetry, the masked volume will cover the whole complex (Figure 4).

The situation is further complicated when there are combinations of crystallographic and non-crystallographic symmetries, or of proper and improper symmetries. Another problem occurs when a non-crystallographic symmetry element maps a cell edge onto itself - in this case the volume of matching density is infinite, and arbitrary limits must be placed upon the mask along one crystal axis.

Additional complications arise in the case of flexible proteins. In this case it may be necessary to average only part of the molecule. In other case it may be necessary to apply multi-domain averaging: in this case the protein is divided up into rigid domains which can appear in differing orientations - each domain must then have a separate mask and set of averaging matrices.

4.1. SKELETONISATION

Skeletonisation was originally developed as a tool to assist model building by tracing high ridges in the electron density to describe the connectivity in the map [5]. This skeleton is far easier to interpret to the human eye than wire-frame density contours.

Skeletonisation has more recently been adapted to the problem of density modification [2]. A skeleton is constructed by tracing the ridges in the map. The skeleton is then 'pruned' to select for protein-like features, and then a new map is built by adding backbone-like density along the links of the skeleton.

A variant of skeletonisation is 'peak-picking'. This involves placing atoms in the map in order to account for the electron density. This sort of method was originally applied as the final stage of a direct-methods calculation in order to fit a model to the final map, However, it has recently been employed as a density modification technique as a part of the 'Shake'n'Bake' approach of [13].

4.2. SAYRE'S EQUATION

Sayre's equation is a set of equations relating each structure factor to every other (or equivalently each density value to every other), and is obeyed by any map which is composed of equal, resolved, atoms [9].

Sayre's equation has been used widely and successfully in the solution of small molecule crystal structures. It is limited in proteins by presence of solvent regions which do not display resolved atoms, and the lower resolution of the diffraction data.

When applied to protein structures at lower resolutions the missing high resolution terms mean that the equations no longer hold. Forcing the equations to agree in the absence of the missing terms introduces errors in those terms which are present. Despite these limitations, Sayre's equation often has some power to refine phases, even at protein resolutions.

4.3. GENERAL COMMENTS

In our experience, solvent flattening and histogram matching can safely be applied in almost every case. Averaging is arguably the most powerful constraint on density, if the mask and matrices can be defined. In extreme cases of high order NCS then it is sometimes possible to solve the structure from the coordinates of a few strong scatterers alone. Sayre's equation is time consuming but worth trying when resolution is better then 2.0Å. Skeletonisation works well in some cases but not others, and so is worth trying when other methods fail.

5. Phase Weighting and Combination

Once the map has been modified it is back transformed to give a set of modified magnitudes and phases. This is similar to the case of a partial structure model, which gives rise to a set of structure factors which differ from the structure factors of the complete structure in both magnitude and phase.

If we assume that the difference structure factors arising from atoms missing from the model are independent from the model structure factors, then we can estimate the phase error between the partial and completed models from the mean difference in the magnitudes. This is the foundation of Sim weighting [10], or through a fuller treatment Sigma-a weighting [8].

There is a difficulty when applying this approach in direct methods. In the case of an atomic model, as the model approaches correctness, the model

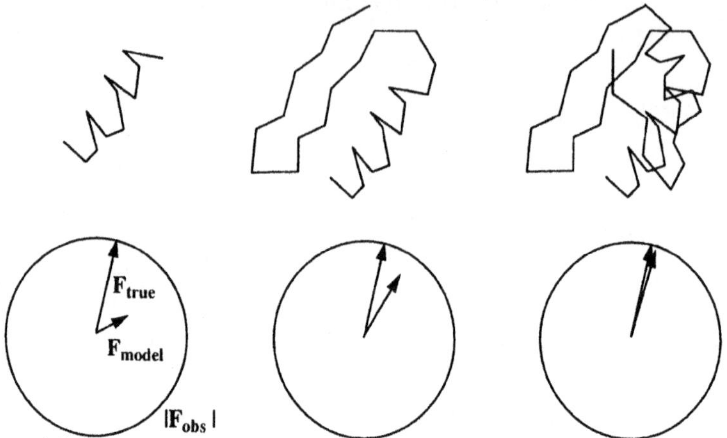

(a) When the F_{calc}'s are from a model, the phases approach their true values with the magnitudes.

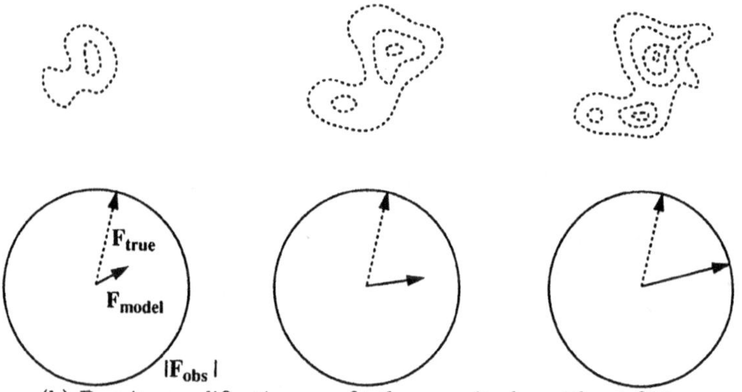

(b) Density modification can fit the magnitudes without fitting the phases.

Figure 5. Sim phase combination.

structure factors approach the true values both in magnitude and phase. In the case of density modification, since the density constraints are typically under determined it is possible to achieve an arbitrarily good agreement between the model magnitudes and their observed values without improving the phases (Figure 5). As a result, phase weights from density modification are typically overestimated.

A number of approaches have been employed to reduce this problem. One is to exclude some reflections from the calculation of the initial map - these reflections then provide a better indication of the power of the density modification to restore magnitudes and phases. This approach is applied in 'dm' in the 'Free Sim' and 'Reflection Omit' schemes [4].

Another approach is adopted by J. P. Abrahams [1] by 'exaggerating' the changes to the map through solvent flipping in order to reduce the dependence between the initial and modified magnitudes. He has shown this to be closely linked with the reflection-omit approach. Vellieux, Hunt, Roy and Read have also developed an effective ad-hoc approach which is applied in the DEMON package [11].

6. Example

As an example, density modification was applied to the structure 5-Carboxy methyl-2-Hydroxymuconate Isomerase, solved by Wigley *et al* [14]. MIR phases were available to 3.7Åwith SIR information to 2.6Å. Density modification was used to improve and extend phases to the limit of the data at 2.1Å. The structure includes 3-fold non-crystallographic symmetry.

A poor section from the initial MIR map is shown, with the resulting maps after phase improvement. The MIR map (Figure 6 (a)) shows low resolution, with poor main chain connectivity and only the largest side-chains visible. Phase improvement was then carried out using solvent flattening, histogram matching, and reflection-omit phase combination, giving (Figure 6 (b)). This map shows considerably better connectivity and some side chains. Finally, the phase improvment calculation was repeated including the 3-fold averaging. The final map (Figure 6 (c)) shows good connectivity with all the side chain density clearly defined.

7. Multi-Crystal Averaging

A related situation is when several crystal forms of the same protein are available. In this case it is possible to run a phase improvement calculation simultaneously on all crystal forms, averaging the protein density between all copies of the protein across all crystal forms. This can be very effective for improving phases, even when phase information is only available in one crystal form. A software package ('dmmulti') for performing this sort of calculation is available from the author.

8. Common Errors

There are a number of common errors which can be made. The first is not understanding the difference between 'modifying a map' and 'improving phases'. The processes described here are for phase improvment, using density modification. It is sometimes desirable to modify a map without performing a phase improvement calculation (e.g. we may wish to start building in the averaged density of a protein rather than in any individual copy).

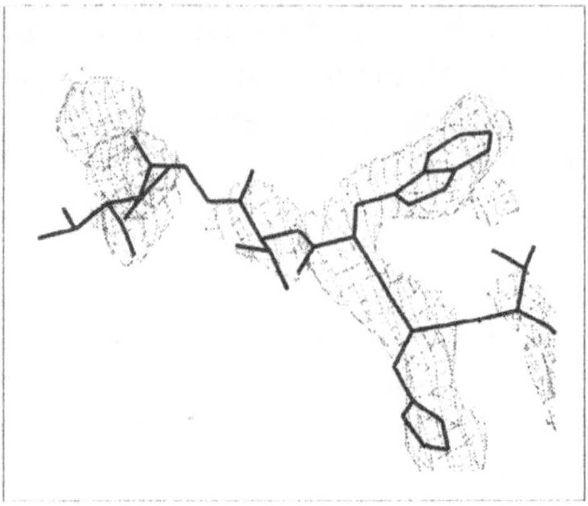

(a) Initial MIR map.

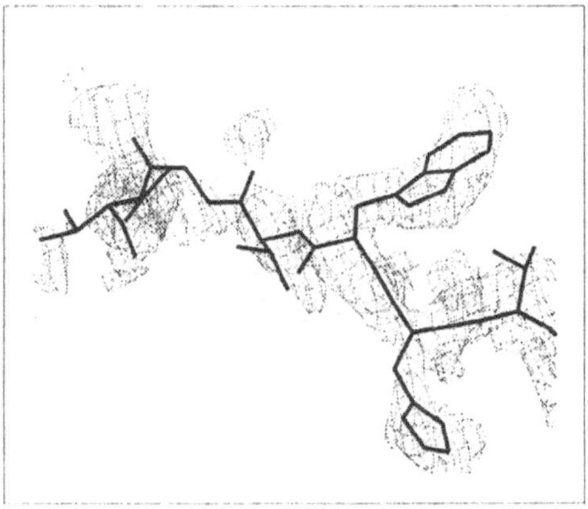

(b) After phase improvement using solvent flattening and histogram matching.

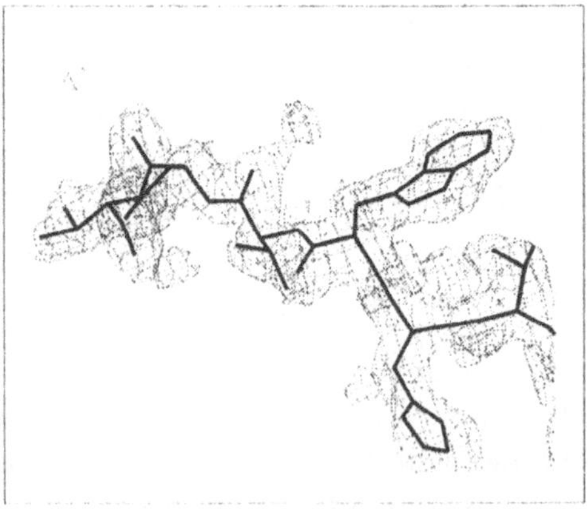

(c) After phase improvement using solvent flattening, histogram matching, and 3-fold averaging.

Figure 6. Maps from CHMI before and after phase improvement (generate using Quanta)

It is also important to be aware of the distinction between solvent and averaging masks - they are not interchangeable. A solvent mask may defined over an asymmetric unit or unit cell, and shares the same symmetry and repeat as the crystal structure. An averaging mask is local - it is defined once in the whole of crystal space, and does not in general obey any crystallographic symmetry or repeat. The definition of an averaging mask is also dependent on the choice of non-crystallographic symmetry operators.

A number of authors have applied phase improvement techniques to improve maps obtained from molecular replacement calculations. In this case it is necessary to generate weights for the model phases using the observed and model magnitudes by a procedure such as Sigma-a. Setting the weights to an arbitrary constant value does not work.

References

1. Abrahams J. P. (1995) Methods used in the structure determination of bovine mitochondrial F1 ATPase *Acta Cryst*, **D52**, pp. 30–42
2. Baker D., Bystroff C., Fletterick R., Agard D. (1994) PRISM: Topologically constrained phase refinement for macromolecular crystallography. *Acta Cryst*, **D49**, pp. 429–439
3. Bricogne G. (1974) Geometric redundancy in intensity data and their use for phase determination *Acta Cryst*, **A30**, pp. 395–405
4. Cowtan K. D., Main P. (1996) Phase combination and cross validation in iterated density modification calculations. *Acta. Cryst.*, **D52**, pp. 43–48
5. Greer J. (1985) Computer skeletonization and automatic electron-density map analysis *Methods in Enzymology*, **115**, pp. 206–224
6. Navaza J. (1994) AMoRe: an Automated Package for Molecular Replacement *Acta Cryst*, **50**, pp. 157–163
7. QUANTA, Molecular simulation inc., 9685 Scranton road, San Diego, CA 92121-3752
8. Read R.J. (1986) Improved Fourier coefficients for maps using phase from partial structures with errors. *Acta. Cryst.*, **A42**, pp. 140–149
9. Sayre D. (1974) *Acta Cryst*, Least squares phase refinement **A30**, pp. 180–184
10. Sim G. A. (1959) *Acta Cryst*, The distribution of phase angles for structures containing heavy atoms **12**, pp. 813–815.
11. Vellieux F. M. D. A. P., Hunt J. F., Roy S., Read R. J. (1995) DEMON/ANGEL: a suite of programs to carry out density modification. *J. Appl. Cryst.*, **28**, pp. 347–351
12. Wang B. C. (1985) Resolution of phase ambiguity in macromolecular crystallography *Methods in Enzymology*, **115**, pp. 90–112
13. Weeks C. M., Hauptman H. A., Chang C. S., Blessing R. H. Teeter M. M., Miller R. (1995) Crambin - a direct solution for a 400-atom structure *Acta Cryst*, **D51**, pp. 33–38
14. Wigley D.B., Roper D.I., Cooper R.A. (1989) Preliminary crystallographic analysis of 5-carboxymethyl-2-hydroxymuconate isomerase from escherichia-coli *J. Mol. Biol.*, **210**, pp. 881–882
15. Zhang K. Y. J., Main P. (1990) The use of Sayre's equation with solvent flattening and histogram matching for phase extension and refinement of protein structures *Acta Cryst*, **A46**, pp. 377–381

AN INTEGRATED GRAPHICAL USER INTERFACE FOR CRYSTALLOGRAPHIC SOFTWARE

JOHN J. IRWIN and ERIC DE LA FORTELLE
MRC Laboratory of Molecular Biology
Hills Road, Cambridge CB2 2QH England

Abstract

A graphical user interface[1] (GUI) for the crystallographic data processing program SHARP[2] has been designed and built from scratch using the hypertext transfer protocol (http). The advantages of this design are as follows. The scripts are easy to write, debug and maintain. The interface is easy for beginners because it uses familiar software such as a web browser and the graphics program "O", as well as extensive hyperlinks which provide context-sensitive help while preparing input and examining output. Results such as graphs and electron density maps can be presented graphically using the mouse, making it a productivity tool for novices and experts alike. In particular, the systematic use of the residual map (log-likelihood gradient map) speeds up and facilitates the improvement of the heavy-atom model by identification of minor sites and anisotropy. Since the software uses the client-server model, it can be run from any networked computer allowing remote control and remote viewing of crystallographic calculations on distant and nearby computers alike.

1. Introduction

Most macromolecular crystallographers do not want to be computer geeks. Instead they want task-oriented software that allows them to solve their structures as quickly as possible without getting bogged down memorising arcane commands or learning a new scripting language. Our desire to make modern task-oriented software led us to design and build a graphical user interface (GUI) following the client-server model, using a web browser as the cornerstone tool.

While there are some macromolecular crystallographers who use computers regularly and can afford to accumulate a detailed knowledge of the software they use, many more are only occasional users who end up re-learning the software each time they use it. We have tried to build our software so that it is both easy to use for beginners, yet unencumbering for productive daily use by "power-users".

This article describes the GUI, highlights some of its novel features, compares our approach with some of the other ways we could have designed it, examines some of the limitations of the current design, and concludes by suggesting how the interface could be improved.

2. Description of the Interface

The Simple to Use SHARP Interface (SUSHI) consists of scripts written mostly in perl[3] and UNIX c-shell (csh). The shell scripts and hypertext markup language (HTML) pages are accessed using a browser as client connecting to a hypertext server such as Apache[4]. The server and its associated applications can be run on any computer running UNIX. The client may be any computer that supports a browser, with the restriction that some of the helper applications such as "O" are only supported on a limited number of platforms.

S. Fortier (ed.), Direct Methods for Solving Macromolecular Structures, 181–186.

From the user's point of view, the main parts of the interface as viewed through the browser are the Control Panel, the input preparation forms, and the logfiles.

2.1 THE CONTROL PANEL

The Control Panel is the first and central page of the Interface. From it, all the main facilities of the software can be accessed. In addition to input preparation and output examination which will be described later, the following facilities are available: job control, providing stop, resume and restart capabilities; setting of preferences on a user-by-user basis to customise the interface to each individual's tastes; a bug reporting facility; a list of frequently asked questions; on-line documentation; tutorials containing examples of both MAD and SIRAS phasing; information about the SHARP discussion group, an on-line bulletin board with over 100 participants who can exchange tips about the program.

The Control Panel also contains links to the UNIX directories `datafiles`, `cardfiles` and `logfiles`. The `datafiles` directory contains the user's file of experimental observations in CCP4's MTZ[5] format. The mtz files are prepared off-line before using SHARP with the help of the CCP4 suite of software. The `cardfiles` directory contains the ASCII input instruction command files for SHARP that are normally not viewed by themselves but entirely prepared and processed by the Interface. The `logfiles` directory contains one sub-directory for each calculation. The project name is suffixed by a monotonically increasing integer to distinguish successive runs on the same project. e.g. The fifth run of a project called PROTEIN would be stored in a directory called PROTEIN.5.

2.2 INPUT PREPARATION FORMS

The input preparation forms are used to produce input for SHARP. The user may elect to begin completely from scratch, or to use the parameters from a previous calculation. In this way a considerable amount of time may be saved by simply making a small change to an existing parameter set . For example, a new heavy-atom site may be added, or an existing one allowed to become anisotropic as a result of examining the residual map of a previous calculation.

The input is managed in a series of HTML forms. Each form or page includes one or more hyperlinks directly to the documentation providing a context-sensitive help facility. The first page contains general information such as project name, space group and type of calculation to be performed. The second page contains a list of all geometric sites (G-sites) to be used in the heavy atom model. Unlike its predecessors, SHARP uses abstract geometric points to which a chemical identity may later be assigned. In this way, only a unique copy of the geometric coordinates of a heavy atom site shared by more than one derivative or at more than one wavelength needs to be refined. Subsequent pages allow the user to navigate through a tree structure which successively defines compounds (C-sites, in which a chemical identity is assigned to geometric sites); crystals (T-sites, in which the "tunable" parameters of occupancy and B-factor are assigned to a chemical site); wavelengths, in which the resolution range of a dataset may be specified; and batches[6] (in which scale, nonisomorphism, and anomalous signal parameters may be specified). This hierarchy is summarised in Figure 1. At the bottom of the batch page, the user may elect to branch to create a new batch at this wavelength, or a new wavelength for this crystal, or a new crystal for this compound, or a new compound altogether. Alternatively, if the heavy-atom model description is complete, the user may choose to end the input preparation and submit the calculation.

Tree Structure	Description/comment
Global Level G-Site-1 X1 Y1 Z1 G-Site-2 X2 Y2 Z2 .... G-Site-n Xn Yn Zn	Geometric Sites (G-Sites) abstract points in space with no chemical identity
Compound Level C-Site-1 = G-Site-i & Atom type C-Site-2 = G-Site-j & Atom type	Chemical identity assigned to point in space In this example, a compound is described containing two heavy-atom sites. A native dataset with no heavy atom would define zero.
Crystal Level	"Tunable" parameters occupancy and B-factor assigned to chemical site.
Wavelength Level	Apply resolution cutoffs to data
Batch Level	Set up parameters for scale, non-isomorphism and scattering behaviour

Figure 1 - The Hierarchical Tree Structure of the SHARP is echoed in the structure of the SUSHI GUI

Before submission takes place, the user is presented with a summary of the input that has just been prepared and is given three choices: submit the calculation now, save these parameters for later editing or submission, or restart the tree traversal in order to make changes. This last option is necessitated by the currently linear fashion in which the tree is navigated using a series of HTML pages. This shortcoming and our planned remedy are discussed further in the "Drawbacks" section.

As each calculation is submitted a new directory is created for it in the `logfiles` directory by monotonically augmenting an integer suffix to the project name. All the files produced by SHARP for this calculation are stored in this UNIX directory. The main output listing is called LIST.html, which is a symbolic link to the most recent output whose name takes the form LIST.<n>.html, where <n> is the number of the run.

2.3 OUTPUT LISTING

The output main listing contained in the LIST.html consists of six parts. 1. Identification of program by version number, date and a literature citation; 2. Confirmation of input as read from the command file and the datafiles; 3. Preparation of data including initial scaling and parameter estimation and calculation of statistics; 4. Output of maximum likelihood refinement, if performed; 5. Residual map calculation if requested; 6. Electron density map calculation and solvent flattening[7] option if requested. A detailed description of each of these sections is beyond the scope of this article. Instead, some of the salient features of the output will now be highlighted.

The output listing contains many hyperlinks and a few HTML forms. The hyperlinks marked "explanation" form a context-sensitive help facility linking directly to the user manual. There are also many hyperlinks containing details of the calculation that it may not be necessary to consult every time the program is run. In this way the main output listing LIST.html is kept as short as possible and contains only essential information. At the same time, many detailed diagnostics are produced which may be easily consulted if desired. Furthermore, some hyperlinks marked "Plot" or "graphical summary" trigger the external helper application `plotmtv`, which is used to display graphs of various parameters, such as non-isomorphism as a function of resolution or Wilson plots.

The output also contains several forms, distinguished by HTML buttons, text input fields and pop-up menus. Using these buttons, residual maps may be displayed, or solvent flattening calculations launched.

3. Summary of features of the Interface

The SUSHI interface makes use of off-the-shelf components that are widely ported, independently maintained, and either free or available at low cost to academics. These third-party applications have a familiar interface that facilitates the adoption of the SUSHI GUI.

The client-server model (Figure 2) allows for the centralisation of calculations on a server, which in turn may dispatch the calculations over a heterogeneous network of UNIX workstations using a queuing system such as DQS[8]. In this way a large computing network may be used efficiently. Because of the client-server character of the Interface, SHARP calculations may be set up and started from any browser on the departmental intranet, or, indeed, any browser on the Internet, if access is permitted by local security measures. Any suitably configured graphics workstation may be used to view the results, allowing for flexible use of departmental graphics resources.

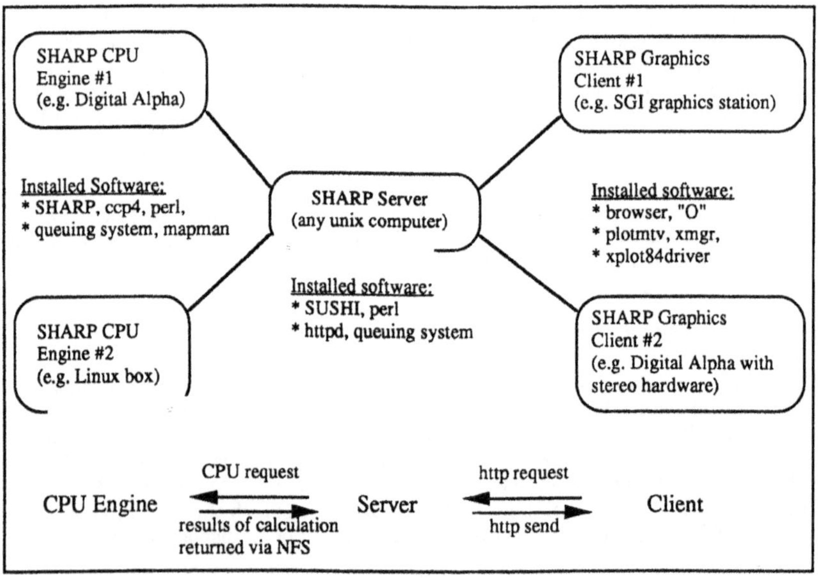

Figure 2 - The Client-Server Architecture of SHARP/SUSHI

The hierarchical structure of the input file is managed by a hierarchical succession of HTML forms which help avoid book-keeping mistakes that might arise if the user were to edit the SHARP input file by hand.

Log-likelihood gradient maps to detect deficiencies in the current heavy atom model can be used systematically and with little effort because all the tedious work of feeding the maps and macros to "O" is done by the Interface. The residual maps are also useful for detecting systematic problems with the data such as incorrect scaling or systematically missing data.

There are also considerable advantages for developers: debugging can be done faster; users may use new version via http prior to its release. When the software is used by remote beta testers connecting via the Internet to the developers' computers, the developers can examine the files produced in great detail, make patches that can be used in minutes without a large binary to transport, thereby accellerating the code/test/debug cycle.

4. Comparison of our approach with some other interface designs

There are a number of other ways in which the SHARP interface could have been built. This section explains what other designs were considered, and ponders the merits of each.

4.1 SHELL SCRIPT INTERFACE

Programs such as CCP4 and TNT are used using scripts that are edited by hand. The advantage of this approach is that editing can be done quickly and the calculation resubmitted rapidly. Indeed, the current design of SUSHI cannot match the speed at which this can be done. See the "Drawbacks" section of below.

The drawback of this approach compared with ours is that users need to write their own scripts, which makes for a steep learning curve. This means that in practice the residual map will only be seen if the user makes an explicit effort to write a script to produce the map, i.e. it will seldom be done.

4.2 ELABORATE COMMAND LANGUAGE

Programs such as X-PLOR[9] and "O", which contain their own scripting language, offer a lot of power to the user and are very popular.

While these systems are ideally suited to daily users, they can be quite difficult for beginners. Indeed, the O program has recently introduced default pull-down menus with prepared scripts while CNS, the successor to X-PLOR, is being designed with a GUI.

4.3 NATIVE INTERFACE IN TCL/TK, MOTIF, OR X11

We could have written an interface in a language that gives more control over the design of the users interface. In particular, HTML forms are notoriously limited in the type of user interaction supported. This is changing rapidly, however, due to immense industry pressure to develop HTML capabilities and the associated technologies JavaScript and Java. When we first wrote SUSHI, neither Java nor JavaScript were available, and HTML did not support mouse events. This has all changed, so that it should now be possible to write an interface that has all the user interface elements of a native application written in Motif, X11 or Tcl/Tk in several different ways.

5. Drawbacks of the current design and how it can be improved

The most notable disadvantages of the current design are the infexible input preparation cascade, the absence of a mechanism for feeding back features in the residual map into the next cycle of heavy-atom model refinement, the lack of control of solvent flattening calculations, and the lack of interactivity currently requiring HTML POST transactions for screen refreshes.

Of these drawbacks, the most widely reported by our users is the first one, i.e. the rigid way in which the Interface requires the complete retraversal of the input, even if only one small detail in the input has changed. We acknowledge this limitation and are actively seeking to redesign this area of the program using Java, HTML frames, JavaScript, and HTML 4.0. We hope to release a new version of the Interface soon that allows for direct access, viewing, and modification of input files. We also have plans to simplify updating the heavy-atom model and controlling solvent flatting more tightly from the GUI.

6. Current Status

SHARP is available free to academics from the BUSTER Development Group site on the World-Wide Web, http://Lagrange.mrc-lmb.cam.ac.uk.

Acknowledgements

This work was carried out on grants from the Howard Hughes Medical Institute (HHMI), Pfizer Inc. and Glaxo-Wellcome plc to Gérard Bricogne. We are grateful to Gérard Bricogne for his guidance and discussions in the course of this work. We would like to thank Digitial Equipment Corporation for a generous loan of state-of-the-art computers.

References

[1]Irwin, J. J and Bricogne, G. (1995) Collected Abstracts of the American Crystallographic Association Annual Meeting, Buffalo, USA; Irwin, J. J. and LaFortelle, E. de (1997) Collected Abstracts of the American Crystallographic Association, Buffalo, USA

[2]LaFortelle, E. de and Bricogne, G (1997) In *Methods in Enzymology*, Vol 276 (C. W. Carter and R. M. Sweet, eds) Chapter 7, pp472-494, Academic Press

[3] Wall, L. and Schwartz, R. L.(1991) *Programming Perl*, O'Reilly and Associates, Sebastopol, CA, USA

[4]Apache is free software available from the Apache Group at http://www.apache.org.

[5]Collaborative Computing Project Number (1994) The CCP4 program Suite: Programs for Protein Crystallography *Acta Cryst.* **D50**, 760-763

[6] A batch is a subset of the data collected in a time slice where all parameters are assumed to be constant

[7]Abrahams, J. P. and Leslie, A. G. W. (1996) Methods used in the Determination of F1ATPase *Acta Cryst.* **D52**, 30-42

[8] Green, T. P & Supercomputer Computations Research Institute (1996) Distributed Queuing System, Florida State University, Tallahassee, Florida, USA

[9]Brünger, A. T. (1992) X-PLOR, A System for X-Ray Crystallography and NMR, Yale University, New Haven, CT USA

ISOMORPHOUS REPLACEMENT: A SURVEY OF RECENT DEVELOPMENTS

A. D. Podjarny, L. Moulinier, F. Tete-Favier* and J. Cavarelli
UPR de Biologie Structurale, IGBMC, B.P.163, 67404, Illkirch, C.U. de Strasbourg, France
**Present address: LCMMMB,BP239,54506 Vandoeuvre-lès-Nancy,France*

1. Introduction

The heavy atom replacement method [1] remains the major tool to phase new macromolecular structures. A derivative is created by adding a heavy atom compound to the macromolecule, while keeping the crystal lattice. This is possible in the case of macromolecular crystals, since they contain solvent channels which allow the diffusion of heavy atom salts. In this case, the electron densities add, and if the reciprocal lattices are sampled at the same points (isomorphous crystals) the following relation is valid:

$$\mathbf{Fph} = \mathbf{Fp} + \mathbf{Fh} \qquad (1)$$

where **Fp**=Fp. exp (iαp) is the macromolecular structure factor, **Fh**= Fh. exp (iαh) is the heavy atom structure factor and **Fph**=Fph. exp (iαph) is the derivative structure factor. To apply the method, Fp and Fph are measured experimentally. A Patterson map is then calculated with the coefficients |Fph-Fp |; it shows the peaks corresponding to the heavy atoms plus noise. These peaks are readily identifiable, provided their signal is significantly higher than the noise. They lead to the calculation of the heavy atom positions and, after the refinement of the heavy atom parameters, to the determination of **Fh**. Equation (1) can then be solved using the amplitudes Fph and Fp. As shown in Figure 1, there are two solutions, symmetrical around **Fh**, which cause a two fold ambiguity for αp (native phase). This ambiguity can be solved by a second derivative.

Because of errors, the circles corresponding to different derivatives will not intersect in a single point. To solve this problem, Blow and Crick [2] proposed a probability distribution for the phase αp ,

$$P(\alpha p) = \exp\left(-\varepsilon(\alpha p)^2/2E^2\right) \qquad (2)$$

based on the assumption that all errors can be treated as residing in Fph, and that the rms value E can be estimated from centrosymmetric projections. This allowed to treat several

S. Fortier (ed.), Direct Methods for Solving Macromolecular Structures, 187–197.

derivatives simultaneously by multiplying the corresponding probability distributions. The technique was initially applied with success to soluble proteins of intermediate size; it works best when the following criteria are met:

1) derivatives are readily obtained;
2) the native and derivative crystals are isomorphous to a resolution around 2-3 Å;
3) the signal of the heavy atom is significantly stronger than the measurement noise;
4) more than one derivative is measured to solve the phase ambiguity.

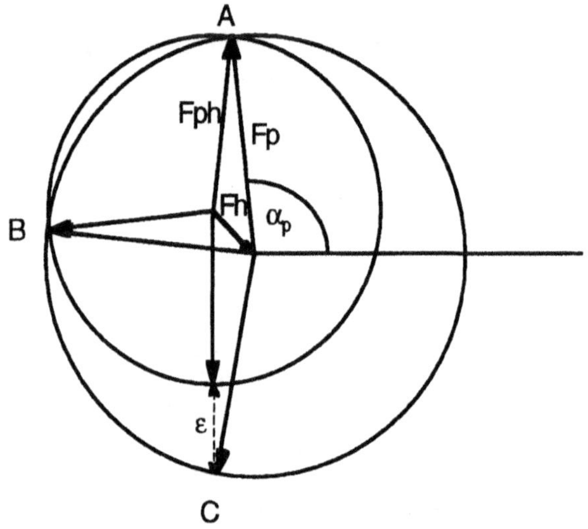

Figure 1. Vector diagram showing the relationship between **Fh**, Fp and Fph. The two circles with radii Fp and Fph intersect in points A and B, giving the two possible solutions of equation (1). At all other points (e.g. C) the triangle formed by **Fph**, **Fh** and **Fp** does not close. The lack of closure difference ε =Fph -|**Fh+Fp**| is a function of the phase αp.

These criteria were met in a large number of cases for the first applications of the method. Since then, the large variety of macromolecules studied pushed the method to its limits, leading to several problems. Specific efforts have been done to solve them.

2. Problems and solutions.

We can qualify the difficulties broadly in two categories, experimental and methodological, and we will give some examples of solutions. Experimental questions concern the preparation of samples, while methodological questions concern the computational algorithms used to find the heavy atoms, refine their parameters and obtain the native phase.

2.1 EXPERIMENTAL

2.1.1 Problem: Difficulty of obtaining derivatives

Solution: Genetically substituted cysteines [3]

This method is based on the affinity of mercury derivatives to the free cysteines in the surface of the molecule. It consists of replacing a surface residue by a cysteine through site-directed mutagenesis . The replaced residue should be structurally analogous to cysteine and likely to be at the surface; serine is a residue of choice, and glutamine, threonine, aspartic and glutamic acid have also been used. This creates a site for mercury derivatives. It also helps further map interpretation, since the site marks a specific residue in the sequence.

2.1.2 Problem: Poor isomorphism

Solution: Use of xenon [4 and references therein]

This methods consists in using xenon, a heavy inert gas, by linking the capillary to a pressurized xenon container with relatively low pressure ($10\text{-}20 \times 10^5$ Pa). Since xenon is soluble in water, it can rapidly diffuse through the solvent channels. Xenon generally occupies pre-existing cavities bound by hydrophobic residues. The interaction with these residues is through weak van der Waals forces, and therefore the derivative is very isomorphous. The occupancy of the heavy atom varies with the applied pressure.

2.1.3 Problem: Poor diffraction power of the derivative, relative to the size of the native molecule

Solution: Use of large clusters [5,6].

This problem has appeared in the solution of very large complexes, like the nucleosome and the ribosome. Clusters of heavy atoms, such as TAMM, undecagold or multi-tungsten compounds were used to boost the diffracting power of the derivative. In the case of the nucleosome, the use of TAMM led to an interpretable map (T. Richmond, personal communication). In the case of the ribosome, the use of W30 clusters led to a SIR map at 12 Å resolution.

2.2 METHODOLOGICAL

2.2.1 Problem: Find the heavy atom positions

This problem appears specially in the following cases:

1) multiple site derivatives;
2) space groups without Harker planes;
3) marginally isomorphous derivatives.

Several algorithms have been developed to deal with this problem, by automating the interpretation of the difference Patterson maps. Some examples follow:
-*HASSP* [7]. This program searches for solutions by systematically moving a trial position in real space and looking at the Patterson value in the corresponding positions in vector space. It does so for one and two trial positions, and correlates the findings in a matrix of self and cross-vectors. A probability is assigned to every solution based on the values of the Patterson function. Table 1 shows such a matrix for the case of aldose reductase, in P1 with 12 Hg atoms per unit cell.
-*GROPAT* [8]. This program also searches systematically for solutions by moving a trial position. Several possibilities are available to assess the quality of the solutions, including the mean, product, minimum, and RMS of pixel values at predicted peaks, as well as probability estimations. The pairwise combination is checked against a list of crossvectors. The general philosophy is to consider initially a large number of ranked solutions and then to apply increasingly stringent filters.
-*SHELXS-90* [9]. This program uses a Patterson vector superposition minimum function, which uses a sharpened Patterson with coefficients $[E^3F]^{1/2}$, where E is a normalized structure factor. To calculate this function, the sharpened Patterson is shifted by a vector **U** (corresponding to a peak position). At each point, the value of the shifted Patterson is compared with the value of the unshifted one, and the lowest one is chosen. The resulting map has peaks only where both Pattersons had peaks, and consists of 2N peaks corresponding to the heavy atom distribution plus its inverted image, with an origin shift. This shift is then determined by imposing the condition that the heavy atom distribution has the space group symmetry. The program then calculates a crossword table which summarizes the sites and the self and cross-vectors, and figures of merit. It is interesting to note that this approach led to the *ab-initio* solution of Crambin by locating the sulfurs as heavy atoms in the native Patterson function.

2.2.2 Problem: Refine the heavy atom parameters [10 and references therein].

The heavy atom structure factor **Fh** is a function of the heavy atom positions $\mathbf{r}_j$, occupancies q_j and temperature factors B_j, as follows:

$$\mathbf{Fh} = \Sigma_j\, q_j \exp(-B_j s^2) \exp(2\pi i\, \mathbf{h}.\mathbf{r}_j)$$

To obtain accurate values of these parameters, a function $R(Fph, Fp, \mathbf{Fh}(q_j, B_j, \mathbf{r}_j), k_i)$ is minimized (k_i represents additional parameters, like the overall scale and temperature factors). Several schemes have been proposed, as listed below.

Heavy atom amplitude refinement : minimizes the difference between $Fh(q_j, B_j, \mathbf{r}_j)$ and $|\mathbf{Fph}-\mathbf{Fp}|$

These methods need an estimate of |**Fph-Fp**|, which can be done restricting the refinement to centric reflections or using the anomalous contribution (FHLE method [11]). The method of using centric zones is quite accurate, since **Fph** and **Fp** are parallel, but it needs an adequate space group and uses only a part of the information. Furthermore, the possibility that **Fph** and **Fp** are of opposite sign (crossover) must be contemplated. The use of the anomalous contribution allows an estimate of |**Fph-Fp**| for acentric reflections, but it depends on the quality of the anomalous data.

Phased refinement: minimizes the lack of closure difference between Fph and $|\mathbf{Fh}(q_j,B_j,\mathbf{r}_j)+\mathbf{Fp}|$

Phased refinements need an estimate of the native phase. Since this phase is dependent on the heavy atom parameters, this leads to bias in the refinement. Classically, this has been treated by phasing with one derivative while refining others; however, the bias can still be introduced by alternative cycles of phasing and refinement or common sites.

Furthermore, all errors were originally grouped in a « lack of closure » (figure 1), and the phase probability distribution was considered a normal distribution of the lack of closure. A more detailed treatment can improve the method by differentiating between different sources of errors, such as: 1) measurement errors in Fp and Fph; 2) lack of isomorphism errors ; 3) errors in the heavy atom model.

Of these, the most important is the lack of isomorphism error. An approach to its estimation has been developed by Dumas [12]. This approach gives a statistical estimate of the contribution to the isomorphous differences of the signal due to the heavy atoms and the noise due to the lack of isomorphism, before the heavy atoms sites have been determined.

The different error sources were analyzed by Read [13], and a maximum likelihood approach was developed [13,14,15,16,17] which in a first approximation solves the problem of the bias introduced by the native phase by integrating over all its possible values[17] leading to a first implementation (MLPHARE). This maximum likelihood approach was then fully developed and programmed by Bricogne and de la Fortelle[18](SHARP). This program has the further advantage of providing an accurate map for improving the heavy atom model. It also links automatically the isomorphous replacement procedure with a density flipping algorithm (SOLOMON) [19].

Another possibility of solving the problem of the native phase is the use of external phases for refinement. For example, a set of external phases can be obtained from a partial molecular replacement solution or from the refinement of an original MIR set with density modification [20] or non crystallographic symmetry averaging methods [21].

Patterson refinement: minimizes the difference between the observed and calculated heavy atom Pattersons

The comparison of observed (coefficients $[(Fph-Fp)^2-<(Fph-Fp)^2>]$) and calculated (coefficients $[(Fh)^2-<(Fh)^2>]$)) origin-removed heavy atom Pattersons is particularly useful in the absence of centric zones. It has been programmed in reciprocal space in the program HEAVY [22]. It does not have the bias of the phased refinement, but it refines a single derivative. Tickle [23] has proposed to compare the peaks in Patterson space (program VECREF), since the (Fph-Fp) Patterson map contains the Fh Patterson with a weight of 0.5 plus a considerably reduced native Patterson and random noise. By restricting the comparison to the peaks most of the noise is discarded.

2.2.3 Problem: Calculate the native phases and accurate estimates of its error

Once the heavy atom model is obtained, the overall probability distribution for the native phase can be derived by multiplying the individual derivative probability distributions (formula 2). From this probability distribution, a centroid phase α_b and a figure of merit m can be derived from the equation

$$m.Fp.\exp(i\alpha_b) = \int Fp.\exp(i\alpha_p)P(\alpha_p)d\alpha_p \qquad (3)$$

This treatment has been improved by the maximum likelihood approaches which generalize the treatment of errors [13,16,18]. In this case, the probability distribution becomes two-dimensional and formula (3) becomes

$$m\,.F_b.\exp(i\alpha_b) = \int Fp.\exp(i\alpha_p)P(Fp,\alpha_p)\,d\,Fp\,d\alpha_p \qquad (4)$$

where $F_b\exp(i\alpha_b)$ represents the best estimate of the native structure factor. A mean figure of merit value above 0.5 usually indicates an interpretable map. However, this value is dependent on the program. Another estimate of the phase quality is the phasing power, defined for each derivative as the ratio of the RMS heavy atom amplitude to the RMS lack of closure error . The phasing power measures the usefulness of a derivative; values of 1.0 are marginal, while values above 2.0 indicate a good derivative.

3. Example SIR: Aldose Reductase in the apoenzyme form [24]

This enzyme crystallizes in space group P1, with four monomers per asymmetric unit, with local symmetry initially assigned as 222. Only one derivative could be obtained. We faced therefore a SIR multisite case without centric zones.

3.1 DERIVATIVE DATA. HEAVY ATOM SEARCH

A search for heavy atom derivatives was conducted. Numerous compounds were tried. Since aldose reductase from pig lens has six cysteines, mercurials were logical candidates. The only effective derivative was indeed a mercury salt, sodium phenylmercurysalicylate (PMS). Since in the space group P1 self-vectors fall in the origin, heavy atom sites have to be identified from cross-vectors. The difference Patterson could not be manually interpreted, since there are 30 peaks above 3σ and 142 peaks above 2σ. Therefore the automatic search procedure HASSP [7] was used. Table 1 shows the result of this search, in which 9 correlated positions were found. All 36 cross-vectors are in positive zones of the Patterson, and 50% of them are above 2σ, with a maximum of 5.4σ. Two other positions were located by residual maps, and the last one was assigned using the 222 local symmetry.

TABLE 1. Result of the cross-vector search on aldose reductase. The values indicate the height of the Patterson at the cross-vector position, scaled to 1σ=1000. Peak numbers following the final assignment.

Site	1	4	5	6	7	8	10	11	12
1	-	2697	2422	4205	1671	5275	4432	2928	3637
4	2697	-	1393	2277	1242	68	1903	1377	1694
5	2422	1393	-	2043	1231	5418	1027	970	850
6	4205	2277	2043	-	4858	2484	3867	1679	2957
7	1671	1242	1231	4858	-	840	1496	1478	1385
8	5275	68	5418	2484	840	-	2698	1433	1717
10	4432	1903	1027	3867	1496	2698	-	3040	3847
11	2928	1377	970	1679	1478	1433	3040	-	2235
12	3637	1694	850	2957	1385	1717	3847	2235	-

3.2 HEAVY ATOM REFINEMENT

These heavy atom positions were refined using the program HEAVY, which is specially designed for the SIR case, and the result is shown in Table 2, values Occ_{hea} and B_{hea}.

The heavy atom parameters were also refined with the programs MLPHARE and SHARP, to test their ability to overcome the problems of the phased refinement in the SIR case. The results in terms of occupancy and temperature factor are given in Table 2, which shows that both programs provide reasonable estimates of the occupancy and of the temperature factors of the strongest sites. However, the behavior of these two programs was different. SHARP was able to refine occupancies, then B's, and then both simultaneously. MLPHARE needed the two refinements to be kept separate; if not, the values diverged. This problem is probably linked to the limited resolution of the data, which makes difficult to obtain reliable temperature factors for weak sites.

TABLE 2. Values of occupancy (Occ, in an arbitrary scale) and temperature factor (B) obtained using the program HEAVY (hea), MLPHARE (mlp) and SHARP (shar), and after re-refinement against averaged phases (re-ref). **The last ones are taken as the best estimation**. The sites are classified by monomer (ABCD) and ordered from strongest (1) to weakest (3).

Monomer	Site	Occ hea	Occ mlp	Occ shar	**Occ re-ref**	B hea	B mlp	B shar	**B re-ref**
A	1	13.1	11.8	12.8	**13.7**	21.3	13	22.9	**17.8**
	2	5.5	4.7	4.3	**5.6**	22.5	10.5	10	**18.2**
	3	3.5	3.2	3.3	**3.4**	16.7	15.1	16.7	**15.0**
B	4	8.3	7.3	7.3	**8.2**	16.6	17.6	14	**18.0**
	5	6.3	4.2	5.3	**4.8**	22.4	28.8	49.2	**20.2**
	6	6.5	6.6	7	**6.5**	15.0	15.8	21.1	**14.5**
C	7	9.3	9.7	8.9	**11.1**	18.3	17	16.7	**17.3**
	8	5.4	4.4	4.5	**4.6**	17.5	28.9	47.5	**20.0**
	9	3.6	3.9	5.25	**4.2**	20.7	18.6	33.1	**17.5**
D	10	9.9	7.1	8.8	**6.8**	21.3	2.2	22.2	**21.4**
	11	5.1	4.8	5	**5.3**	20.2	23.5	24.7	**20.0**
	12	3.3	5.5	5.6	**5.6**	14.5	16.9	24.5	**17.7**

3.3 SIR PHASING AND HEAVY ATOM RE-REFINEMENT

The heavy atom data and parameters were used to calculate phases in the 30-3.5 Å range using the program PHARE from the CCP4 suite [25]. *A-posteriori* comparison of the SIR phases with the current model showed that the mean phase error was 70.6°. Phases were then improved by non crystallographic symmetry averaging, which led to a partially interpretable map. Using the partial model as a rigid body, refinement of the four independent monomers was performed using data between 15 and 6 Å resolution. This refinement displaced significantly the positions and orientations of the local axes and changed it from an exact 222 symmetry to a pseudo $2_t2_t2_t$ one, where the axes are not exact 2 folds, they have a translation component in the direction of the axis and they do not intersect.

A new averaging procedure was started with the new symmetry and original heavy atom parameters from HEAVY. In order to improve the starting point, map averaging at 3.5 Å was combined with re-refinement of heavy atom parameters against averaged phases [19]. This changed the initial heavy atom parameters, in particular the occupancies (see table 2, values $Occ_{re\text{-}ref}$ and $B_{re\text{-}ref}$), and decreased the SIR phase error by 2°. However, the effect is more important after averaging. The use of the re-refined SIR parameters decreased the phase error of the averaged phases from 43° to 38°, and the maps became more continuous. Finally, a model was built and refined to an R-factor of 0.22 at 2.5 Å resolution.

4. Example MIR with mutated cysteines :
The case of *Staphylococcus aureus* epidermolytic toxin A (ETA)

ETA crystallises in space group $P2_1$ with unit cell dimensions a = 49.09 Å, b=66.40 Å, c= 81.77 Å, β=93.92° and two molecules in the asymmetric unit. For solving the phase problem, many heavy atom derivatives were tested without any statistical differences with the native data set, except for two of them (see below). To overcome this crucial step, two mutated proteins have been purified and crystallized in very similar conditions: one where Ser195 has been substituted by Cys (Ser195Cys), the other one where Ser147 has been substituted by Cys (Ser147Cys).

Initial phases were determined by multiple isomorphous replacement using 2 derivatives (Mersalyl and K2AuCl4) for the native protein, 2 derivatives for the mutant Ser195Cys (EtHgCl, DiHgCL) and one derivative (PCBMS) for the mutant Ser147Cys (Table 2). Heavy atom binding sites were determined by difference Patterson and Fourier difference maps. Four derivatives (Mersalyl, K2AuCl4, EtHgCl, DiHgCl) have a common major binding site which corresponds to the serine protease active site (Ser195, Asp120, His72). Refinement of the heavy atom parameters was performed by the maximum likelihood approach as coded in the program SHARP [17]. The calculated phases gave an overall figure of merit of 0.622 for data between 27.0Å-3.3 Å resolution Similar results were obtained using the program MLPHARE. The SHARP result was improved by solvent flipping using the program SOLOMON [18]. This map was already interpretable. Further phase refinement was done using solvent flattening, and a model was built and refined (R-factor = 0.18, R-free= 0.23).

Figure 2 shows the phase error (a) and the correlation of phase error and figure of merit (b) before (curve A) and after (curve B) the application of SOLOMON. Even when there are 5 derivatives, as 4 of them share a major binding site the MIR phasing quality is not sufficient for direct map interpretation. The application of the solvent flipping algorithm improves the situation, both in terms of phase error and quality of the FOM as an estimate of phase error. A second run of SHARP and SOLOMON at 3.5Å resolution was done as a test case; the mean phase error improved between the two runs by 3.3° after SHARP and by 7.3° after SOLOMON (fig. 2, curve C).

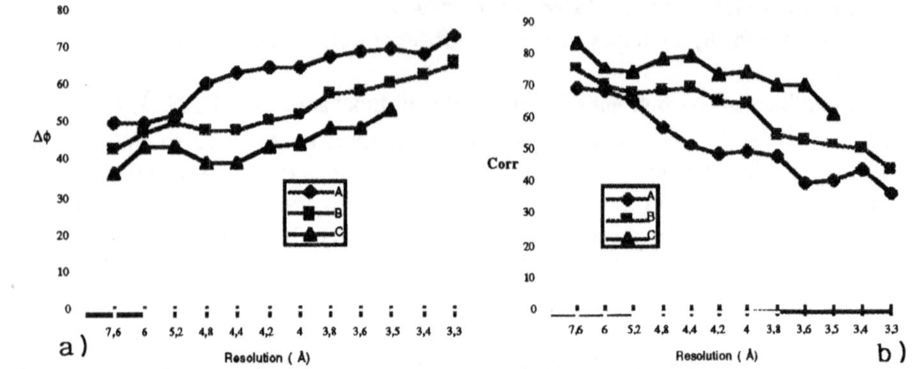

Figure 2. a) Phase error in ° (compared with the current model) after the first run of SHARP (curve A), the first run of SOLOMON (curve B) and the second run of SHARP plus SOLOMON (curve C).
b) Correlation (Corr) in %, between phase error and figure of merit, for the same cases as (a).

5. Conclusions

While the basis of the isomorphous replacement method remains the same as 40 years ago, the method is still evolving in order to tackle larger and more difficult problems. The crystallographer has now a panoply of possible techniques to be adapted to the particularities of their problem. It is also clear that the MIR technique provides a starting set of phases, which are then refined by diverse density modification or averaging procedures. In this sense, the accurate estimate of the MIR phases and MIR phase errors becomes critical, as a small improvement in phase quality can be amplified by the refinement procedures, *which also rely on the error estimates.* A large part of the methodological effort has been aimed at estimating the errors properly. Nevertheless, the main problem remains obtaining derivatives, and the main limit to the phase quality is the lack of isomorphism. The MAD techniques, treated elsewhere in this book, are providing the answer.

Acknowledgments

We thank A. Urzhumtsev for his careful reading of the manuscript and his interest, and B.Rees and D.Moras, for useful discussions and their continuing interest and encouragement. This work was supported by the Centre National de la Recherche Scientifique (CNRS) through the UPR 9004, by the Institut National de la Santé et de la Recherche Médicale, by the Centre Hospitalier Universitaire Régional.

References

1. Green, D.W., Ingram, V.M. and Perutz, M.F. (1954) The structure determination of hemoglobin : IV. Sign determination by the isomorphous replacement method, *Proc. R. Soc. London*, **A225**, 287-307.
2. Blow, D.M. and Crick, F.H.C. (1959) The treatment of errors in the isomorphous replacement method, *Acta Cryst.* **12**, 794-802.
3. Dao-Pin, S., Alber, T., Bell, J.A., Weaver,L. and Mathews, B.W. (1987) Use of site-directed mutagenesis to obtain isomorphous heavy-atom derivatives for protein crystallography: cysteine-containing mutants of phage T4 lysozyme, *Protein Engineering* **1**, 115-123.
4. Schiltz, M., Prangé, T. and Fourme, R. (1994) On the preparation and X-ray data collection of isomorphous Xenon derivatives, *J. Appl. Cryst.* **27**, 950-960.
5. Thygesen, J., Weinstein, S., Franceschi, F. and Yonath, A. (1996) The suitability of multi-metal clusters for phasing in crystallography of large macromolecular assemblies, *Structure* **4**, 513-518.
6. O'Halloran, T.V., Lippard, S.J., Richmond, T.J. and Klug, A. (1987) Multiple heavy atom reagents for macromolecular X-ray structure determination: application to the nucleosome core particle, *J. Mol. Biol.* **194**, 705-712.
7. Terwilliger, T. C. and Eisenberg, D. (1987) Generalized method of determining heavy atom positions using the difference Patterson function, *Acta Cryst.* **A43**, 1-5.
8. Jones, Y. and Stuart, D. (1991) Locating heavy atom sites by automatic Patterson search-GROPAT, in W. Wolf, P.R. Evans and A.G.W. Leslie (eds.), *Isomorphous Replacement and Anomalous Scattering*, CCP4, SERC, Daresbury, pp. 39-48.
9. Sheldrick, G.M. (1991) Heavy atom location using SHELXS-90, in W. Wolf, P.R. Evans and A.G.W. Leslie (eds.), *Isomorphous Replacement and Anomalous Scattering*, CCP4, SERC, Daresbury, pp. 23-38.
10. Evans, P.R. (1991) Refinement of heavy atom parameters and isomorphous phasing , in W. Wolf, P.R. Evans and A.G.W. Leslie (eds.), *Isomorphous Replacement and Anomalous Scattering*, CCP4, SERC, Daresbury, pp. 49-59.
11. Dodson, E.J. (1976) A comparison of different heavy atom refinement procedures, in F.R. Ahmed (ed.), *Crystallographic Computing Techniques*, Munskgaard, Copenhagen, pp. 259-268.
12. Dumas, P. (1994) The heavy atom problem: a statistical analysis. I. A Priori determination of best scaling, level of substitution, lack of isomorphism and phasing power. *Acta Cryst.*. **A50**, 526-537.
13. Read, R.J. (1991) Dealing with imperfect isomorphism in multiple isomorphous replacement, in W. Wolf, P.R. Evans and A.G.W. Leslie (eds.), *Isomorphous Replacement and Anomalous Scattering*, CCP4, SERC, Daresbury, pp. 69-79.
14. Einstein, J.R. (1977) An improved method for combining isomorphous replacement and anomalous scattering diffraction data, *Acta Cryst.* **A33**, 75-85.
15. Terwilliger, T. C. and Eisenberg, D. (1987) Isomorphous replacement: effect of errors on the phase probability distribution, *Acta Cryst.* **A43**, 6-13.
16. Bricogne,G. (1991) A maximum likelihood theory of heavy atom parameter refinement in the isomorphous replacement method , in W. Wolf, P.R. Evans and A.G.W. Leslie (eds.), *Isomorphous Replacement and Anomalous Scattering*, CCP4, SERC, Daresbury, pp. 60-68.
17. Otwinowski, Z. (1991) Maximum likelihood refinement of heavy atom parameters, in W. Wolf, P.R. Evans and A.G.W. Leslie (eds.), *Isomorphous Replacement and Anomalous Scattering*, CCP4, SERC, Daresbury, pp. 80-86.
18. de la Fortelle, E. and Bricogne, G. (1997) Maximum likelihood heavy atom parameter refinement for multiple isomorphous replacement and multiwavelength anomalous diffraction methods. in C.W. Carter Jr. and R.M. Sweet, *Methods in Enzymology Vol 276,* Academic Press, New York , pp. 472-493.
19. Abrahams, J.P. and Leslie, G.W. (1996) Methods used in the structure determination of bovine mitochondrial F1 ATPase, *Acta Cryst.* **D52**, 30-42.
20. Cura, V., Khrishnaswamy, S. and Podjarny, A. D. (1992) Heavy atom refinement against solvent-flattened phases, *Acta Cryst.*, **A48**, 756-764.
21. Tête-Favier, F., Rondeau, J.-M., Podjarny, A.D. and Moras D.(1993) The structure determination of aldose reductase: joys and traps of local symmetry averaging, *Acta Cryst.* **D49**, 246-256
22. Terwilliger, T.C. and Eisenberg, D. (1983) Unbiased three dimensional refinement of heavy atom parameters by correlation of origin removed Patterson functions, *Acta Cryst.* **A39**, 813-817.
23. Tickle, I.J (1991) Refinement of single isomorphous replacement heavy atom parameters in Patterson vs. reciprocal space, in W. Wolf, P.R. Evans and A.G.W. Leslie (eds.), *Isomorphous Replacement and Anomalous Scattering*, CCP4, SERC, Daresbury, pp. 87-95.
24. Rondeau, J-M., Tete, F., Podjarny, A. D., Reymann, J.M., Barth, P., Biellmann, J.F and Moras, D. (1992), Novel NADPH binding domain revealed by the crystal structure of Aldose Reductase, *Nature* **355**, 469-47 .
25. Collaborative Computational Project, Number 4 (1994) *Acta Cryst.*, **D50**, 760-763.

A THEORETICAL STUDY ON THE MEASURABILITY OF BIJVOET DIFFERENCES IN MACROMOLECULAR CRYSTALS

S.PARTHASARATHY AND V.SWARNALAKSHMI
Department of Crystallography & Biophysics
University of Madras, Guindy Campus
Madras 600 025, INDIA.

1. Introducation

The Anomalous Scattering (AS, hereafter) method is well suited to proteins, since protein crystals are always non-centrosymmetric and, in principle, it is possible to prepare suitable heavy atom derivatives of proteins. The success of AS method strongly depends on the measurability of Bijvoet differences (MBD, hereafter). A theoretical study of MBD helps us in two ways: (i) It enables us to understand the relative importance of the various factors that influence the measurability. (ii) It also enables us to know ***a priori*** (i.e. before data collection) the percentage of reflections whose Bijvoet ratios could be expected to be greater than a given value for any heavy atom derivative (HAD, hereafter) with a radiation of any desired wavelength. This, in turn, helps us in the proper choice of the heavy atom for preparing the HAD and for choosing an appropriate wavelength for data collection.

2. Definition of MBD at Any Required Level

We shall use ***PH*** to denote a HAD in which the protein part is denoted by ***P*** and the heavy atom part by ***H***. Let n_P be the number of light atoms (excluding the hydrogens) and let n_H be the number of heavy atoms per protein molecule which we take to be the asymmetric unit.The scattering factor of

S. Fortier (ed.), Direct Methods for Solving Macromolecular Structures, 199–209.

the heavy atom under anomalous scattering can be written as $f = f^{o} + f' + if''$ where f^{o} is the normal scattering factor and f' and f'' are the real and imaginary dispersion corrections. The intensities of the inverse reflections **h** and -**h** become unequal under AS and their difference ΔI [$= I(\mathbf{h}) - I(-\mathbf{h})$] is called the Bijvoet difference. The Bijvoet ratio X for the inverse reflections **h** and **-h** is defined to be

$$X = |\Delta I|/\overline{I} \tag{1}$$

where $\overline{I}$ ($=0.5[I(\mathbf{h}) + I(-\mathbf{h})]$) is the mean intensity of the inverse reflections. In the measurability studies, only the magnitude of the Bijvoet ratio is relevant. Theoretical Expression for ΔI and $\overline{I}$ can be shown to be [1]

$$\Delta I = 4\,|F_P(\mathbf{h})|\,|F''_H(\mathbf{h})|\sin\psi \tag{2}$$

$$\overline{I} = |F'_{PH}(\mathbf{h})|^2 + |F''_H(\mathbf{h})|^2 \tag{3}$$

where $F_P(\mathbf{h})$, $F'_{PH}(\mathbf{h})$ and $F''_H(\mathbf{h})$ are respectively the contributions to the structure factor of the reflection **h** due to (i) the P-atoms, (ii) total real part of the atomic scattering factors of the atoms in the PH-group and (iii) the imaginary part of the scattering factor of the H-group of atoms. ψ is the angle between $F_P(\mathbf{h})$ and $F'_H(\mathbf{h})$.

Using (2) and (3) in (1) and the result $|F''_H| = k|F'_H|$ which is valid when all the anomalous scatterers are of the same type, we get

$$X = \frac{4k\sigma_1\sigma_2 y'_H y_P\,|\sin\psi|}{\sigma_2^{\,2} y_P^{\,2} + \sigma_1^{\,2}(1+k^2)y_H'^{\,2} + 2\sigma_1\sigma_2 y'_H y_P\cos\psi} \tag{4}$$

where

$$k = f''_H/(f^{o}_H + f'_H) \tag{5}$$

$$\sigma_1^{\,2} = 1 - \sigma_2^{\,2} = \langle|F'_H|^2\rangle / \langle|F'_{PH}|^2\rangle \tag{6}$$

$$y'_H = |F'_H|/<|F'_H|^2>^{1/2} \; , \; y_P = |F_P|/<|F_P|^2>^{1/2} \qquad (7)$$

Let y_t be the threshold value (i.e. the least observable value) of the normalized structure factor magnitude in a given $(\sin\theta)/\lambda$-range. A Bijvoet ratio can be experimentally measured only if the intensities of the reflections **h** and **-h** are both above the threshold of observation. If y_{min} is the minimum of the quantitites $y_{PH}(\mathbf{h})$ and $y_{PH}(-\mathbf{h})$, then the above requirement for the measurability of the intensities can be written as $y_{min} > y_t$.

Let X_o be a particular value of X. Let G be the set of reflections for which $X > X_o$ and at the same time $y_{min} > y_t$. That is, $G = \{(X > X_o) \cap (y_{min} > y_t)\}$ where the symbol $\cap$ denotes the set-theoretic intersection operation. The probability of occurrence of the event G represents the fractional number of reflections whose intensities can be measured and for which the Bijvoet ratios are greater than X_o and we shall refer to this probability as the MBD at the level X_o. We shall denote this probabiliity by the symbol $M(X_o, y_t)$. That is,

$$M(X_o, y_t) = \Pr\{(X > X_o) \cap (y_{min} > y_t)\} = \Pr(G) \quad (8)$$

Since a study of the normalized intensity data of a few crystals shows that y_t can be chosen to be 0.3 [2] and since a Bijvoet ratio of 0.1 can be easily measured, we shall refer to M(0.1, 0.2) as the MBD in the rest of this article and denote it by the symbol m. That is,

$$m = \Pr\{(X > 0.1) \cap (y_{min} > 0.3)\} \qquad (9)$$

3. Evaluation of m for Any Situation

In the Foster-Hargreaves classification of crystals of the triclinic, monoclinic and orthorhombic systems [3] into 7 space group categories (SGC, hereafter), only the categories 1, 3 and 5 are relevant to protein crystals. The theoretical expressions of $M(X_o, y_t)$ for these have been obtained in the literature and these expressions involve the parameters k and σ_1^2. These expressions also differ for different values of n_H and different SGC. These expressions are multidimensional integrals which could be evaluated only by the Monte Carlo method. Sekar and Parthasarathy [4] have obtained Tables of values of $M(X_o, y_t)$ as a function of k and σ_1^2 for different values of X_o for these three SGC and for n_H = 1, 2, 3 and 4.

3.1 PROPERTIES OF σ_1^2 AND k

From (6) we obtain

$$\sigma_1^2 = \frac{n_H \, (f_H^o + f_H')^2}{n_H \, (f_H^o + f_H')^2 + \Sigma_{j=1}^{P} \, f_{pj}^2} \tag{10}$$

From (10) it is seen that σ_1^2 depends on the size of the protein (i.e. n_p) , the number of heavy atoms/asymmetric unit (i.e. n_H), the wavelength of the radiation (through f_H') and sin θ/λ (through f_H^o and f_{pj}). Since σ_1^2 increases with sin θ/λ, a mean value $<\sigma_1^2>$ is to be used to characterize a given HAD for a given resolution and such a mean value for the reflections in $0 < \sin\theta/\lambda < S_{max}$ can be obtain by numerical integration.

For proteins of molecular weight (MW, hereafter)more than 25000 daltons, the values of $<\sigma_1^2>$ are less than 0.2. For a given protein, the value of σ_1^2 can be increased by using a heavy atom of high atomic number (e.g.U) to prepare the HAD and / or by choosing a HAD in which a large number of sites are occupied by heavy atoms.

From (5) it is seen that the value of k depends on the heavy atom type and λ. Since k increases with sin θ/λ, a mean value <k> is to be obtained for a given heavy atom type for any particular λ and resolution. The mean value of k for reflections in the ragne $0 < \sin\theta/\lambda < S_{max}$ can be obtained by numerical integration. Values of <k> for typical elements are given in Table 1.

TABLE 1. Average values of k (in %) for typical elements corresponding to 2Å resolution

(i) Values (in %) for CuKα

Fe	Co	I	Cs	Ba	Pr	Sm	Gd	Pt	Au	Hg	Pb	U
16.7	19.0	16.1	18.2	19.3	23.3	27.0	27.2	11.2	11.6	12.0	13.0	18.5

(ii) Values (in %) for the radiation of wavelength (λo in Å) at which the value of <k> is the largest

	Co	Cs	Pr	Sm	Gd
k	32.0	58.3	10.64	85.9	116.6 %.
λo	1.605	2.4722	2.0789	1.8448	1.7104 Å

It is found that Sm and Gd have the largest value of <k> for CuKα and that the value of <k> can be made close to 0.9 for Pr, Sm and Gd by choosing a proper wavelength with Synchrotron radiation (SR, hereafter). The value of <k> can be increased by choosing an appropriate heavy atom and/or wavelength.

3.2 EVALUATION OF m FOR ANY REQUIRED SITUATION

Since m is an implicit function of sin θ/λ, it is necessary to obtain an average value of m (<m>, say) appropriate to the resolution. The value of <m> can be obtained using bilinear interpolation on the values of m available in the literature as a function of k and σ_1^2 [5].

4. Effect of Various Factors on MBD

4.1 EFFECT OF σ_1^2

For a given k, MBD steeply increase with σ_1^2 in the region $0 < \sigma_1^2 < \alpha$ (where $\alpha = 0.15$ for $k = 0.25$ and $\alpha = 0.05$ for $k = 0.8$) and the increase is not significant in the region $\alpha < \sigma_1^2 < 0.5$ [see Figure 1(i)].

4.2 EFFECT OF k

For a given σ_1^2, MBD increases steeply with k in the region $0 < k < 0.3$ and thereafter the increase is at a much smaller rate. [see Figure 1(ii)].

4.3 EFFECT OF SPACE GROUP SYMMETRY

(i) When $n_H = 1$, MBD is the largest for SGC 1 (the other conditions being the same); it becomes smaller but similar for the SGCs 3 and 5 [see Figure 2(i)]. (ii) When $n_H \geq 2$, MBD is practically independent of the SGC [see Figure 2(ii)].

4.4 EFFECT OF NON-RANDOMNESS IN THE ATOMIC DISTRIBUTION

So far we have considered the P-atoms to be randomly distributed in the unit cell. However protein moelcules have rich secondary structure (e.g. α-helix, β-sheet) and this might affect the MBD. No theoretical result is available for such a situation. However the values of m available as a function of the degree of centrosymmetry of the molecule for a crystal of space group P1 in which $n_H = 1$ shows that (for the typical situation $\sigma_1^2 = 0.05$, $k = 0.24$) even if 40% of the atoms in the molecule form a single ideally centrosymmetric group, the reduction in MBD is only marginal [6]. We may extrapolate this result and state that the internal ordering in the protein molecule in the from of secondary structure may not reduce MBD in a significant way.

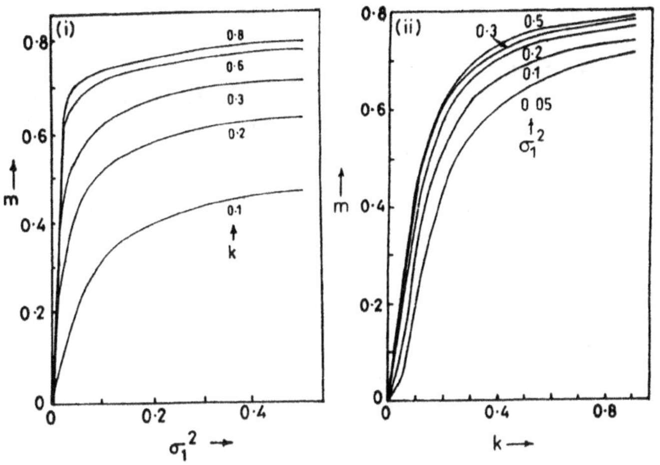

Figure 1: Variation of m as functions of σ_1^2 and k. (i) m as a function of σ_1^2 for different fixed values of k. (ii) m as a function of k for different fixed values of σ_1^2.

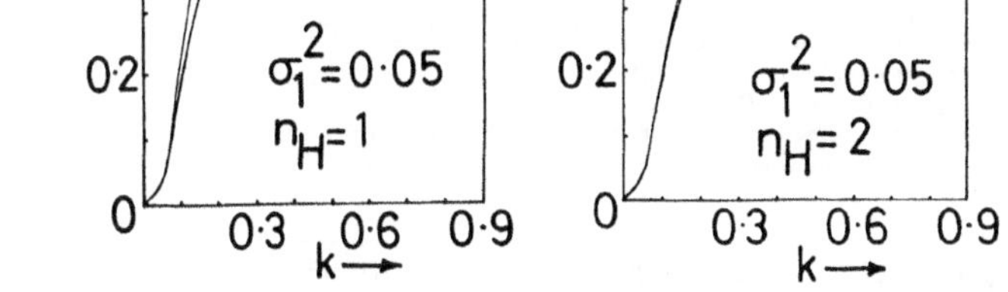

Figure 2 : Variation of m as a function of k for SGCs 1, 3 and 5 corresponding to σ_1^2 = 0.05. (i) - curves for n_H = 1. (ii) - curves for n_H = 2.

5. MBD Expected with CuKα Radiation

From the theoretical values of m for HADs of proteins calculated for CuKα radiation for Fe, Co, I, Cs, Pr, Sm, Pt, Au, Hg, Pb and U (see Table 2), the following results are obtained : (i) For a given number of anomalous scatterers per protein Sm, Gd and U derivatives have the largest measurability and the measurabilities for these derivatives are practically the same. (ii) Pt, Au, Hg and Pb are not good anomalous scatterers for CuKα radiation. (iii) A measurability close to 50% may be obtained with CuKα radiation even for a large protein, provided a HAD conatining more than two atoms of U (or Sm or Gd) per protein molecule could be prepared.

TABLE 2. MBD (in %) expected in heavy atom derivatives of typical proteins containing n_H =1,2,3 or 4 heavy atoms: Radiation:CuKα, Resolution: 2Å, SGC:3.

Protein	n_H	Fe	Co	I	Cs	Ba	Pr	Sm	Gd	Pt	Au	Hg	Pb	U
I	1	21	26	44	48	50	55	59	58	41	42	43	46	57
	2	28	33	50	54	56	61	64	64	44	46	46	49	59
	3	39	43	56	60	61	66	69	69	49	50	51	54	64
	4	40	44	56	60	61	65	68	68	48	49	50	52	65
II	1	21	25	30	35	38	45	50	49	29	31	33	36	50
	2	19	23	38	43	46	53	57	56	37	38	40	42	55
	3	23	28	48	52	55	60	63	63	44	45	47	49	61
	4	25	29	49	53	55	60	64	63	44	45	46	49	61
III	1	21	25	20	25	27	34	40	40	19	21	23	26	42
	2	18	23	26	32	35	43	49	48	27	28	30	33	49
	3	21	26	37	42	45	51	56	55	35	37	39	42	56
	4	19	23	39	44	46	53	57	56	37	39	40	43	56
IV	1	21	25	20	24	26	33	38	39	14	15	16	19	34
	2	18	23	20	25	28	36	41	41	21	22	24	27	42
	3	20	25	29	35	37	45	50	49	29	30	32	36	52
	4	19	23	33	38	40	47	52	52	31	33	35	38	52

Data for proteins I - IV are taken from Protein Data Bank (Bernstein et al., 1977).
I - Lectin IgE receptor (Mouse) (MW: 12084)
II - Blood coagulation factor XA (Human) (MW: 29820)
III - Human Growth Hormone complexed with its receptor (EC domain) (MW: 60918)
IV - Lipase (Triacyl Glycerol Hydrolase) (Horse pancreas) (MW: 92596)

6. MBD Expected under Pronounced Anomalous Scattering Effect Obtained with SR

Using the measured dispersion correction data of Templeton and coworkers [8] - [10] for Cs, Pr, Sm and Gd, the values of m were computed for typical HADs under pronounced AS effect obtained with SR (Table 3). It is found that, for a given n_H, the

TABLE 3. MBD (in %) expected in HADs of typical proteins
SGC:3, $n_H = 1$, Resolution:2Å

Protein	Cs			Pr			Sm			Gd		
	(i)	(ii)	p	(i)	(ii)	p	(i)	(ii)	p	(i)	(ii)	p
I	48	65	35	55	81	47	59	72	22	58	88	52
II	35	60	71	45	79	76	50	67	34	49	86	76
III	25	60	140	34	79	132	40	67	68	40	86	115
IV	24	60	150	33	79	139	38	67	76	39	86	121

(i) – Values for CuKα. (ii) – Values for a radiation obtainable with SR at which the value of k is the largest

p =Percentage increase in MBD for (ii) relative to (i)
=100*[MBD (ii)-MBD (i)] /MBD (i)

increase in m (relative to CuKα) occurring under pronounced AS effect obtained with SR becomes larger as the size of the protein increases. Thus, purely from the point of view of MBD, the advantage of SR is more for a large protein than for a small one.

It is found that the percentage of reflections with large Bijvoet ratio ($X > 0.5$, say) increases systematically as k increases (see Table 4). Thus a large value of k increases the power of the AS technique of phase determination.

Table 4. Values of m_{10} and m_{50} (in %) as functions of k and σ_1^2
$m_{10} = P[(X > 0.1) \cap (Y_{min} > 0.3)]$, $m_{50} = P[(X > 0.5) \cap (Y_{min} > 0.3)$

σ_1^2	m	k: 0.2	0.3	0.5	0.6	0.9
0.05	m_{10}	44	55	68	71	72
	m_{50}	1	5	22	30	34
0.10	m_{10}	51	62	71	73	74
	m_{50}	4	11	31	39	42
0.15	m_{10}	56	65	73	75	76
	m_{50}	7	17	36	43	46

7. Conclusion

From the above considerations, it is clear that, by a judicious choice of the type of heavy atom as well as increasing their number per asymmetric unit, a measurability as high as 50% may be obtained with CuKα radiation even in a large protein. Furthermore, by obtaining enhanced AS effect with SR, the phase determination procedure may be made more effective due to the availability of a sufficiently large percentage of reflections with large values of Bijvoet ratio.

8. References

1. Srinivasan, R. and Parthasarathy, S. (1976) ***Some Statistical Applications in X-ray Crystallography,*** Pergamon Press, Oxford.
2. Ponnuswamy, M.N. and Parthasarathy, S. (1977) Quantitative statistical criteria for use in the cumulative, semi-cumulative and distribution function tests for centrosymmetry, ***Acta Cryst.*** **A33,** 838-844.
3. Foster, F. and Hargreaves, A. (1963) ***The use of moments of X-ray intensity in space-group determination. I. Derivation of theoretical moments, Acta Cryst.*** **16,** 1124-1133.
4. Sekar, K. and Parthasarathy, S. (1987) Measurability of Bijvoet differences in triclinic, monoclinic and orthorhombic crystals. Part II, ***Acta Cryst.,*** **A43,** 653-655.
5. Sekar, K. (1991) ***Ph.D. Thesis,*** University of Madras.
6. Velmurugan, D. (1984) ***Ph.D. Thesis,*** University of Madras.
7. Bernstein, F.C., Koetzle, T.F., Williams, G.J.B., Meyer, D.F., Jr. Brice, M.D., Rodgers, J.R., Kennard, O, Shimanouchi, T. and Tasumi, M. (1977) ***J.Mol.Biol.,*** **112,** 535-542.

8. Templeton, D.H., Templeton, L.K., Philips, J.C. and Hodgson, K.O. (1980) Anomalous scattering of X-rays by Cesium and Cobalt measured with Synchrotron radiation, ***Acta Cryst.*** **A36,** 436-442.

9. Templeton, L.K., Templeton, D.H., and Phizacherley, R.P. (1980) L_3-Edge anomalous scattering of X-rays by Praseodymium and Samarium, ***J. Am. Chem. Soc.***, **102,** 1185-1186.

10. Templeton, L.K., Templeton, D.H., Phizacherley, R.P. and Hodgson, K.O. (1982), L_3-Edge anomalous scattering by Gadolinium and Samarium measured at high resolution with Synchrotron radiation, ***Acta Cryst.***, **A38,** 74-78.

MULTIWAVELENGTH ANOMALOUS DIFFRACTION IN MACROMOLECULAR CRYSTALLOGRAPHY *

JANET L. SMITH

Department of Biological Sciences
Purdue University
West Lafayette, Indiana 47907 USA

1. Introduction

Multiwavelength anomalous diffraction (MAD) is the fastest growing method of structure determination in macromolecular crystallography. At least twenty-five new structures solved with MAD were published in the past year. Many factors contribute to the growth of MAD, and its future is extremely bright. The experience gained over the past several years is now being generalized to make MAD more accessible. This paper aims to present a practical overview of MAD. I first review the observational equation for MAD and describe the basis of the phasing signal and how it is estimated for specific problems. This is followed by a discussion of the design of a MAD experiment, schemes for data analysis and phasing, and considerations in solving the anomalous-scatterer partial structure. Finally, there is a discussion of selenomethionine as a phasing vehicle. More comprehensive reviews of MAD have been published by W. A. Hendrickson, who pioneered its development and application in macromolecular crystallography [1,2]. Several short papers covering various aspects of MAD are included in the proceedings of the 1997 CCP4 Study Weekend [*Recent Advances in Phasing*, K. Wilson, G. Davies, and S. Bailey, eds., SERC Daresbury Laboratory, Warrington, U.K, 1997, in press].

2. Theoretical Basis

Electrons bound in atomic orbitals have specific resonant frequencies corresponding to allowed transitions. Anomalous scattering is the manifestation in X-ray diffraction of these resonance effects. The resonant frequencies of most chemical elements in biological macromolecules are far below the energies used for diffraction experiments, and their anomalous scattering is thus negligible. However, elements of atomic number 24 through 92 have resonant frequencies between 6 keV (λ = 2Å) and 40 keV (λ = 0.3Å), which give rise to detectable effects in X-ray scattering from macromolecular specimens labeled with these elements. Information about the phase of

S. Fortier (ed.), Direct Methods for Solving Macromolecular Structures, 211–225.

the scattered X-rays can be derived from the resonance effects, or anomalous scattering. Anomalous scattering is an atomic property and thus enters the equations for X-ray diffraction in the expression for the atomic scattering factor (f), which is the sum of "normal" atomic scattering factor f^0 and a complex "anomalous" correction having real (f') and imaginary (f") components:

$$f = f^0 + f' + if''.$$

The breakdown of Friedel's law caused by the imaginary component of anomalous scattering (f") has been used for many years as a source of phase information in macromolecular crystallography. Wavelength-tunable synchrotron radiation allows the real component (f') to be used as well, providing the opportunity for direct phasing through combination of the orthogonal effects of f' and f". MAD exploits differences in the observed diffraction intensities caused by differential f' and f" values at different X-ray wavelengths to achieve such direct phasing.

The formulation of the MAD observational equation used here is based on that of Karle [3] as modified by Hendrickson *et al.* [4].

$$\begin{aligned} |^{\lambda}\mathbf{F}_{obs}{}^{\pm}|^2 = |\mathbf{F}_T|^2 &+ a_\lambda|\mathbf{F}_A|^2 \\ &+ b_\lambda|\mathbf{F}_T||\mathbf{F}_A|\cos(\phi_T-\phi_A) \\ &\pm c_\lambda|\mathbf{F}_T||\mathbf{F}_A|\sin(\phi_T-\phi_A), \end{aligned} \quad (1)$$

where $a_\lambda = (f''^2_\lambda + f'^2_\lambda)/(f^0)^2$,

$b_\lambda = 2f'_\lambda/f^0$

and $c_\lambda = 2f''_\lambda/f^0$.

This formulation is distinguished from many others relating phases to anomalous scattering by Karle's insight that the real ($|^{\lambda}\mathbf{F}_A'|$) and imaginary ($|^{\lambda}\mathbf{F}_A''|$) structure amplitudes, due to f' and f", respectively, can be expressed as products of scattering factor ratios and normal structure amplitudes, due to f^0:

$$|^{\lambda}\mathbf{F}_A'| = (f'_\lambda/f^0)|\mathbf{F}_A|$$

and

$$|^{\lambda}\mathbf{F}_A''| = (f''_\lambda/f^0)|\mathbf{F}_A|.$$

Wavelength-dependence and structure-dependence are thus separated into different quantities. All wavelength dependence is in the anomalous scattering factors, f'_λ and f''_λ, which do not depend on atomic positions, and all structure dependence is in the normal structure factors $\mathbf{F}_T$ and $\mathbf{F}_A$, which do not depend on wavelength. The structure factor $\mathbf{F}_T$ represents normal scattering from the total structure, and $\mathbf{F}_A$ represents normal scattering from the partial structure of anomalous scatterers. An Argand diagram showing the relationships of these structure factors has been published [5]. Eq. 1 describes the case for one type of anomalous scatterer. In general, Eq. 1 will relate experimental observations to unknown quantities whose number equals twice the number of anomalous-scatterer types plus one, here $|\mathbf{F}_T|$, $|\mathbf{F}_A|$ and $(\phi_T-\phi_A)$ for one anomalous-scatterer type.

The MAD observational equation (Eq. 1) involves no approximations, and the accuracy of MAD phases is limited only by the precision of the diffraction data. This is in contrast to isomorphous replacement where phase accuracy is limited most severely by breakdown of the assumption of isomorphism of native and derivative crystals. The new prominence of MAD is due primarily to a significant improvement in the quality of diffraction data in general. This comes from the ability to measure better data faster thanks to widespread adoption of cryocooling techniques and to improvements in synchrotron sources and X-ray detectors.

3. Anomalous Scattering Factors

Anomalous scattering factors in the region of an absorption edge are sensitive to the chemical environment of the absorbing atom, and are significantly enhanced by sharp spectral features in many cases. Therefore, f" and f' for anomalous scatterers in macromolecules cannot be calculated as free-atom anomalous scattering factors [6,7], which are accurate estimates for all chemistries at energies away from absorption edges. Several laboratories have schemes for extracting anomalous scattering factors f' and f" from X-ray spectra, none of which has been published in rigorous detail. However, all exploit the fact that the imaginary component of anomalous scattering f" is proportional to the atomic absorption coefficient μ_a, which can be obtained easily from raw X-ray fluorescence or transmission data. The scheme of Hendrickson *et al.* [8] is described briefly here and illustrated in Fig. 1. The X-ray spectrum of the labeled macromolecule, typically a macromolecule single crystal, is measured as fluorescence through the edge of interest (Fig. 1a). Regions of the experimental spectrum slightly away from the edge are fit to theoretical values using the program XASFIT in order to place the experimental spectrum on an absolute scale (Fig. 1b). Theoretical values are obtained from a program by Don Cromer, modified by Wayne Hendrickson to produce spectra rather than f' and f" at single energies and variously called FPRIME, SPECTRUM or CROMER. Care must be taken to measure enough edge-remote points for reliable fit of the experimental spectrum, which may be quite noisy. A narrow region around the absorption edge is then cut from the scaled experimental spectrum and spliced into the theoretical spectrum. From the hybrid spectrum of f" values thus obtained, f' values are calculated by Kramers-Kronig transformation:

$$f'(E) = \frac{2\delta}{\pi} \sum_{i=0}^{\infty} \frac{E_i f_i''}{E^2 - E_i^2} \tag{2}$$

where E is energy in eV and δ is the energy increment of the f" spectrum being transformed. In practice, the point of singularity for each f' ($E_i = E$) is not included in the summation, and a transformation range of ~500 eV beyond the f' being computed is sufficient to eliminate truncation effects. Splicing and f' calculation (Fig. 1c) are done with the program KRAMIG.

Typical anomalous scattering factors, f''_{max} and f'_{min}, estimated from X-ray spectra of protein crystals taken at MAD experimental stations, are given in Table 1 for several elements. In addition to the electronic environment of the anomalous scatterer, the energy dispersion of the incident X-ray beam also influences the values of anomalous scattering factors in the edge region.

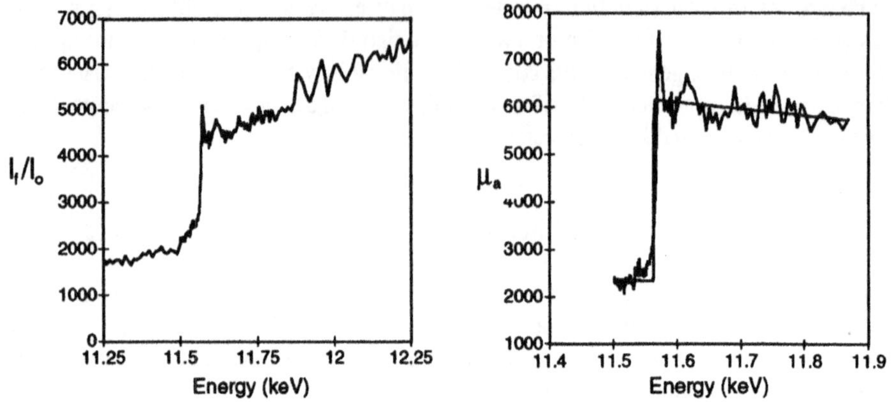

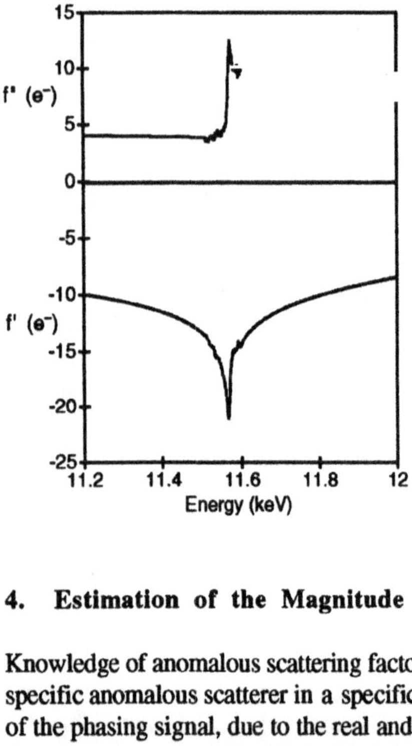

Figure 1. A. Fluorescence spectrum (I/I_0 on an arbitrary scale) through the Pt L_{III} absorption edge from a single crystal of β-hydroxydecanoyl thiolester dehydrase [9]. A single methionine amino acid of the crystalline protein was labeled with Pt by soaking in a solution of K_2PtCl_4.
B. Scaling of fluorescence data to theoretical atomic absorption coefficients (μ_a). The raw fluorescence spectrum was fit to the theoretical spectrum for the Pt L_{III} edge using the program XASFIT. The scaled experimental spectrum is shown superimposed on the theoretical free-atom spectrum.
C. Hybrid f" and f' spectra for the Pt L_{III} edge. Using the program KRAMIG, the edge region has been cut from the experimental spectrum in B and spliced into the theoretical spectrum, μ_a converted to f", and f' calculated from f" by Kramers-Kronig transformation (Eq. 2).

4. Estimation of the Magnitude of the MAD Signal

Knowledge of anomalous scattering factors allows estimation of the MAD signal for a specific anomalous scatterer in a specific macromolecule. The orthogonal components of the phasing signal, due to the real and imaginary anomalous scattering factors f' and f", are estimated separately because both are required for phase determination. The maximum MAD Bijvoet signal is due to Bijvoet differences at the energy of peak absorption, or f''_{max}, and is proportional to $2f''_{max}$ of Table 1. The maximum MAD dispersive signal is due to wavelength differences between structure amplitudes at the energy of the inflection point of the edge (f'_{min}) and at a remote energy (f'_{remote}), and is proportional to $|f'_{min}-f'_{remote}|$.

TABLE 1. Typical anomalous scattering factors

Element	f^0 (e^-)	Edge	λ (Å)	f'_{min} (e^-)	λ (Å)	f''_{max} (e^-)	Ref.
Fe	26	K	1.7402	-9	1.7380	5	[8]
			1.7425	-8	1.7390	4	[10]
Cu	29	K	1.3790	-8	1.3771	4	[11]
Zn	30	K	1.2826	-9	1.2818	4	[12]
Se	34	K	0.9793	-11	0.9792	6	[13]
Br	35	K	0.9207	-7	0.9196	4	[14]
Sm	62	L_{II}	1.6959	-16	1.6952	17	[15]
Ho	67	L_{III}	1.5363	-28	1.5356	20	[16]
Yb	70	L_{III}	1.3857	-33	1.3853	35	[17]
W	74	L_{III}	1.2136	-24	1.2123	19	[18]
Os	76	L_{III}	1.1402	-23	1.1397	20	[19]
Pt	78	L_{III}	1.0720	-21	1.0714	13	Fig. 1c
Hg	80	L_{III}	1.0094	-18	1.0057	10	[20]
			1.0095	-25	1.0063	12	[21]
U	92	L_{III}	0.7213	-21	0.7208	12	[22]

Energy (keV) = 12.39854/λ (Å)

The magnitude of the MAD phasing signal is estimated as the ratio of expected Bijvoet or dispersive difference to expected total scattering of the macromolecule. This is based on calculation of expected structure amplitudes $\langle|\mathbf{F}|\rangle$, where $\langle|\mathbf{F}|\rangle = \sqrt{\sum f_i^2}$ and $\langle|\mathbf{F}|\rangle = \sqrt{N}f$ for N atoms of identical f [23]. The diffraction ratios of interest to MAD [24] are, for the dispersive signal,

$$\frac{\langle\left||\mathbf{F}_{\lambda 1}|-|\mathbf{F}_{\lambda 2}|\right|\rangle}{\langle|\mathbf{F}_T|\rangle} \approx \sqrt{\frac{N}{2}}\frac{|f'_{\lambda 1}-f'_{\lambda 2}|}{\langle|\mathbf{F}_T|\rangle} \tag{3}$$

for N anomalous-scatterer sites with λ1 chosen at f'_{min} and λ2 chosen for $|f'_{\lambda 1}-f'_{\lambda 2}|_{max}$, and, for the Bijvoet signal,

$$\frac{\langle\left||\mathbf{F}^+_\lambda|-|\mathbf{F}^-_\lambda|\right|\rangle}{\langle|\mathbf{F}_T|\rangle} \approx \sqrt{\frac{N}{2}}\frac{2f''_\lambda}{\langle|\mathbf{F}_T|\rangle} \tag{4}$$

with λ chosen at f''_{max}. These diffraction ratios are analogous to the usual calculation of isomorphous signal from experimental data in which

$$\frac{\langle\left||\mathbf{F}_{PH}|-|\mathbf{F}_P|\right|\rangle}{\langle|\mathbf{F}_P|\rangle} \approx \sqrt{\frac{N}{2}}\frac{f^0}{\langle|\mathbf{F}_P|\rangle} \tag{5}$$

where f^0 is for the heavy atom. Values for f^0, f'_{min} and f''_{max} are those in Table 1. The denominator of all diffraction ratios is the expected total scattering of the macromolecule, which can be estimated for $2\theta = 0$ with the expressions in Table 2.

TABLE 2. Estimates of scattering strength for macromolecules, $<|F_T|>$

Macromolecule	NA=#atoms (e^-)	NR=#residues (e^-)	MW=molecular weight (e^-)
Protein	$6.70\ (NA)^{1/2}$	$(346\ NR)^{1/2}$	$(3.14\ MW)^{1/2}$
DNA	$7.20\ (NA)^{1/2}$	$(1128\ NR)^{1/2}$	$(3.87\ MW)^{1/2}$
RNA	$7.26\ (NA)^{1/2}$	$(1183\ NR)^{1/2}$	$(3.89\ MW)^{1/2}$

A hypothetical example illustrates the issue of signal size in MAD *vs.* isomorphous replacement. Consider a 500-residue protein and the MAD signal generated by 10 Se anomalous scatterers. If $f''_{max} = 6\ e^-$, $f'_{min} = -11\ e^-$ and $f'_{remote} = -4\ e^-$, then by Eq. 4 the maximum Bijvoet signal will be ~6% of $|F_{obs}|$ and by Eq. 3 the maximum dispersive signal will be ~4% of $|F_{obs}|$. By comparison, the isomorphous replacement signal generated by one fully occupied Hg site ($f^0 = 80\ e^-$) in the same protein will be ~14% of $|F_{obs}|$ by Eq. 5. For many typical examples the MAD signal is near the noise level of moderate-quality diffraction data sets, whereas the isomorphous replacement signal is easily detectable in data of moderate quality. On the other hand, detection of the MAD signal is limited only by data quality whereas lack of isomorphism will pollute the isomorphous replacement signal with systematic error that cannot be removed. It is clear from the large number of successful MAD experiments that a relatively weak phasing signal is by no means an insurmountable problem.

5. MAD Experimental Design

Three important considerations distinguish the design and execution of a MAD experiment from more familiar monochromatic experiments in macromolecular crystallography. These are wavelength selection, data completeness and data quality.

The largest MAD phasing signal is obtained at energies with the most extreme values of f' and f", which correspond to the sharpest features of the absorption edge. Therefore, it is critical to determine the position of the absorption edge experimentally from the labeled macromolecule at the time of a MAD experiment. Even when the position of the edge is well known, small unanticipated chemical changes in the sample or energy changes in the X-ray beam can reduce the MAD signal very significantly if the sharp edge features are missed in selecting energies for data collection. Energies are selected at the peak of sample absorption just above the edge ("E_{peak}" for f''_{max}) to optimize the Bijvoet signal and at the inflection point of the edge ("E_{dip}" for f'_{min}) to optimize the orthogonal dispersive signal. The dispersive signal is further optimized if a third energy remote from the edge ("E_{remote}") is chosen. The choice of E_{remote} is experiment dependent, although it is typically above rather than below the edge due to the larger Bijvoet signal. E_{remote} may also be chosen to avoid complications from other edges or to obtain data at a wavelength optimal for model refinement.

There has been much debate about the optimal number of data-collection energies for successful phase determination by MAD. In the commonest MAD experiment $|F^+|$ and $|F^-|$ are measured at each of E_{dip}, E_{peak} and E_{remote}. If the difference in f' is large enough to produce a detectable signal, then one could in principle obtain phases from three measurements: $|F^+|$ and $|F^-|$ at E_{peak} and either $|F^+|$ or $|F^-|$ at E_{dip} [25]. However, redundancy is one of the best ways to minimize the effects of measurement error in macromolecular crystallography. In the full three-energy experiment, the Bijvoet signal is redundant because the remote energy is above the edge. The orthogonal dispersive signal is redundant because two measurements are taken at each of E_{dip} and E_{remote}. There are several examples of even more redundant four- or five-wavelength MAD experiments. While greater redundancy is desirable, it should not be gained at the cost of good counting statistics. Unfortunately, considerations of available beam time frequently preclude MAD experiments with more than three energies.

The MAD phasing signal is derived from intensity differences that may be similar in magnitude to measurement errors. Thus a general philosophy in the design of a MAD experiment is to equalize systematic errors among the measurements whose differences will contribute to each phase determination. This is achieved for each single reflection by recording Bijvoet measurements at all wavelengths from the same asymmetric unit of the same crystal at nearly the same time. Bijvoet mates can be recorded simultaneously by alignment of the crystal with a mirror plane perpendicular to the rotation axis, or Friedel images can be recorded in an "inverse beam" experiment. (Friedel images are related by 180° rotation of the crystal about any axis perpendicular to the incident beam, usually the data-collection axis.) If crystal decay is a problem, small blocks of Bijvoet data can be recorded at each of the selected wavelengths before moving to another block of reciprocal space. When such a data collection strategy is followed, the resulting MAD data set will be complete with respect to recording all multiwavelength, Bijvoet measurements for all regions of the reciprocal lattice that are covered in the experiment. Coverage of reciprocal space can be monitored during the experiment by a strategy program, if available, or by reduction of diffraction images to integrated intensities for data from at least one wavelength. Completeness of the MAD data set is at least as important as for any diffraction experiment that will be used for phasing. If data, and hence phase information, are incomplete, it may be difficult to reproduce the same beam and sample conditions during a subsequent experiment, which is likely to occur only after some weeks or months.

Measurement errors are of major importance in all areas of macromolecular crystallography, but are the limiting factor in phase determination by MAD. MAD data should be of high quality by the usual measures (R_{sym}, redundancy, completeness), especially in experiments where the phasing signal is weak. In the hypothetical 500-residue protein with 10 Se anomalous scatterers, a 5% MAD signal will become undetectable as it is exceeded by R_{sym} "noise". Thus data with good counting statistics are of paramount importance. In a carefully designed experiment, the effect of increasing R_{sym} with increasing θ is mitigated somewhat by equalizing systematic errors. Nevertheless, if R_{sym} (I) is 30% for the outer shells of data, there will be virtually no detectable MAD phasing signal for these reflections in the hypothetical example. Disappearance of the phasing signal into R_{sym} noise is the major reason that useful MAD phases generally are not obtained to the diffraction limit of crystals even though anomalous scattering does not fall off with increasing θ.

6. Data Processing and Scaling

Concerns about signal size dominate special schemes for handling MAD data. Special computer programs for scaling MAD data have been developed [8,26]. Two general approaches to data handling for MAD have been employed.

The approach originally proposed by Hendrickson, known as "phase first, merge later," represents the extreme interpretation of the scheme for equalizing systematic errors – the individual observations constituting a multiwavelength Bijvoet set, as determined by the data-collection strategy, are grouped together and scaled as usual, but are merged with redundant measurements only after phases are determined. Error estimates from the phasing or the agreement of redundant phase determinations can be incorporated into weights for averaging, or can be used to reject outliers. This approach involves complicated, experiment-dependent bookkeeping to assemble exactly the correct observations for each unmerged set.

A second approach, "merge first, phase later," is to scale and merge data at each wavelength, keeping Bijvoet pairs separate, and then to scale data at all wavelengths to one another. This is most easily and reliably done by scaling all data against a common standard data set, which can be the unique data from one wavelength with Bijvoet mates averaged. If the data collection followed one of the strategies outlined above, then measurements for each unique reflection are identically redundant, which itself minimizes systematic errors in the amplitude differences used for phasing. The second approach is computationally simpler than the first because it is experiment independent. However, unanticipated, minor experimental disasters may be more difficult to overcome in the "merge first, phase later" approach to data handling.

7. Approaches to MAD Phasing

There are two general approaches to MAD phasing. One is to treat the problem explicitly and solve the MAD observational equation (Eq. 1). This explicit approach is embodied in the MADSYS package from the Hendrickson laboratory [27], in particular in the phasing program MADLSQ. The other approach is to treat MAD phasing as a special case of multiple isomorphous replacement (MIR). The pseudo-MIR approach is discussed in two recent publications [28,29]. Both approaches have been quite successful, and there are no hard-and-fast rules for which sorts of problems are more amenable to which approach, rumors in the community notwithstanding. There are advantages and disadvantages to both approaches.

The explicit approach provides the quantities $|F_T|$, $|F_A|$ and $(\phi_T-\phi_A)$. Estimates of the anomalous scattering factors at the wavelengths of data collection are required to fit the observations to the MAD phase equation. These estimates can be refined within MADLSQ, so they need not be highly accurate. A second calculation is required to obtain ϕ_T from the phase differences $(\phi_T-\phi_A)$. There are two advantages to the explicit approach. First, it is amenable to the "phase first, merge later" scheme of data handling because refinement of the anomalous-scatterer partial structure is entirely separate from phase calculation. In this case redundancies are merged to produce a unique data set at the level of the derived quantities $|F_T|$, $|F_A|$, $(\phi_T-\phi_A)$ and their error estimates. These error estimates or the agreement of redundant phase determinations can be used to weight terms in a Fourier synthesis from $|F_T|$ and ϕ_T. Phase probability coefficients (ABCDs) have been derived from the MAD phase equation [30]. The second principal advantage of the explicit approach is calculation of an experimentally derived estimate of the normal structure amplitude $|F_A|$ for the anomalous scatterer. This is the quantity with which the partial structure of anomalous scatterers is most directly solved and refined, and therefore should be highly sought. However, while MADLSQ is quite successful in the least-squares fit of the MAD phase equation to $|F_{obs}|$ for high-quality data, it is poorly conditioned to extracting $|F_A|$ from noisy data and requires careful pruning of outliers from the $|F_A|$ values produced. A Bayesian method of $|F_A|$ estimation [31] should be more robust than the least-squares procedure.

In the pseudo-MIR approach data at one wavelength are designated as "native" data, which include anomalous scattering, and data at the other wavelengths as "derivative" data. This approach has the advantage that nothing need be known about the anomalous scattering factors prior to phasing. These quantities are incorporated into heavy-atom atomic "occupancies" and refined along with other parameters. Of course, the partial structure of anomalous scatterers must be known, and refinement of the partial structure is concurrent with phasing. In refinement of the "heavy atom" parameters, greater weight is given to the data set selected as "native." This bias should be removed by the new maximum-likelihood refinement of de La Fortelle and Bricogne [32], which treats data at all wavelengths as statistically equivalent. The amplitudes $|F_A|$ are not determined in the pseudo-MIR approach, and the partial structure is solved from Bijvoet differences between $|F^+|$ and $|F^-|$ or dispersive differences between $|F_{\lambda 1}|$ and $|F_{\lambda 2}|$, with wavelengths selected to optimize the signal. The pseudo-MIR approach is used more frequently than the explicit approach due to the greater familiarity of crystallographers with software for isomorphous replacement.

8. Determination of the Anomalous-Scatterer Partial Structure

A prerequisite for MAD-phased electron density, regardless of the phasing technique, is determination of the partial structure of anomalous scatterers. As described above, the optimal quantities for solving and refining the partial structure of anomalous scatterers are the normal scattering amplitudes $|F_A|$. Frequently $|F_A|$ values are not extracted from the MAD measurements, and the largest Bijvoet or dispersive differences are used instead. This involves the usual approximation of representing structure amplitudes

($|F_A|$) as the subset of larger differences ($||F^+|-|F^-||$ or $||F_{\lambda 1}|-|F_{\lambda 2}||$). The approximation is accurate for only a small fraction of reflections because there is no correlation between the phase of F_P and the phase of F_A. However, it suffices for a suitably strong signal and a suitably small number of sites. For virtually all structures determined by MAD, the anomalous-scatterer sites have been located by Patterson methods. However, the problem quickly becomes intractable by Patterson methods when there are more than a handful of sites. This is a current challenge for MAD, where the aim is to solve the macromolecule structure from one MAD data set using any number of anomalous scatterer sites. Statistical direct methods clearly hold the answer to this problem. Recent results are promising in this regard. Bertrand *et al.* [33] have solved a 12-atom Se partial structure in a 437-residue protein by direct methods using $|F_A|$s, and S. Doublié (personal communication) has solved a 15-atom Se partial structure in an asymmetric unit of 108 kDa using dispersive differences, also by direct methods. These results open the door for routine MAD determination of quite large structures with many anomalous scatterer sites. New direct methods techniques, such as described in this volume in papers by G. M. Sheldrick, by R. Miller and C. M. Weeks and by G. Bricogne, hold great promise for a major expansion in the complexity of anomalous-scatterer partial structure that can be solved.

The correct enantiomorph for the anomalous-scatterer partial structure must be determined (ϕ_A *vs.* $-\phi_A$) in order to obtain an electron-density image of the macromolecule. However, it cannot be determined directly from MAD data. The correct hand is chosen by comparison of electron density maps based on both enantiomorphs of the partial structure. Unlike the situation for MIR, the density based on the incorrect hand of the anomalous-scatterer partial structure is not the mirror image of that based on the correct hand and contains no image of the macromolecule. The correct map is distinguished by features such as a clear solvent boundary, positive correlation of redundant densities, and a macromolecule-like density histogram. If the anomalous scattering centers form a centric array, then the two enantiomorphs are identical and both maps are correct.

9. Selenomethionine

The most successful MAD phasing vehicle to date has been selenium in the form of selenomethionine (SeMet). This particularly clever experiment was devised by Wayne Hendrickson [24], who also pioneered its use [34,35]. Briefly, proteins are labeled with Se by biological substitution of SeMet for methionine amino acids. This is achieved by blocking methionine biosynthesis in the cells in which the protein is produced and substitution of SeMet for Met in the growth medium. The generality of the labeling scheme for proteins is the root of its success. SeMet labeling technology is discussed in a recent review by Doublié [36].

SeMet incorporation has been done most frequently for proteins expressed in *E. coli* strains that are auxotrophic for Met (strain DL41 [35]; strain B834 [37,38]; strain LE392 [39]; strain MIC88 [40]. Nearly complete incorporation has also been reported in nonauxotrophic bacterial strains (*E. coli* strain BL21 [41]; *E. coli* strain

XA90 [42,43]), in a mammalian cell line [44] and in baculovirus-infected insect cells [45]. Special precautions must be taken to prevent oxidation of SeMet proteins. In almost all cases, somewhat higher-than-normal concentrations of disulfide reducing agents, such as dithiothreitol or β-mercaptoethanol, are sufficient to protect SeMet from air oxidation to the selenoxide [46]. In a few cases, crystallization in an inert atmosphere has been necessary [27,47]. Because Se is a light element, the position of the K absorption edge moves to slightly higher energy upon oxidation, and a mixture of oxidation states in a sample crystal is predicted to obliterate the MAD signal.

Methionine is a particularly attractive target for anomalous scatterer labeling. The hydrophobic side chain of methionine, which carries the sulfur atom to be substituted by selenium, is usually buried in the hydrophobic core of globular proteins and is therefore relatively better ordered than are surface side chains. Evidence for isostructuralism of Met and SeMet proteins comes from the labeling experiment itself. All proteins in the biological expression system have SeMet substituted for Met at levels approaching 100%. The cells are viable, therefore the proteins are functional and isostructural with their unlabelled counterparts to the extent required by function.

The natural abundance of methionine in soluble proteins is approximately one in fifty amino acid residues. The N-terminal Met is not included in this estimate because, if present, it is usually disordered. Using Eqs. 3 and 4, this provides a maximal MAD phasing signal of 4-6% of $|F|$, easily detectable in strongly diffracting protein crystals and detectable with careful data collection from crystals of moderate quality. To improve the phasing signal, in a few cases Met has been substituted for other amino acids by site-directed mutagenesis [37,48-50].

SeMet labeling is now part of the repertoire of protein crystallography, and has broader applicability than for MAD phasing alone. This comes from the relative ease of incorporation of the SeMet label, from the remarkable structural similarity of SeMet and wild type proteins, and from the uniformity and completeness of labeling. Crystals of SeMet proteins are usually isomorphous with those of the wild type, and consequently can be used as isomorphous derivatives. The isomorphous signal comes from the excess of 18 electrons in Se relative to S, making the SeMet isomorphous phasing signal (~10% of $|F|$, Eq. 5) about twice as strong as the SeMet MAD phasing signal (4-6% of $|F|$). In most cases SeMet derivatives are more isomorphous, and certainly more rationally produced, than are heavy-atom derivatives produced by the usual soaking procedures. Prior knowledge of exactly how Se labels the protein is itself a powerful tool. For example, the SeMet mutation is an extremely useful amino acid label for fitting a protein sequence to electron density. Also, noncrystallographic symmetry operators usually can be defined more reliably from Se positions in SeMet protein than by heavy-atom positions in conventional derivatives due to the uniformity and completeness of labeling [51].

An analogous label is available for nucleic acids in the form of brominated bases, particularly 5-bromouridine, which is isostructural with thymidine. Iodinated bases are commonly used as isomorphous derivatives (f^0 = 53 e^-) for nucleic acids, but the X-ray edges of I (λ = 0.38Å for K, λ = 2.56-2.72Å for L) occur at energies less favorable for accurate macromolecular data collection than does the K edge of Br (λ = 0.92Å).

10. Conclusion

Why is the enthusiasm for MAD so high today? There are three primary reasons. First, cryocrystallography has improved data quality to the point that the precision required for MAD is usual rather than exceptional. Second, new synchrotron sources and new beamlines provide intense, reliably tunable X-ray beams and the instruments to exploit them. Third, MAD works extremely well and very quickly. For many problems, the experimentally phased electron density is of stellar quality. Crystallographers are only beginning to appreciate the value of nearly error-free, model-independent phases [52]. The remaining challenges are in two areas. The greatest impediment to growth of MAD today is access to suitable experimental facilities. This non-technical problem may be solved only be a concerted effort of the community. The greatest technical challenge is to develop methods for solving large partial structures of anomalous scatterers. Here recent results with statistical direct methods are very promising, and MAD applied to large macromolecules no longer seems such a heroic undertaking. MAD has at last taken its place as a standard tool of macromolecular crystallography.

11. Acknowledgment

Work in the author's laboratory has been supported by grants from the U.S. National Institutes of Health (DK42303), and from the Lucille P. Markey Foundation to the Structural Studies Group at Purdue University. Collaboration with the scientific staffs at synchrotron facilities is gratefully acknowledged, especially A. W. Thompson of the European Synchrotron Radiation Facility, and S. E. Ealick of the Cornell High Energy Synchrotron Source.

* Previously published in Proceedings of the CCP4 Study Weekend, January 1997
(Re-published with permission of the Council for the Central Laboratory of the Research Councils.)

12. References

1. Hendrickson, W.A. (1991) Determination of macromolecular structures from anomalous diffraction of synchrotron radiation, *Science* **254**, 51-58.
2. Hendrickson, W.A. and Ogata, C.M. (1997) Phase determination from multiwavelength anomalous diffraction measurements, *Meth. Enzymol.* **276**, 494-523.
3. Karle, J. (1980) Some developments in anomalous dispersion for the structural investigation of macromolecular systems in biology, *Int. J. Quantum Chem.: Quantum Biol. Symp.* **7**, 357-367.
4. Hendrickson, W.A., Smith, J.L., and Sheriff, S. (1985) Direct phase determination based on anomalous scattering, *Meth. Enzymol.* **115**, 41-55.
5. Smith, J.L. (1991) Determination of three-dimensional structure by multi-wavelength anomalous diffraction, *Curr. Opinion Struc. Biol.* **1**, 1002-1011.
6. Cromer, D.T. and Liberman, D. (1970a) Relativistic calculation of anomalous scattering factors for X-rays, *J. Chem. Phys.* **53**, 1891-1898.
7. Cromer, D.T. and Liberman, D. (1970b) Relativistic calculation of anomalous scattering factors for X-rays, *Los Alamos National Laboratory Report LA-4403*.
8. Hendrickson, W.A., Smith, J.L., Phizackerley, R.P., and Merritt, E.A. (1988) Crystallographic structure analysis of lamprey hemoglobin from anomalous dispersion of synchrotron radiation, *Proteins: Struct., Funct., Genet.* **4**, 77-88.

9. Leesong, M., Henderson, B.S., Gillig, J.R., Schwab, J.M., and Smith, J.L. (1996) Structure of a dehydratase-isomerase from the bacterial pathway for biosynthesis of unsaturated fatty acids: two catalytic activities in one active site, *Structure* **4**, 253-264.
10. Smith, J.L., Zaluzec, E.J., Wery, J.-P., Niu, L., Switzer, R.L., Zalkin, H., and Satow, Y. (1994) Structure of the allosteric regulatory enzyme of purine biosynthesis, *Science* **264**, 1427-1433.
11. Guss, J.M., Merritt, E.A., Phizackerley, R.P., Hedman, B., Murata, M., Hodgson, K.O., and Freeman, H.C. (1988) Phase determination by multiple-wavelength X-ray diffraction: crystal structure of a basic "blue" copper protein from cucumbers, *Science* **241**, 806-811.
12. Zhang, G., Kazanietz, M.G., Blumberg, P.M., and Hurley, J.H. (1995) Crystal structure of the Cys2 activator-binding domain of protein kinase Cδ in complex with phorbol ester, *Cell* **81**, 917-924.
13. Wu, H., Lustbader, J.W., Liu, Y., Canfield, R.E., and Hendrickson, W.A. (1994) Structure of human chorionic gonadotropin at 2.6 Å resolution from MAD analysis of the selenomethionyl protein, *Structure* **2**, 545-558.
14. Ogata, C.M., Hendrickson, W.A., Gao, X., Patel, D.J., and Satow, Y. (1989) Structure of a DNA-drug complex determined by MAD phasing from imaging plate data, *Abstr. Amer. Crystallog. Assoc. Mtg., Series 2* **17**, 53.
15. Tomchick, D.R., Smith, J.L., Turner, R.J., and Switzer, R.L. (1996) PyrR, a bifunctional RNA-binding transcriptional attenuation protein and uracil phosphoribosyltransferase, *Acta Cryst.* **S52**, C-163.
16. Weis, W.I., Kahn, R., Fourme, R., Drickamer, K., and Hendrickson, W.A. (1991) Structure of the calcium-dependent lectin domain from a rat mannose-binding protein determined by MAD phasing, *Science* **254**, 1608-1615.
17. Shapiro, L., Fannon, A.M., Kwong, P.D., Thompson, A., Lehmann, M.S., Grübel, G., Legrand, J.-F., Als-Nielsen, J., Colman, D.R., and Hendrickson, W.A. (1995) Structural basis of cell-cell adhesion by cadherins, *Nature* **374**, 327-337.
18. Egloff, M.-P., Cohen, P.T.W., Reinemer, P., and Barford, D. (1995) Crystal structure of the catalytic subunit of human protein phosphatase 1 and its complex with tungstate, *J. Mol. Biol.* **254**, 942-959.
19. Cate, J.H., Gooding, A.R., Podell, E., Zhou, K., Golden, B.L., Kundrot, C.E., Cech, T.R., and Doudna, J.A. (1996) Crystal structure of a group I ribozyme domain: principles of RNA packing, *Science* **273**, 1678-1685.
20. Tesmer, J.J.G., Stemmler, T.L., Penner-Hahn, J.E., Davisson, V.J., and Smith, J.L. (1994) Preliminary X-ray analysis of *Escherichia coli* GMP synthetase: Determination of anomalous scattering factors for a cysteinyl mercury derivative, *Proteins: Structure, Function and Genetics* **18**, 394-403.
21. Krishna, T.S.R., Kong, X.-P., Gary, S., Burgers, P.M., and Kuriyan, J. (1994) Crystal structure of the eukaryotic DNA polymerase processivity factor PCNA, *Cell* **79**, 1233-1243.
22. Glover, I.D., Denny, R.C., Nguti, N.D., McSweeney, S.M., Kinder, S.H., Thompson, A.W., Dodson, E.J., Wilkinson, A.J., and Tame, J.R.H. (1995) Structure determination of OppA at 2.3 Å resolution using multiple-wavelength anomalous dispersion methods, *Acta Cryst.* **D51**, 39-47.
23. Crick, F.H.C. and Magdoff, B.S. (1956) The theory of the method of isomorphous replacement for protein crystals. I, *Acta Cryst.* **9**, 901-908.
24. Hendrickson, W.A. (1985) Analysis of protein structure from diffraction measurement at multiple wavelengths, *Trans. Amer. Crystallogr. Assn.* **21**, 11-21.
25. Peterson, M.R., Harrop, S.J., McSweeney, S.M., Leonard, G.A., Thompson, A.W., Hunter, W.N., and Helliwell, J.R. (1996) MAD phasing strategies explored with a brominated oligonucleotide crystal at 1.65 Å resolution, *J. Synchrotron Rad.* **3**, 24-34.
26. Friedman, A.M., Fischmann, T.O., Shamoo, Y., and Ealick, S. (1994) MADPRB: a new suite of programs for MAD data analysis incorporating robust estimation, maximum likelihood and Bayesian inference, *Abstracts Amer. Crystallogr. Assn. Series 2* **22**, 39.
27. Wu, H. and Hendrickson, W.A. (1996) The analytical approach of phasing by multiwavelength anomalous diffraction (MAD), *Acta Cryst.* **S52**, C-55.

28. Ramakrishnan, V. and Biou, V. (1997) Treatment of multiwavelength anomalous diffraction data as a special case of multiple isomorphous replacement, *Meth. Enzymol.* **276**, 538-557.
29. Terwilliger, T.C. (1997) Multiwavelength anomalous diffraction phasing of macromolecular structures: analysis of MAD data as single isomorphous replacement with anomalous scattering data using the MADMRG program, *Meth. Enzymol.* **276**, 530-537.
30. Pähler, A., Smith, J.L., and Hendrickson, W.A. (1990) A probability representation for phase information from multiwavelength anomalous dispersion, *Acta Crystallogr. sect. A* **46**, 537-540.
31. Terwilliger, T.C. (1994) MAD phasing: Bayesian estimates of F_A, *Acta Cryst.* **D50**, 11-16.
32. de La Fortelle, E. and Bricogne, G. (1997) Maximum-likelihood heavy-atom parameter refinement for multiple isomorphous replacement and multiwavelength anomalous diffraction methods, *Meth. Enzymol.* **276**, 472-494.
33. Bertrand, J., Auger, G., Fanchon, E., Martin, L., Blanot, D., van Heijenoort, J., and Dideberg, O. (1997) Crystal structure of UDP-N-acetylmuramoyl-L-alanine:D-glutamate ligase from *Escherichia coli*, *EMBO J.* **16**, 3416-3425.
34. Yang, W., Hendrickson, W.A., Crouch, R.J., and Satow, Y. (1990) Structure of ribonuclease H phased at 2 Å resolution by MAD analysis of the selenomethionyl protein, *Science* **249**, 1398-1405.
35. Hendrickson, W.A., Horton, J.R., and Lemaster, D.M. (1990) Selenomethionyl proteins produced for analysis by multiwavelength anomalous diffraction (MAD): a vehicle for direct determination of three-dimensional structure, *EMBO J.* **9**, 1665-1672.
36. Doublié, S. (1997) Preparation of selenomethionyl proteins for phase determination, *Meth. Enzymol.* **276**, 523-530.
37. Leahy, D.J., Erickson, H.P., Aukhil, I., Joshi, P., and Hendrickson, W.A. (1994). Crystallization of a fragment of human fibronectin: introduction of methionine by site-directed mutagenesis to allow phasing via selenomethionine, *Proteins: Structure, Function, and Genetics* **19**, 48-54.
38. Doherty, A.J., Ashford, S.R., Brannigan, J.A., and Wigley, D.B. (1995) A superior host strain for the over-expression of cloned genes using the T7 promoter based vectors, *Nucleic Acids Res.* **23**, 2074-2075.
39. Ceska, T.A., Sayers, J.R., Stier, G., and Suck, D. (1996) A helical arch allowing single-stranded DNA to thread through T5 5'-exonuclease, *Nature* **382**, 90-93.
40. Shamoo, Y., Friedman, A.M., Parsons, M.R., Konigsberg, W.H., and Steitz, T.A. (1995) Crystal structure of a replication fork single-stranded DNA binding protein (T4 gp32) complexed to DNA, *Nature* **376**, 362-366.
41. Harrison, C.J., Bohm, A.A., and Nelson, H.C.M. (1994) Crystal structure of the DNA binding domain of the heat shock transcription factor, *Science* **263**, 224-227.
42. Van Duyne, G. D., Standaert, R. F., Karplus, P. A., Schreiber, S. L., and Clardy, J. (1993) Atomic structures of the human immunophilin FKBP-12 complexes with FK506 and rapamycin. *J. Mol. Biol.* **229**, 105-124.
43. Labahn, J., Schärer, O.D., Long, A., Ezaz-Nikpay, K., Verdine, G.L., and Ellenberger, T.E. (1996) Structural basis for the excision repair of alkylation-damaged DNA, *Cell* **86**, 321-329.
44. Lustbader, J.W., Wu, H., Birken, S., Pollak, S., Kolks-Gawinowicz, M.A., Pound, A.M., Austen, D., Hendrickson, W.A., and Canfield, R.E. (1995) The expression, characterization and crystallization of wild-type and selenomethionyl human chorionic gonadotropin, *Endocrinology* **136**, 640-650.
45. Chen, W. and Bahl, O.P. (1991) Recombinant carbohydrate and selenomethionyl variants of human choriogonadotropin, *J. Biol. Chem.* **266**, 8192-8197.
46. Brot, N., Fliss, H., Coleman, T., and Weissbach, H. (1984) Enzymatic reduction of methionine sulfoxide residues in proteins and peptides, *Meth. Enzymol.* **107**, 352-360.
47. Dyda, F., Hickman, A.B., Jenkins, T.M., Engelman, A., Craigie, R., and Davies, D.R. (1994) Crystal structure of the catalytic domain of HIV-1 integrase: similarity to other polynucleotidyl transferases, *Science* **266**, 1981-1986.
48. Leahy, D.J., Aukhil, I., and Erickson, H.P. (1996) 2.0 Å crystal structure of a four-domain segment of human fibronectin encompassing the RGD loop and synergy region, *Cell* **84**, 155-164.

49. Skinner, M.M., Zhang, H., Leschnitzer, D.H., Guan, Y., Bellamy, H., Sweet, R.M., Gray, C.W., Konings, R.N., Wang, A.H., and Terwilliger, T.C. (1994) Structure of the gene V protein of bacteriophage f1 determined by multiwavelength x-ray diffraction on the selenomethionyl protein, *Proc. Natl. Acad. Sci. U.S.A.* **91,** 2071-2075.
50. Tong, L., Qian, C., Massariol, M.-J., Bonneau, P.R., Cordingley, M.G., and Lagacé, L. (1996) A new serine-protease fold revealed by the crystal structure of human cytomegalovirus protease, *Nature* **383,** 272-275.
51. Tesmer, J.J.G., Klem, T.J., Deras, M.L., Davisson, V.J., and Smith, J.L. (1996) The crystal structure of GMP synthetase reveals a novel catalytic triad and is a structural paradigm for two enzyme families, *Nature Structural Biology* **3,** 74-86.
52. Burling, F.T., Weis, W.I., Flaherty, K.M., and Brünger, A. T. (1996) Direct observation of protein solvation and discrete disorder with experimental crystallographic phases, *Science* **271,** 72-77.

THE SOLVENT CONTRAST VARIATION METHOD

Charles W. Carter, Jr.
Department of Biochemistry and Biophysics,
CB #7260 University of North Carolina at Chapel Hill,
Chapel Hill, NC 27599-7260

1. The Molecular Envelope in Macromolecular Crystallography

The distinctly non-uniform distribution of atomic positions created by solvent channels in the asymmetric units of protein crystals is a defining distinction between small molecules and macromolecules. The contrast variation method is a frontal *experimental* attack on the problem of solving the structure of a molecular envelope. It is rooted in the earliest proposal to solve the three-dimensional structures of proteins (Bragg & Perutz, 1952). That paper outlined in all essential details the theoretical basis for using measurements from crystals prepared with different solvent contrasts to enhance the signal of the envelope transform and separate it from that of the internal fluctuations. The method was used explicitly, and combined with information on the molecular envelope obtained from electron microscopy, to give the initial phases for Tomato Bushy Stunt Virus by triangulation akin to that used with isomorphous derivatives (Harrison, 1969; Harrison & Jack, 1975).

2. The Contrast Variation Effect in Real and Reciprocal Space

Masking by solvent in real space. The intuitive content of the contrast variation method is that by changing the solvent electron density relative to the constant electron density of the macromolecule, one can "unmask" those regions, U, that exclude solvent and thereby determine their shape. The real space construction in Figure 1 reminds us that we can define two regions of a macromolecular unit cell of volume V by the scattering behavior of their contents. Those within the macromolecular envelope, $c(x)_U$, have low disorder parameters, and have a characteristic mean electron density. Solvent molecules outside the envelope, in the region denoted here by $c(x)_{V-U}$, have much greater disorder parameters, and their density depends on the solvent composition.

S. Fortier (ed.), Direct Methods for Solving Macromolecular Structures, 227–237.

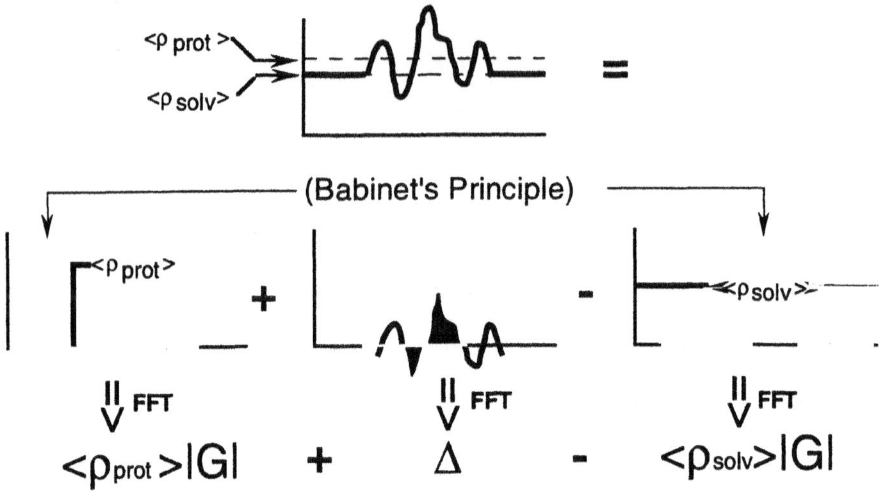

Figure 1. Real space construction of contrast variation effects.

Three contributions to the scattering from the two groups of atoms in the unit cell can therefore be described: scattering arises from a uniform distribution with mean macromolecular density, $\rho_{protein}$, inside the molecular envelope and from a uniform distribution with mean solvent density, $\rho_{solvent}$, outside the envelope. These two contributions arise from contrast at the molecular boundary and are sensitive to the solvent density contrast, and the difference density between two structures with different solvent densities is proportional to the envelope function. By Babinet's principle, the contribution from the mean solvent density is of opposite sign to that from the mean density inside the protein.

Temporally stable differences, $\delta(\mathbf{x})_U$, between the actual macromolecular density inside the molecular envelope and its average value give rise to diffraction that is insensitive to the density contrast, and which is ultimately of interest because it represents the desired structure. At high resolution, the contributions from internal fluctuations dominate, but they are significant even at moderately low resolution (10-15Å). In the special case where the electron density of a crystal mother liquor matches that of the macromolecule, this fluctuation term is all that remains. It is perhaps worth pointing out that this matching occurs approximately for high concentrations of phosphate and/or sulfate salts (see Table 1 below). In contrast, solutions of polyethylene glycol have rather low electron densities. Relative difficulties in interpreting electron density maps may reflect this differential masking; an electron density for a structure crystallized from polyethylene glycol may be easier to interpret simply because the boundary is more sharply contrasted. Thus, the degree to which the solvent masks the protein by contrast matching can contribute to the overall difficulty or ease of solving a structure using experimental phases. By the same token, one can see that solvent density variation provides an experimental means to heighten the contrast, and one can visualize this variation as a basis for extrapolation to infinite contrast.

Table 1. Solvents used in the Tryptophanyl-tRNA Synthetase contrast variation study

Solvent (all pH = 7.5)	**Electron Density**
2.25 M Li_2SO_4, 0.5M NH_4Ac	ρ_{solv} = **0.387 e/Å^3**
3.5 M$(NH_4)_2SO_4$;	ρ_{solv} = **0.403 e/Å^3**
3.5 M$(NH_4)_2SO_4$, 1.0M CsCl;	ρ_{solv}= 0.425 e/Å^3 (non-isomorphous)
4.1 M $(NH_4)_2SeO_4$;	ρ_{solv} = **0.455e/Å^3**
2.9 MK_2SeO_4;	ρ_{solv} = 0.457e/Å^3

Unmasking in reciprocal space. Extrapolation to infinite contrast has a ready correlate in reciprocal space. Various Harker constructions have been given for the effect of contrast variation in reciprocal space (Figure 2). Not all of these suggest the phase and amplitude relationships that give rise to amplitudes for a substructure more amenable to solution by direct methods. A significant problem concerns scaling together the different datasets in a contrast variation series (Bentley *et al.*, 1984; Roth *et al.*,1984). Roth recognized that structure factor difference vectors are proportional to the product of the mean solvent density and the structure factor vectors,$\{G_H\}$, of the interference function. The latter is the transform of the boundary between solvent and macromolecule, and which therefore defines the solvent-accessible molecular envelope. This proportionality means that datasets for the different contrasts can be scaled by assuring that the difference vectors are colinear (Figure 2a). Colinearity assures the same relative phase for each of the contributing difference structure factors shown by black, horizontal vectors. Adjusting the scale factors for the observed structure factor amplitudes (light gray vectors) and for the difference in phase between the contrasts by triangulation leads to a least-squares procedure to estimate the global scale parameters and a phase difference, $\Delta\phi(i\text{-}j)_H$, for each pair of reflection amplitudes measured at different solvent densities i and j.

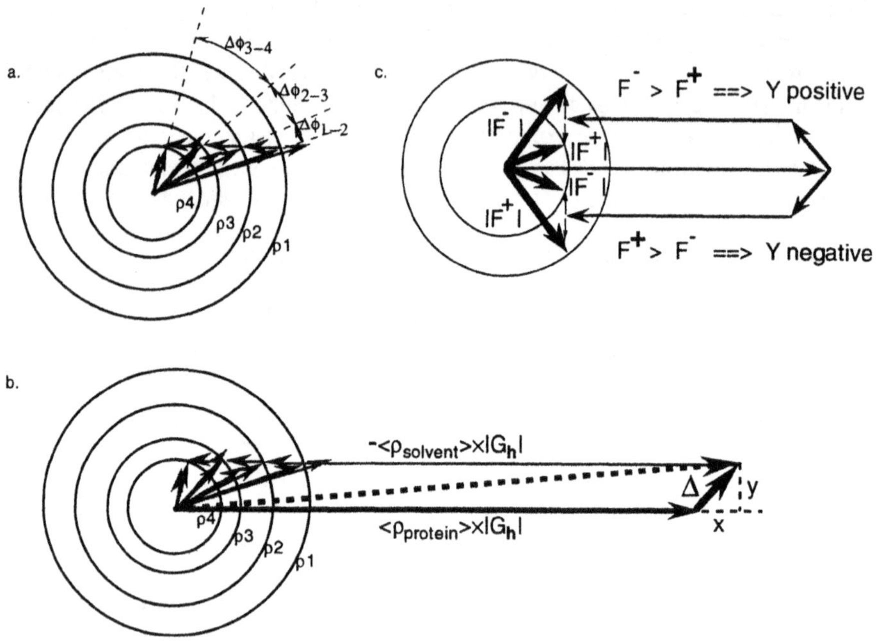

Figure 2. Reciprocal space (Harker) constructions for contrast variation effects. **a.** The construction given by Roth (Roth et al., 1984) for global scaling of contrast variation datasets together and estimating phase differences reflection-by-reflection. **b.**Harker construction for the resolution of $\chi(\mathbf{x})$ from $\delta(\mathbf{x})$ (Bricogne). Contributions to F_{obs} from the envelope are large, parallel, of opposite sign, and proportional to mean electron densities. **c.** Bijvoet differences from anomalously scattering solvent molecules can determine the sign of the imaginary component of Δ.

Different water: D_20 mixtures change the contrast to neutrons over a very wide range, and these differences were used in solving a structure for the nucleosome core particle (Bentley et al., 1984) .

2.1 RESOLVING THE AMPLITUDES, $\{G_H\}$, OF THE ENVELOPE, $\chi(x)_U$, FROM THOSE, $\{\Delta_H\}$, OF THE FLUCTUATION DENSITY, δ.

The phase information present in a contrast variation dataset is always *relative*; the absolute phases of the envelope structure factors remain undetermined. From a direct phasing perspective, distinct advantages are to be gained using an alternative procedure for scaling the data, and which was introduced by Bricogne (Carter *et al.*,1990) (Figure 2b). At low resolution, the envelope transform has considerable contrast, giving rise to large amplitudes in its spectrum. Typically, these are expected to be much larger than either the amplitudes, $\{|F_{obsH}|\}$ or$\{|\Delta_H|\}$ (Figure 2b). A contrast variation series can therefore be used to estimate the amplitudes of the envelope or indicator function, $\{G_H\}$, simultaneously with the amplitude and relative phase (within a sign ambiguity) of $\{\Delta_H\}$, the transform of the fluctuation density.

An additional feature of the alternative construction in Figure 2b is that anomalous scatterers in the solvent provide a way to fix the sign of the imaginary component of the fluctuation term from the sign of the Bijvoet difference (Figure 2c). This aspect of the method has recently been elaborated in the Multiwavelength Anomalous Solvent Contrast (MASC) method (Fourme *et al.*, 1995; *Fourme et al.*, 1996). The latter method appears to have many attractive features, including the fact that in addition to providing a method for envelope determination at low resolution, inclusion of anomalous scatters in the solvent can also lead to specific adsorption onto the protein. In that case, the MAD and MASC methods combine, and a complete phase determination should be possible for all aspects of the structure.

3. Outline of Procedures

The solvent contrast variation method proceeds in an obvious way, with few unusual requirements, as follows:

- Generate appropriate modifications to the solvent.
- Measure isomorphous and/or anomalous $\{|F_{obs,H}|\}$i for mother liquors, i, with different electron density, $P_{solv,i}$.
- Scale together the intensity data for different solvents.
- Estimate amplitudes for the structure factors, $\{|G_H|\}$, and $\{|\Delta_H|\}$ from the variation in $\{F_{obs,i}\}$ with ΔP_{solv}. Computer programs to effect the scaling and parameter estimation redescribed in a subsequent section.
- Determine phases for the envelope amplitudes, $\{|G_H|\}$. Using direct methods to aid the *ab initio* structure determination using contrast variation-derived amplitudes is described in another chapter.

4. Solvent Contrast Variation in Practice

4.1. SELECTION OF SOLVENTS

Solvent electron density can be increased by the addition of salts and soluble nonionic additives that increase the density of water. It can be reduced by adding salts and nonionic additives that decrease the density of water. The principal effect seems to be changes in the water concentration. Polyethylene glycols and lithium sulfate are examples of reagents that dilute water extensively and therefore produce solvent densities substantially lower than the mean protein density. Our experience suggests that choosing solvent systems for contrast variation entails considerations and frustrations similar to those encountered searching for isomorphous derivatives. Solvent systems must induce variation in the solvent electron density without changing either the macromolecular conformation or packing in order that the contrast variation datasets remain isomorphous. The MASC method (Fourme et al.,1995; Fourme et al., 1996) greatly enhances the power of the method, eliminating the absolute requirement for isomorphism. Inert gases, and salts of anomalous scatterers become useful reagents for this purpose.

4.2. SOLVENT ELECTRON DENSITY

Obviously, one must also measure the solvent electron density. This can be done by directly weighing a solution of known composition and obtaining the water concentration by difference. Changes in the concentration of water in a mother liquor is generally the most important contributor to $\Delta\rho_{solv}$. For this reason, gravimetric determination or refractometry is essential. Two useful equations express the water concentration, C_{water} in terms of the specific gravity of the solution and the solvent electron density in terms of the solvent composition:

$$\mathbf{C_{water}} = (1/M_{water}) \times (D - \Sigma M_i \times C_i); \; D = \text{specific gravity, g/L} \quad (1)$$

$$\rho_{solv} = 10^{-27}\,\text{L/Å}^3 \times 6.02 \times 10^{23} \times \Sigma(E_i \times C_i); \; Ei = \text{number of electrons/atom i.} \quad (2)$$

These calculations can and should be checked by measuring the crystal density:

$$\mathbf{D_{cryst}} = (V_{solv}/V) \times \mathbf{D}_{solv} + \{(V_{wat}/V)\times \mathbf{D}_{wat} + (V_{prot}/V)\times \mathbf{D}_{prot}\}; \; \mathbf{D} \quad (3)$$

refers to mass density and Vwat and Dwat refer to the properties of boundwater inaccessible to variation in the bulk solvent.

The contrast variation method was used in a study of the molecular envelope of tryptophanyl-tRNA synthetase (TrpRS) complexed with its substrate, tryptophan (Carter et al., 1990; Crumley, 1989). TrpRS crystals grow in ~2.4M K_2PO_4, for which $\rho_{solv} = 0.412$ e/Å^3. Additional solvents prepared in the study are listed in Table I. Of these, only those in bold face were sufficiently isomorphous, even at low resolution, to be useful in subsequent estimation of the envelope structure factor amplitudes. An interesting feature of this series is the use of both potassium and ammonium selenate salts. These were used at quite different concentrations to obtain solutions with essentially identical solvent electron densities (due again primarily to differences in water concentration). Since selenium has appreciable anomalous electron density, it was originally intended that Bijvoet difference measurements for these two solvents could be used as suggested in Figure 2, to determine the sign of the imaginary component of the fluctuation structure factor. In the event, this was not done, because the one crystal mounted in potassium selenate diffracted poorly and was not sufficiently isomorphous (Crumley, 1989).

It should be noted that Selenium compounds are potentially quite toxic and should be handled with proper care in a fumehood.

4.3. DATA COLLECTION

It is critical to measure all of the lowest resolution data accurately. This requirement makes the contrast variation method somewhat more difficult technically than routine data collection. It is essential to reduce the penumbra of the beamstop shadow to within the radius of the lowest resolution reflections (Figure 3). This generally increases the noise from air scatter markedly. For this reason, it is advantageous to use a helium path between the crystal and detector, and to mount the beamstop carefully on the exit window of the helium path or the entrance window of the detector. The overriding consideration is that the beamstop casts a shadow whose penumbra occludes a much larger angle than does its umbra. Consequently, data out to roughly 15 Å are usually corrupted, giving inconsistent integration and merging.

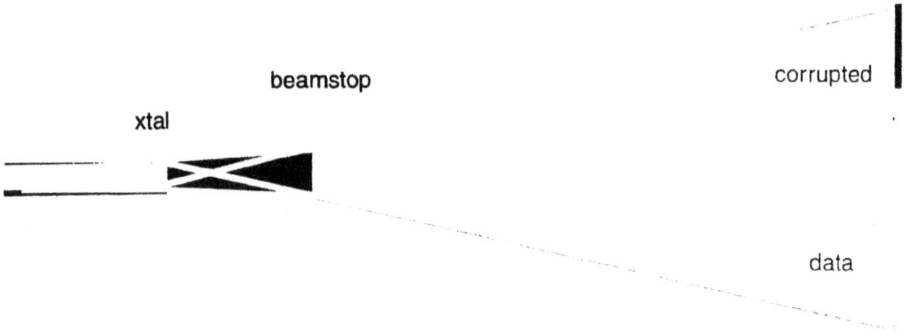

Figure 3. Corruption of low resolution data by the penumbra of the beamstop.

Several strategies can be employed to free these measurements from effects of this penumbra. The simplest modification of the usual setup is to replace the beamstop with a smaller one, which just catches the direct beam, and to move it toward the detector as far as practically possible from the crystal. This procedure requires rather painfully finicky adjustments, however, and a more reliable procedure is to mount the beamstop close to the detector face at the exit window of a conical helium tunnel. The helium path will reduce the intense air scatter generated by the direct beam.

4.4. GFROMF: SCALING AND ESTIMATING THE AMPLITUDE PARAMETERS

GFROMF (Figure 4) is a least-squares computer program written to scale contrast variation data and estimate the $\{|G_H|\}$ and $\{\Delta_H\}$ parameters (Carter et. al., 1990). It uses Bricogne's algorithm (Bricogne, 1982) for estimating partial derivatives to represent the entrainment by the global scale constants of local parameters derived for each reflection. This program was tested extensively using simulations of the contrast variation effect involving a simple and very effective model for the solvent continuum. In this model, each pixel near the molecular boundary was smoothed using a weighted average of the six surrounding pixels that preserved the total volume requirements of molecular and solvent regions. Vector sums of the simulated contrast variation series and envelope and fluctuation structure factors agreed to within ~2% with calculated structure factor vectors for bare protein, in accordance with the models illustrated in Figures 1 and 2. GFROMF recovered estimated amplitudes for the envelope and fluctuation densities with similar fidelity, suggesting that it would work well with real data. Success with real data has been confirmed (Carter et al., 1990; Fourme et al., 1996). Source and simulated data are available from the author (carter@med.unc.edu).

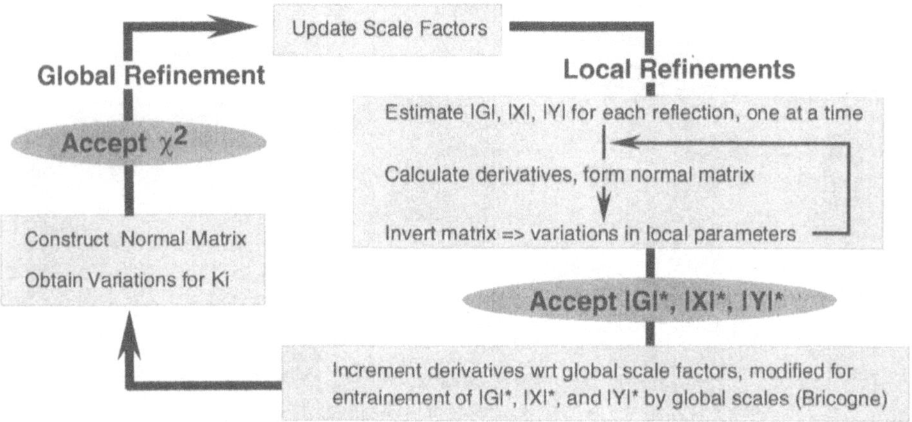

Figure 4. Schematic diagram of the GFROMF program.

One reflection at a time. The contrast variation method has the unexpected virtue of inviting the investigator back into the world of the first protein crystallographers, where each reflection posed an individual puzzle. The two reflections illustrated in Figure 5 are representative of the behavior one is likely to see. According to the sign and magnitude of the real component of the fluctuation term, the behavior of $|F_{obs,\,H}|_i$ with increasing solvent electron density can be ascending, descending, or pass through a minimum. Diagnosing this behavior in advance was helpful in estimating initial values for the x and y parameters indicated in the figure.

$\{|G_H|\}$ values for a macromolecular crystal can be very large. The two examples in Figure 5 both produced $|G_H|$ values nearly an order of magnitude larger than the largest $|F_{obs,H}|_i$. Subsequent work at Lure with real measurements of other solved structures indicate that GFROMF does work well (M. Ramin, personal communication). The SHARP program will eventually provide a stronger framework for dealing with lack of isomorphism, which was a problem in several of the TrpRS datasets.

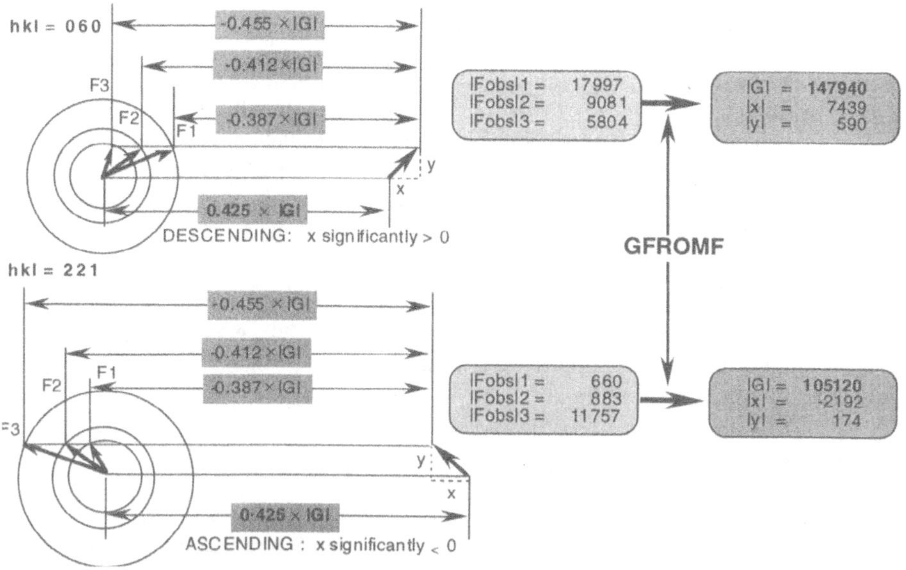

Figure 5. Examples of the fit to two reflections using the contrast variation measurements from the tryptophanyl-tRNA synthetase study (Carter et al.,1990).

Amplitudes of the envelope transform are a source of useful phase information. The coupling between structure factors expected from a spherical envelope transform are discussed elsewhere in this volume (Chapman, 1988) . Experimentally determined values for the actual envelope transform represent considerable potential improvement over the spherical approximation. The algebraic formalism for utilizing these coupling patterns in probabilistic direct phasing have been outlined by Bricogne (Bricogne, 1988) .

$\{|G_H|\}$ and direct phase determination. Resolving $\rho(\mathbf{x})$ for the bare protein into the molecular envelope function, $\chi(\mathbf{x})$, and the fluctuation density inside the envelope, $\delta(\mathbf{x})$ should provide a dataset of amplitudes that are more amenable to direct phase determination than are the $\{|F_{obs}|\}$. In principle and in practice, the envelope transform, $\{|G_H|\}|$, has three characteristic features that encourage this hope:

1. Their large amplitudes lead directly to large E values and important triplet phase invariant probabilities. This statement oversimplifies reality, because the random atom model is not the right probability model for the binary valued envelope

function. Nevertheless, the coupling induced by the strong values should make an appropriate probabilistic approach more effective.

2. The complexity of the envelope does not scale with N. The pattern of solvent channels in a macromolecular crystal does not change dramatically as the unit cell dimensions climb by orders of magnitude.

3. The envelope is proportional to its square. Sayre-equation relations between all structure factor vectors therefore exist and are essentially complete at quite low resolution. Simulations suggest that a significant resolution limit for significant $|G_H|$ values occurs at around 5 Å. This value is consistent with the onset of significant scattering from the solvent continuum at around 4.5 Å. The actual practical limit probably occurs at much lower resolution, say 8-10 Å.

These considerations suggest that the $\{G_H\}$ should be amenable to direct probabilistic phase determination at low resolution. Generation of a molecular envelope by *ab initio* use of direct methods using amplitudes determined experimentally for the envelope transform is described in another chapter (Carter, 1998).

5. References

1. Bentley, G., Lewit-Bentley, A., Finch, J. T., Podjarny, A. D. & Roth, M.(1984). Low Resolution Structure of the Nucleosome Core Particle of Chromatin Analyzed by Neutron Solvent Contrast Variation. *Journal of Molecular Biology* **176**, 55-75.
2. Bragg, W. L. & Perutz, M. F. (1952). The External Form of the Hemoglobin Molecule. I. *Acta Crystallographica* **5**, 277-289.
3. Bricogne, G. (1982). Multiple Isomorphous Replacement: The problem of parameter refinement from acentric reflexions. In *Computational Crystallography* (Sayre, D., ed.), pp. 223-230. Clarendon Press, Oxford.
4. Bricogne, G. (1988). A Bayesian Statistical Theory of the Phase Problem. A Multichannel Maximum-Entropy Formalism for Constructing Generalized Joint Probability Distributions of Structure Factors. *Acta Crystallographica* **A44**, 517-545.
5. Carter, C. W., Jr. (1998). *Ab initio* Determination of a Low Resolution Molecular Envelope by Direct Methods. In *Direct Methods for Solving Macromolecular Structures* (Fortier, S., ed.), pp. In Press. Kluwer Academic Publishing, Dordrecht.
6. Carter, C. W., Jr., Crumley, K. V., Coleman, D. E., Hage, F. & Bricogne, G.(1990). Direct Phase Determination for the Molecular Envelope of *Bacillus stearothermophilus* Tryptophanyl-tRNA Synthetase by X-ray Contrast variation. *Acta Crystallographica* **A46**, 57-68.
7. Chapman, M. (1998). Use of non-crystallographic symmetry in phasing. In *Direct Methods for Solving Macromolecular Crystal Structures* (Fortier, S.,ed.). Kluwer Academic Publishers.
8. Crumley, K. V. (1989). X-ray Solvent Contrast Variation: Obtaining A Molecular Envelope Structure for *Bacillus stearothermophilus*Tryptophanyl-tRNA Synthetase. Masters, University of North Carolina at Chapel Hill.
9. Fourme, R., Shepard, W., Kahn, R., l'Hermite, G. & de La Sierra, I. L.(1995). The Multiwavelength Anomalous Solvent Contrast (MASC) Method in Macromolecular Crystallography. *Journal of Synchrotron Radiation* **2**, 36-48.
10. Fourme, R., Shephard, W. & Kahn, R. (1996). Application of the Anomalous Dispersion of X-rays to Macromolecular Crystallography. *Progress in Biophysics and Molecular Biology* **64**, 167-199.

11. Harrison, S. C. (1969). Structure of Tomoto Bushy Stunt Virus I. The Spherically Averaged Density. *Journal of Molecular Biology* **42**,457-483.
12. Harrison, S. C. & Jack, A. (1975). Structure of Tomato Bushy Stunt VirusIII. Three-dimensional X-ray Diffraction Analysis at 16 Å *Resolution. Journal of Molecular Biology* **97**, 173-191.
13. Roth, M., Lewit-Bentley, A. & Bentley, G. A. (1984). Scaling and Phase Difference Determination in Solvent Contrast Variation Experiments. *Journal of Applied Crystallography* **17** (77-84).

OBTAINING PHASES FROM ELECTRON MICROSCOPY FOR SOLVING PROTEIN STRUCTURES

Tribute to Boris Vainshtein (1921-1996)

Sven Hovmöller
Structural Chemistry,
Stockholm University, S-106 91 Stockholm, Sweden

This paper is given in honour of Professor Boris Vainshtein, one of the founders of electron crystallography. He was an invited speaker at this NATO Advanced Study Institute on Electron Crystallography, but regrettably died half a year before it took place. Already in his classical book *Structure Analysis by Electron Diffraction* from 1964 [1] he wrote: "There is no doubt now that electron diffraction may be used for the complete analysis of crystals whose structure is unknown".

1. The phase problem in crystallography

The phase problem is one of the most intriguing phenomena in X-ray crystallography. Among all mathematical problems, the phase problem may be one of the most studied, in terms of man-years. Thousands of crystallographers have tried to solve the phase problem. None has succeeded but many have been successful. Still today there is no general and complete solution to the phase problem in X-ray crystallography, but partial solutions, or solutions of special cases, have been essential parts of at least 8 Nobel prizes in chemistry and physics over the years. The two latest of these were awarded in chemistry in 1982 and 1985. Aaron Klug at the MRC Laboratory of molecular biology in Cambridge received the Prize in 1982 "for his development of crystallographic electron microscopy and his structural elucidation of biologically important nuclei acid-protein complexes"[2]. In 1985 it was awarded to Herbert Hauptman and Jerome Karle "for their outstanding achievements in the development of direct methods for the determination of crystal structures"[2]. These direct methods revolutionised the field of structure determination of organic molecules with up to and even above 100 atoms.

It is easy for X-ray crystallographers to get into the idea that these phases are mere mathematical abstractions, since the crystallographic structure factor phases cannot be directly determined experimentally in an X-ray diffraction pattern. The use of words such as "real" and "imaginary" for phases contribute to this feeling. Actually, the phases are very concrete physical entities, as hopefully will come clear in the following. It is quite the other way with the unit cell - it is a concept which is easily understood although it is

S. Fortier (ed.), Direct Methods for Solving Macromolecular Structures, 239–250.

an abstraction - there are no small boxes in the crystal where the molecules are put inside! One remarkable advantage of electron crystallography is that the enigmatic phases can actually be observed and measured in the electron microscope. This paper will discuss the phase problem and relations between the scattering of X-rays and electrons, real and reciprocal space, images and diffraction. The questions are: Are the phases measured from electron microscopy images the same as those of the X-ray structure factors and can they be used for helping to solve the phase problem in protein X-ray crystallography?

2. Dualism in crystallography

A fascinating aspect of the life of crystallographers is that we tend to spend half our time in real space and the other half in reciprocal space. These dual worlds, also called direct space and Fourier space, are at the same time totally different and closely related. In terms of information content, they may even be considered identical. The phases have a key role in relating these dual spaces.

There are many other important pairs in crystallography, some of which will be the subject of this review, namely:

electrons - X-rays
particles - waves
images - diffraction
wave front phases - structure factor phases
strong interaction - weak interaction
small objects - large objects
low resolution - high resolution
molecular biology - inorganic chemistry
amorphous - crystalline

3. Electrons and X-rays

If we want to determine the positions of atoms, we need a probe which is able to detect distances of the order of 1 Ångström (10^{-10}m). The three kinds of radiation used in crystallography are X-rays, electrons and neutrons. It is possible to tune the wavelength of each of them, such that an optimal wavelength is used for the experiment. By far most crystallographic studies are carried out using X-rays, with neutrons on second place, at least if we talk about determination of atomic co-ordinates. Neutrons are especially useful because they, unlike X-rays, interact relatively strongly with hydrogen, which is an extremely important element in biology. However, here we will concentrate on electrons and X-rays and their interrelations.

X-rays are created by bombarding a metal plate with electrons accelerated to some 40 kV. If the metal is copper, then the main wavelength (λ) of the X-rays is CuK_α at 1.54 Å, incidentally the same as the length of the C-C single bond, the most important bond for life. When X-rays pass through matter, they interact with the electrons surrounding the atomic nuclei, but not with the nuclei themselves. Yet, since most electrons are located very close to the atomic nuclei, X-rays give information mainly about the positions of atoms, rather than about the more diffuse shapes of valence electrons. The interaction is rather weak - the mean free path length is in the order of a millimetre. If the wavelength is shorter (for example MoK_α with $\lambda = 0.71$ Å), the interaction becomes even weaker.

If the accelerated electrons are produced in an electron microscope, they can themselves be used to study the structure of matter. The most common accelerating voltages for electron microscopes today are 100 kV to 400 kV. The electrons reach a velocity of about half the speed of light and have a wavelength of about 0.04 to 0.02 Å. The electrons interact with the electrostatic potential in matter, *i.e.* they sense the positive and negative charges. All atoms are made up of positive nuclei surrounded by negative electrons. Since electrons are themselves negatively charged, they interact very strongly with all these charges when passing through an object. The electrons are attracted by the positive nuclei and repelled by the electrons. The distribution of charges in solids is such that the positive charges are concentrated in the extremely small nuclei, with most of the negatively charged electrons also located close to these nuclei, partially shielding the positive charges. The electrostatic potential is thus a set of rather sharp positive peaks on a low and relatively smooth background. The mean free path length is very short for electrons, in the order of nanometers. Because of the strong interaction of electrons with matter, extremely small samples can be studied. The optimal crystal thickness is less than 100 Å. A crystal structure determination can be carried out on a crystal of less than 100 unit cells, as compared to 10^{12} to 10^{15} needed for single crystal X-ray diffraction [3]. In fact, even single atoms can be studied by electron microscopy, opening up the world of crystal defects and even amorphous materials for study at the atomic level. As with X-rays, if the wavelength is shorter (higher accelerating voltage), the energy becomes higher but the interaction becomes weaker.

In spite of the facts that X-rays and electrons are so different and interact with different features in matter, in both cases it all boils down to sensing the positions of the centers of atoms. We may therefore expect similar results from structure investigations carried out with X-rays and electrons, and a close interaction between the scientific fields of X-ray crystallography and electron microscopy. For various practical and historical reasons, some of which will be discussed in the following, this is not really the case. But there are strong reasons why science in general and crystallography in particular would gain from a stronger interaction between these two fields.

4. Images, diffraction and phases

The mathematical definition of an image involves a one-to-one relation between the points in an object and its image. One fascinating aspect of the Fourier transform (FT) of an object is that it is not at all an image, yet contains the same information as the image. Each point in an object contributes to every point in its FT and *vice versa.* The FT of a periodic object is especially useful, in that the continuous (analogue) function of the object is condensed into a discrete (digital) function; the Fourier coefficients.

Only the first Fourier coefficient is non-zero on average - it gives the average value of the function. In the case of an X-ray diffraction study it is the F(000) term which is the average electron density within the unit cell. For a protein this term is 0.38 electrons/Å^3. All the other Fourier coefficients are cosine waves, only modulating the average electron density, so as to concentrate the density into sharp peaks (where the atomic nuclei are) on a rather smooth background of some 0.01 electrons/Å^3. Each Fourier coefficient is characterised by three numbers; the index (h,k,l) giving the direction and periodicity of the cosine wave, an amplitude and a phase. The phase is the position of the maximum of the wave relative to the origin of the co-ordinate system.

If a function is known, the calculation of its FT is straight-forward, although very time-consuming. The FT of a 2D image consisting of 256 x 256 points (pixels) also consists of 256 x 256 points. Each point in the FT is the sum of every point in the image after having been multiplied by a 2D cosine function. Even at the rate of making one such calculation per second, it would take more than a life-time to calculate a 256 x 256 FT by hand. Thus image processing involving Fourier transforms did not emerge as a science until the computers became fast enough. In the late 1960ies, when Klug and co-workers started image processing, a 256 x 256 FT was an over-night job at the single main-frame computer of the university. Today such FTs can be calculated in a fraction of a second on a personal computer. The inverse FT gives back the original function.

The bright spots seen in a diffraction pattern (often called the reflections) are the squares of the Fourier coefficients. From the positions of the reflections we can immediately get the indices. The intensities are equal to the squares of the amplitudes. However, we can not calculate the inverse Fourier transform, since there is no information about where the maxima of the cosine waves are positioned relative to each other or to the origin of the unit cell. The phase information is lost in a diffraction pattern. If the atomic co-ordinates are known, the FT (with phases) can easily be calculated. The crux is that we do not know the atomic co-ordinates when we start a crystal structure determination.

An image, for example an electron micrograph, is a two-dimensional function. Its FT is easily calculated on modern computers and ***does contain the phases***. That this is really the case should be clear from Figure 1. It is worth emphasising that here the word *phase* is used for crystallographic structure phases. The reason why the crystallographic phases are present in EM images, but not in electron diffraction or X-ray diffraction patterns, is that the different diffracted beams are focused into an image by the lenses in the EM. Each diffracted beam carries information about the relative atomic positions on Bragg planes through the crystal. In the image all the contributing reflections interact, with phases, giving rise to a projection of the crystal structure. Since there are no lenses for X-rays, no X-ray microscope can be constructed and thus we only have X-ray diffraction patterns, where the phase information is lost.

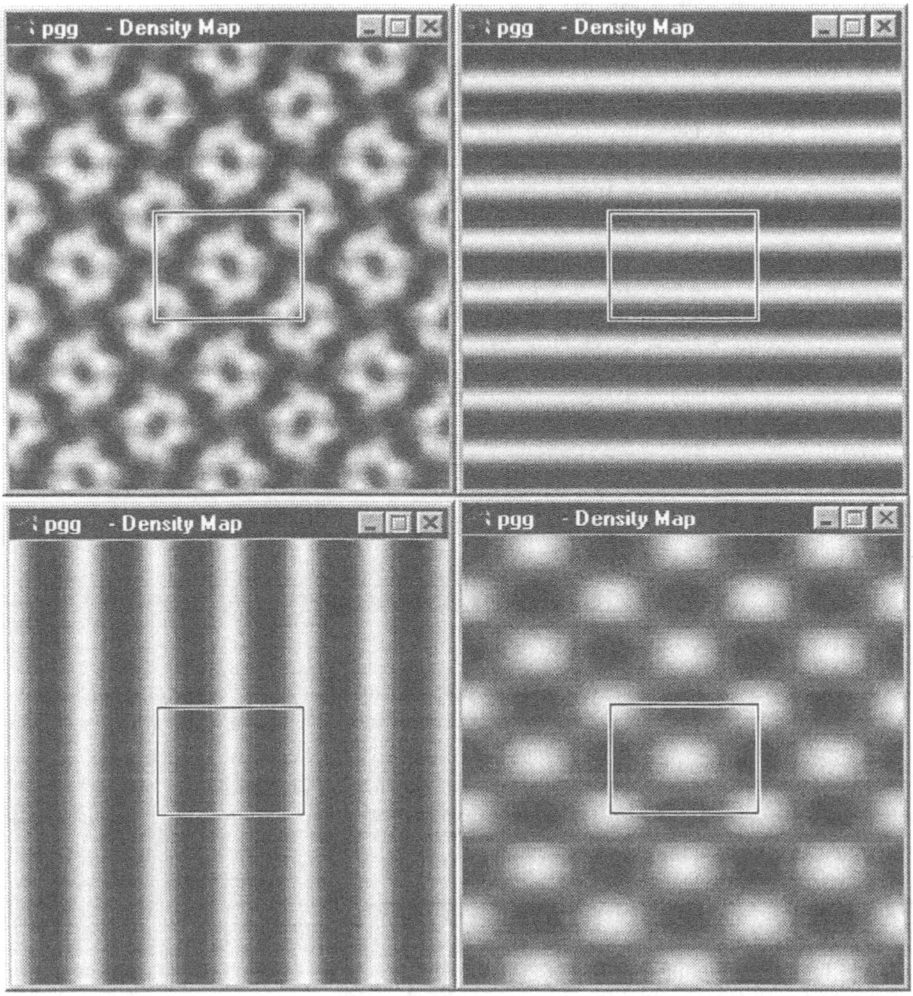

Figure 1. ***The physical meaning of phases***

Top left: An EM image of a negatively stained membrane crystal of cytochrome reductase [17]. One unit cell is indicated; a = 174Å, b = 137Å. Protein is white. At this resolution (25 Å) the protein complex is seen as a white ring. The internal features (α-helices and β-sheets etc.) cannot be seen, since the resolution is too low. The crystal is orthorhombic with a projected centro-symmetric symmetry pgg. There are glide planes g both vertically and horizontally, giving rise to a zigzag arrangement of the molecules. Each ring is a dimer. There are two dimers in each unit cell; one in the centre and one in the corner.

Top right: The (0 2) reflection has a phase of 180° since the electron density of the protein is at its minimum at the origin of the unit cell. (The minimum of the cosine function is -1, obtained for cos180.)

Bottom left: The phase of (2 0) is 0°.

Bottom right: The pair of two symmetry-related reflections (1 1), going \\ and (1 -1) going //, are shown together. Their phases are both 0°, since they have maximum density (white) at the origin of the unit cell.

This electron microscopy image is composed of 26 unique reflections. If their phases were not known from an EM image, each of these 26 reflections could have any of the two phase values allowed for centro-symmetric crystals; 0° or 180°, resulting in as many as $2^{26} = 64 \cdot 10^6$ *possible combinations.*

5. The phase confusion problem

It has frequently been claimed that there should be a phase problem also in electron microscopy. [4,5] It may be more correct to say that there is no phase problem, only a phase confusion problem in electron microscopy. The phase confusion problem has arisen since the word phase is used for (at least) two very different physical entities in the field of electron microscopy. When crystallographers speak about phases, they mean the ***crystallographic structure factor phases***, while physicists usually mean the ***wave front phases***. In order to resolve the phase confusion problem, it is necessary to define clearly which phases are meant. Some of the most prominent differences between the structure factor phases and the wave front phases are shown in the Table 1 below. A full description of the process crystal structure → structure factors → electron wave → lenses → image formation → structure restoration with mathematical and physical formulas has been published by Zou [6].

TABLE 1. A comparison of some central properties of crystallographic structure factor phases and wave front phases.

Entity	Crystallographic structure factor phases	Wave front phases
The waves are	stationary standing waves	moving - propagating
Wave length	≈ 1 - 20 Å	≈ 0.03 Å
Physical entity	electron density or potential	electromagnetic wave
Phases relative to	symmetry elements in crystal	incident electron wave front
Phase information	present in EM images	lost in EM images
Method of retrieval	Fourier transform of EM image	holography, focal series EM
Interrelations	modulates the wave front	records the structure factor

6. Small and large objects at high and low resolution

It is illuminating to compare the sizes of objects studied by electron microscopy and X-ray crystallography during different periods of the 20th century. In spite of the much shorter wavelength of electrons (0.02 - 0.04 Å) compared to that of X-rays (0.5 - 2 Å), electrons are actually best suited for studies of large objects at low resolution, while X-rays are ideal for structure determination of smaller structures at atomic resolution.

X-ray crystallography started in 1912 with the determination of exact atomic co-ordinates for inorganic crystals with small unit cells and then gradually developed to handle larger and larger molecules and unit cells. Electron microscopy, on the other hand, started off 20 years later, with a resolution comparable to that of a light microscope, which is useful for studying large systems such as whole cells, bacteria and organelles. As the electron optical lenses were improved, the resolution of electron microscopes improved. It was only in the early 1970-ies that the EM resolution reached values around 5 Å, allowing the visualisation of details within a unit cell of an inorganic crystal. The historical development of X-ray crystallography and electron microscopy can be considered as running in opposite directions, as shown in Table 2.

TABLE 2. Main new categories of structures studied by X-rays and electrons during different periods of the 20th century

Period	**X-ray crystallography**	**Unit cells (in Å)**	**Electron microscopy**	**Resolution (in Å)**
1910-1930	Inorganic crystals	3-10	-	-
1930-1950	Organic molecule	5-20	Whole cells and organelles	50
1950-1970	Proteins	30-100	Viruses and ribosomes	15
1970-1990	Viruses	200-400	Proteins, inorganic crystals	5-2
1990-	Ribosomes	600-1000	Atomic resolution	1-2

X-ray crystallography has had a tremendous impact on chemistry and has been the main method providing us with details of how the atoms are arranged in molecules and crystals. The solving of the double helix DNA structure by Watson and Crick and haemoglobin and myoglobin by Perutz and Kendrew in the early 1950-ies marked the birth of molecular biology. At this time, electron microscopy had shown that the inside of a cell was much more complex and ordered than had been thought before. There was not just a nucleus and a soup of cytoplasm, but a large number of highly specialised and structured organelles within the cell, such as the mitochondria, the endoplasmic reticulum, the Golgi apparatus and so on. This marked the birth of another science: cell biology.

7. Membrane protein structures solved by electron microscopy

By 1975, about 200 water soluble proteins had been determined to (near) atomic resolution by X-ray crystallography, but very little was known about the structures of the hydrophobic membrane proteins, because they were so difficult to purify and crystallise. It was then the revolutionary structure determination of the membrane protein bacteriorhodopsin (= purple membrane) was completed by Henderson and Unwin [7]. They took advantage of the natural tendency of membrane proteins to insert themselves into lipid bilayers. If lipids and membrane proteins are mixed in a ratio close to 1:1, the protein molecules sometimes aggregate into a 2D crystal. This happens *in vivo* with the bacteriorhodopsin of *Halobacterium halobium.* Such a membrane crystal is only 40-100 Å thick which is far too thin for single crystal X-ray diffraction, but ideal for electron microscopy. Once Unwin and Henderson had overcome several difficult technical problems, especially how to take EM images of the tremendously radiation sensitive unstained proteins, they obtained a 3D structure at 7Å resolution, showing rods of density perpendicular to the plane of the membrane. This was correctly interpreted as being α-helices composed of 20-25 hydrophobic amino acids. We now know that this is a very general building principle of membrane proteins, and it gave inspiration to the hydrophobicity plots, where amino acid sequences of unknown proteins are searched by computers for stretches of 20-25 consecutive hydrophobic amino acids which, if found, give a good indication that the protein is membrane bound. It took another ten years until Hartmut Michel managed to make 3D crystals of a membrane protein suitable for X-ray crystallography and together with Deisenhofer and Huber solve its 3D structure to atomic resolution. For this work on the photosynthetic reaction centre, they received the Nobel Prize in chemistry 1988.

Two other membrane proteins have been solved in 3D to atomic resolution by electron microscopy; light harvesting complex (Kühlbrandt and Wang, [8]) and porin by Jap *et al.* [9], see below.

8. The large mitochondrial membrane complexes

Until just a couple of years ago, all we knew about the 3D structures of the huge mitochondrial membrane protein complexes, numbered I-V, came from electron microscopy investigations of 2D crystals stained with heavy metal salts {(I) NADH: ubiquinone oxidoreductase, (III) cytochrome reductase, (IV) cytochrome oxidase and (V) the F_1F_0-ATPase}. Unfortunately, severe mistakes were made on the structure interpretations of complex II and cytochrome oxidase.

Crystals that were for a long time claimed to be of the entire complex I [10] turned out only to be from a much smaller sub-complex. This was clear from a calculation of the molecular mass of the protein per asymmetric unit. While the unit cell (a = b = 153 Å, c = about 105Å, giving a total cell volume of 2465 nm^3) could possibly contain two molecules of complex I, the asymmetric unit cell when taking the crystallographic symmetry $p42_12$ into account is only 1/8th of the whole unit cell. Assuming 50% protein in the crystal, the volume of the protein complex is then 2465 000 Å^3/16 = 154 000 Å^3. The volume per non-hydrogen atom in proteins is 20 Å^3, and the average molecular mass per non-H atom is 14.4, resulting in a volume of 20/14.4 = 1.4 Å^3/dalton. Thus, the molecular mass of the crystallised protein had to be about 110 kDa; much smaller than that of complex I which was estimated to be at least 600 kDa. The crystals were subsequently identified as made up of only a small subcomplex of complex I [11]. However, Leonard *et al.* later succeeded in crystallising the entire complex I [12].

Cytochrome oxidase was the first membrane protein to be crystallised and studied by electron microscopy, but all interpretations of its 3D structure were wrong since they are based on the hypothesis by Henderson and Capaldi that the crystals formed in Triton were composed of two membranes per unit cell, with the cytochrome oxidase inserted very asymmetrically into the membrane[13], whereas in fact there is only a single membrane in the unit cell passing through the middle of the proteins[14]. Consequently, the crystals made using deoxycholate as the detergent are not monomers[15], but dimers. This interpretation was recently confirmed by the atomic resolution structure of cytochrome oxidase solved by X-ray diffraction [16], where the membrane passes through the centre of the molecule.

The 3D structure of Complex III, cytochrome reductase was solved from electron microscopy images of membrane crystals to 25 Å resolution by Weiss and Leonard [17].

Already in the 1960ies electron microscopy images of the inner mitochondrial membrane had shown protruding "lollipops" which were interpreted as the water soluble part (also called F_1) of the ATPase complex V. The correctness of this interpretation of low-resolution data was confirmed when the complete 3D structure of F_1 to 2.8 Å resolution was solved by John Walker *et al.* [18], using X-ray crystallography.

9. Ribosomes

The ribosomes are among the largest and most complicated of all biological complexes. All we know today about the 3D structures of ribosomes has come from electron microscopy, but in this case mainly from studies of single particles rather than crystals. Randomly oriented ribosomes have been rapidly frozen into ice and studied by cryomicroscopy without adding any stains. Thousands of individual ribosomes have been added after their relative orientations have been found. These procedures make use of advanced multi-variate statistics and have been developed mainly by Joachen Frank and Marin van Heel [19]. It has also been possible to crystallise these huge molecular complexes into 3D crystals which diffract X-rays to high resolution, as shown by Ada Yonath *et al.* [20], but the tremendous task of solving the ribosome structure to atomic resolution will most likely not be finished until next century.

10. Electron crystallographic studies of organic and inorganic compounds

Although the scattering of X-rays and electrons are so different, in effect they both probe positions of atomic nuclei. It is now nearly half a century since Vainshtein and Pinsker in 1950 localised hydrogen atoms in paraffin by electron diffraction data [21]. The Moscow group succeeded in the structure determination of several inorganic compounds from electron diffraction data already in the 1950ies. One advantage of electrons pointed out by Vainshtein was that the relative scattering power of lighter elements, including hydrogen, is higher for electrons than for X-rays. The heavy elements are not so heavy for electrons.

There was an unfortunate and long period of almost no structure determinations by electron crystallography outside the Soviet Union, mainly because of exaggerated fears for the effects of multiple scattering of electrons within the crystals. One of the very few who pursued such investigations was Douglas Dorset whose book [22] is the only modern complete treatise on electron crystallography on organic molecules. Dorset has solved numerous lipids and other organic molecules from electron diffraction data and also shown that it is possible to phase such data by the direct methods used in X-ray crystallography.

The spectacular results on membrane proteins led to a restart of electron crystallography studies also of inorganic compounds, but now using the full power of electron microscopy and image processing rather than only electron diffraction that had been used by the Moscow group. Several metal oxides were solved by electron microscopy and image processing, giving atomic co-ordinates of the metal atoms accurate to within about 0.1 Å [3]. The first refinement of an inorganic structure against selected area electron diffraction data to better than 1 Å resolution, reaching an accuracy of 0.02 Å was recently completed by Weirich *et al.* [23]. After nearly one hundred years, the circle is closed and electrons and X-rays can be used for crystal structure determination in all fields of chemistry, from inorganic solids over organic molecules to proteins, with more or less similar accuracy.

11. Comparison of phases obtained by electron microscopy and X-ray diffraction

As shown above, electron microscopy can give structure factor phases. If these phases are similar to the X-ray structure factor phases, they may be used as a starting set for phase extension, for example by direct methods or maximum entropy. Even if only a few low-resolution reflections can be found, this can be very helpful in determining the packing of proteins in the unit cell. If reliable phases could be determined to intermediate resolution, 5-7 Å or so, then the complete solution of a protein structure should be straight-forward.

We must ask: how similar are the amplitudes and phases obtained from the two techniques X-ray and electron crystallography? Can phases obtained by electron microscopy be used for phasing X-ray diffraction data of proteins? Strangely enough, it turns out that very little work has been carried out trying to answer these questions.

Already in 1975, Unwin and Henderson studied thin crystal plates of the protein catalase by electron microscopy and obtained a projection map to 9Å resolution [24]. Later, several different crystal forms of catalase have been determined by X-ray crystallography, but none of these were from the same crystal form as the one studied by electrons. Only a few other proteins have been studied to high resolution by both electrons and X-rays. Unfortunately, none of these structure determinations were on the same crystal form, as would be necessary if the electron microscopy phases were to be used directly for phasing the X-ray diffraction amplitudes. However, in the case of inorganic structures, it has been shown that indeed the phases obtained from electron microscopy images, are nearly identical to the ones obtained after refinement of X-ray crystallography data on the same crystal, at least to 2.5 Å resolution [25].

The membrane protein porin has been studied by both electron microscopy and X-ray crystallography. The structure is very interesting and beautiful. As had been predicted from optical dichroism measurements, the structure is almost entirely built up from β-pleated sheets. A barrel of 16 anti-parallel β-strands is embedded in the membrane. Unlike the completely hydrophobic α-helices, these β-strands contain many hydrophilic amino acid side chains. In this barrel, alternate side chains point out towards the lipids and in towards the middle of the protein. All the hydrophilic amino acids are on the inside, where they form a hydrophilic channel, allowing transport of specific water-soluble molecules across the membrane. All side chains pointing out towards the lipid bilayer of the membrane are hydrophobic. Again the crystal forms studied by X-rays [26], and electrons [9], are not the same. However, the projected structures of porin obtained independently by the two methods are extremely similar.

A similar case is the structure of the protein streptavidin [27], which also shows very good correspondence between X-ray and electron crystallographic results, but again the molecules crystallised in different space groups, making a direct numerical comparison of phases difficult.

What reasons can there be that so few, if any, proteins crystals have been studied by both X-rays and electrons? One reason may be that only a few laboratories master both of these two techniques. Another may be that the crystal forms ideally suited for electrons are extremely thin plates of one or just a few unit cells, whereas the crystals that can be studied by single crystal X-ray diffraction must be at least 1000 unit cells thick, corresponding to about 10 μm. The chunky crystals that are optimal for X-rays are not suitable for electron diffraction. Early attempts [28], to fix thick 3D protein crystals for example in tannic acid and then cut them with an ultramicrotome seems not to have been pursued.

Cryo-microscopy is today the main technique for studying proteins at atomic resolution by electrons. This has the advantage over the classical method of negatively stained samples (using metal salts) that not only the envelope but also internal features of the protein are visualized. Perhaps small 3D crystals could be deep frozen in ice and then crushed into smaller pieces, thin enough to be transparent to electrons. Considering the amount of work involved in phasing proteins, for example making one or several isomorphous heavy atom derivatives, it should be worth developing this combined approach of electrons and X-rays.

12. References

1. Vainshtein, B.K. (1964) *Structure Analysis by Electron Diffraction.* Pergamon Press, Oxford.
2. http://www.nobel.se/ Home page of the Nobel Foundation, Sweden
3. Hovmöller, S. (1992) Structure Determination of Inorganic Structures by Electron Microscopy and Crystallographic Image Processing. In: Electron Crystallography, Trans. of the American Crystallographic Association. Ed. D.Dorset. **28**, 95-103.
4. Van Dyck, D. and de Beeck, M. (1992) *Scanning Microscopy Supplement* **6**, 115.
5. BRITE-EURAM Project No. BE-3322 (1996) *Ultramicroscopy* **64** Entire volume.
6. Zou, X.D. (1995) *Electron Crystallography of inorganic structures - Theory and practice*
   Chem. Commun. 5. Stockholm University. Doctoral thesis.
7. Henderson, R. and Unwin, P.N.T. (1975) Three-dimensional model of purple membrane obtained by electron microscopy. *Nature* ***257***, 28-32.
8. Kühlbrandt, W., Wang , D.N., Fujiyoshi, Y. (1994) Atomic model of plant light-harvesting complex by electron crystallography *Nature* **367**, 614-621.
9. Jap, B.K., Walian, P.J. and Gehring, K. (1991) Structure of porin *Nature* **350**, 167-
10. Boekema, E.J., Van Heel, M.G. and Van Bruggen, E.F.J. (1984) *Biochem. Biophys. Acta* **787**, 19-26.
11. Brink,J., Hovmöller, S, Ragan, C.I., Cleeter, M.W.J., Boekema, E.J and van Bruggen,E.F.J. (1988) The Structure of NADH:Ubiquinone Oxidoreductase from Beef Heart Mitochondria. Crystals Containing an Octameric Arrangement of Iron-Sulphur Protein Fragments. *Eur. J. Biochem.* **166**, 287-294.

12. Leonard, K., Haiker, H. and Weiss, H. (1987) Three-dimensional structure of NADH:Ubiquinone reductase (Complex I) *J.Mol.Biol.* **194**, 277-286.
13. Henderson, R., Capaldi, R.A. and Leigh, J.S. (1977) Arrangement of cytochrome oxidase molecules in two-dimensional vesicle crystals. *J.Mol.Biol.* **112**,631-648.
14. Hovmöller, S. (1990) Structure of cytochrome oxidase. An alternative interpretation. Manuscript submitted to *J. Mol. Biol.* but rejected.
15. Fuller, S.D., Capaldi, R.A. and Henderson, R. (1979) Structure of Cytochrome c Oxidase in Deoxycholate-derived Two-dimensional crystals. *J. Mol. Biol.* **134**, 305-327.
16. Tsukihara, T., Aoyama, H., Yamashita, E., Tomizaki, Yamaguchi, H., Shinzawa-Itoh, K., Nakashima, R., Yaono, R. and Yoshikawa, S. (1995) Structures of metal sites of oxidized bovine heart cytochrome c oxidase at 2.8 Å. *Science* **269**, 1069-1074.
17. Leonard, K., Wingfield, P., Arad, T. and Weiss, H. (1981) Three-dimensional structure of Ubiquinol:cytochrome c reductase from Neurospora mitochondria determined by electron microscopy of membrane crystals. J.*Mol.Biol.* **149**, 259-274.
18. Abrahams, J.P., Leslie, A.G.W., Lutter, R. and Walker, J. (1994) Structure at 2.8 Å resolution of F_1- ATPase from bovine heart mitochondria. *Nature* **370**, 621-628.
19. Van Heel, M. and Frank, J. (1981) Use of multivariate statistics in analysing the images of biological macromolecules. *Ultramicroscopy* **6**, 187-194.
20. Evers, U., Franceschi, F., Boeddeker, N., Yonath, A. (1994) Crystallography of Haliphilic Ribosomes: the Isolation of an Internal Ribonucleoprotein Complex, *Bioph. Chem.*, **50**, 3-16.
21. Vainshtein, B.K. and Pinsker, Z.G. (1950) Determination of the hydrogen positions in the crystal structure of a paraffin. *Dokl. Akad. Nauk SSSR* **72**, 53-56.
22. Dorset, D. (1995) *Structural Electron Crystallography,* Plenum Press, New York.
23. Weirich,T., Ramlau,R. Simon,A., Zou,X.D. and Hovmöller,S. (1996) A crystal structure determined to 0.02 Å accuracy by electron microscopy. *Nature* **382**, 144-146.
24. Unwin, P.N.T. and Henderson, R. (1975) Molecular structure determination by electron microscopy of unstained crystalline specimens. *J. Mol. Biol.* **94**, 425-440.
25. Zou, X.D., Sundberg, M., Larine, M. and Hovmöller, S. (1996) Structure projection retrieval by image processing of HREM images taken under non-optimum defocus conditions. *Ultramicroscopy* **62**, 103-121.
26. Weiss, M.S., Abele, U., Weckesser, J. Welte, W. Schiltz, E. and Schulz, G.E. (1991) Molecular architecture and electrostatic properties of a bacterial porin. *Science* **254**, 1627-1630.
27. Avila-Sakar, A.J. and Chiu W. (1996) Visualization of β-sheets and side-chain clusters in two-dimensional periodic arrays of streptavidin on phospholipid monolayers by electron crystallography, *Biophysical Journal* **70**, 57-68.
28. Akey, C.W., Spitsberg, V. and Edelstein, S.J. (1983) Electron microscopy of beef heart F_1-ATPase crystals. *J. Biol. Chem.* **258**, 3222-3229.

HALF TITLE PAGE

3. APPLICATIONS OF DIRECT METHODS TO MACROMOLECULAR STRUCTURES

SOLVING A HEAVY-ATOM SUBSTRUCTURE BY DIRECT METHODS

Charles W. Carter, Jr.
Department of Biochemistry and Biophysics,
CB #7260 University of North Carolina at Chapel Hill,
Chapel Hill, NC 27599-7260

1. Isomorphous Displacement and the Simplification of Complex Structures

Strategies for solving a new, unknown crystallographic phase problem can generally be divided into two categories. Statistical coupling between structure factors within a single dataset can be used to refine phases from random starting phase sets or from a partial structure obtained, for instance by Patterson methods. Success with this "direct" approach requires high redundancy in the data. The requisite redundancy can arise either from atomic resolution datasets or from very high noncrystallographic symmetry (Rossmann, 1972), and is necessary to sufficiently "over determine" the structure.

Most macromolecular structures do not afford the high redundancy required for successful application of direct methods. In these cases the successful strategy invariably involves simplifying the structure by isolating amplitude data for a solvable subset of the structure from measurements on multiple crystals or at multiple wavelengths. Once solved, this known subset affords an "isomorphous displacement" of multiple datasets by a known vector of structure factors corresponding to the transform of the known substructure. This isomorphous displacement vector reduces the complexity of the macromolecular phase problem and leads to phases for the entire structure by triangulation (Figure 1).

S. Fortier (ed.), Direct Methods for Solving Macromolecular Structures, 253–273.

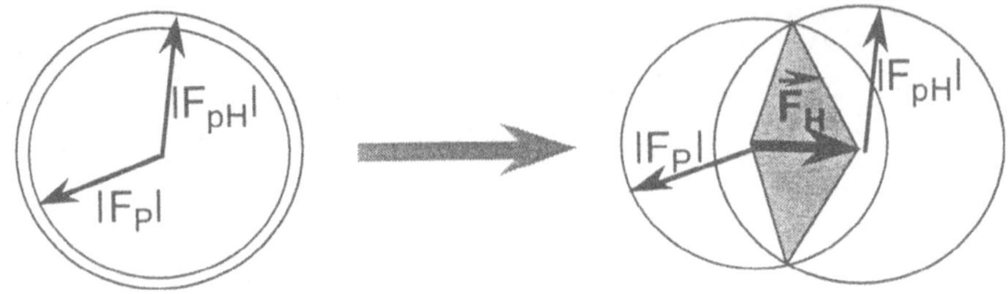

Figure 1. Isomorphous displacements and macromolecular phase determination. A substitution structure that has been located in the asymmetric unit provides a displacement vector of known amplitude and phase that must connect the centers of phase circles for amplitudes from isomorphous crystals whose unit-cell contents differ by the substitution. In the displacement simplifies the macromolecular phase problem by reducing it to a triangulation involving the phased displacement vector and the two known amplitudes.

Except in cases where an extensive partial structure is afforded by molecular replacement, this "displacement" method is stronger than partial structure or "heavy-atom" methods, because of the power of reflection-by-reflection triangulation. This applies to SIR, MAD and MIR methods, as well as to solvent contrast variation (Carter, 1998). It is an important "Achilles' tendon" of the molecular replacement method that the corresponding isomorphous displacement vector is unknown and essentially impossible to determine independently of solving the structure. Although they are used with increasing frequency, borrowed phases derived from a known structure using the molecular replacement method are vulnerable to the errors represented by these unknown displacement vectors. These problems arise from the strong bias introduced by the fact that the dominant source of structural information in reciprocal space is the phases (Read, 1997). They are difficult to detect and/or correct because of the intensely multimodal properties of the crystallographic R-factor and the serious limitations of conventional least-squares used in model refinements.

Experimental macromolecular phasing methods thus generally require Δ|F| values approximating the Fourier spectrum of a simpler, solvable structure. Complexity can be reduced in this manner by two limiting strategies, wherein differences between multiple datasets can be related to the scattering from simpler, potentially solvable structures:

- Changing the scattering characteristics of the solvent continuum by the solvent contrast variation (Carter *et al.*,1990) or MASC (Fourme *et al.*, 1995; *Fourme et al.*, 1996) method. These methods are described in other chapters in this book (Carter 1998a and Carter 1998b).
- Introducing and/or exploiting anomalous dispersion from strong, discrete scatterers.

In any isomorphous displacement method, the substructure must be "solved" before phased structure factor vector displacements can be determined. Since the substructure represented by difference amplitudes is simpler than that of the macromolecule, solution of its phase problem is also simpler. It is often also more amenable to solution using probabilistic direct methods. Thus, for most macromolecular problems, the most immediate practical relevance of direct methods is the extent to which they can expedite solving structures that provide these known isomorphous displacement vectors.

2. Conditional Probability Relationships From Direct Methods and the Solution of Discrete, Heavy-Atom Substructures.

I will focus on two aspects of this topic. First, I will review the applications of traditional "direct methods" (Mukherjee *et al.*, 1989; Navia & Sigler, 1974; Neidle, 1973; Sheldrick, 1993; Steitz, 1968). Then, I will describe work in which the "Bayesian" paradigm advocated by Gérard Bricogne (Bricogne, 1997) was used to improve an inadequate set of MIR phases to the point where a complicated substructure of nine selenium atoms could be determined using isomorphous difference Fourier maps (Doublié *et al.*,1995; Doublié *et al.*, 1994).

Several aspects of the substructure problem suggest that it should be a nearly ideal context in which to exploit the strengths of direct methods. The problem is, in general, quite highly over determined, as there are usually data to a resolution well beyond that required to separate individual heavy atoms in a macromolecular crystal. Moreover, the heavy atoms are distributed essentially randomly throughout the unit cell, in accordance of the fundamental postulate from which the conditional probability distributions of direct methods are derived. On the other hand, one lacks accurate values for the structure factor amplitudes of the unknown structure, because these must be estimated from the $\Delta|F|$data.

This disadvantage notwithstanding, it appears that direct methods offer a route to solving the heavy atom substructure that is currently underutilized. Use of direct methods actually fills a natural gap between situations where a small number (1-2) of strong substitution sites can be solved by difference Patterson methods, and solution of an entire structure. Increasing use is being made of selenomethionine substitution, for which there are now quite general methods (Doublié, 1997), and multiwavelength anomalous scattering techniques(Smith, 1998). This combination will often create situations where there are quite good estimates for the amplitudes of a substructure having ~ten atoms per protein molecule, and which therefore are quite difficult to solve using Patterson methods. In these cases, direct methods have begun to prove quite superior to other approaches (Doublié *et. al.*, 1997).

2.1. OVERVIEW

Direct methods, *sensu stricto*, mean using conditional probability distributions to extract phases from amplitudes alone. The fundamental match between the foregoing aspects of the substructure problem and probabilistic direct methods were recognized and exploited quite rapidly following the first implementations of direct methods (Steitz, 1968). Sayre's equation

$$s(F_H)=\sum_H(F_K F_{H-K}) \tag{1}$$

(Sayre, 1952): was imposed on reflections from the centrosymmetric projections of carboxypeptidase from a starting point of two signs to fix the origin plus four additional reflections to give sixteen different starting sets. Imposition of the squaring equation (1) was iterated to convergence. For each of three derivatives, the correct solution was identified by comparing peaks in the map with those in a map phased with MIR phases. Thus, although the direct method produced a correct solution, identifying that correct solution from among those produced was not accomplished independently of the protein structure solution. This aspect of direct phasing is a significant and recurrent theme in the efforts summarized in Table 1.

Table 1. Direct methods solutions for heavy-atom substructures

Authors	Application	Methods	Comments
Steitz, 1968	Centosymmetric zones of Carboxypeptidase	Sayer's equation, symbolic addition	2.0 Å data
Steitz, 1973	Centosymmetric zones of hexokinase	Sayer's equation, symbolic addition	4-9 sites found using 3.0Å and 6.0Å data
Neidle, 1973	Three-dimensional data for elastase derivatives	Tangent formula refinement	Three-dimensional, acentric data to 3.5Åand to 6.0 Å
Navia and Sigler, 1974	3D data to 6 Å for $tRNA^{fMet}$	Hybrid: solutions in projection improved by tangent formula refinement	Recycling and elimination of redundant sites at each heavy-atom site
Mukherjee, Helliwell and Main, 1989	locate anomalous scatterers	MULTAN	
Sheldrick, ~1990	Tutorials on locating heavy atoms	SHELX86	
Doublié (Doublié et al.,1994)	Locate nine selenium atoms from difference Fourier maps	MICE	Phase improvement by phase and envelope permutation
Doublié, 1997	Locate 15 selenium atoms by Multiple solution direct methods	SHELX86	$\|Fa\|$ derived from MAD data set

This pioneering effort was soon extended by a logical succession of applications, which are summarized in Table 1. Multisolution methods based on the tangent formula (Karle & Hauptman, 1956) and using three-dimensional data were in use by the early 1970s (Navia & Sigler, 1974; Neidle, 1973, Steitz *et al.*, 1973). As other phasing methods developed utilizing anomalous scattering, they were accompanied by applications of direct methods to the solution of anomalously scattering substructures (Mukherjee et al.,1989). An important tutorial on these methods was provided by Sheldrick (Sheldrick, 1993).

2.2. PROTOCOLS

The foregoing and subsequent applications of traditional direct methods used a common protocol for structure solution. Three steps can be distinguished:

1. Estimate and normalize amplitudes from $\Delta|F|$ data. The ΔF data are generally treated as if they were $|F_{obs}|$ data in the subsequent scaling and normalization steps to give$\{|E_\Delta|\}$. As noted (Navia & Sigler, 1974), the term ΔE is to be avoided, as it implies that the normalized amplitudes were obtained from differences between E values for protein and derivative.
2. Generate enough phase sets to insure that correct structure is represented, if approximately, and expand them to estimate phases for a sufficient number of reflections. Usually this is done by magic integer permutation and tangent formula refinement.
3. Select correct structure from among the others.

Normalization. Normalization of the data is carried out in order to assemble a list of amplitudes, $\{|E_H|\}$, identifying those that are unusually stronger ($|E_H| >> 1.0$) or unusually weaker ($|E_H| << 1.0$) than an "average" reflection ($|E_H| \sim 1.0$). Corrections are usually made to compensate the diffraction effects of smearing of atomic electron distributions and of various sorts of disorder. This process is sometimes called "sharpening". It is described in other chapters (Blessing, 1998a; Blessing, 1998b).

Generation of phase invariants. Direct phasing proceeds from this list of normalized structure by the processes of generating triplet (and possibly higher-order) phase invariants. This process is purely arithmetic, and consequently is essentially automatic.

Convergence mapping. Reflections are then sorted in reverse order, according to the number and strength of the phase invariants to which they contribute. This process is called convergence mapping (Germain & Woolfson, 1968) because it systematically eliminates less well-coupled reflections until only the strongest remain. This sorted list is sometimes called a "phasing path". Reflections from this end-point of the convergence map constitute a set from which reflections are selected for permutation. The course of phase determination is sometimes highly dependent on the phasing path given by convergence mapping. Since this, in turn, can depend sensitively on the normalization parameters alternative normalizations can sometimes lead to solutions where the default path fails.

Phase set generation. Origin fixing reflections are generally selected from the last remaining reflections in the convergence map. Depending on the potential difficulty of the structure, a small number of additional reflections are selected by the same criteria. Since their phases are unknown but are not arbitrary, once the origin has been fixed, they are permuted by factorial design (either a complete factorial or a sampling design like that of magic integers (Main, 1977)).

Tangent formula refinement. Each separate phase set is then "refined" by iterative application of the tangent formula (Fan, 1998):

$$\tan(\phi_H) \sim \frac{\sum_K \kappa_{HK} \sin(\phi_H + \phi_{H-K})}{\sum_K \kappa_{HK} \cos(\phi_H + \phi_{H-K})} \qquad (2)$$

This process expands the conditional probability relationships implied by the particular "basis set" reflections to the rest of the dataset.

Scoring. A number of figures of merit have been described which purport to "score" the results from tangent formula refinement (Gilmore, 1988). It is at this stage that the traditional process is weakest. None of the scores evaluated from tangent formula refinement actually perform very well, and selection of the correct solution most often depends on judicious application of prior information about the structure.

Summary of results. Published accounts of direct methods solutions for heavy-atom substructures generally concur on the following major conclusions:

- Correct solutions are generated, often when a (noisy) $\Delta|F|^2$ difference Patterson is not readily interpretable.
- Correct solutions are not reliably indicated by any of the figures of merit accumulated for the different solutions.
- Correct solutions must be confirmed by:
  - prior evidence; ie., from a known structure
  - consistency with $\Delta|F|^2$ difference Patterson
  - consistency of subsequent atomic parameter refinement.

2.3. ADAPTATIONS TO LOW RESOLUTION.

An especially interesting discussion of the use of direct phasing methods is (Navia & Sigler, 1974). The authors did not use the multi-solution strategy outlined above. Rather, they solved using projection data, and formulated lists of likely sites by combing results from different projections. The hand ambiguity had to be resolved when combining projections, and this provided the strongest impetus to work with 3D data. Atomic positions from this process were used to estimate phases in three-dimensions, which were then refined using the tangent formula.

The radius of convergence was reproducible with respect to the accuracy of the starting coordinates. For 6 Å data, sites within ~3 Å of the correct positions could be refined in this manner. Sites further than this from correct positions deteriorated during refinement, and could be discarded from the phasing model. Thus, many incorrect sites can be removed by tangent formula refinement, provided one site is correct.

This observation suggested combining real and reciprocal space improvements. A constellation of atoms was introduced around those sites identified from Emap peaks. Each possible site was provided with six satellite sites (Figure 2) with the idea that the incorrect sites would refine away, leaving only the correct ones. Recycling would eliminate up to ten incorrect positions, provided a correct site was among them. Recycling between map interpretation and direct phase refinement is essentially the formalism that recently made possible the much more substantial successes of the "Shake and Bake" procedure (DeTitta *et al.*, 1994; Hauptman, 1993; Weeks *et al.*, 1994).

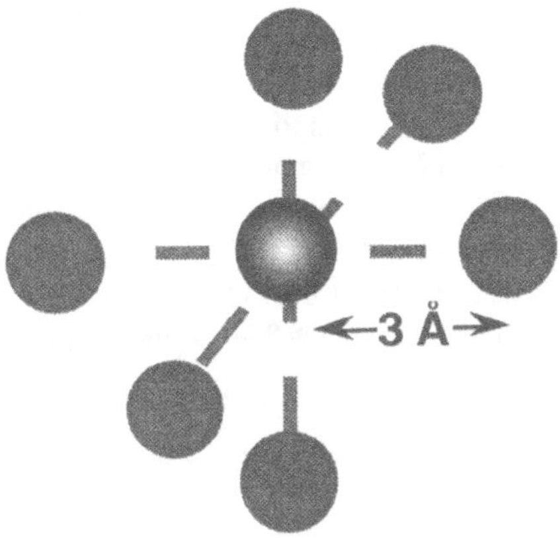

Figure 2. Recycling strategy of (Navia & Sigler, 1974). Sites identified from combining projections (center atom, light gray shading) were supplemented by six additional positions separated from it by 3 Å. That distance had been determined to be approximately equal to the radius of convergence for indirect positional refinement via phase improvement by the tangent formula. Generally, the correct site would emerge at the expense of the six erroneous locations.

2.4. ADVANTAGES AND DISADVANTAGES

Recent successes locating large numbers of selenium atoms in large (~100 Kd) proteins augur well for an increased role of traditional direct methods in similar, future applications. Patterson superposition (Sheldrick, 1997; Sheldrick *et al.*, 1993) is in many respects a complementary method. However, both methods work well in direct proportion to the quality and resolution of the data. Reasons to consider using a direct methods program such as SHELX97 (http://linux.uni-ac.gwdg.de/SHELX) include the fact that direct methods are useful when high substitution, high symmetry, or poor data quality create problems solving a Patterson by inspection.

The principal disadvantage arises from the fact that ΔF data are reliable only to estimate large F_{PH} values. Consequently, only the triplet phase invariants, which rely exclusively on large normalized amplitudes, can be estimated from$\{|E_{\Delta}|\}$. Higher order invariants (ie., negative quartets), which can strengthen direct methods and especially figures of merit, require small amplitudes, and cannot be estimated. This disadvantage is less serious with MAD data, from which improved estimates can be obtained for the amplitudes of the heavy atom substructure. In general, these methods are probably underutilized at present.

3. The Bayesian Paradigm, Phase Improvement, and a Nine-Atom Selenium Substructure.

The Bayesian paradigm (Bricogne, 1988a; Bricogne, 1993a; Bricogne, 1997) provides a conceptually parallel approach to traditional direct methods, but one that is implemented primarily in real space. This approach was recently used in determining a substructure containing nine selenium atoms per asymmetric unit in the course of phasing tryptophanyl-tRNA synthetase (Doublié et al., 1995; Doublié et al., 1994). The strategy was actually not an *ab initio* phase determination for the substructure, but rather the use of conditional probability methods for model-independent phase improvement of an inadequate MIR phase set for the entire structure. It made use of direct phase determination for 28 reflections by permutation method shaving much in common with the outline described for traditional direct methods in §2.2.

3.1. JPDS IN RECIPROCAL AND DIRECT SPACES

The Bayesian paradigm is rooted in the same context as direct methods because it addresses the same question, that of conditional probability (Figure 3). Conditional probability, P(A|B), is always formulated using the paradigm: "What is the probability of A, given that B has a particular value or property". Confusion can stem from the fact that the Karle-Hauptman determinant is formulated in reciprocal space in order to evaluate joint probability distributions for structure factors, while conditional maximum entropy distributions are evaluated in real space to evaluate JPD of atomic positions.

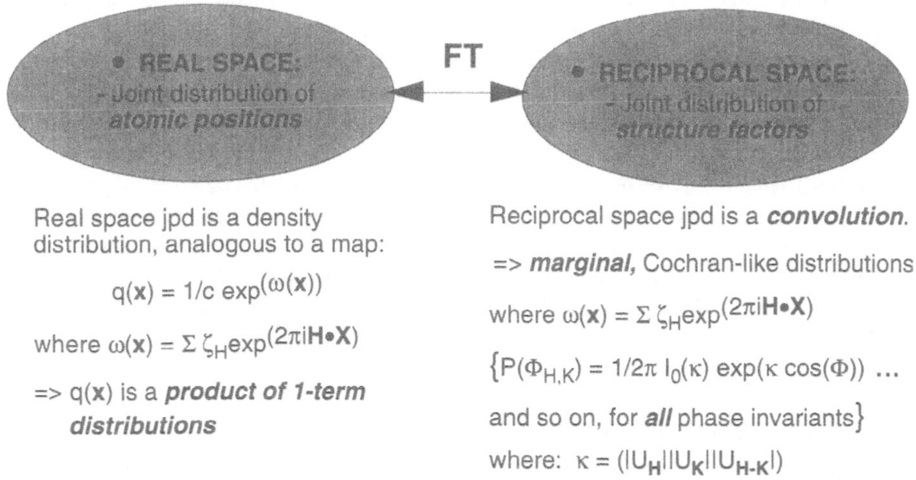

Figure 3. Schematic comparison between the traditional implementation of conditional probability in direct methods (right) and the role of entropy maximization in the Bayesian paradigm (Bricogne, 1988a; Bricogne, 1993b; Bricogne, 1998). In the objective of real-space entropy maximization corresponds to simultaneously evaluating all triplet and higher-order probability distributions in reciprocal space.

In crystallography, we are interested in the probable distribution of atomic positions, *given that* certain conditions are satisfied. In small molecule applications (Woolfson, 1987), probability calculations are conditional first on the assumption that the distribution of atomic positions is positive everywhere. Generally, assumptions are also made concerning the phases of a small number of reflections (§2.3, phase set generation).

Constraints are potentially richer for macromolecules than they are for small molecules because in addition to positivity, macromolecular crystals have solvent channels where the density is approximately constant and noncrystallographic symmetry redundancies. These, respectively, lead to the following physical constraints: (i) $\rho(\mathbf{x})$ outside a presumed boundary $=\langle\rho_{solvent}\rangle$ and (ii) equivalent subunits in asymmetric unit have $\rho(x) = \langle\rho(x)_{subunit}\rangle$. Moreover, since the macromolecular problem is so much more difficult, no *ab initio* structure determination has been attempted, and applications to date have involved the assumption that phases are known for the strongest ~30% of the reflections.

Bricogne (Bricogne, 1984) showed that the real space equivalent to the full, joint probability distribution (JPD) for the atomic positions could be calculated under most any reasonable set of constraints by finding the constrained maximum entropy distribution, $q^{ME}(x) = \dfrac{1}{C}\exp\omega(x)$ where $\omega(\mathbf{x})$ is a Fourier synthesis over the basis set reflections. The Fourier coefficients of $\omega(\mathbf{x})$ are coordinates of a "saddlepoint", where the constraints are satisfied to the greatest extent possible with the least possible restriction in the distribution's entropy, giving rise to the possibly misleading phase "maximum entropy".

The maximum entropy distribution, $q^{ME}(\mathbf{x})$, is *conditional* because it is constrained to include the basis set reflections in its Fourier spectrum; it is a *probability distribution* owing to the exponentiation. Moreover, its Fourier spectrum includes estimates for many reflections outside the basis set, which arise through the process of maximum entropy extrapolation. These estimates provide the basis for an optimal figure of merit, called the Log-likelihood gain (LLG; (Bricogne, 1988a; Bricogne, 1988b; Bricogne, 1993a)) The likelihood itself gives the probability of the data, given the constraints placed on the conditional probability. Since these constraints are generally hypotheses about phases, or about other aspects of the structure such as envelopes, the likelihood can be used to evaluate how much more probable the experimentally observed data outside the basis set become as those hypotheses better approximate the actual structure. The LLG is the improvement in likelihood of a given hypothesis over that for the null hypothesis, and is therefore an excellent figure of merit for comparing different hypotheses.

3.2. THE BAYESIAN PARADIGM IN ACTION

Figure 4 illustrates the essential features of the real-space conditional probability implementation in the Bayesian paradigm. It is a summary from a tutorial available on the WWW (http://russell.med.unc.edu/~carter/bayes). A model, one-dimensional structure of the linear triatomic molecule, $HgCl_2$, is solved, step by step, using entropy maximization and likelihood scoring (Carter, 1994; Carter & Xiang, 1997; Carter *et al.*, 1997). Both real and reciprocal space are shown in their entirety, as maps and phased histograms of structure factor vectors. The centroid electron density is shown for two successive "basis sets", differing only in the presence or absence of a strong, phased high-resolution reflection in **b.** Maximum-entropy extrapolation from only one reflection (**a**) provides a reasonable low resolution picture of the molecule. Extrapolation to high resolution is absent from its Fourier spectrum because none of the higher-order reflections lie in the second neighborhood of the basis set reflection, h=1. In contrast, adding information from one strong, phased high-resolution reflection, h=11(**b**) allows maximum entropy extrapolation to fill in the remaining information necessary to solve the structure. The phase of the latter reflection can be determined by permuting its phase and comparing the log-likelihood gain for each permuted phase.

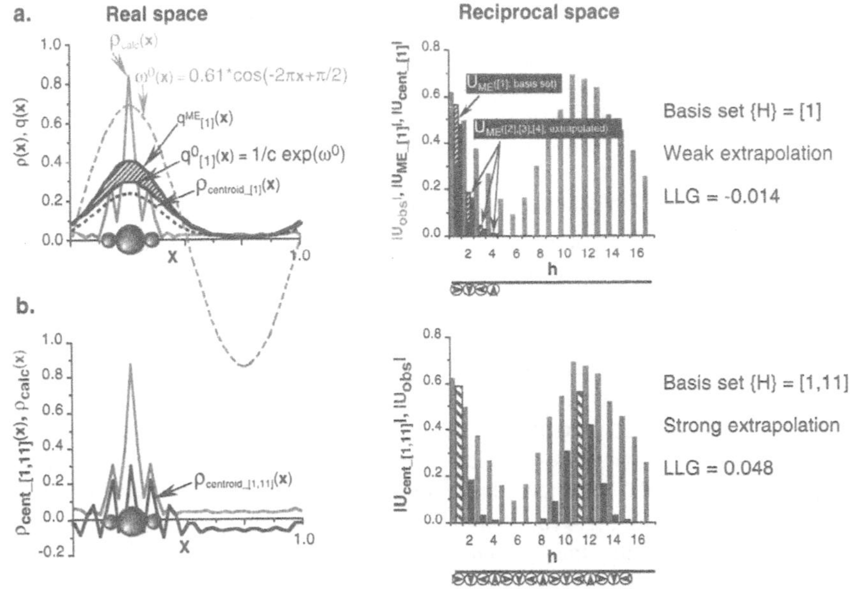

Figure 4. The Bayesian paradigm in action. Real and reciprocal space representations of two successive "nodes" in the phase determination for A model, one-dimensional structure of the linear triatomic molecule, $HgCl_2$. Target values for the map and its structure factors are shown in light gray, superimposed on results from the respective conditional probability calculations, indicated in dark gray. In the dials underneath the two structure factor amplitude histograms represent their phases. **a.** Conditional probability calculation for a one-reflection basis set consisting of the h=1 reflection. **b.** Conditional probability calculation for a two reflection basis set including h=1 and h=11. In both **a** and **b** ρ_{cent} refers to the centroid electron density map calculated with Sim-like weights.

3.3. ADVANTAGES OF WORKING IN REAL SPACE

- The distribution, $q^{ME}(\mathbf{x})$, is a constrained exponential model for $\rho(\mathbf{x})$ which has several optimal properties.
- It is *recentered* away from the random-atom model, and is hence the most accurate which can be derived from the given reciprocal space constraints.
- Since it is implemented in real-space it is possible to represent the complete set of marginal reciprocal space phase probability distributions.
- Because it represents a correct, simultaneous implementation of triplet and all higher-order phase invariants evaluation of $q^{ME}(\mathbf{x})$ is both valid and useful in all resolution ranges. It is thus an optimal approximation to the conditional probability distribution. Moreover, it has been shown to effect an optimal coordination of all available phase information (Bricogne, 1991).
- Maximum entropy extrapolation from the exponential model is sufficiently accurate that hypotheses about phases can be compared with confidence, using the LLG as the figure of merit (Doublié et al.,1994; Gilmore, 1998; Gilmore *et al.*, 1991).

These decisive advantages of the real space evaluation are related to the fact that the real space JPD is a product of simple exponential distributions, while the JPDs in reciprocal space are convolutions, and hence much more difficult to evaluate. In practice, this means that the real space JPDs can be *recentered*, and hence are much more accurate, that they are complete, and that hypotheses about non-random prior distributions (e.g., molecular envelopes) can be evaluated in the same way as are hypotheses about phases (Doublié et al., 1994).

3.4. PHASE PERMUTATION AND A SELENIUM SUBSTRUCTURE

During the course of solving the tryptophanyl-tRNA synthetase structure (Doublié et al., 1995; Doublié et al., 1994), it became apparent that making heavy-atom derivatives invariably induced lattice-preserving non-isomorphism. We have subsequently begun to understand this from the fact that subsequent structure solutions revealed translational shifts in the tetragonal packing of up to 0.5 Å when different ligands are bound. The non-isomorphism left us with approximate, but rather poor phases for the protein structure. We turned to substitution with selenomethionine as an alternative source of phases. At the time, it was still somewhat premature to carry out a successful MAD experiment for such a large and radiation sensitive problems, so we decided to use the Cu Kα selenomethionine data as an isomorphous derivative. TrpRS has 10 methionine residues, so it was not possible to solve the selenium substructure using the difference Patterson. Moreover, the MIR phases were not good enough to clearly identify the selenium positions above the noise in a difference Fourier map.

We had previously shown that maximum entropy solvent flattening (MESF) provided a superior density modification algorithm at 3 Å resolution (Xiang *et al.*, 1993). However, MESF alone did not produce an interpretable map, particularly in the crucial region formed by the β-sheet of the Rossmann dinucleotide-binding fold. Thus, we began a more active process of model-independent map improvement, hoping to be able to locate the selenium atoms. This process succeeded, and the phase information from the selenomethionine derivative contributed substantially to the structure determination. In many respects, this procedure followed the outline given in §2.2. At each stage, we evaluated the MESF centroid map using Sim-like figures of merit as representative of the current position.

Extrapolation and $\{|U_{\mathbf{H}}|_{renorm}\}$. For the list of extrapolated structure factors outside the basis set (and occasionally from within the basis set) we estimated the amplitude of the difference structure factor vector between the observed and maximum entropy models. These estimates, obtained by a cosine law approximation (Figure 5), are called renormalized structure factor amplitudes, $\{|U_{\mathbf{H}}|_{renorm}\}$.

Identification of reflections to permute. Proceeding by analogy to convergence mapping, we selected from a sorted list of these $\{|U_{\mathbf{H}}|_{renorm}\}$ those reflections most likely to enhance the current basis set. We used this construction to estimate from the extrapolation pattern and the observed data just where the current hypothesis was weakest (had the most potential noise reduction for the smallest enhancement of the basis set). Since these reflections generally had large observed amplitudes but quite small extrapolated amplitudes, they represented, in miniature, a phase determination problem.

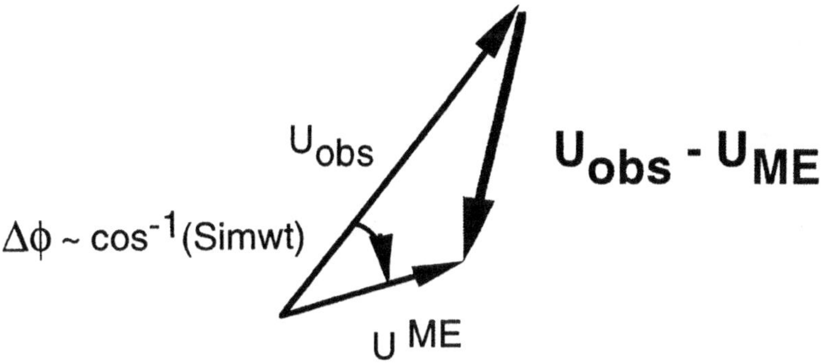

$$|U_{obs} - U_{ME}|^2 \sim |U_{obs}|^2 + |U_{ME}|^2 - 2|U_{obs}||U_{ME}| \text{ Simwt}$$

Figure 5. Cosine law approximation to the amplitude, $|U_H|_{renorm}$, of the missing information represented by a particular (usually unphased) structure factor.

Generation of phase hypotheses. Using a sampling procedure previously developed for sampling crystal growth conditions (Carter & Carter, 1979), we proceeded to permute phases for these reflections, generating a set of "nodes" analogous to those encountered in the small molecule multi-solution approach. Acentric reflections were given values according to quadrant permutation $\left(\frac{\pi}{4}, \frac{3\pi}{4}, \frac{5\pi}{4}, \text{and } \frac{7\pi}{4}\right)$. Samples were chosen randomly and were balanced. They included approximately $\frac{\sqrt{N}}{2}$ from the N experiments in a full factorial design covering all possibilities. This sampling is considerably more sparse than that employed in the magic integer strategy (Main, 1977). This *incomplete* factorial design allows averaging of balanced samples of each phase bit. It permits the main effects (ie,. the signs) of many phase bits to be determined simultaneously. In our hands, 10-12 phase bits can be explored in a single design of ~24 nodes.

Maximum entropy expansion of nodes. This process, analogous to tangent formula refinement of different nodes in a multisolution small molecule determination, involves application of MESF to each of the nodes in the permutation design. Its sole purpose is to obtain the LLG associated to each node and hence to each of the different phase (and/or envelope) hypotheses in the design.

Table 2. Regression model for the Log-Likelihood Gain scoring gives highly significant phase indications for 7 permuted reflections from 24 nodes. In the squared multiple R for this model was 0.97; the standard error of the estimated score, 10.3

VARIABLE	COEFFICIENT	STD ERROR	Students t	P(2 TAIL)
CONSTANT	2098.6	2.035	.10E+04	0.10E-14
Re 6 4 48	26.194	2.408	10.8	0.84E-08
Re 10 9 26	3.120	2.503	1.3	0.23
Im 10 926	-7.116	2.736	-2.6	0.02
Re 9 8 28	17.304	2.189	7.9	0.65E-06
Im 9 8 28	17.734	2.327	7.6	0.10E-05
Sign 2 054	22.239	2.205	10.1	0.24E-07
Sign 0 0 48	-5.359	2.490	-2.2	0.05
Sign 9 0 35	21.222	2.708	7.8	0.72E-06

Analysis of Variance

SOURCE	SUM-OF-SQUARES	DF	MEAN-SQUARE	F-RATIO	P
REGRESSION	62631.2	9	6959.0	65.3	0.65E-10
RESIDUAL	1704.6	16	106.5		

Analysis of variance and Student t-testing. The LLG values for each node represent a sample of the behavior of this scoring function over the complete set of permutations in the design. To interpolate from these values to the optimal set of phases from the complete factorial design, one must estimate parameters for some model of the LLG over the entire space. We made use of the simplest multivariate model of the LLG, namely the linear combination of signs for each phase bit. The contribution of each phase bit was estimated as the coefficient of this linear model by multiple regression methods. Stepwise regression methods were used to eliminate insignificant contributors to the calculated LLG, and an appropriate model was then evaluated by the analysis of variance. An example is given in Table 2 (Doublié et. al., 1994)

New phase information. By coordinating the results from the ensemble of nodes, this model provides significant phase indications transcending what one would achieve simply by choosing the best node, which is rarely correct in every phase bit. It implies that the LLG for any combination of phase bits can be calculated as the sum:

$$LLG_{calc} = \text{Constant} + \sum_{i=1}^{M} \beta_i F_i \tag{3}$$

where the constant and coefficients are those in the second column. Nearly 98% of the variation in the observed scores is reproduced by the model (from the squared multiple R value). The standard error in the calculated LLG, 0.10, is only ~0.5%. Each phase bit is associated with a Student t-value whose probability under the null hypothesis is given in the far right column. These P values indicate the statistical significance of the indicated coefficient.

For centric reflections, where only a sign is to be determined, the sign of the coefficient indicates the probable phase. For acentric reflections, indications are somewhat more accurate than the sampling interval of 90°. Cases with significant indications for both the real and imaginary components of the phase can be phased by taking the inverse tangent of the ratio of the two coefficients. For example, for **H** = [9 8 28] both real and imaginary components are highly significant and the inverse tangent of their ratio gives a phase of 46°. In the actual phase for this reflection from the final structure refinement is 86°. Overall, for roughly 60 reflections we have phased in this way for three structures, the average phase error is ~30°.

It is interesting to note that the constant term in the linear model, which is the average LLG for the experiments in Table 2, is 100 times greater than the variation in LLG induced by different phase hypotheses. This sensitivity to the correct phase bits is a satisfying confirmation of the appropriateness of the LLG as a score. A noted previously (Gilmore et al., 1991), it shows the innate superiority of the LLG over all other figures of merit used in traditional direct methods.

Envelope permutation. In principle, hypotheses can be framed for any aspect of a crystal structure for which there are evident ambiguities preventing structure solution. In addition to phase ambiguities, we have found it very useful to determine which of various hypotheses is correct, regarding the molecular envelope used in enforcing solvent flatness (Doublié et. al., 1994).

4. Practical Importance of the Bayesian Paradigm

Direct Phase determination of ~30 strong, but poorly phased reflections improved MIR phases sufficiently to locate 9 Selenium atoms in an isomorphous difference Fourier map. Only four of the nine positions were above the general noise level in the corresponding difference map calculated with the best MIR phases.

The missing information recovered for TrpRS by the application of hypothesis permutation, evaluated by constructing joint probability distributions and likelihood scoring was quite significant. In the overall map improvement we achieved by combining all sources of phase information, including the corrected envelope, the selenomethionine $\{|\Delta F|\}$, another lead derivative, and phase recombination is illustrated for the weakest portion of the map in Figure 6a and 6b. For comparison, the effect of including the 28 reflections phased by permutation and likelihood scoring, with all other effects at the same level (the corrected envelope, without the selenium and lead isomorphous differences) is shown in Figure 7a and 7b. The 28 additional reflections clearly effected a comparable transformation in the map to the overall effect. The likely reason for this is that by using the construction in Figure 5 for $\{|U_H|_{renorm}\}$, we could identify the most important missing information at each stage of map improvement. Of special relevance is the relationship noted by Bricogne between large $|U_H|_{renorm}$ values and the r.m.s. noise in a map (Bricogne, 1997). We have found subsequent to the work described here that the strategy of identifying missing information via these outlying reflections and recovering such information by phase permutation has been crucial in removing model bias introduced in structures phased by molecular replacement (V. Ilyin, unpublished results).

In the Bayesian paradigm for expressing conditional probability relationships between structure factors and for probing the effects of missing information during structure determination represents an important addition to the arsenal of macromolecular crystallography (Bricogne, 1988b; Bricogne, 1991; Bricogne, 1992; Bricogne, 1997). In the permutation experiments described in §3.4 demonstrate convincingly that optimal implementation of the underlying conditional probability relationships in macromolecular crystallography can be useful in solving quite difficult structures at well below atomic resolution. These results may be relevant to the longer term unification of macromolecular phasing methods (Bricogne, 1988a) under the rubric of conditional probability first envisioned by the pioneers of direct methods (Hauptman & Karle, 1953; Sayre, 1952; Woolfson, 1954).

a.

b.

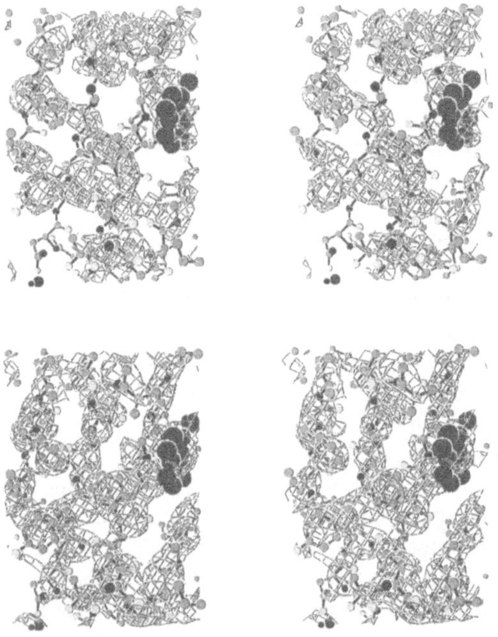

Figure 6. The overall map improvement achieved in the solution of the TrpRS structure. Maps here and in Figure 7 show the beta sheet of the Rossmann fold, the region where the MIR map was least interpretable, in retrospect. **a.** In the initial MIR map phased by two non-isomorphous derivatives. **b.** The map calculated with the final experimental phase set obtained by a process of model-independent map improvement.

Figure 7. The isolated effect of including 28 reflections phased by permutation and likelihood scoring. **a.** Map obtained by MESF using the final envelope and a basis set selected from the best MIR phases used in the map shown in Figure 5**a**. **b.** Map calculated after MESF using the same envelope and basis set as in **a**, but with the addition of 28 additional reflections.

a.

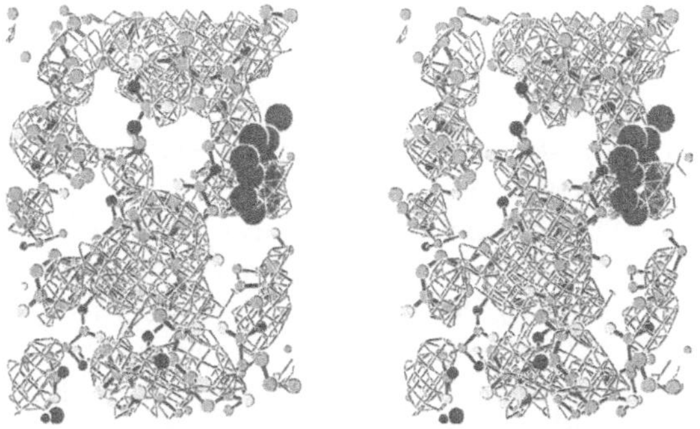

b.

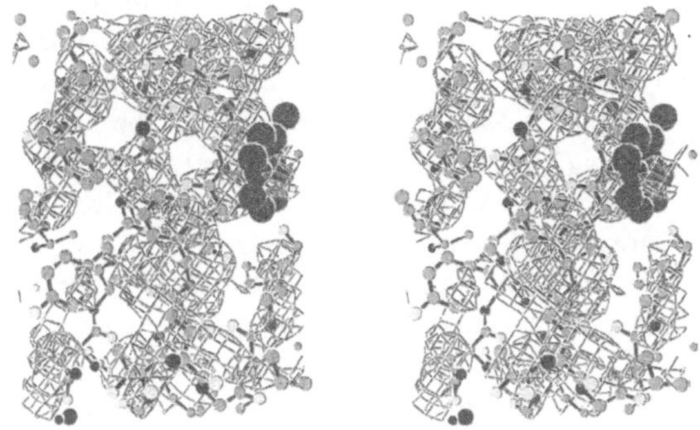

References

1. Blessing, R. (1998a). Intensity statistics and normalization: application to protein crystals. In *Direct Methods for solving Macromolecular Structures* (Fortier, S., ed.). Kluwer Academic Publishers, Dordrecht.
2. Blessing, R. (1998b). Intensity statistics and normalization: Basic Principles. In *Direct Methods for Solving Macromolecular Structures*(Fortier, S., ed.). Kluwer Academic Publishers, Dordrecht.
3. Bricogne, G. (1984). Maximum Entropy and the Foundations of Direct Methods. *Acta Crystallographica* **A40**, 410-445.
4. Bricogne, G. (1988a). A Bayesian Statistical Theory of the Phase Problem. A Multichannel Maximum-Entropy Formalism for Constructing Generalized Joint Probability Distributions of Structure Factors. *Acta Crystallographica* **A44**, 517-545.
5. Bricogne, G. (1988b). Maximum Entropy Methods in the X-ray Phase Problem. In *Crystallographic Computing 4* (Isaacs, N. W. & Taylor, M. R., eds.), pp. 60-79. IUCR and Oxford University Press, Oxford, UK.
6. Bricogne, G. (1991). Maximum Entropy as a Common Statistical Basis for all Phase Determination Methods. In *Crystallographic Computing V* (Moras, D., Podjarny, A. D. & Thierry, J. C., eds.), pp. 257-297. Clarendon Press, Oxford.
7. Bricogne, G. (1992). A Statistical Formulation of the Molecular Replacement and Molecular Averaging Methods. In *Molecular Replacement* (Dodson, E. J., Gover, S. & Wolf, W., eds.), pp. 62-75. Science and Engineering Research Council, Daresbury Laboratory, Daresbury, Warrington, UK.
8. Bricogne, G. (1993a). Direct Phase Determination by Entropy Maximization and Likelihood Ranking: Status Report and Perspectives. *Acta Crystallographica* **D49**, 37-60.
9. Bricogne, G. (1993b). Fourier transforms in crystallography: theory, algorithms, and applications. In *International Tables for Crystallography* (Shmueli, U., ed.), Vol. B, pp. 22-104. Kluwer Academic Publishers, Dordrecht.
10. Bricogne, G. (1997). Bayesian Statistical Viewpoint on Structure Determination: Basic Concepts and Examples. *Methods in Enzymology* **276**, 361-423.
11. Bricogne, G. (1998). Maximum entropy methods. In *Direct Methods for solving Macromolecular Structures* (Fortier, S., ed.). Kluwer Academic Publishers, Dordrecht.
12. Carter, C. W., Jr. (1994). Entropy Maximization, Permutation and Likelihood scoring Methods for Improving Macromolecular Electron Density Maps. In *From First Map to Final Model* (Hubbard, R., Waller, D. & Bailey, S., eds.), pp. 41-58. Science and Engineering Research Council, Daresbury Laboratory, Daresbury, UK.
13. Carter, C. W., Jr. (1998a). The solvent contrast variation method. In *Direct Methods for Solving Macromolecular Structures* (Fortier, S., ed.), pp. In Press. Kluwer Academic Publishing, Dordrecht.

14. Carter, C. W., Jr. (1998b). *Ab initio* determination of a low resolution molecular envelope by direct methods. In *Direct Methods for Solving Macromolecular Structures* (Fortier, S., ed.), pp. In Press. Kluwer Academic Publishing, Dordrecht.
15. Carter, C. W., Jr. & Carter, C. W. (1979). Protein Crystallization Using Incomplete Factorial Experiments. *Journal of Biological Chemistry* **254** (December 10 1979), 12219-12223.
16. Carter, C. W., Jr., Crumley, K. V., Coleman, D. E., Hage, F. & Bricogne, G.(1990). Direct Phase Determination for the Molecular Envelope of *Bacillus stearothermophilus* Tryptophanyl-tRNA Synthetase by X-ray Contrast Variation. *Acta Crystallographica* **A46**, 57-68.
17. Carter, C. W., Jr. & Xiang, S. (1997). Phase Improvement Using Conditional Probability Methods: Maximum Entropy Solvent Flattening and Phase Permutation. *Methods in Enzymology* **277**, In press.
18. Carter, C. W., Jr., Xiang, S., Bricogne, G. & Gilmore, C. (1997). Solving Macromolecular Crystal Structures using the Bayesian Paradigm. In *Likelihood, Bayesian Inference, and Their Application to the Solution of New Structures* (Bricogne, G. & Carter, C. W., Jr., eds.), Vol. 30, pp. 41-58. American Crystallographic Association, Buffalo, NY.
19. DeTitta, G. T., Weeks, C. M., Thuman, P., Miller, R. & Hauptman, H. A.(1994). Structure Solution by Minimal-Function Phase Refinement and Fourier Filtering. Theoretical Basis. *Acta Crystallographica* A50, 203-210.
20. Doublié, S. (1997). Preparation of Selenomethionyl Proteins for Phase Determination. *Methods in Enzymology* **276**, 523-530.
21. Doublié, S., Bricogne, G., Gilmore, C. J. & Carter, C. W., Jr. (1995). Tryptophanyl-tRNA synthetase crystal structure reveals an unexpected homology to Tyrosyl-tRNA synthetase. *Structure* **3**, 17-31.
22. Doublié, S., Tabor, S., Long, A., Richardson, C. C. & Ellenberger, T.(1997). Crystal structure of bacteriophage T7 DNA polymerase complexed to a primer-template, a nucleoside, triphosphate, and its processivity factor, thioredoxin. *Nature* **In press**.
23. Doublié, S., Xiang, S., Gilmore, C. J., Bricogne, G. & Carter, C. W. J.(1994). Overcoming Non-Isomorphism by Phase Permutation and Likelihood scoring: Solution of *Bacillus stearothermophilus* Tryptophanyl-tRNASynthetase. *Acta Crystallographica* **A50**, 164-182.
24. Fan, H.-F. (1998). Sayre's Equation, Tangent Formula, and Saytan. In *Direct Methods for Solving Macromolecular Structures* (Fortier, S., ed.).Kluwer Academic Publishers, Dordrecht.
25. Fourme, R., Shepard, W., Kahn, R., l'Hermite, G. & de La Sierra, I. L.(1995). The Multiwavelength Anomalous Solvent Contrast (MASC) Method in Macromolecular Crystallography. *Journal of Synchrotron Radiation* **2**, 36-48.
26. Fourme, R., Shephard, W. & Kahn, R. (1996). Application of the Anomalous Dispersion of X-rays to Macromolecular Crystallography. *Progress in Biophysics and Molecular Biology* **64**, 167-199.

27. Germain, G. & Woolfson, M. M. (1968). On the use of phase invariants for structure determination: getting off to a good start. *Acta Crystallographica* **B24**, 91-96.
28. Gilmore, C. J. (1998). Figures of merit for evaluating possible phase solutions. In *Direct Methods for Solving Macromolecular Structures* (Fortier, S., ed.). Kluwer Academic Publishers, Dordrecht.
29. Gilmore, C. J., Henderson, A. N. & Bricogne, G. (1991). A Multisolution Method of Phase Dermination by Combined Maximization of Entropy and Likelihood. V. In the Use of Likelihood as a Discriminator of Phase Sets Produced by the Saytan Program for a Small Protein. *Acta Crystallographica* **A47**, 842-846.
30. Hauptman, H. (1993). Phasing Macromolecular Structure via Structure-Invariant Algebra. *Acta Crystallographica* **D49**, 3-8.
31. Hauptman, H. & Karle, J. (1953). *In the Solution of the Phase Problem: I. The Centrosymmetric Crystal.* American Crystallographic Association Monograph, 3, Polycrystal Book Service, Pittsburg.
32. Karle, J. & Hauptman, H. (1956). In the tangent formula. *Acta Crystallographica* **9**, 635-651.
33. Main, P. (1977). On the Application of Phase Relationships to Complex Structures. XI. A Theory of Magic Integers. *Acta Crystallographica* **A33**, 750-757.
34. Mukherjee, A. K., Helliwell, J. R. & Main, P. (1989). In the Use of MULTAN to Locate the Positions of Anomalous Scatterers. *Acta* **Crystallographica A45**, 715-718.
35. Navia, M. A. & Sigler, P. B. (1974). In the Application of Direct Methods to the Analysis of Heavy Atom Derivatives. *Acta Crystallographica* **A30**, 706-712.
36. Neidle, S. (1973). On the determination of heavy-atom positions in various elastase derivatives. *Acta Crystallographica* **B29**, 2645-2647.
37. Read, R. J. (1997). Model Phases: Probabilities and Bias. *Methods in Enzymology* **277**, In Press.
38. Rossmann, M. G., Ed. (1972). In the Molecular Replacement Method: A Collection of Papers on the Use of Non-Crystallographic Symmetry. New York: Gordon Breach.
39. Sayre, D. (1952). In the Squaring Method: A New Method for Phase Determination. *Acta Crystallographica* **5**, 60-65.
40. Sheldrick, G. M. (1993). Tutorial on Automated Patterson interpretation to find heavy atoms. In *Crystallographic Computing 5: From Chemistry to Biology* (Moras, D., Podjarny, A. D. & Thierry, J.-C., eds.), pp. 145-157. Oxford University Press, Oxford, UK.
41. Sheldrick, G. M. (1997). Patterson Superposition and *ab* initio phasing. *Methods in Enzymology* **276**, 628-641.
42. Sheldrick, G. M., Dauter, Z., Wilson, K. S., Hope, H. & Sieker, L. C. (1993). In the application of direct methods and Patterson interpretation to high resolution protein data. *Acta Crystallographica* **D49**,18-23.
43. Smith, J. (1998). Multiwavelength anomalous diffraction in macromolecular crystallography. In *Direct Methods for Solving Macromolecular Structures* (Fortier, S., ed.). Kluwer Academic Publishers, Dordrecht.

44. Steitz, T. A. (1968). A New Method of Locating Heavy Atoms Bound to Protein Crystals. *Acta Crystallographica* **B24**, 504-507.
45. Steitz, T. A., Fletterick, R. J. & Hwang, K. J. (1973). Structure of Yeast Hexokinase II. A 6 Å Resolution Electron Density Map showing Molecular Shape and Heterologous Interaction of Subunits. *Journal of Molecular Biology* **78**, 551-561.
46. Weeks, C. M., DeTitta, G. T., Hauptman, H. A., Thuman, P. & Miller, R. (1994). Structure Solution by Minimal-Function Phase Refinement and Fourier Filtering. II. Implementation and Applications. *Acta Crystallographica* **A50**, 210-220.
47. Woolfson, M. M. (1954). Structure Determination by the Method of Permutation Synthesis. *Acta Crystallographica* **7**, 65-67.
48. Woolfson, M. M. (1987). Direct Methods: From Birth to Maturity. *Acta Crystallographica* **A43**, 593-612.
49. Xiang, S., Carter, C. W., Jr., Bricogne, G. & Gilmore, C. J. (1993). Entropy Maximization Constrained by Solvent Flatness: a New Method for Macromolecular Phase Extension and Map Improvement. *Acta Crystallographica* **D49**, 193-212.

DIRECT METHODS IN RELATION TO ISOMORPHOUS DATA

R. PESCHAR
Laboratory for Crystallography, Amsterdam Institute for Molecular Studies.
University of Amsterdam, Nieuwe Achtergracht 166, 1018 WV Amsterdam.
The Netherlands

Abbreviations

c.f.	Characteristic function
j.p.d.	Joint probability distribution
c.p.d.	Conditional probability distribution
(p.)r.v.	(Primitive) random variable
(d.)s.f.	(Difference) structure factor
DM	Direct methods
SIR(N)AS	Single isomorphous replacement with(out) anomalous scattering effects
SAS	Single-wavelength anomalous scattering
MAD	Multi-wavelength anomalous dispersion

1. Introduction

In the last decades Direct Methods (DM) have played an important role in the structure determination of small molecules, say up to 100 independent atoms in the asymmetric unit. Larger structures up to 600 atoms have been determined by the DM-based Shake-and-Bake technique [1] but only at the expense of an enormous computing effort. The success rate of DM drops dramatically if the amount of independent atoms becomes larger so at first sight a DM application to macromolecules seems to be impossible. On the other hand, a long tradition exists in solving macromolecular structures by means of multiple isomorphous replacement data. Recently, also algebraic techniques have been developed that exploit single and multiple anomalous scattering (SAS, MAD). In the MAD case this led to a solvable set of equations in wavelength-independent quantities [2, 3, 4] and several macromolecular structures [5, 6] have been determined successfully by the program system MADSYS [4] that is based upon this technique.

In case of SAS or single isomorphous replacement (SIR) data, exact algebraic solutions are not feasible and therefore the theoretical basis of the current DM will be discussed from the point of view of *ab initio* structure determination using these type of data.

2. Fundaments of Direct Methods

DM try to solve the crystallographic phase problem via phase relations, structure (sem-)invariants, that exist among the structure factors (s.f.'s). In particular, triplet phase sum relations ψ_{123},

$$\psi_{123} = \varphi_1 + \varphi_2 + \varphi_3 \tag{2.1}$$

between the phases φ_1, φ_2 and φ_3 of three reflections H_1, H_2 and H_3 (= $-H_1-H_2$) are used to express all unknown φ_H in terms of a few starting phase values only. Since in general the ψ_{123} are not accessible experimentally, their numerical values have to be

S. Fortier (ed.), Direct Methods for Solving Macromolecular Structures, 275–284.

determined otherwise, e.g. inferred from the observed s.f. magnitudes.

Although exact relations between s.f.'s have been derived, and also other mathematical tools such as inequalities [7] and determinants [8] have been applied, especially in the early days of DM, their practical use has been limited because numerical phase values are still needed. Since s.f. magnitudes are related quantities, the perception arose that the relations between s.f.'s should be described on basis of multivariate statistics. This idea, pioneered by Hauptman and Karle, led to the method of setting up joint probability distributions (j.p.d.'s) of s.f.'s from which phase sum relations like (2.1) can be estimated. If the latter are estimated sufficiently reliably, DM phase processing may lead automatically to a structure solution, see e.g. [9, 10] for details on DM procedures and theory. The estimation of the ψ_{123} depends directly on the j.p.d. of three s.f.'s. To gain insight in the effectiveness and limitations of DM, its derivation will be discussed in some detail, in particular the assumptions involved.

3. Joint probability distributions of structure factors in space group P1

A j.p.d. of three s.f.'s F_ν (= $F(H_\nu)$) can be set up via the j.p.d. of the real and imaginary parts A_ν and B_ν respectively as a six-fold integral

$$P(A_1,A_2,A_3,B_1,B_2,B_3) = (2\pi)^{-6}\int_{-\infty}^{\infty}\cdots\int_{-\infty}^{\infty}\exp\left[-i\left\{\sum_{\nu=1}^{3}(A_\nu v_\nu + B_\nu w_\nu)\right\}\right]\cdot$$

$$\cdot\; C(v_1,v_2,v_3,w_1,w_2,w_3)\,dv_1dv_2dv_3dw_1dw_2dw_3 \tag{3.1}$$

while the characteristic function (c.f.) $C(v_1,..,w_3)$ can be expressed,

$$C(v_1,v_2,v_3,w_1,w_2,w_3) = \left\langle\exp\left[+i\left\{\sum_{\nu=1}^{3}(A_\nu v_\nu + B_\nu w_\nu)\right\}\right]\right\rangle \tag{3.2}$$

Once the c.f. is available, the j.p.d. is directly obtained because of the one-to-one Fourier-transform relationship that exists between the j.p.d. and its c.f. After a change to polar coordinates,

$$v_\nu = \rho_\nu\cos(\Theta_\nu) \quad\text{and}\quad w_\nu = \rho_\nu\sin(\Theta_\nu) \quad\text{for } \nu = 1,2,3$$

$$A_\nu = R_\nu\cos(\Phi_\nu) \quad\text{and}\quad B_\nu = R_\nu\sin(\Phi_\nu) \quad\text{for } \nu = 1,2,3 \tag{3.3}$$

with R_ν and Φ_ν the r.v.'s associated with $|F_\nu|$ and φ_ν, $P(A,B)dAdB$ changes into $P(R,\Phi)RdRd\Phi$, $C(v,w)dvdw$ into $C(\Theta,\rho)\rho d\Theta d\rho$ and $A_\nu v_\nu + B_\nu w_\nu$ becomes $\rho_\nu R_\nu\cos(\Theta_\nu-\Phi_\nu)$. As a result (3.1) is changed into (see e.g. [11])

$$P(R_1,R_2,R_3,\Phi_1,\Phi_2,\Phi_3) = R_1R_2R_3(2\pi)^{-6}\cdot$$

$$\int_0^\infty\int_0^\infty\int_0^\infty\int_0^{2\pi}\int_0^{2\pi}\int_0^{2\pi}\rho_1\rho_2\rho_3\exp\left[-i\left\{\sum_{\nu=1}^{3}\left[\rho_\nu R_\nu\cos(\Theta_\nu-\Phi_\nu)\right]\right\}\right]\cdot$$

$$C(\rho_1,\rho_2,\rho_3,\Theta_1,\Theta_2,\Theta_3)\,d\Theta_1d\Theta_2d\Theta_3d\rho_1d\rho_2d\rho_3 \tag{3.4}$$

If for simplicity, the space group P1 is taken, the trigonometric expressions of A_ν and B_ν can be inserted. Expressing now $C = \exp[\mathrm{Ln}(C)]$, the c.f. becomes

$$C = \exp\left\{\mathrm{Ln}\left[\left\langle\exp\left[i\left\{\sum_{j=1}^{N}\sum_{\nu=1}^{3}\rho_\nu f_{j\nu}\cos(2\pi\underline{H}_\nu\cdot\underline{r}_j-\Theta_\nu)\right\}\right]\right\rangle\right]\right\} \tag{3.5}$$

4. Primitive random variables

The average in (3.5) is to be carried out under the assumption of a prior distribution of the r.v.'s associated with A_ν and B_ν or, equivalently, the R_ν and Φ_ν. S.f.'s are functions of both reciprocal and direct space coordinates and therefore so-called primitive random variables (p.r.v.'s) are introduced, associated with either choice of coordinates (direct or reciprocal space). P.r.v.'s associated with direct or reciprocal space coordinates imply different probabilistic models. The latter option implies that the crystal structure is regarded to be fixed, though unknown. $P(\Phi_1,..,R_3)d\Phi_1..dR_3$ becomes now the probability that a particular combination of phase values and magnitudes occur for a certain set of reciprocal space coordinates, i.e. the indices. In contrast, if p.r.v.'s are associated with direct space coordinates the crystal structure is regarded to be variable. Now $P(\Phi_1,..,R_3)d\Phi_1..dR_3$ is the probability that a particular combination of phase values and magnitudes occurs for a certain set of atomic coordinates. This model rationalizes the DM process as to pick out the (one and only) structure that fits the observed data the best, given the ensemble of structures that is allowed for by the prior (assumption of) the p.r.v.'s,. An important advantage of this direct-space approach over the reciprocal-space related p.r.v.'s is that structure-specific or partial structure information can be used to limit the ensemble of structures from which the crystal structure should be selected.

4.1 UNIFORM PRIOR: UNIFORMLY AND INDEPENDENTLY DISTRIBUTED ATOMS

The average in (3.5) can be expressed as an integration of the form

$$\langle g(\underline{r}_1,\ldots,\underline{r}_N)\rangle = \int g(\underline{r}_1,\ldots,\underline{r}_N)\; p(\underline{r}_1,\ldots,\underline{r}_N)d\underline{r}_1..d\underline{r}_N \tag{4.1}$$

The integration can be carried out if the prior distribution $p(\underline{r}_1,..,\underline{r}_N)$ of the p.r.v.'s is available. If nothing is known about the structure, except the contents of the unit cell, and no preference can be stated concerning the atomic positions, all positions are equally likely. This leads to the concept of uniformly and independently distributed atoms,

$$p(\underline{r}_1,\ldots,\underline{r}_N) = p(\underline{r}_1).p(\underline{r}_2)\ldots p(\underline{r}_N) \text{ with}$$

$$p(\underline{r}_j) = 1 \quad \text{for } j = 1,\ldots,N \text{ and } \underline{r}_j \in [0,1] \tag{4.2}$$

This assumption changes $\langle\exp[i\sum(..j)]\rangle$ in (3.5) into $\prod_j \langle\exp[i(..j)]\rangle$ and since $\mathrm{Ln}\{\prod_j \langle\exp[i(..j)]\rangle\} = \sum_j \{\mathrm{Ln}\langle\exp[i(..j)]\rangle\}$ the c.f becomes,

$$C = \exp\left[\sum_{j=1}^{N}\left\{\mathrm{Ln}\left\langle\exp\left[i\left\{\sum_{\nu=1}^{3}\rho_\nu f_{j\nu}\cos(2\pi\underline{H}_\nu.\underline{r}_j-\Theta_\nu)\right\}\right]\right\rangle_{\underline{r}_j}\right\}\right] \tag{4.3}$$

To calculate the averages in (4.3), the exponential expression inside the $\langle..\rangle$ is expanded using a Taylor series expansion. This gives a series in the integration variables ρ_ν and Θ_ν with the so-called moments of the distribution as coeffients,

$$C = \exp\left[\sum_{j=1}^{N}\mathrm{Ln}\left\{\sum_{nmax=0}^{\infty} U_{nmax}\left[m^{\beta_1,\ldots,\beta_3}_{\alpha_1,\ldots,\alpha_3}\right]\right\}\right] \tag{4.4}$$

$$U_{nmax} = \sum_{\substack{\alpha_1,\ldots,\beta_3=0\\ \alpha_1+..+\beta_3=nmax}}^{nmax}\cdots\sum^{nmax}\prod_{\nu=1}^{3}\left\{\frac{(i\rho_\nu)^{\alpha_\nu+\beta_\nu}}{2^{\alpha_\nu+\beta_\nu}\alpha_\nu!\beta_\nu!}.\exp[i\theta_\nu(\beta_\nu-\alpha_\nu)]\right\}.m^{\beta_1,\ldots,\beta_3}_{\alpha_1,\ldots,\alpha_3}$$

$$m^{\beta_1,\ldots,\beta_3}_{\alpha_1,\ldots,\alpha_3} = \prod_{\nu=1}^{3} |f_{j\nu}|^{\alpha_\nu+\beta_\nu} \exp[i\delta_{j\nu}(\alpha_\nu-\beta_\nu)]\cdot$$

$$\left\langle\exp\left[2\pi i\left\{\sum_{\nu=1}^{3}(\alpha_\nu-\beta_\nu)\underline{H}_\nu.\right\}.\underline{r}_j\right]\right\rangle \tag{4.5}$$

The integration in (4.5) gives $\sum_\nu (\alpha_\nu-\beta_\nu)\underline{H}_\nu = \underline{0}$ as the condition for the non-zero moments with $\sum_\nu \alpha_\nu+\beta_\nu$ = nmax defining the N-dependence. In order to end up with an integrable c.f., the non-zero moments at nmax ≤ 2 are important. Expression (4.5) shows that $U_0 = 1$ and all $U_1 = 0$, the latter corresponding with all $\langle F_\nu\rangle = 0$. At nmax = 2, for each s.f. a non-zero moment exists ($\alpha_\nu = \beta_\nu = 1$) that is of order N^0. At nmax = 3 triplet terms (order $N^{-\frac{1}{2}}$) occur for $\alpha_1 = \alpha_2 = \alpha_3 = 1$ and $\beta_1 = \beta_2 = \beta_3 = 1$, at nmax = 4 quartet terms (order N^{-1}) etc.

The subsequent steps that lead to the c.f. are quite simple. The series-expansion of the logarithm in (4.3) leads to a second series in the integration variables ρ_ν and Θ_ν though with different coefficients, the cumulants, which can be expressed in terms of the moments. This process is called the moments-cumulants transformation. After adding the cumulant terms of the individual p.r.v.'s the c.f. is obtained. In case of the j.p.d. of three s.f.'s, the c.f. becomes,

$$C = \exp\left[-\sum_{\nu=1}^{3}(\rho_\nu^2/4)Z_{\nu\nu} - i|Z_{123}|(\rho_1\rho_2\rho_3/4)\cos(\Theta_1+\Theta_2+\Theta_3+\Delta_{123})\right] \tag{4.6}$$

$$Z_{\nu\nu} = \sum_{j=1}^{N}|f_{j\nu}|^2 \quad ; \quad Z_{123} = \sum_{j=1}^{N} f^*_{j1}f^*_{j2}f^*_{j3} = |Z_{123}|\exp(i\Delta_{123})$$

Expression (4.6) can be integrated easily and the resulting j.p.d is,

$$P(\Phi_1,\Phi_2,\Phi_3,R_1,R_2,R_3) = (R_1R_2R_3/Z_{11}Z_{22}Z_{33}\pi^3).\exp\left[-R_1^2Z_{11}^{-1} - R_2^2Z_{22}^{-1}\right.$$
$$\left.- R_3^2Z_{33}^{-1} + 2|Z_{123}|R_1R_2R_3(Z_{11}Z_{22}Z_{33})^{-1}\cos(\Phi_1+\Phi_2+\Phi_3+\Delta_{123})\right] \tag{4.7}$$

Finally, a r.v. $\Psi_{123} = \Phi_1+\Phi_2+\Phi_3$ can be introduced, associated with the triplet ψ_{123}. Then, from (4.6) a conditional probability distribution (c.p.d.) can be derived of Ψ_{123} given the numerical values of the s.f. magnitudes associated with R_1, R_2 and R_3,

$$P(\Psi_{123}|R_1,R_2,R_3) \propto \exp\left[2G_{123}\cos(\Psi_{123}+\Delta_{123})\right] \tag{4.8}$$

$$G_{123} = |Z_{123}|(Z_{11}Z_{22}Z_{33})^{-1}|F_1F_2F_3| = |Z_{123}|(Z_{11}Z_{22}Z_{33})^{-\frac{1}{2}}|E_1E_2E_3|$$
$$|E_\nu|^2 = |F_\nu|^2Z_{\nu\nu}^{-1}$$

Expression (4.7), which has been derived first by Cochran [12] in a different way, is a so-called Von Mises distribution and is commonly referred to as the Cochran distribution. The Cochran distribution is the basis of many DM programs. In particular, the tangent formula [13, 14], a popular expression for phase processing DM programs, is obtained easily from a generalized form of (4.8).

The form of (4.8) and its underlying assumptions lead to some conclusions:
(*a*) The distribution is a function of normalized s.f.'s ($|E_\nu| = |F_\nu|Z_{\nu\nu}^{-\frac{1}{2}}$) with $Z_{\nu\nu}$ arising from $\langle|F_\nu|^2\rangle$ as calculated under the assumption of the p.r.v. model.
(*b*) In absence of anomalous scattering, $\Delta_{123} = 0$ so the most likely value of Ψ_{123} will be 0 mod(2π). This expresses that, in general, the Cochran distribution is not

enantiomorph specific, although exceptions due to symmetry exist in chiral space groups. If (some) atoms scatter anomalously, $\Delta_{123} \neq 0$ occur frequently and this can be used successfully in DM applications with SAS and MAD data.
(*c*) In case of equal-atom structures, the sharpness of the Cochran distribution is determined by $|E_H E_K E_{H-K}| N^{-\frac{1}{2}}$. If N becomes larger, a decrease in $N^{-\frac{1}{2}}$ will not be compensated by a simultaneous increase of $|E_H E_K E_{H-K}|$ because with an increase of the structure the chances for structure-specific large normalized s.f.'s reduce as well. As a result, the distributions will become more flat at larger N and, consequently, DM that process phase information via triplets will break down.

In order to get sharper distributions some conditions should be fulfilled: (i) A smaller N, thus dependence on less p.r.v.'s, (ii) a smaller $Z_{\nu\nu}$ thus larger $|E_\nu|$ and (iii) $\Delta_{123} \neq 0$. One way to fulfil these conditions is to exploit a non-uniform prior of $p(\underline{r}_1,..,\underline{r}_N)$. Another way is to make use of isomorphous data (SAS, SIR).

5. Ab initio DM using isomorphous protein data.

A long-standing tradition exists in expressing isomorphous replacement and/or anomalous scattering data algebraically in the phase differences and magnitudes of the heavy-atom/ anomalous scatterer substructure. More recently, procedures have been reported intending to fuse this algebraic approach with DM without solving the substructure first. For example, combining the Bijvoet difference technique with double Patterson based quantities, Kroon et al. [15] were the first to estimate triplet sine phase sums.

An alternative to the algebraic approach is to derive j.p.d.'s of the appropriate s.f.'s involved, the obvious advantage being that it is not necessary to enter prior knowledge of the heavy-atom/ anomalous scatterer substructure. DM have been applied often successfully in the determination of the heavy-atom substructure of macromolecules (e.g. [16] and references therein), although as yet they did not succeed in replacing the Patterson techniques.

5.1 JOINT PROBABILITY DISTRIBUTION OF ISOMORPHOUS STRUCTURE FACTORS

In spite of their physical differences, both isomorphous replacement and anomalous scattering lead to the existence of isomorphous data. Two s.f.'s F^1_ν and F^2_ν, in which the "1" and "2" denote that they belong to different data sets, are isomorphous if the trigonometric form of their s.f. expressions is the same. This model encompasses, for example, the SIR case with F^1_ν and F^2_ν being s.f.'s from the native crystal and the heavy-atom derivative. Also the SAS case is included because F(H) and the complex conjugate of F(-H) are isomorphously related [17].

Two or more isomorphously related s.f.'s lead to extra non-zero moments in (4.4). For simplicity, now only the case of two isomorphously related s.f.'s will be considered. Then, under the assumption that the p.r.v.'s are independent and uniformly distributed in the interval [0,1], for every two isomorphous s.f.'s up to order N^0 (nmax = 2) the non-zero moments in (4.4) lead to the presence of an extra term in the c.f,

$$\exp\left[-\frac{\rho_1\rho_2}{2}|Z_{12}|\cos(\Theta_1+s_{12}\Theta_2+\Delta_{12})\right] \tag{5.1}$$

$$Z_{12} = |Z_{12}|\exp[i\Delta_{12}] = \sum_{j=1}^{N}|f_{j1}||f_{j2}|\exp[-i(\delta_{j1}+s_{12}\delta_{j2})] \tag{5.2}$$

$$s_{12} = -1 \text{ if } H_1 = H_2, \text{ while } s_{12} = +1 \text{ if } H_1 = -H_2$$

After integration of the c.f., the j.p.d. becomes

$$P(\Phi_1,\Phi_2,R_1,R_2) \propto \exp\left[\left[Z_{11}Z_{22}-|Z_{12}|^2\right]^{-1}\left\{-R_1^2 Z_{22}-R_2^2 Z_{11}\right\}\right].$$

$$\exp\left[2R_1R_2|Z_{12}|\left[Z_{11}Z_{22}-|Z_{12}|^2\right]^{-1}\cos(\Phi_1+ s_{12}\Phi_2+ \Delta_{12})\right] \quad (5.3)$$

Thus, the j.p.d. of two isomorphously related s.f.'s leads to the presence of a doublet phase sum term $\Phi_1+ s_{12}\Phi_2$ that is of order N^0.

In a similar way expressions can be derived for the triplet phase sums present among sets of isomorphous s.f. For example, in the SIRNAS and SAS cases, with for each reflection two isomorphous choices, eight different types of triplets are possible: ψ_{123}^{111}, ψ_{123}^{112}, ψ_{123}^{121}, ψ_{123}^{122}, ψ_{123}^{211}, ψ_{123}^{212}, ψ_{123}^{221} and ψ_{123}^{222} in which the superscript (abc) denotes that reflection H_1 is from data set *a*, reflection H_2 from data set *b* and reflection $H_3 = (-H_1- H_2)$ from data set *c*. Probabilistic expressions have been secured for the SIR case [18, 19], the SAS case [20, 21, 22] while also a complete probabilistic integration of DM with any type and number of isomorphous data has been established [23], encompassing the above-mentioned SIR and SAS expressions as marginal distributions. For space-saving reasons we will not repeat these expressions here, but an analysis of (5.2)-(5.3) shows some common characteristics:

(*i*) Isomorphous s.f.'s lead to doublet phase sums that are of order N^0. Doublets play an important role because via the doublet estimates all phases of the second data set that occur in triplets can be expressed in terms of those of the first data set only, as will be discussed in §7. This process to improve phase sum estimates of triplets using invariants of lower order is quite similar to that of estimating quartets (order N^{-1}). The latter process requires cross-term phases to be removed, e.g. by using cross-term phase estimates on basis of triplets (order $N^{-\frac{1}{2}}$) that are of lower order than the quartets [24].

(*ii*) In spite of the isomorphism, all the probabilistic quantities still depend explicitely on all N p.r.v.'s.

(*iii*) In the SAS case doublet and triplet estimates anywhere in the interval $(-\pi,\pi)$ are possible.

The doublet phase sum has been the subject of some recent studies. It has been pointed out that in the SAS case the doublets tend to have the same (positive) sign [25, 26]. The doublet sign ambiguity known to exist in the SIRNAS case has been analysed using cluster-analysis [27, 28]. As a solution, an analysis of all possible triplets given the eight possible sign combinations of the doublets in the SIR and SAS cases has been suggested but it was observed that triplets cluster around a few values and a two-fold sign ambiguity could not be avoided [28].

6. Difference structure factor technique

It has been mentioned in §5 that the probabilistic quantities still depend on all N p.r.v.'s when isomorphous s.f.'s are used as r.v.'s. Recently, it was recognized that the applicability of DM in *ab initio* structure determination with isomorphous data may be improved by reducing explicitely the amount of p.r.v.'s that should be allowed to vary uniformly. This reduction can be achieved by defining the difference structure factor (d.s.f.) between two isomorphous s.f.'s as a r.v. [29, 30]

$$F^d_\nu = F^1_\nu - F^2_\nu = \sum_{j=1}^{N} (f^1_{j\nu} - f^2_{j\nu})\exp(2\pi i\underline{H}_\nu.\underline{r}_j) = \sum_{j=1}^{n} f^d_{j\nu}\exp(2\pi i\underline{H}_\nu.\underline{r}_j) \quad (6.1)$$

The individual isomorphous s.f.'s F^1_ν and F^2_ν depend on the complete set of *N* atoms but, in contrast, the d.s.f. F^d_ν depends only on those *n* atoms for which the atomic scattering factors in F^1_ν and F^2_ν are different. For the purpose of deriving j.p.d.'s F^d_ν can be considered formally to be a s.f. to which all j.p.d. techniques developed so far apply. For example, the normalized j.p.d. of a d.s.f in space group P1 can be expressed as:

$$P(R^d_\nu, \Phi^d_\nu) = 2R^d_\nu (Z^d_\nu)^{-1} \exp[-(R^d_\nu)^2 (Z^d_\nu)^{-1}] \quad (6.2)$$

$$Z^d_\nu = \sum_{j=1}^{n} |f^d_{j\nu}|^2$$

By taking the squared magnitude of (6.1), the relation to the doublet phase sum becomes visible,

$$|F^d_\nu|^2 = |F^1_\nu|^2 + |F^2_\nu|^2 - 2|F^1_\nu||F^2_\nu|\cos(\psi^d_\nu) \quad (6.3)$$

$$\psi^d_\nu = \varphi^1_\nu + s^d_\nu \varphi^2_\nu \quad ; \quad s^d_\nu = +1 \text{ (SAS) or } -1 \text{ (SIRNAS, SIRAS, MAD)}$$

From (6.2) it follows that $<|R^d_\nu|^2> = Z^d_\nu$. Combining this with (6.3) while assuming that $|F^1_\nu|$ and $|F^2_\nu|$ are exactly known and replacing the exact doublet phase sum $\cos(\psi^d_\nu)$ by the estimate $<\cos(\Psi^d_\nu)>$, with Ψ^d_ν the r.v. for ψ^d_ν, we get,

$$<\cos(\Psi^d_\nu)> = (|F^1_\nu|^2 + |F^2_\nu|^2 - Z^d_\nu)/(2|F^1_\nu||F^2_\nu|) \quad (6.4)$$

The above expressions can also be derived algebraically and it has been shown [31] that from a new difference Patterson synthesis employing cosine doublet estimates like (6.4),

$$P(\underline{u}) = \sum_{H} |F^1_\nu - F^2_\nu|^2 \cos(2\pi \underline{H}.\underline{u}) \quad (6.5)$$

interatomic vector information can be derived which in turn may improve the initial estimates of $\cos(\Psi^d)$. In this respect, the difference Patterson [32] can be considered a special case of (6.5) with all $\cos(\Psi^d) = 1$.

7. The probability distribution of the triplet phase sums present among three difference structure factors

The joint probability theory of s.f.'s while allowing for complex-valued atomic scattering factors is directly applicable to the d.s.f.'s of isomorphous s.f.'s. Let us define R^d_1, R^d_2 and R^d_3 to be r.v.'s for the magnitudes $|F^d_1|$, $|F^d_2|$ and $|F^d_3|$ respectively and Ψ^d_{123} to be the r.v. of the d.s.f. triplet ψ^d_{123},

$$\Psi^d_{123} = \Phi^d_1 + \Phi^d_2 + \Phi^d_3 \quad (7.1)$$

for which, analogous to (4.8), the c.p.d. can be expressed,

$$P(\Psi^d_{123}|R^d_1, R^d_2, R^d_3) \propto \exp\left[2C^d_{123}\cos(\Psi^d_{123} + \Delta^d_{123})\right]$$

$$C^d_{123} = R^d_1 R^d_2 R^d_3 |Z^d_{123}| \quad (7.2)$$

$$Z^d_\nu = \sum_{j=1}^{N} |f^1_{j\nu} - f^2_{j\nu}|^2 \quad (\text{for } \nu = 1,2,3) \quad (7.3)$$

$$Z^d_{123} = |Z^d_{123}|\exp[i\Delta^d_{123}] = \left[z^d_1 z^d_2 z^d_3\right]^{-1} \sum_{j=1}^{N} \prod_{\nu=1}^{3} \left[f^1_{j\nu} - f^2_{j\nu}\right]^* \quad (7.4)$$

with * denoting the complex conjugation.
Expression (7.4) is a function of both data sets. In order to arrive at an expression that is a function of variables of one data set only, we follow now a procedure, introduced in [29]. The product of the r.v.'s R^d_1, R^d_2, R^d_3 and $\exp[i\Psi^d_{123}]$ in terms of the s.f.'s is

$$R^d_1 R^d_2 R^d_3 \exp[i\Psi^d_{123}] = \prod_{\nu=1}^{3} (F^1_\nu - F^2_\nu) \tag{7.5}$$

The right-hand side of (7.5) can be written out as a sum of the eight contributing isomorphous triplets,

$$\Psi^{abc}_{123} = \Phi^a_1 + \Phi^b_2 + \Phi^c_3 \qquad (a,b,c = 1,2) \tag{7.6}$$

Each of these eight terms is now expressed exclusively in one of the eight isomorphous triplets. After replacing in (7.5) the doublet phase sums $\exp(i\psi^d_\nu)$ by the estimates $<\exp(i\Psi^d_\nu)>$, with Ψ^d_ν the r.v. for the doublet phase sum

$$<\exp(i\Psi^d_\nu)> = <\exp[i(\Phi^1_\nu + s^d_\nu \Phi^2_\nu)]> = \exp(i\lambda^d_\nu) \tag{7.7}$$

$$\lambda^d_\nu = \cos^{-1}\left[\frac{|F^1_\nu|^2 + |F^2_\nu|^2 - z^d_\nu}{2|F^1_\nu||F^2_\nu|}\right]$$

Since in this way the r.v. Φ^2_ν can be expressed in Φ^1_ν and λ^d_ν, all eight triplets in (7.5) can be expressed completely in terms of the triplet of data set 1 reflections only,

$$(F^1_1 - F^2_1).(F^1_2 - F^2_2).(F^1_3 - F^2_3) =$$

$$\exp[i\Psi^{111}_{123}]\Big\{|F^1_1F^1_2F^1_3| - |F^1_1F^1_2F^2_3|\exp[-i\lambda^d_3]$$

$$- |F^1_1F^2_2F^1_3|\exp[-i\lambda^d_2] + |F^1_1F^2_2F^2_3|\exp[-i(\lambda^d_2 + \lambda^d_3)]$$

$$- |F^2_1F^1_2F^1_3|\exp[-i\lambda^d_1] + |F^2_1F^1_2F^2_3|\exp[-i(\lambda^d_1 + \lambda^d_3)]$$

$$+ |F^2_1F^2_2F^1_3|\exp[-i(\lambda^d_1 + \lambda^d_2)] - |F^2_1F^2_2F^2_3|\exp[-i(\lambda^d_1 + \lambda^d_2 + \lambda^d_3)]\Big\} \tag{7.8}$$

The term between {...} in (7.8) does not depend on the triplet Ψ^{111}_{123} itself and can be expressed as $A^{111}_{123}\exp[i\Lambda^{111}_{123}]$ so

$$|R^d_1R^d_2R^d_3|\exp[i\Psi^d_{123}] = A^{111}_{123}\exp[i(\Psi^{111}_{123} + \Lambda^{111}_{123})] \tag{7.9}$$

After combining (7.9) with (7.2), the distribution of Ψ^{111}_{123} becomes

$$P(\Psi^{111}_{123}|R^d_1, R^d_2, R^d_3) \propto \exp\left[2C^{111}_{123}\cos(\Psi^{111}_{123} + \Lambda^{111}_{123} + \Delta^d_{123})\right]$$

$$G_{123}^{111} = A_{123}^{111} |Z_{123}^{d}| \tag{7.10}$$

The same methodology can also be applied to quartets [30] and sem-invariants (Peschar, to be published) and it forms the basis of the computer program SIMPROT (Seljée et al., to be published) for ab initio structure determination using isomorphous data (SAS, SIR, etc.) that is under development

References

1. Miller, R., Gallo, S.M., Khalak, H.G. & Weeks, C.M. (1994) SnB: Crystal structure determination *via* Shake-and-Bake, *J. Appl. Cryst.* **27**, 613-621
2. Karle, J. (1980) Some developments in anomalous dispersion for the structural investigation of macromolecular systems in biology, *Int. J. Quantum Chem.*, 7, 357-367.
3. Karle, J. (1983) A simple rule for finding and distinguishing triplet phase invariants with values near 0 or π with isomorphous replacement data, *Acta Cryst.* **A39**, 800-805.
4. Hendrickson, W.A. (1988) Determination of macromolecular structures from anomalous diffraction of synchrotron radiation, *Science*, **254**, 51-58.
5. Hendrickson, W.A., Pähler, A., Smith, J.L., Satow, Y., Merrit, E.A. and Phizackerley, R.P. (1989) Crystal structure of core streptavidin determined from multiwavelength anomalous diffraction of synchrotron radiation, *Proc. Natl. Acad. Sci. USA.* **86**, 2190-2194
6. Guss, J.M., Merritt, E.A., Phizackerley, R.P., Hedman, B., Murata, M., Hodgson, K.O. & Freeman, H.C. (1988) Phase determination by multiple-wavelength X-ray diffraction: crystal structure of a basic "blue" copper protein from cucumbers, *Science*, **241**, 806-811.
7. Harker, D. & Kasper, J.S. (1948) Phases of fourier coefficients directly from crystal diffraction data, *Acta Cryst.* **1**, 70-75.
8. Karle, J. & Hauptman, H. (1950) The phases and magnitudes of the structure factors, *Acta Cryst.* **3**, 181-187.
9. Giacovazzo, C. (1980) *Direct Methods in Crystallography,* Academic Press, London.
10. Schenk, H. (1991) (Ed.) *Direct Methods for solving crystal structures*, Plenum, New York.
11. Karle, J. & Hauptman, H. (1958) Phase determination from new joint probability distributions: Space group P1, *Acta Cryst.* **11**, 264-269.
12. Cochran, W. (1955) Relations between the phases of structure factors, *Acta Cryst.* **8**, 473-478.
13. Karle, J. & Hauptman, H. (1956) A theory of phase determination for the four types of non-centrosymmetric space groups 1P222, 2P22, $3P_12$, $3P_22$, *Acta Cryst.* **9**, 635-651.
14. Karle, I. & Karle, J. (1964) An application of the symbolic addition method to the structure of L-arginine dihydrate, *Acta Cryst.* **17**, 835-841.
15. Kroon, J., Spek, A.L. & Krabbendam, H. (1977) Direct phase determination of triple products from Bijvoet inequalities, *Acta Cryst.* **A33**, 382-385.
16. Fortier, S. (1991) Direct methods for macromolecular crystallography, in H. Schenk (ed.), *Direct Methods of solving crystal structures,* Plenum, New York, pp.311-334.
17. Fortier, S. & Nigam, G.D. (1989). On the probabilistic theory of isomorphous data sets: General joint distributions for the SIR, SAS and partial/complete structure cases, *Acta Cryst.* **A45**, 247-254.
18. Hauptman, H. (1982) On integrating the techniques of direct methods and isomorphous replacement. I. The theoretical basis, *Acta Cryst.* **A38**, 289-294.
19. Giacovazzo, C., Cascarano, G. & Zheng Chao-De (1988). On integrating the techniques of direct methods and isomorphous replacement. A new probabilistic formula for triplet invariants, *Acta Cryst.* **A44**, 45-51.
20. Heinerman, J.J.L., Krabbendam, H., Kroon, J. & Spek, A.L. (1978) Direct phase determination of triple products from Bijvoet inequalities. II. A probabilistic approach, *Acta Cryst.* **A34**, 447-450.

21. Hauptman, H. (1982) On integrating the techniques of direct methods with anomalous dispersion. I. The theoretical basis, *Acta Cryst.* **A31**, 632–641.
22. Giacovazzo, C.(1983) The estimation of two-phase and three-phase invariants in P1 when anomalous scatterers are present, *Acta Cryst.* **A39**, 585–592.
23. Peschar, R. & Schenk, H. (1991) The joint probability distribution of structure factors incorporating anomalous-scattering and isomorphous-replacement data, *Acta Cryst.* **A47**, 428–440.
24. Peschar, R. (1987) Automatic derivation of joint probability distributions of normalized structure factors and optimal symbolic phase determination, Ph.D. Thesis, Univ. of Amsterdam, The Netherlands
25. Guo, D.Y. (1990) Sign distribution of two-phase structure invariants, *Acta Cryst.* **A46**, 942–944.
26. Guo, D.Y., Blessing, R.H. & Hauptman, H. (1991) On integrating the techniques of direct methods with anomalous dispersion. II. Statistical properties of the two-phase structure invariants, *Acta Cryst.* **A47**, 340–345.
27. Fortier, S. Moore, N.J. & Fraser, M.E. (1985) A direct-methods solution to the phase problem in the single isomorphous replacement case: Theoretical basis and initial applications, *Acta Cryst.* **A41**, 571–577.
28. Fortier, S., Fraser, M.E. & Moore, N.J. (1986) On the number of ambiguities in direct methods – anomalous scattering estimates of the two- and three-phase structure invariants, *Acta Cryst.* **A42**, 149–156.
29. Kyriakidis, C.E., Peschar, R. & Schenk, H. (1993) On direct-methods phase information from differences between isomorphous structure factors, *Acta Cryst.* **A49**, 557–569.
30. Kyriakidis, C.E., Peschar, R. & Schenk, H. (1996) The estimation of four-phase structure invariants using the single difference of isomorphous structure fcators, *Acta Cryst.* **A52**, 77–87.
31. Kyriakidis, C.E., Peschar, R. & Schenk, H. (1993) On the doublet phase sums of isomorphous data sets, *Acta Cryst.* **A49**, 359–369.
32. Rossmann, M.G. (1961) The position of anomalous scatterers in protein crystals, *Acta Cryst.* **A29**, 77–82.

INTEGRATION OF DIRECT METHODS WITH SIR TECHNIQUES

C. GIACOVAZZO[+] & D. SILIQI[*]
[+]*Istituto di Ricerca per lo Sviluppo di Metodologie Cristallografiche CNR, c/o Dipartimento Geomineralogico, Campus Universitario, Via Orabona 4, 70125 Bari, Italy.*
[*]*Department of Inorganic Chemistry, Tirana University, Tirana, Albania*

1. Notation

$F_p = \|F_p\|\exp(i\varphi)$	Structure factor of the protein
$F_d = \|F_d\|\exp(i\psi)$	Structure factor of the isomorphous derivative
$F_H = F_d - F_p$	Structure factor of the heavy-atom structure (i.e. the atoms added to the native protein)
$\delta = \psi - \phi$	
$\Phi = \varphi_{\mathbf{h}} - \varphi_{\mathbf{k}} - \varphi_{\mathbf{h-k}}$	
$E_p = R\exp(i\varphi)$	Normalized structure factor of the protein
$E_d = S\exp(i\psi)$	Normalized structure factor of the isomorphous derivative
N	Number of non-H atoms in the primitive unit cell for the native protein
$\sigma_i = \sum_{j=1}^{N} Z_j^i$	Z_j = atomic number of the *j*th atom
$N_{eq} = \sigma_2^3 / \sigma_3^2$	(Statistically equivalent) number of atoms in the primitive unit cell
$\left[\sigma_2^3 / \sigma_3^2\right]_p$	Value of N_{eq} for the native protein
$\left[\sigma_2^3 / \sigma_3^2\right]_H$	Value of N_{eq} for the heavy atom structure
$[\sigma_2]_p, [\sigma_2]_H, [\sigma_2]_d = \sum_j Z_j^2$	The sum is extended to the native protein atoms, to the heavy atom structure and to the derivative atoms
$D_i(x) = I_i(x) / I_0(x)$	I_i = modified Bessel function of order *i*

B_p, B_d: overall thermal factors of the protein and of the derivative as obtained by Wilson plot

K_p, K_d: scale factor of the protein and of the derivative as obtained by Wilson plot

$\sigma = [\sigma_2]_H / [\sigma_2]_p$

S. Fortier (ed.), Direct Methods for Solving Macromolecular Structures, 285–295.

2. Introduction

The integration of direct methods with isomorphous replacement techniques (SIR case) started with Hauptman [1]. His paper aimed at deriving a probabilistic theory of two and three phase structure invariants for isomorphous pairs. In particular Hauptman derived:

a) the joint probability distribution

$$P(\phi_{\mathbf{h}}, \psi_{\mathbf{h}}, R_{\mathbf{h}}, S_{\mathbf{h}}),$$

from which the conditional distribution

$$P(\delta_{\mathbf{h}} | R_{\mathbf{h}}, S_{\mathbf{h}}) \tag{1}$$

was derived;

b) The joint probability distribution function ($\mathbf{h} + \mathbf{k} + \mathbf{l} = 0$)

$$P(\phi_{\mathbf{h}}, \phi_{\mathbf{k}}, \phi_{\mathbf{l}}, \psi_{\mathbf{h}}, \psi_{\mathbf{k}}, \psi_{\mathbf{l}}, R_{\mathbf{h}}, R_{\mathbf{k}}, R_{\mathbf{l}}, S_{\mathbf{h}}, S_{\mathbf{k}}, S_{\mathbf{l}}) \tag{2}$$

which involves eight types of three-phase structure invariants:

$$\Omega_1 \equiv \Phi = \phi_{\mathbf{h}} + \phi_{\mathbf{k}} + \phi_{\mathbf{l}}, \quad \Omega_2 = \phi_{\mathbf{h}} + \phi_{\mathbf{k}} + \psi_{\mathbf{l}}, \quad \Omega_3 = \phi_{\mathbf{h}} + \psi_{\mathbf{k}} + \phi_{\mathbf{l}},$$

$$\Omega_4 = \psi_{\mathbf{h}} + \phi_{\mathbf{k}} + \phi_{\mathbf{l}}, \quad \Omega_5 = \phi_{\mathbf{h}} + \psi_{\mathbf{k}} + \psi_{\mathbf{l}} \quad \Omega_6 = \psi_{\mathbf{h}} + \phi_{\mathbf{k}} + \psi_{\mathbf{l}}$$

$$\Omega_7 = \psi_{\mathbf{h}} + \psi_{\mathbf{k}} + \phi_{\mathbf{l}} \quad \Omega_8 = \psi_{\mathbf{h}} + \psi_{\mathbf{k}} + \psi_{\mathbf{l}}$$

The algebraic basis of the distributions lies in the neighborhood principle [2] and in the theory of representation [3], [4]. From (2) the conditional probability distributions

$$P(\Omega_i | R_1, R_2, R_3, S_1, S_2, S_3) \cong [2\pi I_0(A_i)]^{-1} \exp(A_i \cos\Omega_i) \tag{3}$$

may be derived by standard techniques. The most important for practical applications is $P(\Omega_1 | R_1, R_2, R_3, S_1, S_2, S_3)$ which involves the protein phases only. Distribution (3) has been checked [5] on error-free (calculated) diffraction data from the protein cytochrome C_{555} from Paracoccus denitrificans and after [6] interpreted in terms of parameters directly connected to the diffraction experiment.

Giacovazzo, Cascarano and Zheng [7] revised the method in order to:

a) prove that results are independent of the chosen primitive random variables (i.e., the reciprocal vectors $\mathbf{h}$ or the atomic positional parameters $\mathbf{r}_j$);

b) take full account of the resolution effects on the distribution parameters β and β_j , j=0,1,2,3. The conclusive distribution so obtained proved to be more effective than the original Hauptman's formula but was also much more cumbersome.

c) derive a simple and effective formula for the case "native-heavy atom derivative". This last effect brought to a von Mises distribution

$$P(\Phi) \cong [2\pi I_0(A)]^{-1} \exp(A \cos\Phi) \tag{4}$$

where

$$A = 2[\sigma_2/\sigma_3^{3/2}]_p R_1 R_2 R_3 + 2[\sigma_2/\sigma_3^{3/2}]_H \Delta_1 \Delta_2 \Delta_3 \tag{5}$$

where

$$\Delta = (F_d - F_p)/\Sigma_H^{1/2}$$

is a pseudo-normalized difference (with respect to the heavy-atom structure). According to (5) the concentration parameter A is the sum of two contributions: a) the Cochran term calculated for the native and a central term which depends on the pseudo-normalized differences Δ. Since the number of heavy atoms in the unit cell is much smaller then the number of the protein atoms

$$\left[\sigma_2/\sigma_3^{3/2}\right]_p \ll \left[\sigma_2/\sigma_3^{3/2}\right]_H .$$

The consequence is that for usual proteins the first contribution to A is negligible with respect to the second. This is quite a relevant result: it is shown that the supplementary prior knowledge of the derivative diffraction data reduces the complexity of the phasing problem from the order $[N_{eq}]_p$ (number of protein atoms in the unit cell) to the order N_H (number of heavy atoms in the unit cell). It may be observed:

a) the formula is more effective when the number of heavy atoms is small;

b) the sign of the triplet cosine depends mainly on the sign of the product $\Delta_1\ \Delta_2\ \Delta_3$: positive values of the product indicate positive cosine values. Negative values of $\Delta_1\ \Delta_2\ \Delta_3$ are also frequent and suggest Φ values close to π.

The distribution (4) is the probabilistic (and more general) counterpart of the algebraic rule R_{iso} stated by Karle [8] according to which: "If the sign of the product $(F_{d\mathbf{h}} - F_{p\mathbf{h}})(F_{d\mathbf{k}} - F_{p\mathbf{k}})(F_{d\mathbf{l}} - F_{p\mathbf{l}})$ is plus the expected value of the invariant $(\phi_{\mathbf{h}} + \phi_{\mathbf{k}} + \phi_{\mathbf{l}})$ with $\mathbf{h} + \mathbf{k} + \mathbf{l} = 0$ is close to zero, if the sign of the product is minus the expected value of the invariant is close to π." Indeed the signs of a triplet cosine predicted by Karle's rule and by (4) coincide provided the Cochran-type term may be neglected. However (5) states that reliability values do not depend on $(F_{d\mathbf{h}} - F_{p\mathbf{h}})(F_{d\mathbf{k}} - F_{p\mathbf{k}})(F_{d\mathbf{l}} - F_{p\mathbf{l}})$ but on the product $\Delta_{\mathbf{h}}\ \Delta_{\mathbf{k}}\ \Delta_{\mathbf{l}}$.

From (4) a weighted tangent formula may be derived

$$\tan\varphi_{\mathbf{h}} = \sum_j \beta_j \sin\left(\varphi_{\mathbf{k}_j} + \varphi_{\mathbf{h}-\mathbf{k}_j}\right) \Big/ \sum_j \beta_j \cos\left(\varphi_{\mathbf{k}_j} + \varphi_{\mathbf{h}-\mathbf{k}_j}\right) = T_{\mathbf{h}}/B_{\mathbf{h}} \qquad (6)$$

where β_j is defined by the equation

$$D_1(\beta_j) = D_1(A)D_1\left(\alpha_{\mathbf{k}_j}\right)D_1\left(\alpha_{\mathbf{h}-\mathbf{k}_j}\right)$$

and

$$\alpha_{\mathbf{h}} = \left(T_{\mathbf{h}}^2 + B_{\mathbf{h}}^2\right)^{1/2} .$$

The reliability parameter $\alpha_{\mathbf{h}}$ of any determined phase $\varphi_{\mathbf{h}}$ is modified according to the agreement between the calculated and the expected value of $\alpha_{\mathbf{h}}$.

3. The procedure

The use of (6) is not straightforward and requires a supplementary insight into the role of the parameters involved in the phasing process. A successful procedure has been described in the following papers [9], [10], [11], [12], [22]. Such procedure will be shortly described in steps.

STEP 1 - **Normalization**. This step is very critical since it provides scale and vibrational parameters necessary to the estimation of the Δ's. These last quantities are generally small, with size order often comparable with the experimental errors on the measurements and with the effects of the lack of isomorphism. Large errors in scale and vibrational parameters, combined with lack of isomorphism, can condemn to the failure any trial for solving the structure. A satisfactory normalization is achieved as follows:

a) The standard Wilson method is applied to native protein data (up to native resolution) to obtain the scale factor K_p and the overall thermal factor B_p. Estimates of the corresponding factors for the derivative are obtained by a differential Wilson plot [14] through the equation

$$\ln\left[\left(\Sigma_p + \Sigma_H\right)\left\langle F_p^2\right\rangle / \left(\Sigma_p\left\langle F_d^2\right\rangle\right)\right] = \ln\left(K_p / K_d\right) + 2\left(B_d - B_p\right)\sin^2\theta / \lambda^2 \qquad (7)$$

The use of (7) does not guarantee a good rescaling especially when the derivative resolution is equal to or lower than 4 Å. A remarkable improvement is obtained by exploiting the P(Δ) distribution ([11], [13]): since the experimental distribution of the Δ's is modeled by different sources of error, it seems reasonable to constrain it to comply with the theoretical distribution expected for that pair of isomorphous structures. We show in Fig.1 the P(Δ) distribution for selected value of σ. The problem may be solved by histogram techniques: the experimental curve is transformed into the theoretically expected P(Δ) distribution at the chosen σ value. The resulting Δ values will then be introduced into (5) for obtaining more accurate triplet invariant estimates.

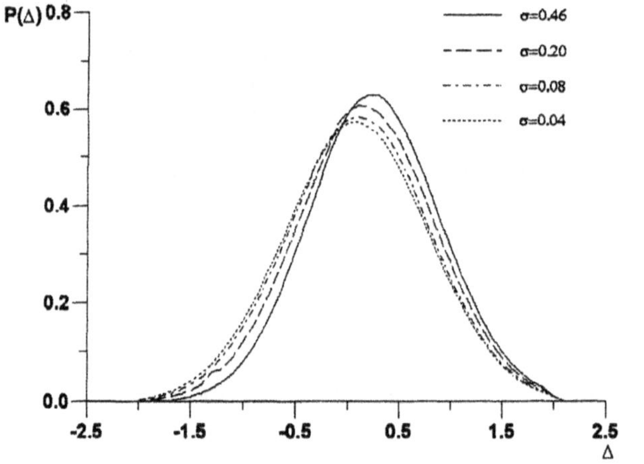

Figure 1 -P(Δ) distribution for select values of σ

STEP 2 - **Triplet calculations**. There are two possible strategies for the phase determination: a) to simultaneously involve a large number of reflections into the phasing process (e.g., almost all the reflections up to derivative resolution); b) to select a relatively small number of reflections among which the triplets are found. Once phases have been given, use such a phase set as a seed for phase expansion. Since the first alternative would require the calculation of several tens of millions of triplets, large storage and computing time, we chose the second alternative.

STEP 3 - **Phasing step**. The NLAR reflections with large values of Δ and R are chosen (they constitute the batch 1):random phases are given to NLAR/2 reflections, and expansion and refinement is performed *via* (7).

STEP 4 - **Selection of the correct solution**. Suitable figures of merit have been designed to pick up the correct solution (see [14], [15] and [9]). They are integrated in the combined figure of merit CFOM: the larger is CFOM, the highest is the probability of the trial solution to be correct. In Table 1 we show for some test structures the values of CFOM and the average error relative to the batch 1 for the best trial solutions as ranked by CFOM. The bold character emphasizes the correct solutions.

TABLE 1 FOM values for the 'best' trial solutions as ranked by CFOM for the various test structures

APP			BPO			E2		
Trial	CFOM	ERR	Trial	CFOM	ERR	Trial	CFOM	ERR
14	**0.49**	**30**	**18**	**0.63**	**29**	**24**	**0.76**	**27**
28	**0.49**	**30**	6	0.48	84	**1**	**0.76**	**27**
7	0.47	82	19	0.48	83	**22**	**0.76**	**27**
29	0.46	83				9	0.74	86
24	0.43	84				16	0.73	86
						31	0.46	78

MFABP			NOX		
Trial	CFOM	ERR	Trial	CFOM	ERR
24	**0.44**	**39**	**61**	**0.45**	**52**
12	0.39	63	**65**	**0.44**	**52**
6	0.38	83	**93**	**0.43**	**53**
			66	0.42	63

STEP 5 - **Phase expansion**. Among the various trials provided by the multisolution approach, the most probable one (on the basis of the figures of merit) is used as a seed for phasing the remaining reflections. Batches of about 200 reflections, chosen in decreasing order of |Δ|, are progressively phased *via* a phase extension procedure from the batch number one, until a phase value is given to all the reflections up to derivative resolution.

4. About the quality of the phases

The quality of the phases obtained by application of the above procedure may be assessed by calculating the average phase error (ERR) between direct methods phases and true phases, and the correlation factor (CORR) between the actual and the published (refined) electron density map:

$$\text{CORR} = \frac{\langle \rho\, \rho_{\text{mod}} \rangle - \langle \rho \rangle \langle \rho_{\text{mod}} \rangle}{\left(\langle \rho^2 \rangle - \langle \rho \rangle^2 \right)^{1/2} \left(\langle \rho^2_{\text{mod}} \rangle - \langle \rho_{\text{mod}} \rangle^2 \right)^{1/2}} \tag{8}$$

The values of ERR and CORR for the test structures are shown in Table 2.

Table 2 Mean phase error (ERR) for the test structures up to derivative resolution. NREF is the number of phased reflections up to derivative resolution. CORR is the correlation factor between direct methods map (derivative resolution) and "true" map (native resolution). The last two columns show ERR and CORR after having exploited the heavy-atom structure.

Structure Name	NREF	ERR(Weighted)	CORR	ERR(Weighted)	CORR
APP	1850	61 (57)	0.3927	58 (53)	0.4667
BPO	12774	57 (52)	0.4490	57 (48)	0.4525
E2	6575	57 (52)	0.5121	56 (47)	0.5026
M-FABP	5456	64 (61)	0.3733	64 (59)	0.3992
NOX	4066	73 (69)	0.3129	74 (67)	0.2939

Sections of BPO and E2 maps from direct methods phases are shown in Figs. 2 and 3 respectively, and are compared with the corresponding sections of the true map. A deeper analysis of the maps suggest that protein and solvent regions are not well separated: however E2 and BPO maps are interpretable, while M-FABP and APP maps are hardly interpretable, and NOX is completely uninterpretable. A comparison with traditional SIR techniques showed that direct methods constitute a valid alternative to them, often offering more useful structure details.

How to improve the effectiveness of direct methods procedures? There are several possible ways:

a) to introduce the representation theory [3], [4] within the framework of the probabilistic theory of isomorphous structures;

b) to exploit the information (available at a certain stage) on the heavy atom structure,

c) to combine direct methods with solvent flattening techniques.

All the points will be discussed shortly in the next sections.

5. Representation theory and isomorphous structures.

Formulas for the estimation of the quartet invariants have been recently derived [17], [18] (see also [19]). Let us denote

$R_1\exp(i\phi_1) = R_{\mathbf{h}}\exp(i\phi_{\mathbf{h}})$, $\quad S_1\exp(i\psi_1) = S_{\mathbf{h}}\exp(i\psi_{\mathbf{h}})$,

$R_2\exp(i\phi_2) = R_{\mathbf{k}}\exp(i\phi_{\mathbf{k}})$, $\quad S_2\exp(i\psi_2) = S_{\mathbf{k}}\exp(i\psi_{\mathbf{k}})$,

$R_3\exp(i\phi_3) = R_{\mathbf{l}}\exp(i\phi_{\mathbf{l}})$, $\quad S_3\exp(i\psi_3) = S_{\mathbf{l}}\exp(i\psi_{\mathbf{l}})$,

$R_4\exp(i\phi_4) = R_{\mathbf{m}}\exp(i\phi_{\mathbf{m}})$, $\quad S_4\exp(i\psi_4) = S_{\mathbf{m}}\exp(i\psi_{\mathbf{m}})$,

$$R_5\exp(i\phi_5) = R_{\mathbf{h+k}}\exp(i\phi_{\mathbf{h+k}}), \quad S_5\exp(i\psi_5) = S_{\mathbf{h+k}}\exp(i\psi_{\mathbf{h+k}}),$$
$$R_6\exp(i\phi_6) = R_{\mathbf{h+l}}\exp(i\phi_{\mathbf{h+l}}), \quad S_6\exp(i\psi_6) = S_{\mathbf{h+l}}\exp(i\psi_{\mathbf{h-l}}),$$
$$R_7\exp(i\phi_7) = R_{\mathbf{k+l}}\exp(i\phi_{\mathbf{k+l}}), \quad S_7\exp(i\psi_7) = S_{\mathbf{k+l}}\exp(i\psi_{\mathbf{k+l}})$$

The joint probability distribution $P(\phi_1,\psi_1,\ldots,\phi_7,\psi_7,R_1,S_1,\ldots,R_7,S_7)$ was first derived from which the conclusive conditional formula

$$P(\Phi_4|R_1,\ldots,R_7,S_1,\ldots,S_7) \cong \left[2\pi I_0(A_4)^{-1}\right]\exp\{A_4\cos\Phi_4\} \tag{9}$$

has been obtained, where

$$A_4 \cong \frac{2}{N_H}\frac{\Delta_1\Delta_2\Delta_3\Delta_4}{1+B}\left\{1+\langle L_5'\rangle+\langle L_6'\rangle+\langle L_7'\rangle\right\},$$
$$B = \frac{1}{2N_H}\Big[\langle L_1'\rangle\langle L_2'\rangle\langle L_5'\rangle+\langle L_3'\rangle\langle L_4'\rangle\langle L_5'\rangle$$
$$+\langle L_1'\rangle\langle L_3'\rangle\langle L_6'\rangle+\langle L_2'\rangle\langle L_4'\rangle\langle L_6'\rangle$$
$$+\langle L_1'\rangle\langle L_4'\rangle\langle L_7'\rangle+\langle L_2'\rangle\langle L_3'\rangle\langle L_7'\rangle\Big],$$
$$\langle L_i'\rangle = \left(S_i'^2 + R_i'^2 - 2R_i'S_i'D_{1i}\right) - 1. \tag{10}$$

The main features of the formula may be so described:

a) the phase relation is of the order N_H^{-1}. Since N_H is usually small, quartets are expected to be reliable (at least in principle).

b) the sign of A_4 is determined by the product of two factors: the first is $\Delta_1\Delta_2\Delta_3\Delta_4$, which may be positive or negative, the second is the term $\left[1+\langle L_5'\rangle+\langle L_6'\rangle+\langle L_7'\rangle\right]$ which again may be positive or negative.

A formula estimating triplets invariants from their second representation has been recently derived [21]. The conclusive formula estimating the triplet invariant Φ may be written as

$$P_{10} = \left[2\pi I_0(A_{10})\right]^{-1}\exp(A_{10}\cos\Phi) \tag{11}$$

where

$$A_{10} = 2\left[\sigma_3/\sigma_2^{3/2}\right]_p R_1R_2R_3 + 2\frac{\Delta_1\Delta_2\Delta_3'}{\sqrt{N_H}}\left\{1+\sum_{\mathbf{k}} CORR_{\mathbf{k}}\right\} \tag{12}$$

$CORR_{\mathbf{k}}$ is a correction term which modulates the first representation estimate. For the sake of shortness we do not give the detailed expression of $CORR_{\mathbf{k}}$: the reader is referred to the original paper.

The role of quartets and of the second representation estimate of the triplet invariants in the phasing procedures is not well established. More efforts will be needed to define it.

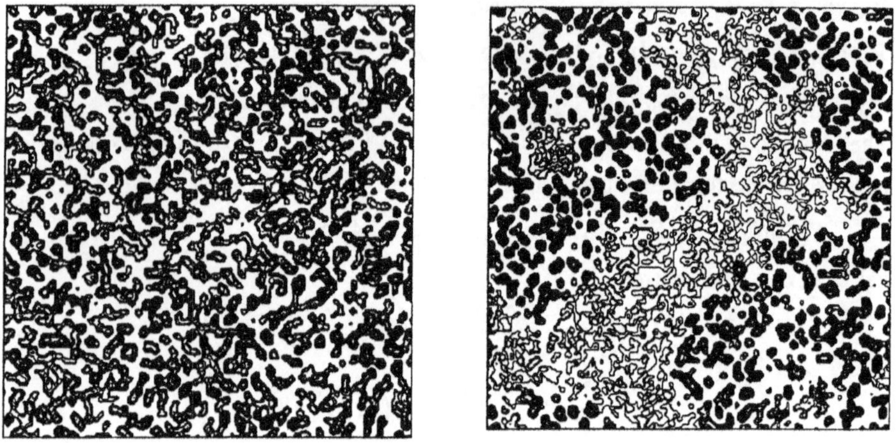

a) b)

Fig. 2 -BPO a) section y=0.8 of the map obtained by Direct Methods. b) section y=0.8 of the true (obtained from the published model) map.

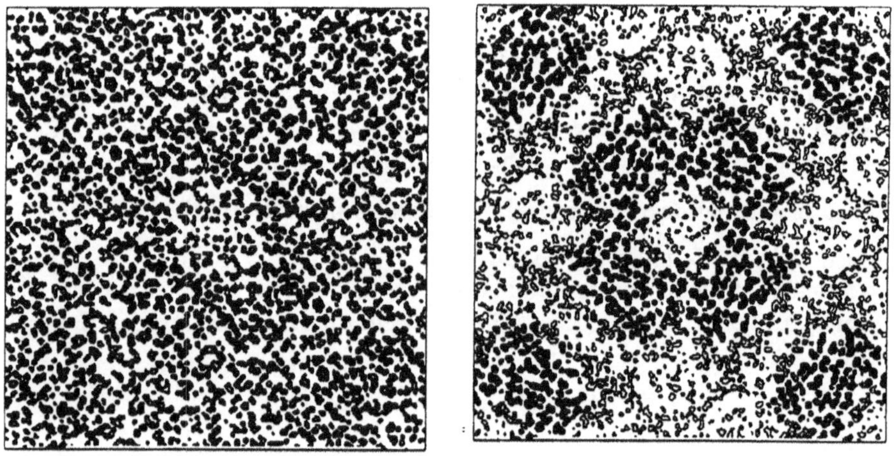

a) b)

Fig. 3 -E2 a) section y=0.8 of the map obtained by Direct Methods. b) section y=0.8 of the true (obtained from the published model) map.

6. About the use of the heavy-atom structure in direct procedures.

The phasing process is able to provide useful electron density maps without any information on the heavy-atom positions. However, once phases are available, a difference Fourier synthesis with coefficients $\left(F_d - F_p\right)\exp\left(i\phi_p\right)$ may provide

approximate heavy-atom structure parameters, which can then be refined by standard techniques. The scaling of derivative on native data is performed by exploiting the P(Δ) curves corresponding to the new $[\sigma_2]_H/[\sigma_2]_p$ values. New Δ's arise, which are then used in the phasing process. This time we do not need to enter into a multisolution procedure: it suffices to refine by a tangent process the phases obtained *via* the old Δ's. The results [22] are in the last two columns of Table 2.
Remarkable improvement of ERR and/or CORR values are obtained for APP and M-FABP. For these structure the new $[\sigma_2]_H/[\sigma_2]_p$ values are very different from those used in absence of information: consequently the corresponding P(Δ) curves are also quite different (see Fig. 1). For BPO and E2 the refined values of $[\sigma]_H/[\sigma]_p$ do not substantially modify the previous situation, while for NOX it becomes slightly worse.
Alternative proposals for incorporating the heavy-atom structure information into the triplet phase distributions were made by various authors. The most significant are those suggested by Fortier, Moore & Fraser [19] and by Klop, Krabbendam & Kroon [20]. Both are based on the following background. Once the heavy-atoms have been located, the cosine moduli of the doublet invariants $\delta_i = \psi_i - \phi_i$, i=1, 2, 3, can be estimated *via* the Carnot relation

$$\cos\delta = \left(|F_d|^2 - |F_p|^2 - |F_H|^2\right)/2|F_dF_p|. \tag{13}$$

Accordingly the distribution

$$P\left(\phi_1,\phi_2,\phi_3,\delta_1,\delta_2,\delta_3 \middle| \left\{R_i',S_i',i=1,2,3\right\}\right) \tag{14}$$

can be derived by a simple change of variable from the distribution

$$P\left(\phi_1,\phi_2,\phi_3,\psi_1,\psi_2,\psi_3 \middle| \left\{R_i',S_i',i=1,2,3\right\}\right).$$

In Fortier *et al.* method the signs of the δ_i's were asumed unknown: then from (14) the conditional distribution

$$P\left(\Phi \mid \left\{R_i',S_i',|\delta_i|,i=1,2,3\right\}\right) \tag{15}$$

may be obtained as a weighted sum of the eight distributions $P\left(\Phi \mid \left\{R_i',S_i',|\delta_i|,i=1,2,3\right\}\right)$ corresponding to the eight sign combinations of the doublet invariants δ_1, δ_2 and δ_3.
Klop, Krabbendam & Kroon proposal tries to exploit also the δ_i signs: indeed estimates of them are available when ϕ_p and heavy-atom structure are known. The conditional distribution

$$P\left(\phi_1,\phi_2,\phi_3,\delta_1,\delta_2,\delta_3 \middle| \left\{R_i',S_i',i=1,2,3\right\}\right) \tag{16}$$

may then be used.
Our tests show that formulas (15) and (16) work well on calculated but not experimental data. Thus the improvement we are able to introduce *via* the knowledge of the heavy-atom structure basically corresponds to the difference between the last two columns of Table 2 and the columns 3 and 4 of the same table.

7. The application of FLEX, a new solvent procedure

Each reflection is phased by direct methods with a given reliability parameter. It occurs that such a parameter is often a good measure of the accuracy of the assigned phase. Such a situation is very promising for application of solvent flattening techniques. We have designed a solvent flattening procedure [22] (from now on denoted FLEX) which may be applied automatically to phases provided by direct methods: it is quite effective for reducing the phases error and providing improved electron density maps. Specific features of the procedure are: a) the envelope is not used in the first stages of the refinement; b) small fractions of the electron density map are inverted; c) very small averaging radius R may be used; d) the solvent region is not set to a constant value. A weight 0.5 is associated to pixels in the solvent region: this retards the solvent flattening process but allows to improve the envelope if, in subsequent cycles, some prominent structural features of the solvent can become part of the protein region. In Table 3 we show, for the test structures, the mean phase error corresponding to initial phases (ERRD) up to derivative resolution and the final phase error (ERRP) corresponding to refined (up to native resolution). CORRD and CORRP are the correlation factors between the final map and the current map [calculated via (8)].

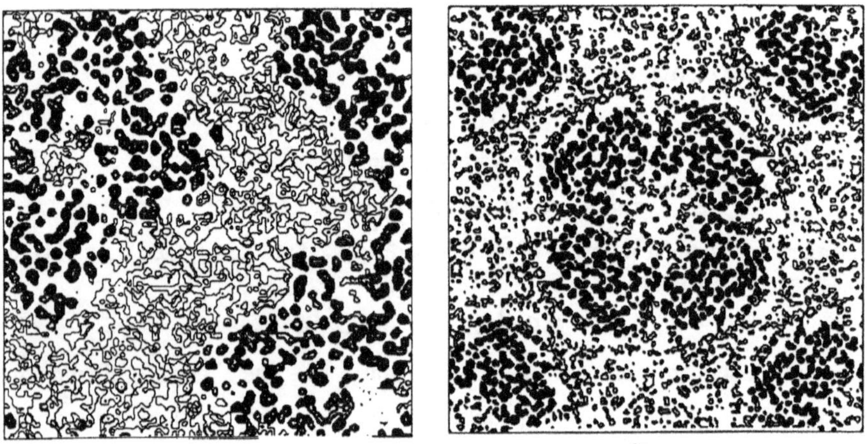

a) b)

Figure 4 -a) BPO: section y=0.8 for the map obtained by applying FLEX to our direct methods map; b) E2: section y=0.8 for the map obtained by applying FLEX to our direct methods map

The efficiency of the technique may be appreciated from Fig 4, where we show for BPO and E2 the same sections emphasied in Figs 2 and 3, but now obtained by using phases refined by FLEX. The solvent and the protein regions are well defined ; the high correlation factors (except for NOX, for which strong lack of isomorphism occured) suggest that all maps are immediately interpretable.

TABLE 3 Number of phased reflections (NREFD), mean phase error (ERRD) and correlation factor (CORRD) for data up to derivative resolution (phase value provided by direct methods). Number of phased reflections (NREFP), mean phase error (ERRP) and correlation factor (CORRP) for data up to native resolution as extended and refined by FLEX procedures.

	APP	BPO	E2	M-FABP	NOX
NREFD	1863	12673	6556	5630	3858
ERRD (weighted)	59 (55)	57 (52)	56 (52)	64 (60)	73 (69)
CORRD	0.4565	0.4452	0.4968	0.4069	0.3020
NREFP	17058	23956	10391	7589	4619
ERRP (weighted)	54 (48)	53 (46)	41 (38)	54 (48)	72 (66)
CORRP	0.7780	0.7391	0.8761	0.6991	0.4175

8. Conclusions

Direct methods can solve protein structures without any prior knowledge of the heavy-atom structure. However they may improve their effectiveness when such information becomes available. Phases so obtained are particularly suitable for phase extension and refinement by solvent flattening procedures like FLEX.

One of the authors (D. Siliqi) undertook this work with the support of the "ICTP Programme for Training and Research in Italian Laboratories, Trieste, Italy".

9. References

1. Hauptman, H. (1982). Acta Cryst. **A38**, 289-294.
2. Hauptman, H (1976). Acta Cryst. **A32**, 934-940.
3. Giacovazzo, C. (1977) Acta Cryst. **A33**, 934-944
4. Giacovazzo, C. (1980) Acta Cryst. **A36**, 362-373
5. Hauptman, H., Potter, S. & Weeks, C. M. (1982). Acta Cryst. **A38**, 294-300
6. Fortier, S., Weeks, C. M., Hauptman, H. (1984). Acta Cryst. **A40**, 544-548
7. Giacovazzo, C., Cascarano, G. & Zheng, C.-D. (1988). Acta Cryst. **A44**, 45-51.
8. Karle J. (1983). Acta Cryst. **A39**, 800-805.
9. Giacovazzo, C., Siliqi, D. & Spagna, R. (1994). Acta Cryst. **A50**, 609-621.
10. Giacovazzo, C., Siliqi, D. & Zanotti, G. (1995). Acta Cryst. **A51**, 177-188.
11. Giacovazzo, C., Siliqi, D. & Gonzalez-Platas, J. (1995). Acta Cryst. **A51**, 811-820.
12. Giacovazzo, C., Siliqi, D., Gonzalez-Platas, J. Hecht, H., Zanotti, G. & York, B. (1995). Acta Cryst. **D52**, 813-825.
13. Blundell, T.L. & Johnson, L.N. (1976). Protein Crystallography, p. 336, London: Academic Press.
14. Cascarano, G., Giacovazzo, C. & Viterbo, D. (1987). Acta Cryst. **A4843**, 22-29.
15. Cascarano, G., Giacovazzo, C. & Guagliardi, A. (1992b). Acta Cryst. **A48**, 859-865.
16. Giacovazzo, C. & Siliqi, D. (1996). Acta Cryst. **A52**, 133-142
17. Giacovazzo, C. & Siliqi, D. (1996). Acta Cryst. **A52**, 143-151
18. Kiriakidis, C.E., Peschar, R. & Shenk, H. (1996) Acta Cryst. **A52**, 77-87
19. Fortier, S., Moore, N. J. & Fraser, M. E. (1985). Acta Cryst. **A41**, 571-577.
20. Klop, E. A., Krabbendam, H. & Kroon, J. (1987). Acta Cryst. **A43**, 810-820.
21. Giacovazzo, C., Siliqi, Cascarano. G., D., Caliandro, R. & Melidoro, A. (1997). Submitted
22. Giacovazzo, C. & Siliqi, D. (1997). Acta Cryst. **A53**, 000-000.

DIRECT METHODS OF SOLVING PHASE AMBIGUITIES IN MACRO-MOLECULAR STRUCTURES

FAN HAI-FU
Institute of Physics, Chinese Academy of Sciences
Beijing 100080, P. R. China

1. Notation

h the reciprocal-lattice vector, which corresponds to the diffraction index *hkl*
$\mathbf{F}_\mathbf{h}$ the structure factor at the reciprocal-lattice node defined by **h**
$F_\mathbf{h}$ the magnitude of $\mathbf{F}_\mathbf{h}$
$\varphi_\mathbf{h}$ the phase of $\mathbf{F}_\mathbf{h}$
$\mathbf{E}_\mathbf{h}$ the normalised structure factor corresponding to $\mathbf{F}_\mathbf{h}$
$E_\mathbf{h}$ the magnitude of $\mathbf{E}_\mathbf{h}$
$\mathbf{F}^\circ_\mathbf{h}$ the structure-factor contribution from the non-anomalous scatterers
$\mathbf{F}'_{\mathbf{h},A}$ the structure-factor contribution from the real-part scattering of the anomalous scatterers, *i.e.*

$$\mathbf{F}'_{\mathbf{h},A} = \sum_{A=1}^{N_A} (f_A + \Delta f'_A) \exp(i2\pi \mathbf{h} \cdot \mathbf{r}_A)$$

$\mathbf{F}''_{\mathbf{h},A}$ the structure-factor contribution from the imaginary-part scattering of the anomalous scatterers, *i.e.*

$$\mathbf{F}''_{\mathbf{h},A} = \sum_{A=1}^{N_A} i\Delta f''_A \exp(i2\pi \mathbf{h} \cdot \mathbf{r}_A)$$

σ_n = $\Sigma_j (Z_j)^n$, Z_j is the atomic number of the jth atom in the unit cell, n is an integer equal to 2 or 3
σ_u = $\Sigma_u (Z_u)^2/\sigma_2$, Z_u is the atomic number of the uth atom, which belongs to the unknown part of the structure

Subscripts:

A the anomalous scattering atoms
R the replacing atoms of an isomorphous pair
N the atoms in the native protein
D the atoms in the heavy-atom derivative of the native protein
p the atoms of the partial structure with known positions in the unit cell
u the atoms of the unknown part in the unit cell

S. Fortier (ed.), Direct Methods for Solving Macromolecular Structures, 297–305.

2. Introduction

Solving a macromolecular structure by using single isomorphous replacement (SIR) or one wavelength anomalous scattering (OAS) data is important when it is difficult to prepare multiple isomorphous derivatives and to collect multiwavelength anomalous diffraction data. However the phase ambiguity intrinsic in SIR or OAS data obstructs the use of them. Various kinds of direct methods have been proposed to solve the problem. [1 - 13]. Some of them have been successfully tested with experimental protein-diffraction data [14 - 18]. In this paper the methods of Fan *et al.* [9, 14] will be discussed in detail.

3. The phase ambiguity in SIR and OAS cases

In the case of SIR for a given reciprocal-lattice vector **h**, we have

$$\mathbf{F}_{\mathbf{h},D} - \mathbf{F}_{\mathbf{h},N} = \mathbf{F}_{\mathbf{h},R} \quad . \tag{1}$$

From experiment we know the magnitudes of $\mathbf{F}_{\mathbf{h},D}$ and $\mathbf{F}_{\mathbf{h},N}$. Accordingly, we can find the positions of the replacing atoms and calculate the vector $\mathbf{F}_{\mathbf{h},R}$. Thus we have two different ways to draw the triangle of (1) leading to an enantiomorphous phase doublet for $\mathbf{F}_{\mathbf{h},D}$ or for $\mathbf{F}_{\mathbf{h},N}$, as shown on the left part of *figure* 1. In the OAS case, we have

$$\mathbf{F}_{\mathbf{h}}^{+} - \mathbf{F}_{\mathbf{h}}^{-*} = 2\mathbf{F}''_{\mathbf{h},A} \quad . \tag{2}$$

The magnitudes of $\mathbf{F}^{+}_{\mathbf{h}}$ and $\mathbf{F}^{-*}_{\mathbf{h}}$ can be obtained from experiment, while $\mathbf{F}''_{\mathbf{h},A}$ can be derived. Hence we have also two ways to draw the triangle of (2) leading to an enantiomorphous phase doublet for $\mathbf{F}_{\mathbf{h}}$, as shown on the right part of *figure* 1.

The phase doublet in both SIR and OAS cases can be expressed as

$$\varphi_{\mathbf{h}} = \varphi'_{\mathbf{h}} \pm |\Delta\varphi_{\mathbf{h}}| \quad . \tag{3}$$

In the SIR case

$$\varphi'_{\mathbf{h}} = \varphi_{\mathbf{h},R} \quad ; \tag{4}$$

$$\Delta\varphi_{\mathbf{h}} = \pm \cos^{-1}(F^2_{\mathbf{h},D} - F^2_{\mathbf{h},N} - F^2_{\mathbf{h},R} / 2F_{\mathbf{h},N} F_{\mathbf{h},R}) \quad , \tag{5}$$

if $\varphi_{\mathbf{h}}$ denotes the phase of a reflection from the native protein;

$$\Delta\varphi_{\mathbf{h}} = \pm \cos^{-1}(F^2_{\mathbf{h},D} + F^2_{\mathbf{h},R} - F^2_{\mathbf{h},R} / 2F_{\mathbf{h},D} F_{\mathbf{h},R}) \quad , \tag{6}$$

if $\varphi_{\mathbf{h}}$ denotes the phase of a reflection from the derivative.

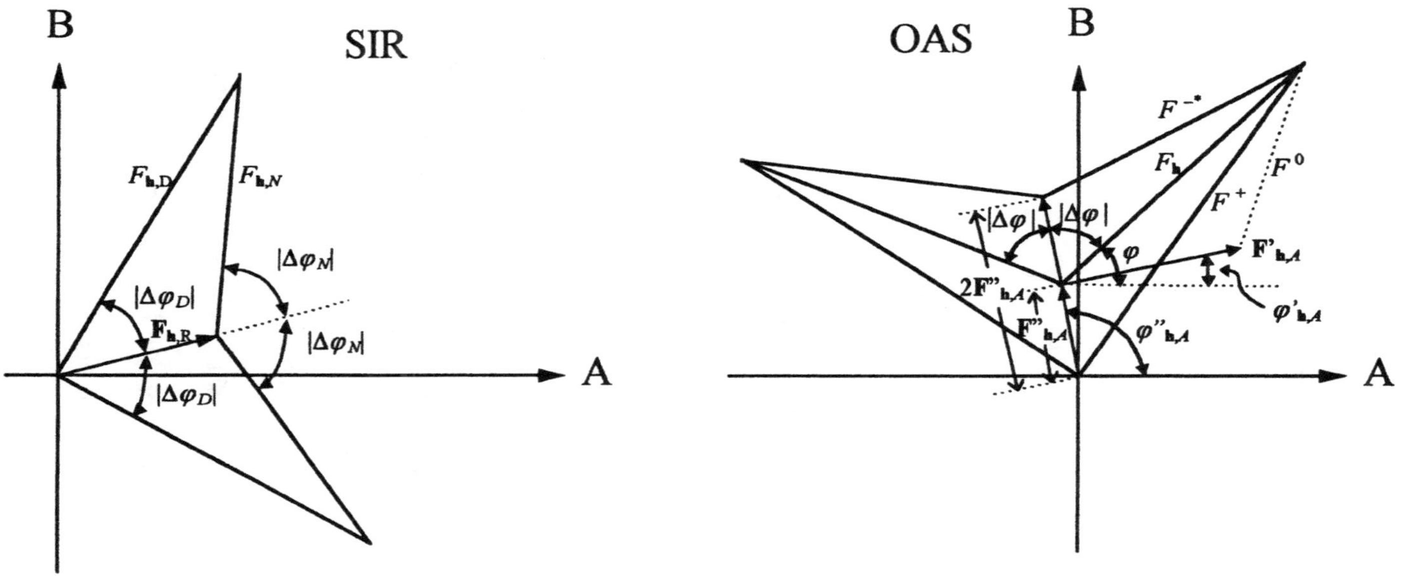

Figure 1. Phase ambiguity (doublets) arising in the single isomorphous replacement (SIR) and one-wavelength anomalous scattering (OAS) cases

In the OAS case

$$\varphi'_{\mathbf{h}} = \varphi''_{\mathbf{h},A} \quad ; \tag{7}$$

$$\Delta\varphi_{\mathbf{h}} = \pm \cos^{-1}(F^{+}_{\mathbf{h}} + F^{-}_{\mathbf{h}} / 2F_{\mathbf{h},A}) \quad . \tag{8}$$

$\varphi_{\mathbf{h}}$ now denotes the phase of

$$\mathbf{F}_{\mathbf{h}} = (\mathbf{F}^{+}_{\mathbf{h}} + \mathbf{F}^{-*}_{\mathbf{h}}) / 2 \quad , \tag{9}$$

where $\mathbf{F}^{-*}_{\mathbf{h}}$ is the conjugate of $\mathbf{F}^{-}_{\mathbf{h}}$.

To summarise, the problem of splitting the above enantiomorphous phase doublets can be converted into that of finding the sign of $\Delta\varphi_{\mathbf{h}}$ given the known values of $\varphi'_{\mathbf{h}}$ and $|\Delta\varphi_{\mathbf{h}}|$. Alternatively, defining

$$A'_{\mathbf{h}} = F_{\mathbf{h}} \cos \Delta\varphi_{\mathbf{h}} \tag{10}$$

and

$$B'_{\mathbf{h}} = F_{\mathbf{h}} \sin \Delta\varphi_{\mathbf{h}} \quad , \tag{11}$$

the problem becomes that of finding the sign of $B'_{\mathbf{h}}$ from the known values of $A'_{\mathbf{h}}$ and $|B'_{\mathbf{h}}|$.

4. The phase difference relations

We start from a modified Sayre's equation [2]:

$$\mathbf{F}_{\mathbf{h}} = (\theta_{\mathbf{h},u} / V) \sum_{\mathbf{h}'} \mathbf{F}_{\mathbf{h}'} \mathbf{F}_{\mathbf{h}-\mathbf{h}'} - \sum_{p} [(\theta_{\mathbf{h},u} / \theta_{\mathbf{h},p}) - 1] \, \mathbf{F}_{\mathbf{h},p} \quad , \tag{12}$$

where θ is an atomic form factor, the subscript u denotes the unknown part of the structure consisting of light atoms, while p denotes the partial structure of known heavy atoms. Replacing $\mathbf{F}_{\mathbf{h}}$ by $F_{\mathbf{h}} \exp(i\varphi_{\mathbf{h}})$ and replacing $\varphi_{\mathbf{h}}$ by $\varphi'_{\mathbf{h}} + \Delta\varphi_{\mathbf{h}}$, we obtain [19]:

$$\begin{aligned} F_{\mathbf{h}} \exp(i \Delta\varphi_{\mathbf{h}}) &= (\theta_{\mathbf{h},u} / V) \sum_{\mathbf{h}'} F_{\mathbf{h}'} F_{\mathbf{h}-\mathbf{h}'} \\ &\quad \times \exp[i(-\varphi'_{\mathbf{h}} + \varphi'_{\mathbf{h}'} + \varphi'_{\mathbf{h}-\mathbf{h}'} + \Delta\varphi_{\mathbf{h}'} + \Delta\varphi_{\mathbf{h}-\mathbf{h}'})] \\ &\quad - \sum_{p} [(\theta_{\mathbf{h},u} / \theta_{\mathbf{h},p}) - 1] \, F_{\mathbf{h},p} \exp[i(\varphi_{\mathbf{h},p} - \varphi'_{\mathbf{h}})] \end{aligned} \tag{13}$$

Taking the imaginary part of (13), it follows

$$F_{\mathbf{h}} \sin \Delta\varphi_{\mathbf{h}} = (\theta_{\mathbf{h},u} / V) \sum_{\mathbf{h}'} F_{\mathbf{h}'} F_{\mathbf{h}-\mathbf{h}'} \sin (\Phi'_3 + \Delta\varphi_{\mathbf{h}'} + \Delta\varphi_{\mathbf{h}-\mathbf{h}'})$$

$$- \sum_p [(\theta_{\mathbf{h},u} / \theta_{\mathbf{h},p}) - 1] F_{\mathbf{h},p} \sin(\varphi_{\mathbf{h},p} - \varphi'_{\mathbf{h}}) \quad , \qquad (14)$$

or equivalently

$$B'_{\mathbf{h}} = (\theta_{\mathbf{h},u} / V) [\sum_{\mathbf{h}'} (A'_{\mathbf{h}'} A'_{\mathbf{h}-\mathbf{h}'} - B'_{\mathbf{h}'} B'_{\mathbf{h}-\mathbf{h}'}) \sin \Phi'_3$$

$$+ 2\sum_{\mathbf{h}'} A'_{\mathbf{h}'} B'_{\mathbf{h}-\mathbf{h}'} \cos \Phi'_3]$$

$$- \sum_p [(\theta_{\mathbf{h},u} / \theta_{\mathbf{h},p}) - 1] F_{\mathbf{h},p} \sin(\varphi_{\mathbf{h},p} - \varphi'_{\mathbf{h}}) \quad , \qquad (15)$$

where

$$\Phi'_3 = -\varphi'_{\mathbf{h}} + \varphi'_{\mathbf{h}'} + \varphi'_{\mathbf{h}-\mathbf{h}'} \quad . \qquad (16)$$

Equation (14) is the 'phase-difference relation', while (15) may be regarded as a 'modified component relation'. In the case of SIR with a centrosymmetric heavy-atom arrangement (15) reduces to one of the original component relations [3]:

$$B_{\mathbf{h}} = (2\theta_{\mathbf{h},u} / V) \sum_{\mathbf{h}'} A_{\mathbf{h}'} B_{\mathbf{h}-\mathbf{h}'} - \sum_p [(\theta_{\mathbf{h},u} / \theta_{\mathbf{h},p}) - 1] B_{\mathbf{h},p} \quad , \qquad (17)$$

while in the case of OAS with a centrosymmetric heavy-atom arrangement, (15) reduces to the other original component relation:

$$A_{\mathbf{h}} = (2\theta_{\mathbf{h},u} / V) \sum_{\mathbf{h}'} (A_{\mathbf{h}'} A_{\mathbf{h}-\mathbf{h}'} + B_{\mathbf{h}'} B_{\mathbf{h}-\mathbf{h}'})$$
$$- \sum_p [(\theta_{\mathbf{h},u} / \theta_{\mathbf{h},p}) - 1] A_{\mathbf{h},p} \quad . \qquad (18)$$

Here $A_{\mathbf{h}}$ and $B_{\mathbf{h}}$ are the real and imaginary components of the structure factor, respectively.

With (14) or (15) it is possible to derive the signs of $\Delta\varphi_{\mathbf{h}}$ or $B'_{\mathbf{h}}$ by a simple procedure like that of using Sayre's equation [20] for a centrosymmetric structure [21]. A special case, which should be mentioned here, is that a centrosymmetric heavy-atom arrangement in a non-centrosymmetric crystal causes phase ambiguity even for small structures. If this occurs with a set of SIR protein data, we get a 'two-fold' phase ambiguity. The solvent flattening technique [22, 23] can not break the SIR phase ambiguity in this case, although it works quite effectively otherwise. On the other hand equation (17) can easily resolve the phase ambiguity under this circumstance, since we know already $A_{\mathbf{h}}$, $B_{\mathbf{h},p}$ and the absolute value of $B_{\mathbf{h}}$. The only problem is to find the signs for $B_{\mathbf{h}}$ by making use of equation (17), which relates the real and the imaginary parts of structure factors, $A_{\mathbf{h}}$ and $B_{\mathbf{h}}$.

5. A tangent-formula among $\Delta\varphi_{\mathbf{h}}$ and the probability of $\Delta\varphi_{\mathbf{h}}$ to be positive

According to Cochran [24], for a set of triplets $\varphi_{\mathbf{h}} - \varphi_{\mathbf{h'}} - \varphi_{\mathbf{h}-\mathbf{h'}}$, the conditional probability distribution of $\varphi_{\mathbf{h}}$ given $E_{\mathbf{h}}$, $E_{\mathbf{h'}}$, $E_{\mathbf{h}-\mathbf{h'}}$. $\varphi_{\mathbf{h'}}$ and $\varphi_{\mathbf{h}-\mathbf{h'}}$ is as follows:

$$P(\varphi_{\mathbf{h}}) = N \exp[\sum_{\mathbf{h'}} \kappa_{\mathbf{hh'}} \cos(\varphi_{\mathbf{h}} - \varphi_{\mathbf{h'}} - \varphi_{\mathbf{h}-\mathbf{h'}})] \quad , \tag{19}$$

where N is a normalizing factor and

$$\kappa_{\mathbf{h},\mathbf{h'}} = 2\sigma_3\sigma_2^{-3/2} E_{\mathbf{h}} E_{\mathbf{h'}} E_{\mathbf{h}-\mathbf{h'}} \quad .$$

With equation (16) and the expression $\varphi_{\mathbf{h}} = \varphi'_{\mathbf{h}} + \Delta\varphi_{\mathbf{h}}$, the triplet phase becomes

$$\varphi_{\mathbf{h}} - \varphi_{\mathbf{h'}} - \varphi_{\mathbf{h}-\mathbf{h'}} = \Delta\varphi_{\mathbf{h}} - (\Phi'_3 + \Delta\varphi_{\mathbf{h'}} + \Delta\varphi_{\mathbf{h}-\mathbf{h'}}) \quad . \tag{20}$$

Substitution of (20) into (19) gives

$$P(\Delta\varphi_{\mathbf{h}}) = N \exp[\sum_{\mathbf{h'}} \kappa_{\mathbf{h},\mathbf{h'}} \cos[\Delta\varphi_{\mathbf{h}} - (\Phi'_3 + \Delta\varphi_{\mathbf{h'}} + \Delta\varphi_{\mathbf{h}-\mathbf{h'}})] \quad . \tag{21}$$

Let

$$\alpha \sin\beta = \sum_{\mathbf{h'}} \kappa_{\mathbf{h},\mathbf{h'}} \sin[\Phi'_3 + \Delta\varphi_{\mathbf{h'}} + \Delta\varphi_{\mathbf{h}-\mathbf{h'}}] \tag{22}$$

and

$$\alpha \cos\beta = \sum_{\mathbf{h'}} \kappa_{\mathbf{h},\mathbf{h'}} \cos[\Phi'_3 + \Delta\varphi_{\mathbf{h'}} + \Delta\varphi_{\mathbf{h}-\mathbf{h'}}] \tag{23}$$

(21) becomes

$$P(\Delta\varphi_{\mathbf{h}}) = [2\pi I_0(\alpha)]^{-1} \exp[\alpha \cos(\Delta\varphi_{\mathbf{h}} - \beta)] \quad , \tag{24}$$

where $I_0(\alpha)$ is the zero-order modified Bessel function of the first kind with α as argument. Equation (24) is the probability distribution of $\Delta\varphi_{\mathbf{h}}$ based on Cochran's distribution. By maximising (24) we have $\beta = \Delta\varphi_{\mathbf{h}}$ and hence from (22) and (23) it follows

$$\tan \Delta\varphi_{\mathbf{h}} = \frac{\sum_{\mathbf{h'}} \kappa_{\mathbf{h},\mathbf{h'}} \sin(\Phi'_3 + \Delta\varphi_{\mathbf{h'}} + \Delta\varphi_{\mathbf{h}-\mathbf{h'}})}{\sum_{\mathbf{h'}} \kappa_{\mathbf{h},\mathbf{h'}} \cos(\Phi'_3 + \Delta\varphi_{\mathbf{h'}} + \Delta\varphi_{\mathbf{h}-\mathbf{h'}})} \quad . \tag{25}$$

This is a tangent formula relating the phase differences.

On the other hand, according to Sim [25], if partial structure information is available, we have

$$P_{Sim}(\theta) = \left[2\pi I_0(x)\right]^{-1} \exp[x\cos(\theta)] \quad , \tag{26}$$

where

$$x = 2\,E_{\mathbf{h}}\,E_{\mathbf{h},p}/\sigma_u \quad , \quad \theta = \varphi_{\mathbf{h}} - \varphi_{\mathbf{h},p}$$

With the expression $\varphi_{\mathbf{h}} = \varphi'_{\mathbf{h}} + \Delta\varphi_{\mathbf{h}}$, (26) becomes

$$P_{Sim}(\Delta\varphi_{\mathbf{h}}) = \left[2\pi I_0(x)\right]^{-1} \exp[x\cos(\Delta\varphi_{\mathbf{h}} - \delta_{\mathbf{h}})] \quad , \tag{27}$$

where

$$\delta_{\mathbf{h}} = \Delta\varphi_{\mathbf{h},p} - \varphi'_{\mathbf{h}} \quad .$$

By multiplying equations (27) and (24), and by introducing the concept of 'best phase relationship' [26], we finally obtain a tangent formula for phase differences $\Delta\varphi_{\mathbf{h}}$, which is based on Cochran's distribution and Sim's distribution and the concept of 'best phases' in protein crystallography:

$$\tan\Delta\varphi_{\mathbf{h}} = \frac{\sum_{\mathbf{h}'} m_{\mathbf{h}'} m_{\mathbf{h}-\mathbf{h}'} \kappa_{\mathbf{h},\mathbf{h}'} \sin(\Phi'_3 + \Delta\varphi_{\mathbf{h}'best} + \Delta\varphi_{\mathbf{h}-\mathbf{h}'best}) + x\sin\delta_{\mathbf{h}}}{\sum_{\mathbf{h}'} m_{\mathbf{h}'} m_{\mathbf{h}-\mathbf{h}'} \kappa_{\mathbf{h},\mathbf{h}'} \cos(\Phi'_3 + \Delta\varphi_{\mathbf{h}'best} + \Delta\varphi_{\mathbf{h}-\mathbf{h}'best}) + x\cos\delta_{\mathbf{h}}} \tag{28}$$

with the corresponding probability

$$P_+(\Delta\varphi_{\mathbf{h}}) = \frac{1}{2} + \frac{1}{2}\tanh\Big\{\sin|\Delta\varphi_{\mathbf{h}}| \times \Big[\sum_{\mathbf{h}'} m_{\mathbf{h}} m_{\mathbf{h}-\mathbf{h}'} \kappa_{\mathbf{h},\mathbf{h}'} \sin(\Phi'_3 + \Delta\varphi_{\mathbf{h}'best} + \Delta\varphi_{\mathbf{h}-\mathbf{h}'best}) + x\sin\delta_{\mathbf{h}}\Big]\Big\} \quad , \tag{29}$$

where

$$\delta_{\mathbf{h}} = \varphi_{\mathbf{h},p} - \varphi'_{\mathbf{h}} \quad , \quad \Delta\varphi_{\mathbf{h}best} = \varphi_{\mathbf{h}best} - \varphi'_{\mathbf{h}} \quad , \quad \Phi'_3 = -\varphi'_{\mathbf{h}} + \varphi'_{\mathbf{h}'} + \varphi'_{\mathbf{h}-\mathbf{h}'} \quad ,$$

$$\tan(\Delta\varphi_{\mathbf{h}best}) = 2(P_+ - \frac{1}{2})\sin|\Delta\varphi_{\mathbf{h}}| \big/ \cos\Delta\varphi_{\mathbf{h}} \quad , \tag{30}$$

$$m_{\mathbf{h}} = \exp(-\sigma_{\mathbf{h}}^2/2)\left\{[2(P_+ - \frac{1}{2})^2 + \frac{1}{2}](1 - \cos 2\Delta\varphi_{\mathbf{h}}) + \cos 2\Delta\varphi_{\mathbf{h}}\right\}^{1/2} \quad . \tag{31}$$

The figure of merit $m_{\mathbf{h}}$ is a measure of reliability of $\Delta\varphi_{\mathbf{h}best}$. $m_{\mathbf{h}}$ includes three factors: (i) $\exp(-\sigma^2_{\mathbf{h}}/2)$, a measure of the sharpness of the phase-doublet distribution;

(ii) $(P_{+} - ½)^2$, a measure of the bias of $\Delta\varphi_{\mathbf{h}}$ towards positive or negative. It reaches the maximum value when P_{+} equals 0 or 1; (iii) $\cos 2\Delta\varphi_{\mathbf{h}}$, a measure of the closeness of the two possible phases, $\varphi'_{\mathbf{h}} + \Delta\varphi_{\mathbf{h}}$ and $\varphi'_{\mathbf{h}} - \Delta\varphi_{\mathbf{h}}$. It reaches the maximum value when $\Delta\varphi_{\mathbf{h}}$ equals 0 or π. Either of the last two factors will have no effect on $m_{\mathbf{h}}$ when the other one reaches the maximum value. A procedure for breaking the SIR or OAS phase ambiguity making use of equations (29), (30) and (31) has been proposed and is implemented in the program OASIS [27].

6. Applications

After a series of test calculations with error-free data, our method was first successfully applied to the OAS data of the protein aPP [14]. The first application to experimental SIR data was done with the protein Rhe [28]. The method was further combined with the solvent-flattening technique showing substantial improvement in dealing with experimental protein OAS and SIR data [15, 18]. More recent results will be discussed in another lecture [29].

References

1. Coulter, C. L. (1965) A possible new method for the determination of crystallographic phases for proteins, *J. Mol. Biol.* **12**, 292-295.
2. Fan, H. F. (1965) The use of sign relationship in the determination of heavy-atom-containing crystal structures I. A sign refinement procedure, *Acta Phys. Sin.* **21**, 1105-1114 (in Chinese). (An English translation can be found in *Chinese Physics* (1965) 1418-1428.)
3. Fan, H. F. (1965) The use of sign relationship in the determination of heavy-atom-containing crystal structures II. Component relationship and its applications, *Acta Phys. Sin.* **21**, 1114-1118 (in Chinese). (An English translation can be found in *Chinese Physics* (1965) 1429-1435.)
4. Karle, J. (1966) Isomorphous substitution and formulas for phase determination, *Acta Cryst.* **21**, 273-276.
5. Hendrickson, W. A. (1971) Some aids for breaking the phase ambiguity in the single isomorphous replacement method, *Acta Cryst.* **B27**, 1474-1475.
6. Hauptman, H. A. (1982) On integrating the technique of direct methods and isomorphous replacement I. The theoretical basis, *Acta Cryst.* **A38**, 289-294.
7. Hauptman, H. A., Potter, S. and Weeks, C. M. (1982) On integrating the technique of direct methods and isomorphous replacement II. The first applications, *Acta Cryst.* **A38**, 294-300.
8. Giacovazzo, C. (1983) The estimation of two-phase and three-phase invariants in *P*1 when anomalous scatterers are present, *Acta Cryst.* **A39**, 585-592.
9. Fan, H. F. and Gu, Y. X. (1985) Combining Direct Methods with Isomorphous Replacement or Anomalous Scattering Data. III. The Incorporation of Partial Structure Information, *Acta Cryst.* **A41**, 280-284.
10. Fortier, S., Moore, N. J. and Fraser, M. E. (1985) A direct-methods solution to the phase problem in the single isomorphous replacement case: theoretical basis and initial applications, *Acta Cryst.* **A41**, 571-577.
11. Klop, E. A., Krabbendam, H. and Kroon, J. (1987) On the direct-methods theory of single isomorphous replacement, *Acta Cryst.* **A43**, 810-820.
12. Giacovazzo, C., Cascarano, G. and Zheng, C. D. (1988) On integrating the techniques of direct methods and isomorphous replacement. A new probabilistic formula for triplet invariants, *Acta Cryst.* **A44**, 45-51.
13. Kyriakidis, C. E., Peschar, R. and Schenk, H. (1993) On the direct methods phase information from differences between isomorphous structure factors, *Acta Cryst.* **A49**, 557-569.
14. Fan, H.F., Hao, Q., Gu, Y.X., Qian, J.Z., Zheng, C.D. and Ke, H. (1990) Combining Direct Methods with Isomorphous Replacement or Anomalous Scattering Data. VII. *Ab Initio* Phasing of One-Wavelength Anomalous Scattering Data from a Small Protein, *Acta Cryst.* **A46**, 935-939.

15. Sha, B.D., Liu S.P., Gu Y.X., Fan H.F., Ke H., Yao J.X. and M.M. Woolfson (1995) Direct Phasing of One-wavelength Anomalous-Scattering Data of the Protein Core Streptavidin, *Acta Cryst.* **D51**, 342-346.
16. Giacovazzo, C., Siliqi, D. and Zanotti, G. (1995) The *ab initio* crystal structure solution of proteins by direct methods. III. The phase extension process, *Acta Cryst.* **A51**, 177-188.
17. Langs, D. A., Guo, D. and Hauptman, H. A. (1995) TDSIR phasing: direct use of phase-invariant distributions in macromolecular crystallography, *Acta Cryst.* **A51**, 535-542.
18. Zheng, X. F., Zheng, C. D., Gu, Y. X., Mo, Y. D., Fan, H.F. and Hao, Q. (1997) Use of single isomorphous replacement data of proteins — resolving the phase ambiguity and a new procedure for phase extension, *Acta Cryst.* **D53**, 49-55.
19. Fan, H. F., Han, F. S., Oian, J. Z. and Yao, J. X. (1984) Combining direct methods with isomorphous replacement or anomalous scattering data I, *Acta Cryst.* **A40**, 489-495.
20. Sayre, D. (1952) The squaring method: a new method for phase determination, *Acta Cryst.* **5**, 60-65.
21. Grant, D. F., Howells, R. G. and Rogers, D. (1957) A method for the systematic application of sign relations, *Acta Cryst.* **10**, 489-497.
22. Wang, B. C. (1981) Protein structure determination by the single isomorphous replacement method with a phase selection and refinement process, *Acta Cryst.* **A37**, C-11.
23. Wang, B. C. (1985) *Diffraction Methods for Biological Macromolecules*, in H. Wyckoff, C.H.W. Hirs and S.N. Timasheff (ed.), *Methods in Enzymology*, Vol. **115**, Academic Press, New York, pp. 90-112.
24. Cochran, W. (1955) Relations between the phase of structure factors, *Acta Cryst.* **8**, 473-478.
25. Sim, G. A. (1959) The distribution of phase angles for structures containing heavy atoms II. A modification of the normal heavy-atom method for non-centrosymmetrical structures, *Acta Cryst.* **12**, 813-815.
26. Fan, H. F., Han, F. S. and Oian, J. Z. (1984) Combining direct methods with isomorphous replacement or anomalous scattering data II. The treatment of errors, *Acta Cryst.* **A40**, 495-498.
27. Hao, Q., Gu, Y. X., Zheng, C. D. and Fan, H. F. (1996) OASIS: a computer program for breaking the phase ambiguity in OAS or SIR protein data, School of Applied Sciences, De Montfort University, Leicester LE1 9BH, England and Institute of Physics, Chinese Academy of Sciences, Beijing 100080, P. R. China.
28. Fan, H. F. & Wang, B. C. (1988). Unpublished work, partial results of which are quoted in Woolfson, M. M. and Fan, H. F. (1995) *Physical and Non-physical Methods of Solving Crystal Structures*, Cambridge University Press, Cambridge, pp. 164-166.
29. Fan, H. F., Hao, Q., Harvey I., Hasnain, S. S., Liu, Y. D., Gu, Y. X., Zheng, C. D. and Ke, H. (1997) Applications of direct methods with single isomorphous replacement or one-wavelength anomalous scattering data, in these *Proceedings*.

PHASE REFINEMENT

H. Schenk, Laboratory for Crystallography, University of Amsterdam
Nieuwe Achtergracht 166, 1018 WV Amsterdam
The Netherlands.

1. Introduction

Years ago it was felt that the only way phases in protein crystallography could be refined was by means of direct methods. In those days in general protein maps were improved through graphical interpretation of Fourier maps. But then Structure Factor Least Squares was developed as a tool for protein refinement by including soft constraints (Waser constraints) and research in the phase refinement through structure invariants became less important. Nevertheless, in order to improve models from proteins and, may be, to set up reasonable models for proteins, phase refinement may prove to be of significant importance.

Phases may be refined by using phase relationships (structure invariants) from which the numerical value is known. In many cases the triplet phase relation

$$\varphi_3 = \varphi_H + \varphi_K + \varphi_{-H-K} \approx 0 \tag{1}$$

with its probability distribution based on the triple product

$$E_3 = N^{-1/2} |E_H E_K E_{-H-K}| \tag{2}$$

is used for refinement of phases, mostly in the form of tangent refinement, using essentially φ_3=0 values for the observed equations.

It is also possible to include observed values for φ_3=a, where a is a value based on the actual probability distribution or on the triple-product (2). In this case the refined phases may well reflect the true phases more reliably than in the case of the tangent refinement, but the calculations involved are more complex.

Then another possibility is to include other phase relationships such as quartets and quintets and use their proper probability distribution as constraints in the refinement.

In all cases the final results are influenced by the symmetry of the space group. This is obvious from the fact that in centrosymmetric structures phases may have just 2 values (0 or π) whereas in noncentrosymmetric structures the full range of values is possible on the interval from $-\pi$ to π.

Many attempts have been reported to use DM for the refinement and extension of macromolecular phase sets, most notably those involving (modified) tangent formulae, determinants and, recently, Sayre's equation.

In spite of the success in phase refinement of small-molecules, the tangent formula and its modifications did not turn out to be the most appropriate refinement tools for macromolecular phase sets. It was noticed that the refinement results deteriorated rapidly with increasing resolution of the extended phases [1] or decreasing resolution of the initial phase set [2].

Determinants [3] have been used successfully, for example, in the expansion and refinement of 1.9 Å observed data of 2 ZN insulin to 1.5 Å [4], but in this area no recent advances seem to have been reported. The most successful DM based technique reported recently seems to be the SQUASH procedure [5], [6], [7] which combines Sayre's equation [8] with a range of real and reciprocal phase constraints (e.g. solvent flattening, non-crystallographic symmetry, histogram matching). From test results on 2 Zn insulin, as tabulated by Zhang [7], it can be concluded that the Sayre equation

S. Fortier (ed.), Direct Methods for Solving Macromolecular Structures, 307–310.

contributes significantly to the phase extension from 3 to 1.5 Å but is less effective in the refinement of the 3 Å data set without the additional constraints.

It is the purpose of this chapter to deal with the above mentioned aspects of refinements.

2. The tangent refinement

A relation commonly used in most DM packages is the tangent expression [9], [10].

$$\tan(\Phi_H) = \frac{\sum_K \kappa_{HK} \sin(\Phi_K + \Phi_{H-K})}{\sum_K \kappa_{HK} \cos(\Phi_K + \Phi_{H-K})} \tag{3}$$

which can be derived in various ways, e.g. from the probability distribution of a triplet (1) by taking the derivative with respect to Φ_H. For a tangent, being unique on the interval 0 to π only, is as such not suitable for phase determination and is therefore always used together with the signs of nominator and denominator. Therefore, when phases are refined by means of the tangent formula, rather its exponential analogue is being used, because phases must be known on the interval $0<\varphi<2\pi$:

$$\exp i\varphi_H = \frac{\sum_K E_3 \exp i(\varphi_H + \varphi_{H-K})}{|\sum_K E_3 \exp i(\varphi_H + \varphi_{H-K})|} \tag{4}$$

The refinement criterion of this exponential expression can be easily deduced to be:

$$\text{Tanres} = \sum_H \sum_K E_3 \sin^2(\tfrac{1}{2}(-\varphi_H + \varphi_K + \varphi_{H-K})) \tag{5}$$

In the expressions (3), (4) and (5) the reflections H all belong to the asymmetric part of reciprocal space, but K and H-K may be found in the entire reciprocal space. Since the phases in the asymmetric part have to be determined, all phases in (3), (4) and (5) outside this part are replaced by phases inside by application of the space-group symmetry rules. E.g. expression (5) then changes into

$$\text{Tanres} = \sum_H \sum_K E_3 \sin^2\{\tfrac{1}{2}(-\varphi_{H^*} + S_K\varphi_{K^*} + S_{H-K}\varphi_{(H-K)^*} + c)\} \tag{6}$$

The starred indices belong to the asymmetric part, S = ±1 follows from symmetry rules, and c is a phase shift occurring in non-symmorphic space groups, i.e. space groups with translational symmetry operations [11]. From this expression it is easily seen that for instance in $P\bar{1}$ (φ=0 or π; $S_K = S_{H-K} = +1$; c=0) the trivial solution with all φ=0 gives the minimum value of Σ_2FOM. Also the values for other solutions may be less reliable, so that Σ_2FOM is unreliable for selecting the correct Σ_2-solution in symmorphic space groups (only unit translations) in which c=0. Another example: in P1 ($S_K = \pm1$, $S_{H-K} = \pm1$, c=0) the most consistent solution with Σ_2FOM=0 is again the one with all phases equal to 0. In general in the noncentric symmorphic space groups (c=0) the use of Σ_2FOM often leads to centrosymmetric solutions. In non-symmorphic polar space groups such as $P2_1$ refinement often results in centrosymmetric phases as well, although here the centric projection reflections may have correct phases. In the other space groups ($P2_1/c$, $P2_12_12_1$) the smallest value of Σ_2FOM may indicate directly the set of correct phases.

It is emphasized here that all FOM's based on the consistency of either the Σ_2 relation or the tangent refinement have the same properties. This implies that:

1. for symmorphic space groups these FOM's are unreliable for selecting the correct phase set.

2. in many noncentrosymmetric space groups extra information is necessary to prevent a starting set from refining to a centrosymmetric solution.

3. The effect of space-group symmetry

In principle a direct method application should lead directly to the image of the structure; in practice however, a few structures remain unsolved. This may occur for several reasons, such as space group symmetry, size of the structure and special features in the atom positions. The effect of symmetry is the most serious reason and will be dealt with here in some more detail.

It can easily be seen that the trivial set of phases with all $\varphi_H=0$ fulfills all triplet relations $\varphi_H+\varphi_K+\varphi_{-H-K}=0$ exactly; this implies that in unfavourable cases no solution of the phase problem will be found. Fortunately in many cases the space group symmetry adds constraints to the phases. E.g. in space group $P2_1/c$ the phase constraints $\varphi_{hk\ell}=0$ or π, $\varphi_{hk\ell}= \varphi_{-h-k-\ell} = \varphi_{-hk-\ell}$ for $k+\ell=2n$ and $\varphi_{hk\ell} = \varphi_{-h-k-\ell} = \pi + \varphi_{-hk-\ell}$ for $k+\ell = 2n+1$ follow from the symmetry and notably the latter condition implies that the set of all phases $\varphi_H=0$ is no longer a solution of the relations (1). In practice, in this space group the most consistent solution under (1) gives nearly always a correct image of the structure. However, also in the triclinic centrosymmetric space group $P\bar{1}$, where no symmetry constraints are present, the triplet relations (1) are reliable enough to carry out a phasing procedure and thus, by using a good starting set of origin defining and symbolic phases, a phase set can be built up containing among other solutions both the trivial and the correct one. Since the triplet relations (1) tend to indicate the trivial solution as the correct one, a FOM based on these relations will not find the correct solution, and this implies that other FOM's must be used for this purpose.

In this chapter it is not possible to deal with the symmetry aspects of the use of triplets (1) in all space groups in detail, for that the reader is referred to [11]. Nevertheless, one other serious problem is relevant for protein refinement, namely that of the refinement of phases in polar non-centrosymmetric space groups. As an example space group $P2_1$ is chosen, in which the origin may lie on any point of the twofold screw axis. This implies that there is an infinite number of correct phase sets instead of just one and they are related by the expression:

$$\varphi_{hk\ell} = a \times k + \varphi_{hk\ell} \tag{7}$$

in which a is a constant related to the shift of the origin. Now, a φ_H derived from a triplet (1) carrying an error, will in general be in error as well. However, when used in further phase extension cycles this wrong phase will be interpreted as being correct with respect to another origin or even as being correct with respect to the enantiomorph of the structure. As a result solutions of the phase problem using the set of triplets (1) often consist of several partial structures with respect to different origins in both enantiomorphs. Because of this one finds that in space groups such as $P2_1$ application of (1) often leads to centric phases, whereas similar refinements in $P2_12_12_1$, where the origin is fixed and eight permissible origins are present only, show stable non-centric phases [12]. It needs no arguing that this behaviour of the triplets (1) in space group $P2_1$ induces extra problems in phase-determination processes.

In summary there exist two categories of centrosymmetric space groups: a) the symmorphic space groups (no translational symmetry, e.g. $P\bar{1}$) for which the most consistent solution under (1) is the trivial one with all $\varphi_H=0$ and b) the other space groups (e.g. $P2_1/c$) for which the most consistent solution usually is the correct one. For the noncentrosymmetric space groups there are three categories: c) the symmorphic space groups (e.g. P1), for which the relations (1) produce also the $\varphi_H=0$ solution, but also less consistent solutions of (1) mostly having centric phases; d) the polar space groups with an infinite number of possible origins (e.g. $P2_1$), for which the phasing process usually ends with centric phases and e) the other space groups (e.g. $P2_12_12_1$) with an origin restricted to a finite number of positions for which the relations produce the correct phases. To overcome the above difficulties, higher invariants play an important role.

4. Enantiomorph-conserving refinements

By using estimated values for triplet-relations refinement procedures have been developed which enable to maintain the proper enantiomorph as soon as it shows up in a refinement procedure.

Also enantiomorph-conserving refinement procedures have been developed employing estimates $|p|$ and $|q|$ for quartet and quintet phase sums respectively. E.g. for quartets a modified tangent formula [13] is given by:

$$\tan \varphi_H = \frac{\sum_K \sum_L E_4 \sin(\varphi_K + \varphi_L + \varphi_{H-K-L} + s|p|)}{\sum_K \sum_L E_4 \cos(\varphi_K + \varphi_L + \varphi_{H-K-L} + s|p|)} \qquad (8)$$

in which s is chosen such that $-\varphi_H+\varphi_K+\varphi_L+\varphi_{H-K-L}+s|p|$ is closest to 0. This refinement will maintain the enantiomorph and introduces random errors in the phases φ_H only. The strongest practical procedure [13] combines the use of triplets, quartets and quintets in a way similar to (8) in the exponential fashion related to (4).

References

1. Reeke, G.N. & Lipscomb, W.N. (1969). On the use of Tangent Formula to Extend the Resolution of Protein Phases. *Acta Cryst.* **B25**, 2614–2623.
2. Olthof, G.J. & Schenk, H. (1982). Phase Extension and Refinement II. Application to Metmyoglobin 2.0 A Data. *Acta Cryst.* **A38**, 117–122.
3. Tsoucaris, G. (1970). A New Method for Phase Determination. The "Maximum Determinant Rule". *Acta Cryst.* **A26**, 492–499.
4. Rango, C. de, Mauguen, Y., Tsoucaris, G., Dodson, E.J., Dodson, G.G. and Taylor, D.J. (1985). The Extension and Refinement of the 1.9 A Spacing Isomorphous Phases to 1.5 A Spacing in 2Zn Insulin by Determinantal Methods. *Acta Cryst.* **A41**, 3–17.
5. Zhang, K.Y.J. & Main, P. (1990). The Use of Sayre's Equation with Solvent Flattening and Histogram Matching for Phase Extension and Refinement of Protein Structures. *Acta Cryst.* **A46**, 377–381.
6. Cowtan, K.D. & Main, P. (1993). Improvement of Macromolecular Electron-Density Maps by the Simultaneous Application of Real and reciprocal Space Constraints. *Acta Cryst.* **D49**, 148–157.
7. Zhang, K.Y.J. (1993), SQUASH-Combining Constraints for Macromolecular Phase refinement and Extension. *Acta Cryst.* **D49**, 213–222.
8. Sayre, D. (1952). The Squaring Method: A New Method for Phase Determination. *Acta Cryst.* **5**, 60–65.
9. Karle, J. & Hauptman, H. (1956). A Theory of Phase Determination for the Four Types of Non-Centrosymmetric Space Groups 1P222, 2P22, $3P_12$, $3P_22$. *Acta Cryst.* **9**, 635–651.
10. Karle, I. & Karle, J. (1964). An Application of the Symbolic Addition Method to the Structure of L-Arginine Dihydrate. *Acta Cryst.* **17**, 835–841.
11. Schenk, H. (1972. The Relation between Space Group Symmetry and the Properties of Direct Solutions Based on the Σ2 Formula. A Method for using Centrosymmetric Σ2 Solutions for Solving Non-centrosymmetric Structures. *Acta Cryst.* **A28**, 412–422.
12. Schenk, H. (1982). Phase Extension and Refinement using Sayre's Equation and Tangent Formula. In *Computational Crystallography*, ed. D. Sayre, Oxford, p. 231–241.
13. Van der Putten, N. and Schenk, H. (1979). Application of Quartet and Quintet Invariants in Phase Determination. *Acta Cryst.* **A35**, 381–387.

ON THE USE OF PARTIAL STRUCTURE INFORMATION IN AB INITIO DIRECT METHODS

R. PESCHAR
Laboratory for Crystallography, Amsterdam Institute for Molecular Studies
University of Amsterdam, Nieuwe Achtergracht 166, 1018 WV Amsterdam
The Netherlands

Abbreviations

j.p.d.	Joint probability distribution
(p.)r.v.	(Primitive) random variable
s.f.	Structure factor
DM	Direct methods
SIR(N)AS	Single isomorphous replacement with(out) anomalous scattering effects
SAS	Single-wavelength anomalous scattering

1. Introduction

Traditionally, the phrase 'ab initio direct methods' has been reserved for processes that intend to derive a crystal structure model from the observed structure factor (s.f.) magnitudes only with as little structural information as possible, often the contents of the unit cell only. This expresses the underlying wish to solve the phase problem exactly in one step only or, as a next best solution, in as few steps as possible. In small-molecule crystallography, with its high resolution data and a large ratio of data/variables, this goal may not seem unrealistic because only a small amount of structural conformations will fit the observed data. In contrast, data of macromolecular compounds are often of lower resolution. Thus, together with the much smaller ratio of data/variables, a much larger ensemble of conformations is feasible, even conformations that differ appreciably. Therefore, to 'solve' a macromolecular crystal structure, i.e to select the most likely conformation(s) given the observed data, any knowledge should be exploited that can limit the conformation(s) to be considered. The initial macromolecular model often consists of a molecular envelope at low resolution, e.g. obtained from (multiple) isomorphous replacement, anomalous scattering or molecular replacement data. The problem of getting a more detailed electron density inside the molecular envelope, using observed s.f. magnitudes of higher resolution, in practice usually carried out by solvent flattening and density modification, resembles closely that of traditional *ab initio* direct methods (DM). One starts from a completely flat (=uniform) electron density map and, after the probability-based DM phase processing, ends up with a structural model or a fragment that can be processed towards a complete structure by other techniques. An example of the latter is the DIRDIF procedure (see [1] and references therein) that processes this partial structural knowledge in the popular DM tangent form.

Therefore, a DM phasing problem will be called *ab initio* if the extension of an initial (low-resolution) structural model can not be arrived at by an iteration of least-squares refinement and (difference) Fourier interpretation.

2. Posterior use of structural knowledge

If, for example, a partial structural model is available in the form of a heavy-atom substructure or correctly positioned groups, the extension to the complete structure is

S. Fortier (ed.), Direct Methods for Solving Macromolecular Structures, 311–316.

expected to be more easy than without this information. However, structural information can be exploited in two essentially different ways. Very popular is the *posterior* use of partial structure knowledge. This means that the structural information is inserted in already existing probabilistic formulae so *after* the latter have been derived.

In connection to SIR and SAS data, it has been shown by various authors that the introduction of heavy-atom phase information in probabilistic SIR and SAS expressions of Hauptman [2, 3] and Giacovazzo [4, 5] may improve their predicting quality. For example, absolute values of the doublets can be calculated if heavy-atom positions are available, leaving only their signs to be determined. Hao Quan & Fan Hai Fu [6] inserted the doublet cosines as calculated from a phase-vector diagram in the above mentioned SIR and SAS expressions of Hauptman. Also Klop et al. [7] used the former expression in combination with algebraically-constructed dispersion-free s.f.'s and doublet phase differences.

3. Prior use of structural knowledge

A theoretically attractive approach, that contrasts the posterior use, is to use structural knowledge *a priori*. This means that the probabilistic expressions are actually derived on basis of the structural knowledge. Obviously, this requires a suitable definition of the *prior* distribution of the primitive random variables (p.r.v.'s).

The use of structural knowledge in DM other than a set of atoms with unknown positions has been initiated by Main [8] who proposed some generalizations of Cochran's expression [9] of the triplet phase sum. The mathematical basis of Main's paper was fully documented in Heinerman [10]. Main and Heinerman considered four types of structural knowledge:

A Atoms with unknown positions.
B One or more groups of atoms with unknown position and unknown orientation.
C One or more groups of atoms with known orientation but unknown position.
D Atoms that are correctly positioned.

Type A has been discussed earlier in this book [11]. It leads to a uniform distribution of p.r.v.'s, in principle over the unit cell area. However, if the unit cell can not be accessed completely by atoms due to space-group symmetry, the uniform distribution of the p.r.v.'s should be modified accordingly [12].

In this contribution we will focus on types B-D. It will be shown that their presence requires only minor modifications of the expressions derived so far on basis of type A.

4. Structure in terms of fragments

From a DM point of view it is very useful to consider a structure as a sum of fragments. In case of *n* fragments, the s.f. $F(H_\nu) = F_\nu$ can be expressed,

$$F_\nu = \sum_{j=1}^{n} g_{j\nu} \exp[2\pi i(\underline{H}_\nu \cdot \underline{r}_j)] \qquad (4.1)$$

If the position vector $\underline{r}_j$ of the fragment coincides with one of the atoms in the fragment, the group scattering factor $g_{j\nu}$ is a function of the interatomic vectors $\underline{r}_k - \underline{r}_j$,

$$g_{j\nu} = \sum_{k} f_{k\nu} \exp[2\pi i(\underline{H}_\nu \cdot (\underline{r}_k - \underline{r}_j))] \qquad (4.2)$$

4.1 FRAGMENTS WITH CORRECT ORIENTATION BUT UNKNOWN POSITION

If the correct orientation of a fragment is available, (4.2) is completely known. The unknown position of a fragment requires that with its group position vector $\underline{r}_j$ a p.r.v. is associated, assumed to be distributed uniformly in the cell. The result is that the same types of non-zero moments, cumulants etc. are obtained as in the model of N uniformly distributed p.r.v.'s. but with the individual $f_{j\nu}$ being replaced by the group scattering factor $g_{j\nu}$. As a result, the expressions derived under the model of uniformly distributed p.r.v.'s also hold for correctly oriented groups with unknown position, provided it is taken into account that:

(*i*) The amount of p.r.v.'s is different. This equals the number of atoms assumed to be uniformly positioned [type A] plus the amount of groups with correct orientation but unknown position [type C]

(*ii*) The group scattering factor $g_{\nu j}$ replaces the individual atomic scattering factors $f_{j\nu}$ for those p.r.v.'s that are associated with the position vector of a group of atoms.

4.2 FRAGMENTS WITH UNKNOWN ORIENTATION AND UNKNOWN POSITION

Exploitation of structural knowledge in the form of fragments that are known *a priori*, i.e. in terms of distances and angles, except for their position and orientation in the cell has been considered by various authors (e.g. [8, 10, 13, 14]. In addition to p.r.v.'s associated with the unknown position now also an extra orientational average must be carried out with respect to the product of group scattering factors of each fragment (see [10])

$$< \prod_{\nu} g_{\nu j}^{\alpha_\nu} (g^*_{\nu j})^{\beta_\nu} >_{\mathrm{orient}} \tag{4.3}$$

If expression (4.3) is written out in terms of the interatomic contributions, terms arise that are associated with the normalization ($\alpha_\nu = \beta_\nu = 1$) and the triplets ($\alpha_1 = \alpha_2 = \alpha_3 = 1$; $\beta_1 = \beta_2 = \beta_3 = 0$ and vice versa),

$$<g_{j\nu} g^*_{j\nu}> = <\sum_k \sum_\ell f_{k\nu} f_{\ell\nu} \exp[2\pi i(\underline{H}_\nu.(\underline{r}_k - \underline{r}_\ell)]>_{\mathrm{orient}} \tag{4.4}$$

$$<g_{j1} g_{j2} g_{j3}> = <\sum_k \sum_\ell \sum_m f_{k1} f_{\ell 2} f_{m3} \exp[2\pi i(\underline{H}_1.(\underline{r}_k - \underline{r}_m) + \underline{H}_2(\underline{r}_\ell - \underline{r}_m)]>_{\mathrm{orient}} \tag{4.5}$$

The calculation of individual orientational averages in (4.4)-(4.5) can be found in earlier literature. Debye [15] was the first to calculate the orientational average of the form $<\exp[2\pi i(\underline{H}.\underline{R})]>$ (with $\underline{R} = \underline{r}_k - \underline{r}_j$), under the assumption that the magnitudes $R = |\underline{R}|$ and $Q = |\underline{H}|$ are available, The resulting expression, $\sin[2\pi QR]/(2\pi QR)$, is often quoted in crystallographic literature. The calculation of $<\exp[2\pi i(\underline{H}_1.\underline{R}_1 + \underline{H}_2.\underline{R}_2)]>$ with $\underline{R}_1 = \underline{r}_j - \underline{r}_m$ and $\underline{R}_2 = \underline{r}_k - \underline{r}_m$ has been undertaken first by Hauptman [13]. He considered two triangular vector sets $\{\underline{H}_1, \underline{H}_2, \underline{H}_1 - \underline{H}_2\}$ and $\{\underline{R}_1, \underline{R}_2, \underline{R}_1 - \underline{R}_2\}$ with the intra-triangular information completely known while the inter-triangular orientation of the vectors is completely arbitrary. Hauptman devised an elegant solution to evaluate the three-fold integral and the final result is attractively simple. The methods of Debye, Hauptman and also Brosius [16] can not be applied in a straightforward way if the fragment consists of more than three atoms. The reason is that an orientational average over a sum of interdependent interatomic contributions, as given in (4.4)-(4.5), can not be approximated by a sum of the individual averages without losing coherence in the structure.

5. Correctly positioned atoms

Various methods have been proposed to exploit correctly positioned atoms (or groups of atoms) in DM but it has been noted (e.g. [17] that essentially all DM expressions derived so far rely on the uniform prior of the p.r.v.'s. For example, Heinerman [10] placed the position vector of the group of correctly positioned atoms on the origin and added the latter to the set of uniformly distributed p.r.v.'s. In effect, this comes down to handling the group of positioned atoms as being correctly oriented only (see previous §). Giacovazzo [18] (see also [19] for related expressions] defined *a priori a* normalized s.f. E_H to be the sum of a fixed term $E_{p,H}$ that comes from the positioned atoms and a variable term $E_{q,H}$ that arises from the atoms with unknown position (assumed to be uniformly distributed).

$$E_H = \frac{F_H}{\left[|F_{p,H}|^2 + \Sigma_q\right]^{1/2}}$$

$$E_{p,H} = \frac{F_{p,H}}{\left[|F_{p,H}|^2 + \Sigma_q\right]^{1/2}} \quad ; \quad E_{q,H} = \frac{F_{q,H}}{\left[|F_{p,H}|^2 + \Sigma_q\right]^{1/2}} \tag{5.1}$$

Fortier & Nigam [20] pointed out that the joint probability distributions (j.p.d.'s) derived on the basis of (5.1) with E_H and $E_{q,H}$ being variables are of the same form as expressions in the SIR and SAS case. Klop et al. [21] combined (5.1) with Heinerman's approach to incorporate structural information and arrived at expressions in the SIR/SAS case that generalizes the corresponding distributions of the previous authors.

6. Correctly positioned atoms as prior in joint probability distributions

A theoretical disadvantage of (5.1) is that a *prior* assumption is made on the form that normalized s.f.'s should have. In the practice of deriving a j.p.d. of s.f.'s normalizing quantities, like $Z_{\nu\nu}$, arise automatically as a result of the *prior* distribution of the p.r.v.'s.

The alternative is to enter an appropiate prior for the positioned atoms in the j.p.d. derivation of s.f's. Suppose now that the position vector $\underline{r}_j$ of p atoms or p groups of atoms lies exactly at $\underline{r}_j(s)$. Then, the prior (probability density) $p(\underline{r}_j)$ of the p.r.v. associated with $\underline{r}_j$ should take on a non-zero value only at $\underline{r}_j(s)$. This can be described by means of a delta function,

$$\delta(\underline{r}_j - \underline{r}_j(s)) = 0 \quad \underline{r}_j \neq \underline{r}_j(s); \quad \int \delta[\underline{r}_j - \underline{r}_j(s)]d\underline{r}_j = 1 \quad j = 1,..,p \tag{6.1}$$

Assuming (6.1) to hold, it can be inferred from expression (4.5) in [11] that non-zero moments exist at nmax=1 for α_ν =1 (and β_ν = 0) and β_ν =1 (and α_ν = 0) respectively,

$$m_1^0(j) = g_{j\nu}\exp[2\pi i\underline{H}_\nu . \underline{r}_j(s)] \qquad j = 1,..,p$$

$$m_0^1(j) = g_{j\nu}^*\exp[-2\pi i\underline{H}_\nu . \underline{r}_j(s)] \qquad j = 1,..,p \tag{6.2}$$

At nmax = 2 now several non-zero moments exist, in particular the moments with α_ν = β_ν =1,

$$m_1^1(j) = |g_{j\nu}|^2 \qquad j = 1,..,n \tag{6.3}$$

After the moments calculation, cumulants are constructed from the moments, taking into account the expansion of the logarithm. Moments-cumulants relations (upto nmax = 2) are:

$$\begin{aligned} k_1^0(\nu, j) &= m_1^0(\nu, j) && j = 1,..,p \\ &= 0 && j = p+1,..,n \end{aligned}$$

$$\begin{aligned} k_1^1(\nu, j) = m_1^1(\nu, j) - m_1^0(\nu, j)m_0^1(\nu, j) &= 0 && j = 1,..,p \\ &= |g_{j\nu}|^2 && j = p+1,..,n \end{aligned} \tag{6.4}$$

It can be shown that the cumulants of p.r.v.'s distributed according to (6.1) are zero, except those at nmax = 1 (Peschar unpublished results).

After summing the cumulants of the n p.r.v.'s that are distributed according to (6.1) and taking the α_ν (=1) and β_ν (=1) terms together, in the characteristic function the term

$+i\rho_\nu |F_{p,\nu}| \cos[\Theta_\nu - \varphi_{p,\nu}]$ arises that depends on a partial s.f. calculated from the p atoms (or atomic groups),

$$F_{p,\nu} = |F_{p,\nu}| \exp[i\varphi_{p,\nu}] = \sum_{j=1}^{p} g_{j\nu} \exp[2\pi i(\underline{H}_\nu . \underline{r}_j(s))] \tag{6.5}$$

Now taking together

$$i\rho_\nu R_\nu \cos(\Theta_\nu - \Phi_\nu) - i\rho_\nu |F_{p,\nu}| \cos[\Theta_\nu - \varphi_{p,\nu}] = i\rho_\nu \Xi_\nu \cos(\Theta_\nu - \xi_\nu)$$

$$\Xi_\nu \exp(-i\xi_\nu) = R_\nu \exp(-i\Phi_\nu) - |F_{p,\nu}| \exp[-i\varphi_{p,\nu}]$$

$$\Xi_\nu^2 = R_\nu^2 + |F_{p,\nu}|^2 - 2R_\nu |F_{p,\nu}| \cos[\Phi_\nu - \varphi_{p,\nu}] \tag{6.6}$$

The remaining $n-p = q$ atomic positions and atomic position vectors that are not known are assumed to be uniformly distributed. Up to nmax = 2 the q uniformly distributed p.r.v.'s lead to same type terms in the characteristic function as given in the expressions (4.6) and (5.2) of [11] in case of non-isomorphous and isomorphous s.f.'s respectively,

$$Z_{\nu\nu'}(q) = \sum_{j=p+1}^{n} |g_{j\nu}| |g_{j\nu'}| \exp[-i(\delta_{j\nu} + s_{\nu\nu'} \delta_{j\nu'})] \tag{6.7}$$

$$= |Z_{\nu\nu'}(q)| \exp[i\Delta_{\nu\nu'}(q)] \qquad \nu, \nu' \in [1,2]$$

$$s_{12} = 1 \text{ if } H_1 = H_2 \text{ and } s_{12} = -1 \text{ if } H_1 = -H_2$$

$$Z_{123}(q) = \sum_{j=p+1}^{n} g_{\nu 1}^{*} g_{\nu 2}^{*} g_{\nu 3}^{*} \tag{6.8}$$

After carrying out the integrations connected with the Fourier-transform of the characteristic function, expressions for the j.p.d's are obtained that are similar to (4.7) and (5.3) in [11]. Thus, the form of the j.p.d. expressions derived for s.f.'s under the model of N uniformly distributed p.r.v.'s remains the same but the parameters that occur in the expressions are different with R_ν being replaced by Ξ_ν, Φ_ν by ξ_ν, $Z_{\nu\nu'}(N)$ by $Z_{\nu\nu'}(q)$ and $Z_{123}(N)$ by $Z_{123}(q)$.

For example, it is easily shown that the j.p.d. of a single F_ν given n atoms positioned according to (6.1),

$$P(R_\nu, \Phi_\nu) = 2R_\nu [Z_{\nu\nu}^{-1}(q)] \exp[-\Xi_\nu^2 Z_{\nu\nu}^{-1}(q)] \tag{6.9}$$

turns out to be the distribution of Sim [22] after inserting (6.6).

As it appears, the presence of a constant term $F_{p,\nu}$ in F_ν does not contribute to normalizing quantities like $Z_{\nu\nu}(q)$. This is in accordance with the interpretation of $Z_{\nu\nu}(q)$ as a variance associated with F_ν [$\sigma^2(F_\nu) = \langle F_\nu^2 \rangle - \langle F_\nu \rangle^2$] given the prior distribution of the p.r.v.'s.

References

1. Beurskens, P.T., Beurskens, G., Bosman, W.P., De Gelder, R., García-Granda, S., Gould, R.O., Israel, R. and Smits, J.M.M. (1996) The DIRDIF-96 program system, Crystallography Laboratory, University of Nijmegen, The Netherlands.
2. Hauptman, H. (1982) On integrating the techniques of direct methods and isomorphous replacement. I. The theoretical basis, *Acta Cryst.* **A38**, 289–294.

3. Hauptman, H. (1982) On integrating the techniques of direct methods with anomalous dispersion. I. The theoretical basis, *Acta Cryst.* **A31**, 632–641.
4. Giacovazzo, C.(1983) The estimation of two-phase and three-phase invariants in P1 when anomalous scatterers are present, *Acta Cryst.* **A39**, 585–592.
5. Giacovazzo, C., Cascarano, G. & Zheng Chao-De (1988). On integrating the techniques of direct methods and isomorphous replacement. A new probabilistic formula for triplet invariants, *Acta Cryst.* **A44**, 45–51.
6. Hao Quan & Fan Hai-Fu (1988) Combining direct methods with isomorphous replacement or anomalous scattering data. VI. Incorporation of heavy-atom information into Hauptman's distributions, *Acta Cryst.* **A44**, 379–382.
7. Klop, E.A., Krabbendam, H. & Kroon, J. (1989) The estimation of triplet invariants from multi-wavelength data, *Acta Cryst.* **A45**, 203–208.
8. Main, P. (1976) Recent developments in the MULTAN system – the use of molecular structure, in F.R. Ahmed (ed.), *Crystallographic Computing Techniques,* Munksgaard, Copenhagen, pp. 97–105.
9. Cochran, W. (1955) Relations with the phases of structure factors, *Acta Cryst.* **8**, 473–478.
10. Heinerman, J.J.L. (1977) The use of structural information in the phase probability of a triple product, *Acta Cryst.* **A33**, 100–106.
11. Peschar, R. (1997) Direct methods in relation to isomorphous data, in S. Fortier (ed.), *Direct methods for solving macromolecular structures*, Kluwer, Dordrecht, This volume.
12. Giacovazzo, C. (1988) New probabilistic formulas for finding the positions of correctly oriented atomic groups, *Acta Cryst.* **A44**, 294–300.
13. Hauptman, H. (1965) The average value of $\exp\{2\pi i(\mathbf{h}.\mathbf{r} + \mathbf{h}'.\mathbf{r}')\}$, *Z. Kristallogr.* **121**, 1–8.
14. Kronenburg, M. (1992) Application of maximum entropy and irreducible cluster integrals to joint probability distributions for x-ray structure determination, PhD thesis, University of Amsterdam, The Netherlands.
15. Debye, P. (1915) Zerstreuung von Röntgenstrahlung, *Ann. Phys. (Leipzig)*, **46**, 809–823.
16. Brosius, J. (1978) The orientational average of $\exp[2\pi i(h.x+ k.y+ \ell.z)]$, *Acta Cryst.* **A34**, 955–959.
17. Bricogne, G. (1988) A Bayesian statistical theory of the phase problem. I. A multichannel maximum-entropy formalism for constructing generalized joint probability distributions of structure factors, *Acta Cryst.* **A44**, 517–545.
18. Giacovazzo, C. (1983) From a partial to the complete crystal structure, *Acta Cryst.* **A39**, 685–692.
19. Camalli, M., Giacovazzo, C. & Spagna, R. (1985) From a partial to the complete crystal structure, *Acta Cryst.* **A41**, 605–613.
20. Fortier, S. & Nigam, G.D. (1989). On the probabilistic theory of isomorphous data sets: General joint distributions for the SIR, SAS and partial/complete structure cases, *Acta Cryst.* **A45**, 247–254.
21. Klop, E.A., Krabbendam, H. and Kroon, J. (1990) The use of structural information in phase determination, *Acta Cryst.* **A46**, 517–520.
22. Sim, G.A. (1959) The distribution of phase angles for structures containing heavy atoms. II. A modification of the normal heavy-atom method for non-centrosymmetrical structures, *Acta Cryst.* **A12**, 813–815.

DIRECT METHODS WITH ELECTRON MICROSCOPE INFORMATION

C.J.Gilmore
Department of Chemistry
University of Glasgow
Glasgow G12 8QQ, Scotland, UK.

1. Introduction

The purpose of this chapter is to examine the process of structure solution when the data have been partitioned into two sets:
(1) The basis set {H} comprising phased reflections where the phase information comes from the Fourier transform of electron microscope images after suitable filtering. Usually the phases so derived correspond to intensities that have a significantly lower resolution than the diffraction data, and one must also remember that there are some significant sources of error in image data:

- At the highest resolutions, more sampled pixels are needed to resolve a detail between two points. Even though averaging over the repeat of the two-dimensional space lattice can be used to minimise the actual radiation dose to the specimen when recording the image, the damage induced by inelastic interactions between electron and sample can be problematic [1].
- Many two-dimensional crystals, especially proteins, contain a curvilinear paracrystalline distortion, probably because of the lipid matrix in which the proteins are embedded. Thus, although the observed electron diffraction pattern might extend to e. g. 3Å, the Fourier transform of a micrograph from a similar area may vanish somewhere in the range from 10 to 6Å. Lattice unbending has been used to restore the higher resolution information [2], but the actual amount of artefact involved is not known.
- There are slight variations of specimen height from the perspective of the microscope objective lens. Although a nearby correction can be made locally for the lens focus before the low-dose image is recorded, the actual transfer function of the micrograph is often unknown.

(2) The non-basis set {K} which is much less problematic, comprising only diffraction intensities.

The problem is clearly one of phasing set {K} from set {H}. Here I will concentrate on the maximum entropy (ME) method [3-5], but they it is clearly not the only available techniques: Dorset, for example, uses the Sayre equation to great effect [6], and the tangent formula can also be used as Fan and co-workers have demonstrated [7].

S. Fortier (ed.), Direct Methods for Solving Macromolecular Structures, 317–327.

Two sorts of problems arise in phase extension:
(1) The basis set is sufficiently large and well distributed in reciprocal space to permit an unambiguous phase extension procedure without recourse to multisolution methods i.e. those that involve phase permutation.
(2) When the basis set is small or in some way inadequate we have a *branching problem.* This problem arises when phases are selected without exploring the relevant phase space in sufficient detail so that what appears to be an unambiguous phase choice is no such thing. For a full description see Bricogne [3].

Sometimes it is not clear which situation is relevant for a given phase extension process; it is important to be constantly on guard. Let us consider examples where branching is probably not a significant problem first.

2. Phase Extension when there is no Branching Problem.

2.1 THE TWO-DIMENSIONAL DATA FROM *HALOBACTERIUM HALOBIUM*

This section describes a trial application [8,9] to the two-dimensional purple membrane data from *Halobacterium Halobium* [2,10]. This data set was used for several reasons:
(1) It is two-dimensional, but of high quality with a maximum resolution of 3.5Å. Dynamical effects have been minimised. It is, however, inaccessible to traditional direct phasing methods having insufficient resolution and being incomplete.
(2) It is a biological macromolecule.
(3) The final map of Baldwin *et al.* [10] acts as a good indicator of the correctness of the ME results.

Nine phased reflections at a resolution less than 15Å were employed as a basis set, and used to produce a U-map (a Fourier map with U magnitudes and phases as coefficients); this is shown in Figure 1a. The same reflections were than used as constraints in an entropy maximisation, and a centroid map generated (Figure 1b) which is based on the 10 constraint reflections plus 39 suitably weighted extrapolates whose resolution exceeds 15Å. This can be compared with a 10Å U-map based on 18 phased reflections using phases published by Henderson *et al.*[2]. Note that we are testing here the extrapolative power of the ME formalism.

The results of the extrapolation process are quite dramatic. Figure 1a gives a very diffuse map with little apparent detail, and only the positions of three of the seven helices can be seen. The centroid map (1b) is much clearer, and displays correct detail that corresponds to the 10Å U-map (1c). In particular the molecular boundary is well defined with no lipid component visible, and with six of the helices correctly indicated. Similar tests using 10Å basis set reflections are shown in Figure 2 with the same significant resolution enhancement and corresponding map clarity, although it is less dramatic. It may be that, from a practical viewpoint, it is only necessary to obtain images at relatively low resolution, and combine them with high resolution diffraction intensities using the ME formalism without recourse to measurement of very high resolution images with all the associated problems

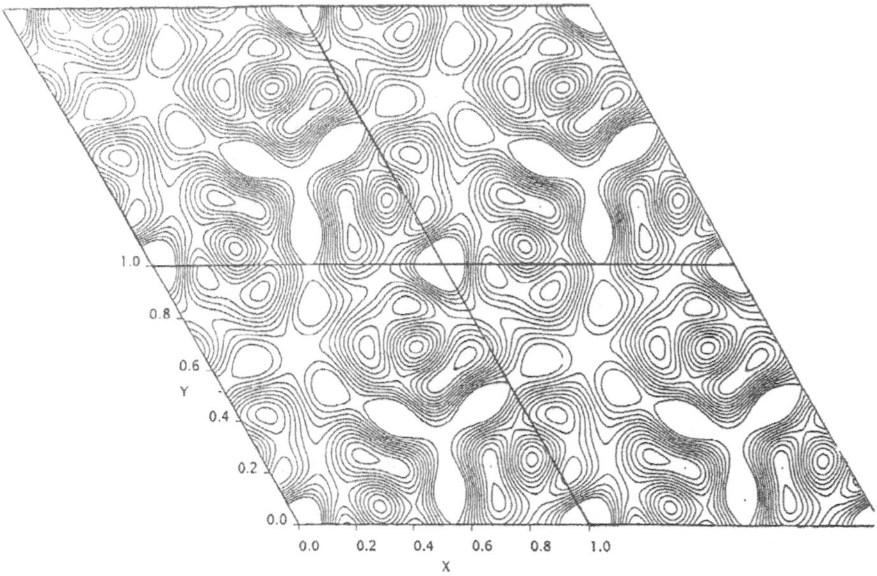
1.0
0.8
0.6
Y
0.4
0.2
0.0
0.0 0.2 0.4 0.6 0.8 1.0
X

(a)

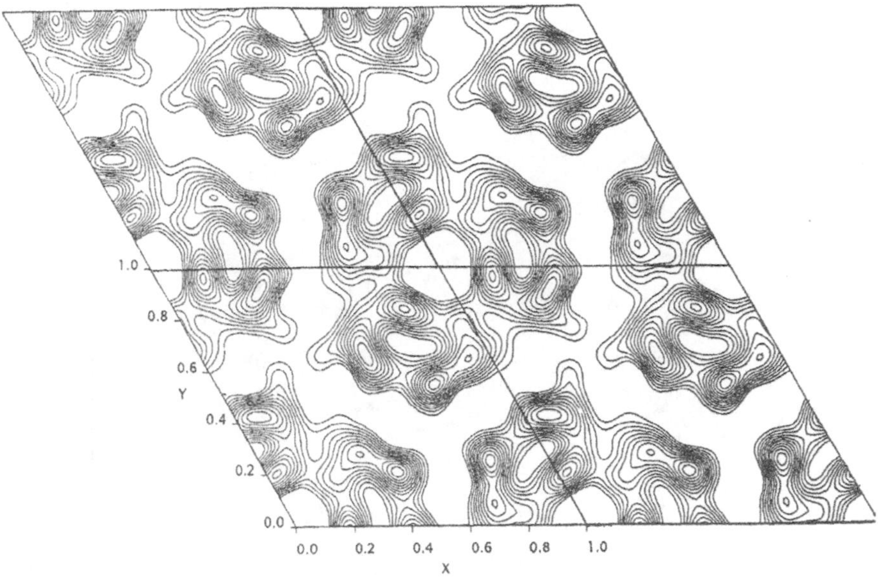
1.0
0.8
0.6
Y
0.4
0.2
0.0
0.0 0.2 0.4 0.6 0.8 1.0
X

(b)

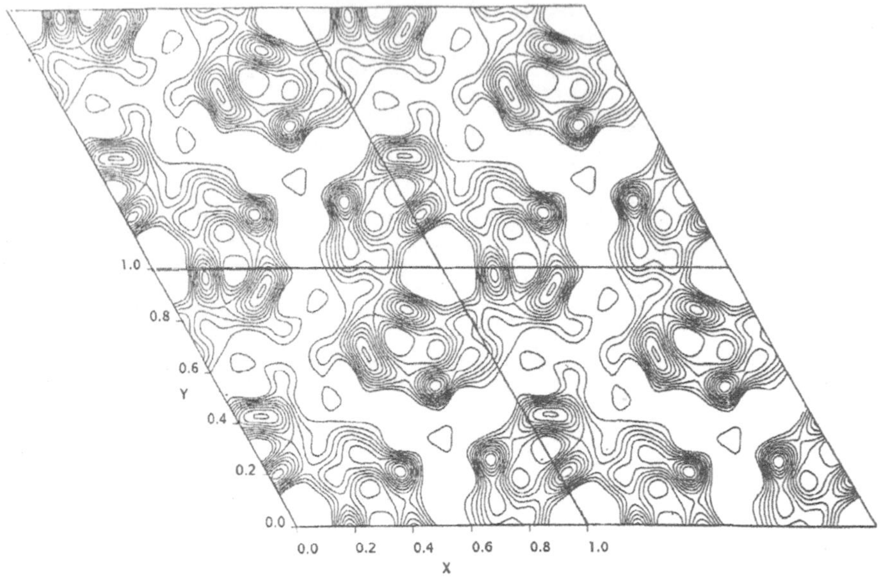

(c)

Figure 1. (a) A U-map based on 9 basis set reflections at 15Å resolution for the data from *Halobacterium Halobium.* (b) a centroid map using the same 9 basis set reflections plus extrapolated phases to 10Å resolution (c) a U-map based on 15 reflections at 10Å.

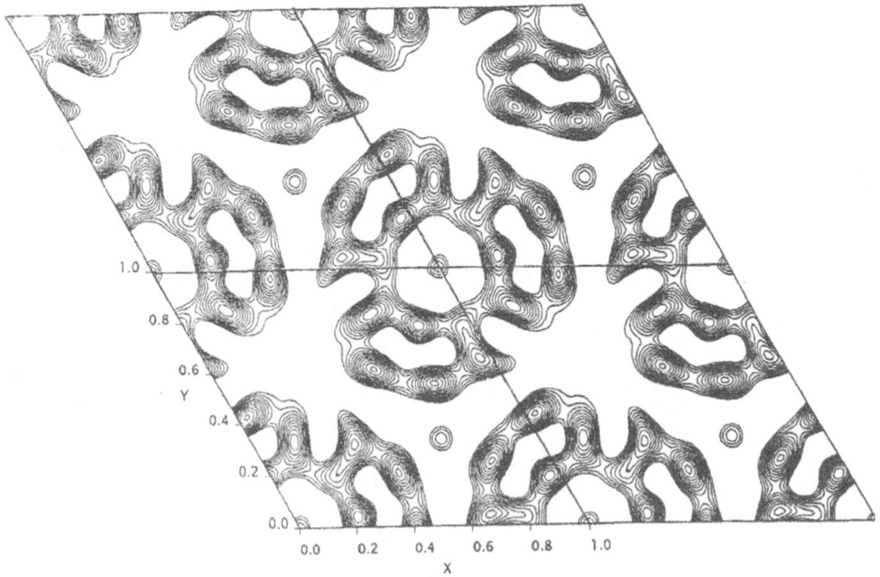

(a)

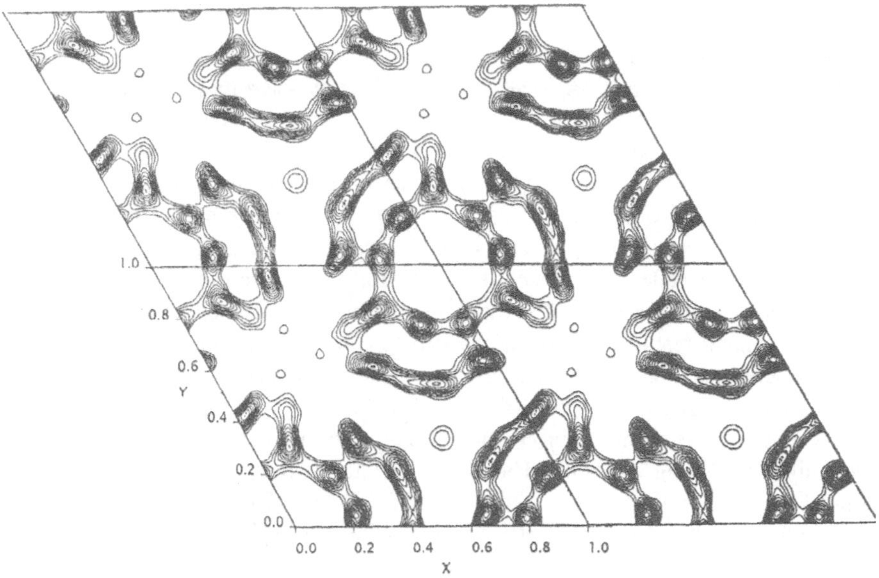

(b)

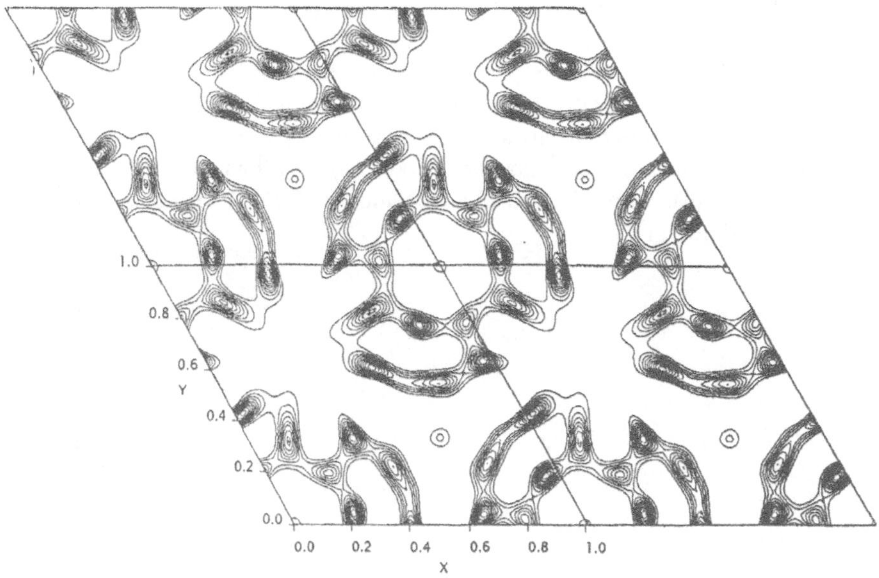

(c)

Figure 2. (a) A U-map based on basis set reflections at 10Å resolution. (b) a centroid map using the 10Å basis set reflections plus extrapolated phases to 8Å resolution (c) a U-map based on reflections phased from image data at 8Å resolution.

2.2 APPLICATION TO SURFACE DIFFRACTION DATA

We are looking here at a two dimensional data set derived from the electron diffraction of the Si(111) 7x7 surface. A total of 193 diffracted intensities are available. It is important to note however, that this data set arises from a reconstructed surface: the bulk Si also diffracts, and these bulk derived reflections have been removed from the data set so that phase extension cannot be taken for granted since there is now a very incomplete sampling of reciprocal space. A total of 41 unique phases were available from image Fourier transform studies of high resolution images [13]. The maximum resolution of these reflections was *ca.* 1.5Å. Most of them have very weak associated U magnitudes. A U map calculated before any entropy maximisation is shown in Figure 3(a). The general features of the surface structure are present although these is some spurious detail and some peaks are absent or coalesced. There is insufficient detail to consider the structure characterised at the atomic level. Accordingly, the 41 reflections were used as a basis set and subjected to entropy maximisation without recourse to any tree building phasing methods. The iterative entropy maximisation algorithm was stopped at the point of maximum likelihood. The LLG was 2.42. The resulting centroid potential map is shown in Figure 3(b) with a resolution of 0.6Å. The whole structure is clearly delineated.

3. Phase Extension when there is a Branching Problem.

As an example, let us look at phase extension from a very small basis set for perchlorocoronene $C_{24}Cl_{12}$ [12]. Figure 4 shows the structure as derived from X-ray crystallography. The electron diffraction data set consists of 39 h0l reflections to a resolution of 1Å . This is a projection which looks edgewise on to the ring. in addition 5 phases were available from images to a resolution of 3.2Å. A map based on these 5 reflections is shown in Figure 5.

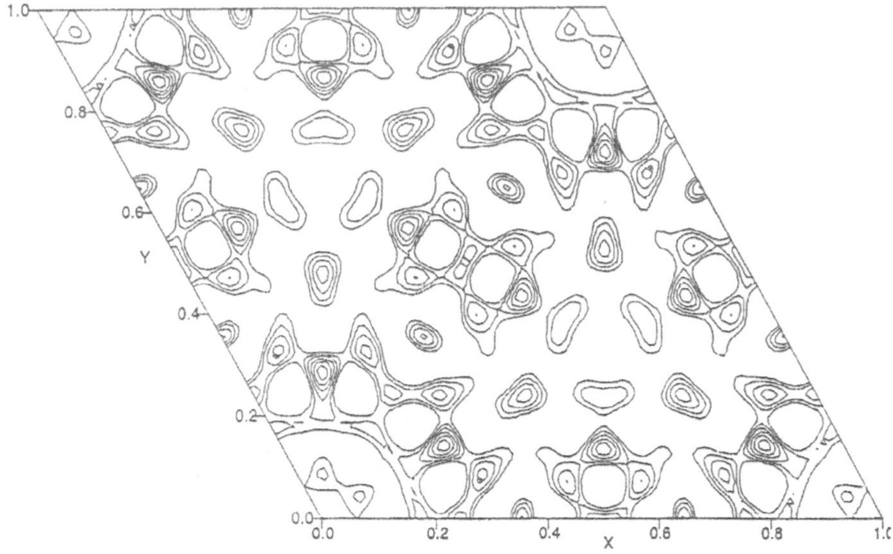

(a)

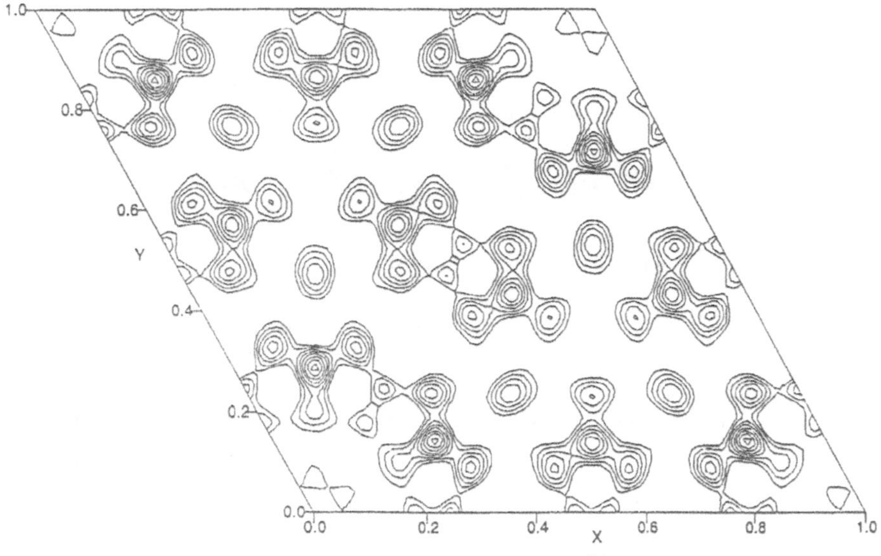

(b)

Figure 3. (a) A U-map based on 41 intensities phased from image data for the Si(111) 7x7 electron diffraction data (b) the corresponding centroid map after entropy maximisation at 0.6Å resolution.

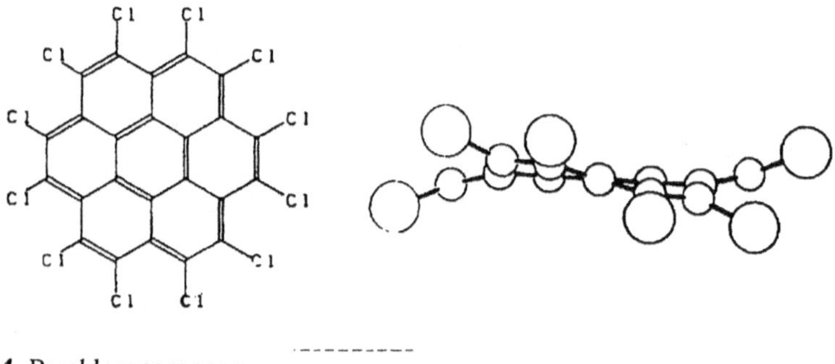

Figure 4. Perchlorocoronene.

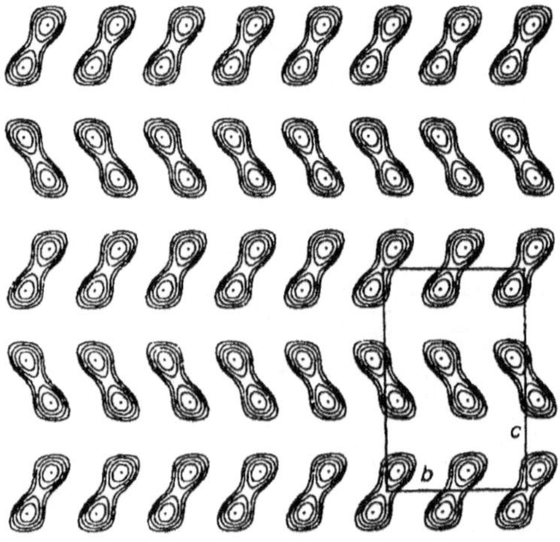

Figure 5. Map calculated from 5 image derived phases.

A 4-level, 53 node phasing tree was constructed; each node was subjected to constrained entropy maximisation and log likelihood gain (LLG) evaluation carried out. Table 1 summarises this procedure. The centroid map corresponding to the LLG preferred node is shown in Figure 6. The agreement with the X-ray structure is excellent.

Node	To node	Entropy	Log likelihood gain
1		−0.21	0.00
2	1	−0.16	−0.10
3	1	−0.19	−0.09
4	1	−0.24	0.00
5	1	−0.24	0.00
6	2	−0.31	1.32
7	2	−0.22	0.56
8	2	−0.32	2.21
9	2	−0.23	0.50
10	3	−0.38	1.12
11	3	−0.17	0.03
12	3	−0.39	1.42
13	3	−0.23	0.06
14	4	−0.26	0.66
15	4	−0.21	0.53
16	4	−0.29	1.61
17	4	−0.21	0.54
18	5	−0.32	0.30
19	5	−0.18	−0.01
20	5	−0.33	0.07
21	5	−0.30	0.20
22	8	−0.25	0.16
23	8	−0.27	0.17
24	8	−0.30	1.23
25	8	−0.24	−0.60
26	16	−0.24	0.26
27	16	−0.28	0.36
28	16	−0.28	1.21
29	16	−0.32	0.16
30	24	−0.28	1.58
31	24	−0.25	0.67
32	24	−0.26	1.07
33	24	−0.23	0.25
34	28	−0.28	2.00
35	28	−0.23	0.61
36	28	−0.25	1·04
37	28	−0.21	0.16
38	30	−0.27	2.76
39	30	−0.30	3.85
40	30	−0.27	2.75
42	30	−0.28	3.18
42	34	−0.27	2.97
43	34	−0.27	3.59
44	34	−0.26	2.53
45	34	−0.26	2.96
46	39	−0.27	4.10
47	39	−0·28	4·17
48	39	−0.29	4·47
49	39	−0.29	4.28
50	43	−0.24	3·13
51	43	−0.25	3.81
52	43	−0.26	3.98
53	43	−0.27	4.59

Table 1. Phasing tree for perchlorocoronene.

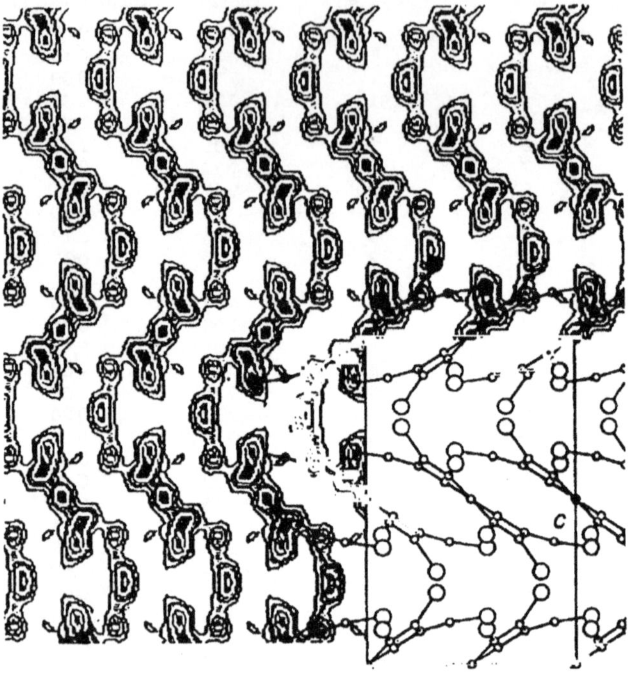

Figure 6. The best centroid map for perchlorocoronene with the molecular structure superimposed.

4. Conclusions

Image data are a very useful source of starting set phases, but branching is a problem that is always present and must be kept in mind. Two questions (unanswered in this chapter) arise:

(1) How much image derived data is necessary?

(2) Are any needed at all?

For a partial answer to (2) see reference [13] where two membrane proteins are solved ab initio without any prior phase information using the ME formalism.

References

1. Glaeser, R. M. (1979) Radiation damage with biological specimens and organic materials (1979)in J.J.Hren, J.I.Goldstein and D.C.Joy (eds.) *Introduction to Analytical Electron Microscopy* pp 24-36 Plenum Press, New York.
2. Henderson, R., Baldwin, J. M., Downing, K. H., Lepault, J. and Zemlin, F. (1986) Structure of purple membrane from *Halobacterium halobium*: recording, measurement and evaluation of electron micrographs at 3.5Å resolution, *Ultramicroscopy.* **19**, 147-178.
3. Bricogne, G. (1984) Maximum entropy and the foundations of direct methods, *Acta Cryst.* **A40**, 410-445.
4. Bricogne, G. and Gilmore, C.J. (1990) A multisolution method of phase determination by combined maximisation of entropy and likelihood. I. Theory, algorithms and strategy, *Acta Cryst.* **A46,** 284-297.
5. Bricogne, G. (1993) Direct phase determination by entropy maximisation and likelihood ranking: status report and perspectives. *Acta Cryst.* **D49**, 37-60.
6. Dorset, D.L., Kopp, S., Fryer, J.F. and Tivol, W.F. (1995) The Sayre equation in electron crystallography, *Ultramicroscopy* **57**, 59-89.
7. Fan, H.F., Xiang, S.B., Li, F.H., Pan, Q., Uyeda, N. and Fujiyoshi, Y. (1991) Image enhancement by combining information from electron density patterns and micrographs *Ultramicroscopy* **36**, 361-373.
8. Gilmore, C.J., Shankland, K. and Fryer, J.R. (1993), Phase extension in electron crystallography using the maximum entropy method and its application to two-dimensional purple membrane data from *Halobacterium halobium*, *Ultramicroscopy*, **49**, 147-178.
9. Gilmore, C.J., Shankland, K. and Bricogne, G. (1993) Applications of the maximum entropy method to powder diffraction and electron crystallography, *Proc.Roy.Soc.Ser. A.* **442**, 97-111.
10. Baldwin, J.M., Henderson, R., Beckman, E. and Zemlin, F. (1988) Images of purple membrane at 2.8Å resolution obtained by cryo-electron microscopy, *J. Mol. Biol.* **202,** 585-591.
11. Gilmore, C.J., Marks, L.D., Grozea, D., Collazo, C., Landree, E. and Twesten R.D, Direct Solutions of the Si (111) 7x7 Structure, *Surface Science*, in press.
12. Dong, W., Baird, T., Fryer, J.F., Gilmore, C.J., MacNicol, D.D., Bricogne, G., Smith, D.J., O'Keefe, M.A. and Hövmoller, S. (1992) Extension of Electron Microscope Resolution to 1Å by Entropy Maximisation and Likelihood Evaluation *Nature, London* **355,** 605-609.
13. Gilmore, C.J., Nicholson, W.V., and Dorset, D.L., (1996) Direct methods in protein electron crystallography: the ab initio structure determination of two membrane protein structures in projection using maximum entropy and likelihood, *Acta Cryst.* **A52,** 937-946.

DENSITY MODIFICATION AND DIRECT METHODS

K. D. COWTAN
Dept. of Chemistry, University of York
Heslington, York, YO1 5DD. United Kingdom

Direct methods have proven very successful in cases where diffraction data is available to atomic resolution, since they generally depend on the atomicity of the structure. Density modification techniques work with much weaker models, and so can only refine existing phases, however these weak models of the density do not depend on atomicity and so can be applied at lower resolution. Combination of the two methods holds promise for improving both.

1. Introduction

Density modification and direct methods, despite the shared initials, have somewhat different origins. Direct methods arose from the ab-initio phasing of small, initially centric, structures through phase relationships in reciprocal space [3]. Density modification owes its modern origins to the resolution of phase ambiguity in proteins by modification of the electron density in real space [10]. However the methods are developing increasing areas of overlap.

Sayre's equation, for a long time the foundation of direct methods, has been applied in a number of authors to phase refinement of proteins [8],[12]. At the same time, recent developments in direct methods, and in particular the Shake'n'Bake procedure, have incorporated real space map-modifications such as peak picking into the direct methods calculation, resulting in a significant increase in the size of structure that can be solved [11].

Both Sayre's equation, and peak picking, are strongly dependent on data resolution. Below 2.0Å resolution, Sayre's equation has very limited power for phase improvement, indeed a point is reached when phase estimates from Sayre's equation are based more strongly on the absence of the high resolution terms than on those terms which are present. Similarly at less than atomic resolution peak picking becomes under-determined, that is it

S. Fortier (ed.), Direct Methods for Solving Macromolecular Structures, 329–337.

become possible to represent a map arbitrarily well by appropriately scaled atoms, thus no information is added.

If the dependence on atomic resolution can be removed, then the potential exists for stronger density modification techniques which can refine much weaker phases, and for direct methods which work at lower resolutions than at present. Some possible approaches are outlined here.

In place of Sayre's equation for phase refinement, a maximum entropy approach is adopted. Bricogne [1] has shown the relationship between the entropy of a distribution of atoms and the joint probability of the structure factors corresponding to that distribution. Thus the distribution with the highest entropy subject to any known information is the most probable distribution - this can be calculated with no assumptions about the values of missing terms, and embodies both the Sayre triplet and higher order phase relationships.

In place of peak picking for map modification, a search for common structural motifs is used. These fragments, after suitable likelihood ranking, can be built into a hypothetical structure to provide additional phase information. A hierarchical search is envisioned, in which large secondary structures are fitted first, and then smaller 1- or 2- residue features, with the possibility of linking these to the ends of the larger fragments. Finally, a pseudo-atom model may be used to account for remaining density [9].

These stages may be combined into a cyclic calculation in a similar manner to traditional density modification methods. However the methods described here are still in the early stages of development, and so no real indication of the practicality of such a scheme may yet be inferred.

2. Maximum entropy phase extrapolation with MIR-like constraints

2.1. PROBLEMS WITH 'DIRECT-METHODS' APPROACHES

Maximum entropy methods have often relied upon the division of reflections into 'phased' and 'unphased' sets - a distinction which has its roots in small molecule direct methods, which in early cases progressed by the phasing of a single new reflection at each stage. However this is not a good representation of the protein phase-improvement problem, where typically there is partial phase information for the majority of reflections. (Note however that Bricogne has also employed phase information in a modified prior probablity distribution).

A more serious difficulty is the computational difficulty of finding the map, consistent with the phased reflections and with unphased magnitudes, which has the highest entropy. Each reflection of known magnitude and unknown phase can be represented by a circle of allowed values in the complex plane. If we combine all the complex planes covered by all reflections, then each reflection describes a hyper-cylindrical surface of allowed values. A valid solution must lie on the intersection of all these hyper-cylinders.

While the entropy is well behaved and has a single global minimum, value of the entropy on the surface of intersection will generally have many local minima. The problem is somewhat analogous to picking a random strand from a ball of wool and trying to guess which way to follow it to reach the centre.

2.2. USING EXPERIMENTAL PHASES

A solution is proposed which resolves both of the above difficulties.

The observed magnitude and phase probability distribution for a reflection is converted into bivariate Gaussians in the complex plane. The long axis of the Gaussian represents the possible variation of the reflection in the phase direction, with the shorter axis representing variation in the magnitude direction.

In the case of well phased reflections, the constraint ellipsoid approaches to a 1-d constraint along a line in the phase direction. For reflections with weak or no phase information, the constraint ellipsoid becomes a circle centred on the origin on the complex plane. Three cases are shown in Figure (1).

Once the constraint ellipsoids are established, χ-squared constraints may be used to ensure an appropriate fit to the constraints. A single χ-squared constraint can be applied, or the constraints may be divided up into sub-classes, each of which must achieve an appropriate fit.

This approach resolves both of the above difficulties. Firstly, it is possible to represent phase probability distributions, rather than simply known and unknown phases. In the case of a bimodal phase probability distribution, it is even possible to constrain a structure factor to a narrow region joining the two peaks.

Secondly, the reflections are only constrained to lie on a surface whose dimension is one or two less than the dimension of the whole space and comparatively simple in shape (a hyper-ellipsoid). Thus local minima should present much less of a problem.

2.3. APPLICATION

Bricogne showed that the Lagrange multiplier solution of the maximum entropy equations with some constrained reflections leads to an exponential model of the density. He modifies this approach to use the Newton Raphson and entropy gradients as alternative search directions in order to find a

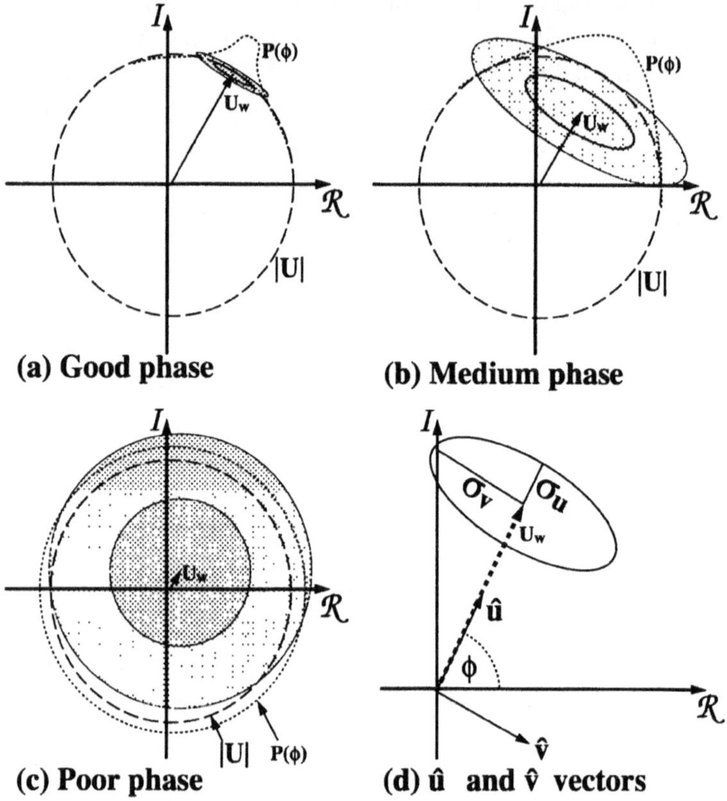

Figure 1. Constraint Gaussian for (a) a good phase, (b) a weaker phase (c) an almost unknown phase. (d) Unit vectors û and v̂ are defined parallel and perpendicular to the complex structure factor.

solution which maximises entropy while achieving satisfactorary constraint agreement.

This formalism is extended here: the weighting of the exponential modelling gradients is altered to reflect the anisotropic constraint of each reflection. The exponential modelling gradients give directions which modify the χ-squared fit, and an entropy gradient is also used to ensure a solution with maximum entropy.

The χ-squared parameter takes the following form:

$$\chi^2 = \frac{1}{2N}\sum_h \left[\left(\frac{(U_{\mathrm{ME}}(h) - U_w(h)).\hat{u}}{\sigma_u}\right)^2 + \left(\frac{(U_{\mathrm{ME}}(h) - U_w(h)).\hat{v}}{\sigma_v}\right)^2\right] \quad (1)$$

where U_w is the FOM-weighted unitary structure factor (the mean of the structure factor integrated over possible phases), U_{ME} is the unitary structure factor on the maximum entropy distribution, and N is the number of reflections. The target value for χ^2 is 1.

Gradient directions may be calculated for both the entropy and the χ^2 values in the exponential model space. Rather than using a plane search to control the step size, the two vectors were orthogonalised by selecting for each gradient the vector closest to the gradient but perpendicular to the other gradient (this is a simple Lagrange multiplier problem). Initially the χ^2 gradient is followed exclusively, but as this value approaches the desired value the step along the entropy direction is allowed to increase until it equals the χ^2 step. This corresponds to a search on the constraint surface for the constrained entropy maxima.

2.4. INITIAL RESULTS AND DISCUSSION

Initial testing suggests that when the elliptical Gaussian constraints are used instead of strict phase constraint of the the best reflections, most of the more weakly constrained magnitudes shrink towards zero, and the centroid map is poor.

Better results are obtained by arbitrarily reducing σ_u by a large factor, narrowing the ellipses and essentially constraining most reflections to lie on lines perpendicular to the current phase directions. Inclusion of a solvent channel and corresponding alterations to the prior distribution leads to similar behaviour.

Some tests have been made with additional constraints to the first and second moments of the distribution of structure factors about the FOM-weighted values. Unfortunately since the constraint surface now has four dimensions less than the entropy function, the constrained entropy once again has many local minima. This problem has not been solved.

The method suggested here have had limited success, since strong imposition of magnitude constraints leads to a search subspace whose topology renders simple search methods impractical, whereas weak constraint of magnitudes gives insufficient structural information. If this is the case then the tree-based phase permutation scheme suggested by Bricogne [2] may be the only effective solution.

3. Fragment location

In his 'Essens' procedure Kleywegt [4] has shown that a 6-dimensional (rotation and translation) search with a secondary structure fragment can give strong indications of the presence of secondary structure features in the map. The fragment is fitted against the electron density in each possible position and orientation. The electron densities at each atomic centre are compared, with the worst densities giving a strong indication of the goodness-of-fit.

This approach rapidly becomes computationally prohibitive as fragment size increases. However, an approximation by Fourier methods is possible for which the computational requirement is independent of fragment size. The conventional Fourier-based translation search method is convolution of the map and the fragment, given by the product of the fragment density and the map density summed over the volume of the fragment. This is simply formed in reciprocal space by the product of the structure factors of the map and the fragment. This is however insufficiently sensitive to regions of disagreement between the fragment and the map.

3.1. METHOD

A simple mathematical function is required whose value reflects the worst agreements between the map and the fragment. A good example of such a function is the mean-squared or root-mean-squared difference between the map and the fragment over the fragment volume, since the largest differences contribute most strongly to the result.

For the mean squared difference to be meaningful, both the fragment density and the map density must be first placed on an absolute scale, and the atomic temperature factors of the fragment should match the overall temperature factor of the map. The difference is only calculated over the volume covered by the fragment, represented by an envelope function.

Let the discriminator be called $t(x)$. The fragment density is $\rho_f(x)$, and the corresponding fragment envelope is $\epsilon_f(x)$. Then the discriminator may be formed from the sum of the mean-squared difference in density between the offset fragment and the the map:

$$\begin{aligned} t(x) &= \sum_y \epsilon_f(y)\,[\rho_f(y) - \rho(x+y)]^2 \qquad (2) \\ &= \sum_y \epsilon_f(y)\rho_f^2(y) - 2\epsilon_f(y)\rho_f(y)\rho(x+y) + \epsilon_f(y)\rho^2(x+y) \end{aligned}$$

The second and third terms on this expression are convolutions which may be efficiently calculated in reciprocal space.

$$t(x) = \sum_y \epsilon_f(y)\rho_f^2(y) + \epsilon_f(x) \otimes \rho^2(x) - 2[\epsilon_f(x)\rho_f(x)] \otimes \rho(x)$$

$\otimes$ represents convolution, performed in reciprocal space.

Possible variants of the discriminator function include higher order difference functions, and functions which perform local scaling of the model

Peak number	Height (/σ's above mean) Convolution	Squared difference
1	4.24172	4.51519
2	4.05857	4.12955
3	3.81619	4.04963
4	3.78091	3.99586
5	3.76134	3.89016
6	3.74526	3.83041
7	3.70356	3.81776
8	3.69312	3.79927
9	3.59330	3.78493
10	3.57736	3.72884

TABLE 1. Heights (in standard deviations) of the top peaks in the discriminator function using convolution/overlap and squared difference discriminators. The correct peak is shaded.

to the map using the mean and/or variance of the map over the envelope region.

3.2. APPLICATION

To test the translation functions a fragment search was performed in a uninterpretable SIR map of RNAse [7]. The search fragment was a correctly oriented alpha-helix of 10 polyalanine residues.

The translation search was performed using an overlap integral (the normal phased translation function), and the mean squared difference function. The heights of the top 10 peaks in the translation function are listed in Table (1), with the correct peak shaded. Using the conventional overlap integral, the correct translation is the second highest peak in the function, and indistinct from the noise. In the mean squared difference function, the correct translation is clearly identifiable.

A genuine rotation and translation search was run using the MIR map O6-methylguanine-DNA methyltransferase [5], which was interpreted only after density modification. A search was made for the same helical fragment, and the best helix orientation at each position in the unit cell was stored.

Matches were found to every helix in the structure, including one of only 7 residues. In most cases the direction of the helix was also determined correctly. Part of the molecule is shown in Figure (2), together with the best fit helices. The 20 best matches were taken from the search function, so several different fits are found for each helix. Note that matches to the

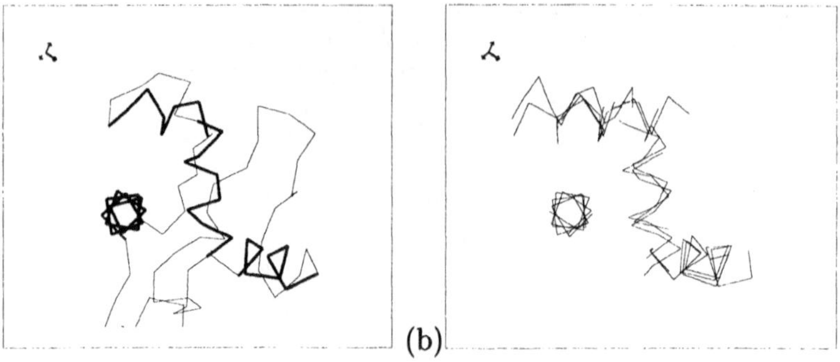

(a) (b)

Figure 2. Cα trace of part of O6-methylguanine-DNA methyltransferase, with the helices highlighted, and the corresponding helix fragments found using the squared-difference discriminator. (Plots generated using QUANTA)

short helix vary somewhat in orientation.

4. Discussion

The improved discriminators described here are effective for the location of small fragments of secondary structure in poor electron density maps. The calculation is time-consuming, but virtually independent of fragment size. While the discriminator function is less versatile than that used in 'Essens', the use of larger fragments allows location of secondary structure features in very poor maps. This is a first stage towards the automatic identification of larger scale features in low quality maps.

References

1. Bricogne G. (1984) Maximum-entropy and the foundations of direct methods *Acta Cryst.*, **A40**, pp. 410–445
2. Bricogne G. (1993) Direct phase determination by entropy maximization and likelihood ranking - status-report and perspectives *Acta Cryst.*, **D49**, pp. 37–60
3. Harker D., Kasper J. S. (1948) Phases of Fourier coefficients directly from crystal diffraction data *Acta Cryst.*, **1**, pp. 70–75.
4. Kleywegt G. J., Jones T. A. (1997) Template convolution to enhance or detect structural features in macromolecular electron-density maps. *Acta Cryst.*, **D**, to be published.
5. Moore M. H., Gulbis J. M., Dodson E. J., Demple B., Moody P. C. E. (1994) Crystal-structure of a suicidal DNA-repair protein - the ADA O6-methylguanine-DNA methyltransferase from Escherichia-coli. *Embo Journal*, **13**, pp. 1495–1501
6. QUANTA, Molecular simulation inc., 9685 Scranton road, San Diego, CA 92121-3752
7. Ševčík J., Both V., Halický P., Zelinka, J. (1981) Study of some properties of ribonuclease SA from streptomyces aureofaciens by physical methods *Biológia*, **36**, pp. 235–239
8. Sayre D. (1974) Least squares phase refinement *Acta Cryst*, **A30**, pp. 180–184
9. Vellieux F. M. D. (1998) A comparison of two algorithms for electron density map improvement by introduction of atomicity: skeletonization, and map sorting followed

by refinement. Direct Methods for Solving Macromolecular Structures: A NATO ASI Series (Springer-Verlag), pp. 519

10. Wang B. C. (1985) Resolution of phase ambiguity in macromolecular crystallography *Methods in Enzymology*, **115**, pp. 90–112
11. Weeks C. M., Hauptman H. A., Chang C. S., Blessing R. H. Teeter M. M., Miller R. (1995) Crambin - a direct solution for a 400-atom structure *Acta Cryst*, **D51**, pp. 33–38
12. Zhang K. Y. J., Main P. (1990) The use of Sayre's equation with solvent flattening and histogram matching for phase extension and refinement of protein structures *Acta Cryst*, **A46**, pp. 377–381

MOLECULAR SCENE ANALYSIS AND ITS ROLE IN PROTEIN STRUCTURE DETERMINATION

Suzanne Fortier*, & and Janice Glasgow &

Depts. of Chemistry and of Computing and Information Science &*
Queen's University, Kingston, Canada K7L 3N6

Abstract

The paper presents an introduction to scene analysis, an area of machine vision concerned with the set of processes associated with the reconstruction, classification and interpretation of complex images. Approaches designed to extend scene analysis to the molecular domain and, in particular, to the analysis and interpretation of protein electron density maps are described. Examples are given of applications to protein maps at medium and low resolution.

1. Introduction

Scene analysis is an activity we engage in almost constantly and, indeed, one in which we excel. From the two-dimensional images perceived by our visual system, we automatically infer three-dimensional scenes and naturally focus our attention at important locations so as to recognize key features within the scene. We are able to put this information together, discern relationships among salient features and gain a holistic understanding of the scene before us. Furthermore, from our analysis of a scene at a given point in time we are often able to infer past and future scenes and construct a plausible scenario of temporal events related to the scene.

Crystal structure analysis, although highly specialised, can be regarded as an exercise in scene analysis. Three-dimensional images are constructed from data "perceived" through the diffraction experiment. An analysis of these images allows us to segment the unit cell into regions, such as protein and solvent regions, and focus on the protein region so as to recognize characteristic structural motifs. Collecting local information and putting all of the pieces together leads us to an understanding of the molecular scene before us, of the conformation of individual molecules and of the important interactions among different molecules. This, in turn, allows us to infer scene events, either past of future, such as molecular recognition and enzyme mechanisms.

Why should one discuss scene analysis in the context of a School on direct methods? The topic may seem out of place since traditionally direct methods have been characterized by their capacity to reconstruct crystal structures with a minimum of structural information and as little structural analysis as possible. Until recently, however, the direct methods approach was limited to small molecule applications and

S. Fortier (ed.), Direct Methods for Solving Macromolecular Structures, 339–359.

few believed that it could ever be extended to macromolecules. Major breakthroughs, made most notably by Hauptman's group[1] and by Sheldrick[2] in the *ab initio* determination of several proteins, have now clearly demonstrated the applicability of direct methods to protein structures. They have also shown the importance of introducing structural information, even at a very coarse level, in the phasing procedure. Direct methods procedures intended for macromolecular applications are therefore usually updated to include some form of map-interpretation and density-modification algorithms in the phasing cycles, allowing frequent feedback between structural and statistical constraints. Thus, as in day-to-day examples, the image reconstruction of a crystal structure as the application of a series of independent sequential steps is only possible in simple cases. For more complex ones, feedback of information from one process to the other is required, making the exercise dynamic, iterative and analogous to scene analysis.

This paper is organized into two sections. The first one provides an introduction to the basic steps of scene analysis, focusing primarily on those components that are pertinent to crystal structure determination and interpretation. The second section will describe a methodology for automated molecular scene analysis and provide examples of applications at low- and medium- level resolution.

2. Scene Analysis: towards Automated Image Interpretation

The concept of scene analysis[3] has emerged from the area of machine vision and refers to the set of processes associated with the reconstruction, classification and interpretation of complex images. Its aim is to provide a computer implementation of the human information processing skills used to translate raw visual data into a fully interpreted scene. Scene analysis is usually broken down into three levels: low, intermediate and high.

2.1 LOW-LEVEL ANALYSIS

At its lowest level, scene analysis centres on local image properties and is concerned with image acquisition, image preprocessing and the detection of local features. One of the first tasks required for machine vision is the transformation of data from the form in which they were originally perceived into one suitable for computation, such as a discrete digital representation model. This model is used, for instance, for the representation of electron density maps, which are usually depicted in the form of a three-dimensional array of intensity values. Another important early task is to clean up the image, that is reduce its noise level, discard unnecessary details and enhance features of interest. Various techniques are used for these purposes, such as thresholding, smoothing, local averaging, multi-copy averaging, background removal and histogram matching. They encompass the density modification techniques commonly used in protein structure determination. Analysis of intensity gradients at the local level is then used to detect local homogeneity and discontinuity which will be used later on to partition the image into regions. As well, the image is examined for the detection of other characteristic local features that can serve as anchors or markers in pattern matching and motif labelling. Finally, in many scene analysis exercises, an important low-level process is the detection of local shading and texture geometry, which is crucial for the derivation of a three-dimensional scene from two-dimensional images.

2.2 INTERMEDIATE-LEVEL ANALYSIS

At its intermediate level, scene analysis is primarily concerned with image segmentation and pattern recognition. These processes address two of the most fundamental questions of image analysis, that is what is present in a scene and where it is. For example in the case of protein electron density maps, at low resolution, i.e. > 5 Å, one may want only to segment the map into two regions and to classify these regions as protein or solvent. At medium resolution, i.e. around 3 Å, one may want to attend to the protein region, segmenting it into main and side chains and classifying main-chain segments in terms of secondary-structure motifs. Segmentation of the main chain into individual residues, followed by their classification into residue classes, and finally their identification can then proceed as resolution increases.

There exists a whole array of techniques designed for image segmentation and different paradigms on which these techniques are based. We note, in particular, techniques based on space clustering and graph constructions, edge linking and skeletonization and region growing. None of the techniques is universally successful and in fact few of the techniques have achieved the desirable level of success in any given domain. One of the main problems lies with the basic notion that an image can be neatly partitioned into separate objects or features of interest. This is rarely the case since most images are perceived in two dimensions and are affected by various factors such as illumination, shading, view angle and movement. This results in numerous problems such as occlusion, overlap and general ambiguity, breaking down the one-to-one relationship between distinct object/feature and clear spatial definition.

In general, pattern recognition seeks to establish similarity and/or equivalence between an unidentified pattern and one previously identified and retrievable from long-term memory. Because it is a very active area of research, it cannot be summarized in a succinct way. Also, many of the approaches and algorithms have been developed with specific applications in mind and would be of little use in the crystallographic domain. We will limit our discussion, therefore, to recognition of images of structured objects, that is objects that are composed of several parts with defined relations among the parts. In such cases, establishing similarity or equivalence between two objects can be achieved through comparing their parts' attributes. To properly capture the structural properties of the object, the process must include comparison of the spatial relationships among the parts. It is also important that the recognition process be applicable to patterns at different granularity level so as to capture both coarse and fine shape properties. Following recognition, it may be useful to position the newly identified object into the broader class of similar objects. If a structural theory already exists for the domain of application, then this theory can be expressed, for example, in the form of rules which are used to establish class membership. In the absence of a theory, one can use various clustering approaches[4] to not only group similar objects together but to also provide a description of the class in the form of a concept that subsumes or represents all members of the class. Clearly the concept of similarity[5] is central to pattern recognition and classification. While the operational definition of similarity may vary from application to application, shape and spatial relationships are recognized as salient components of object/image similarity. One of their key characteristics is their invariance with respect to rotation, translation and scale.

2.3 HIGH-LEVEL ANALYSIS

The aim of high-level scene analysis is to provide a full symbolic description of the scene under consideration. It implies that information gained in the previous lower levels as well as global rules and constraints can be combined together to achieve a holistic view and understanding of the scene. This is clearly an extremely difficult task and one that is still rarely achieved computationally. One of the main hurdles lies with the vast amount of knowledge that is needed to interpret even a fairly simple scene. This knowledge would not only have to be transferred to the computer but also appropriately structured and indexed for rapid access and effective use. In addition, this knowledge has to be appropriately represented so as to capture salient features of the domain under consideration. Knowledge has traditionally been represented using linguistics or rule-based schema. These may not have the desirable semantics, however, for representing images whose main features are best described using a visual and spatial vocabulary rather than a descriptive one.

2.4 CHALLENGES IN SCENE ANALYSIS

Despite a high level of research activity, machine vision is still in its infancy with several major challenges to surmount. One may therefore ask what its contribution could be to an area like protein crystal structure determination, which by contrast is a mature research area with well established and thoroughly tested methodologies. Two comments are warranted. First, many of the current challenges of machine vision are related to problems that either do not exist or are substantially alleviated for molecular scenes. One of the biggest problems in machine vision remains the transition from two to three dimensions. This is clearly not an issue in the crystallographic domain. Also, many images belong to under constrained and general domains. By contrast, the molecular domain has well established rules and constraints and very rich databases of possible structural motifs. In fact, the molecular domain, while still highly challenging, is a good test case for machine vision techniques. Second, there now exists an extensive amount of information on crystal and molecular structures that should be fully exploited in further crystal structure determination. Scene analysis provides a framework for integrating such knowledge into current phasing techniques.

3. Molecular Scene Analysis: Towards Automated Protein Map Interpretation

The interpretation of electron density maps is an important step in protein crystal structure determination, regardless of the phasing approach used. Indeed, this step is normally repeated several times in the course of a protein image reconstruction, when maps of varying degrees of accuracy and resolution are analysed. The task of interpreting protein maps has been greatly facilitated by the advent of powerful graphics stations coupled with the implementation of computer programs, notably FRODO[6] and O[7], designed specifically for the visual analysis of protein electron density maps. Nevertheless, it still relies on significant human intervention and requires considerable time investments. Direct methods procedures usually involve the iterative exploration of several hundreds of possible phasing sets. Integrating map interpretation into such procedures requires, therefore, fully automated and highly efficient algorithms.

Throughout this section, examples will be given using 3 and 5 Å maps generated from either calculated or experimental data. The test structures used are described on Table I. For the calculated maps, structure factors were calculated using protein and solvent atomic coordinates and thermal parameters as stored in the Protein Data Bank (PDB)[8]. From these calculated structure factors, 3 Å resolution electron-density maps were generated. The experimental map of penicillopepsin[9] was generated using all reflections in the resolution range of 10 to 3 Å. Observed structure factor amplitudes and multiple isomorphous replacement phases with their figures of merit were used in the calculation of the map. The phases were derived from eight heavy-atom derivatives and had an overall figure of merit of 0.9 for all observed data to 2.8Å resolution. The experimental map of antifreeze protein (AFP)[10] was generated using all reflections in the resolution range 19.6 to 3 Å. Observed structure factor amplitudes and multiple isomorphous replacement phases with their figures of merit were used in the calculation of the map (using the CCP4[11] package). The phases were derived from three heavy-atom derivatives and had an overall figure of merit of 0.71 for all data observed data to 2.5 Å.

3.1 LOW-LEVEL MOLECULAR SCENE ANALYSIS

As described before, low-level scene analysis is concerned primarily with image acquisition and preprocessing and with the detection of local features that can help in the performance of tasks associated with higher level analysis. Images, in crystal structure determination, are obtained from the transformation of diffraction data into electron density maps. For proteins, the main hurdle in this process remains the solution of the phase problem, which is the topic of this book. A whole array of techniques have been developed for preprocessing electron density maps and are described as well in this book[12]. Here, we focus on the detection of local features in electron density maps.

Analysing intensity gradients is a commonly used approach for detecting local homogeneity and discontinuity in images. For example, small molecule crystallography uses peak picking algorithms for detecting atomic features in electron density maps. A more complete description of the local features of a map can be obtained through locating and characterizing the critical points of the electron density function, $\rho(\mathbf{r})$, thereby providing a full representation of its topological properties. This approach is well known among the theoretical chemistry and accurate electron density distribution research communities[13], where it is applied to high resolution theoretical and experimental maps of small molecules. It has not been used extensively yet in mainstream crystallography despite the fact that its potential usefulness for lower resolution maps of proteins was recognized early on by Johnson[14], who implemented it in the computer program ORCRIT[15]. More recently, a study has been undertaken in our group to evaluate its usefulness for automated map interpretation approaches[16]. As well, the core tracing method, which also draws from the topological properties of electron-density maps, has been proposed by Swanson[17].

Choosing a good approach for local feature detection is important to the higher level scene analysis processes and a key determinant to their effective implementations, as one of its outcomes is the transformation of the initial image representation into one better suited for higher level scene analysis tasks. This latter representation should be able to capture and make explicit those properties of the electron density distribution that help us most in performing the tasks of segmentation, pattern recognitions, classification and global image analysis. These tasks draw primarily from an analysis of shape properties and spatial relationships in the map. Indeed, it is from the local and global topology of the map that structural features such as protein and solvent regions, main chains, helices, individual residues, etc. can be located, segmented and identified and the full image interpreted. While shape and spatial relationship information is contained implicitly in the three-dimensional grid representation of electron-density maps, this is not a form that is suitable for efficient computation. Alternative representations that synthesize and make explicit selected shape and spatial properties of the electron-density distribution have thus been investigated and used. For example, the skeletonization approach introduced by Hildith[18], and adapted to protein crystallography by Greer[19], has been used extensively to capture connectivity in protein maps.

It is important to appreciate that a good representation model should not only capture the features of interest but also discard unnecessary and distracting details. The critical-point mapping approach, for instance, summarizes the shape and spatial properties of the electron-density distribution through its critical points while discarding unnecessary detailed background information. A good test for a representation model is to check it against the criteria established in the context of machine vision by Marr & Nishihara[20] for general shape representation:

- Accessibility - the representation should be derivable from the initial image at a reasonable computing cost
- Scope and uniqueness - the representation should provide a unique description of all possible shapes in the domain of interest
- Stability and sensitivity - the representation should capture both the general (less variant) and the subtle (of finer distinction) properties of shape.

Table I Chemical and structural parameters of the test structures

Name, code	Phospholipase A2 (bovine pancreas), 1BP2
resolutions (Å), R-factor	1.7, 0.170
cell parameters (Å and degrees)	47.070 64.450 38.158 90 90 90
space group	$P2_12_12_1$
solvent/heteroatoms	109 H_2O, $2C_6H_{14}O_2Ca^{2+}$
number of amino acids, disulfide bonds	123, 7
% α, % β, number of turns	54, 8, 10
Name, code	Ribonuclease T1 complex (aspergillus Oryzae), 1RNT
resolution (Å, R-Factor	1.9, 0.180
cell parameters (Å and degrees)	46.810 50.110 40.440 90 90 90
space group	$P2_12_12_1$
solvent/heteroatoms	91 H_2O, $C_{10}H_{14}N_5O_8P$
number of amino acids, disulfide bonds	104, 2
% α, % β, number of turns	16, 35, 10
Name, code	Trypsin inhibitor (bovine pancreas), 4PTI
resolution (Å), R-factor	1.5, 0.162
cell parameters (Å) and degrees)	43.100 22.900 48.600 90 90 90
space group	$P2_12_12_1$
solvent/heteroatoms	60 H_2O
number of amino acids, disulfide bonds	58, 3
% α, % β, number of turns	17, 33, 3
Name, code	Penicillopepsin (fungus) 3APP
resolution (Å), R-factor	1.8, 0.136
cell parameters (Å and degrees)	97.370 46.640 65.470 90 115.4 90
space group	C2
solvent/heteroatoms	320 H_2O
number of amino acids, disulfide bonds	323, 1
% α, % β, number of turns	15, 52, 2
Name, code	Recombiant type III Antifreeze Protein (Marcrozoarces americanus), 1MSI
resolution (Å), R-factor	1.25, 0.198
cell parameters (Å and degrees)	32.330 38.980 45.480 90 90 90
space group	$P2_12_12_1$
solvent/heteroatoms	90 H_2O
number of amino acids, disulfide bonds	66, 0
% α, % β, number of turns	6, 0, 2

3.1.1 *Locating critical points*

The critical points of the three-dimensional electron density distribution, $\rho(\mathbf{r})$, are its points of inflection. They are found at locations for which the *first derivatives of* $\rho(\mathbf{r})$ *vanish* and are characterized as maxima, minima or saddle points by the second derivatives of $\rho(\mathbf{r})$. Critical-point analysis thus requires the computation of the gradients (∇), Hessians (**H**), and Laplacians (∇^2) of the electron density function[13]. The Hessian matrix of a continuous three-dimensional function such as the electron density, $\rho(\mathbf{r})$, is built from its second derivatives:

$$\mathbf{H}(\mathbf{r}) = \begin{vmatrix} \partial^2\rho/\partial x^2 & \partial^2\rho/\partial x\partial y & \partial^2\rho/\partial x\partial z \\ \partial^2\rho/\partial y\partial x & \partial^2\rho/\partial y^2 & \partial^2\rho/\partial y\partial z \\ \partial^2\rho/\partial z\partial x & \partial^2\rho/\partial z\partial y & \partial^2\rho/\partial z^2 \end{vmatrix} \qquad (1)$$

For each critical point, this real and symmetric matrix can be put in a diagonal form, which corresponds to finding a rotation of the original coordinate system that aligns the new coordinate system with the principal axes of the critical points, resulting in :

$$\mathbf{H}'(\mathbf{r}) = \begin{vmatrix} \partial^2\rho/\partial x'^2 & 0 & 0 \\ 0 & \partial^2\rho/\partial y'^2 & 0 \\ 0 & 0 & \partial^2\rho/\partial z'^2 \end{vmatrix} . \qquad (2)$$

The three diagonal elements of **H'**(**r**) are the eigenvalues of **H**. They correspond to the three main curvatures of the electron density function at the critical point **r** while the eigenvectors, resulting from the diagonalization of **H**(**r**), correspond to the three principal axes of curvature. The Laplacian is the trace of **H'**(**r**), that is the sum of the diagonal elements of the Hessian matrix. It gives details about the local concentration (sign < 0) or depletion (sign > 0) of electron density. The rank of **H'** is the number of non-zero eigenvalues, and the signature, *s*, is the algebraic sum of their sign. When the rank of **H'** is 3, as is normally the case for critical points in three-dimensional electron-density distributions, then four possible cases arise:

- $s = -3$ corresponds to a local maximum or *peak*, i.e. the electron-density function adopts maximum values along each of the three principal directions x', y' and z';
- $s = -1$ corresponds to a saddle point or *pass* and there are only two negative eigenvalues;
- $s = +3$ corresponds to a local minimum or *pit*, i.e. the electron-density function adopts minimum values along each of the three principal directions; and
- $s = +1$ corresponds to a saddle point or *pale* characterized by only one negative eigenvalue.

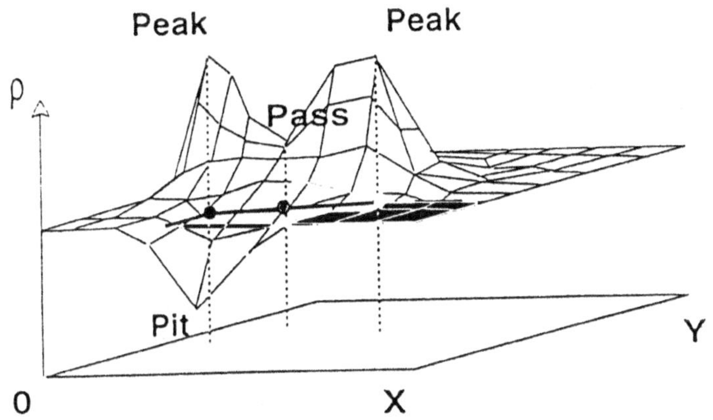

Figure 1 Critical points in an electron density distribution

Peaks, pits and saddle points are depicted in Figure 1. Thus, from computing the gradient, Hessian and Laplacian of the electron-density distribution, one obtains information not only on the location and types of critical points in the function, but also on their local environment. In particular, retaining information on their heights, curvatures, and principal directions of curvature allows a reconstruction of local electron-density distributions and, in turn, of the global distribution. This is important in macromolecular direct methods procedures where the purpose of map interpretation is to achieve density modification and in turn phase refinement through the Fourier transform of direct-space information into reciprocal space.

Locating critical points should correspond ideally to locating important chemical features within the protein. For convenience, protein backbones are commonly represented by a C_α tracing and comparison of models to structures is often done using C_αs. It is important to note, however, that at low and medium resolution, C_αs do not correspond to "important" chemical features, that is they do not correspond to peaks in the electron density function. Indeed, calculations on test proteins have shown that at 3 Å resolution, the peak critical points are located closer to the centre of mass of C_αCON groups than to the C_αs themselves. For the calculated maps, as might be expected, it is found that 100% of the residues are captured by at least one peak critical point. The corresponding figure for both the penicillopepsin and antifreeze protein experimental maps is 98%. Heteroatoms and sulfurs are easily detected among the set of critical points due to their higher peak heights and can therefore serve as anchors in higher level scene analysis.

3.2 MEDIUM-LEVEL MOLECULAR SCENE ANALYSIS

The goal at this level of scene analysis is to provide a local interpretation of the image through its segmentation into separate entities which can then be classified and identified. Although segmentation and pattern recognition are often presented as two separate tasks, it is important to realize that often they are not independent of one another or applied sequentially. For example, analysis of the spatial relationships among various segmented parts may be essential to their identification. Similarly, partial identification can help guide further segmentation.

3.2.1 Segmentation of electron density maps

In his early work, Johnson[14] found that connecting individual critical points through the construction of graphs was an effective way of segmenting the electron density map. From his applications to calculated maps at around 2 Å resolution, he observed that the minimal spanning trees constructed from the collection of critical points showed ...peak-pass-peak-pass-peak.... sequences of adjacent points, which traced out ridge lines following the polymeric protein molecule. Similarly, ...pit-pale-pit-pale-pit...sequences should, in theory, be found to trace out valley lines separating adjacent molecules. We have also used graph construction as a segmentation method. A few terms from graph theory[21], needed to understand the remaining discussion on segmentation, are introduced at this point.

A *graph* is a representation of a set of points and their connectivity. In particular, a *connected graph* consists of a set of points (vertices) in which any pair may be connected by a path formed by one or several edges in the graph. A connected graph containing no *circuit* or *cycle* is called a *tree* and a graph whose components are trees, a *forest.* A subgraph of a connected graph G that is a tree containing all the vertices of G is called a *spanning tree.* A spanning tree can thus be obtained from a connected graph G by removing edges from G in such a way that all circuits are eliminated while all vertices remain connected. In general, a graph may have several possible spanning trees. A *minimum-weight spanning tree* is obtained when graph edges are assigned weights and the edge-removal algorithm minimizes the sum of the edge weights. For example, in ORCRIT minimum spanning trees are obtained through the use of weights derived from attributes of adjacent critical points, such as distances, density heights, types of critical points, and eigenvector projections.

In our applications to calculated maps at 3 Å, because a cut-off ρ_{min} value was used, only peak and pass critical points were considered in the construction of the minimum weight spanning trees. Pits and pales normally have much smaller signal-to-noise ratios and are generally below the cut-off value. We also observed that adjacent critical points tended to outline ...peak-pass-peak-pass-peak...motifs. Furthermore, each test structure had a principal spanning tree with a long ...peak-pass-peak-pass-peak... "branch", with smaller branches jutting out of it, corresponding to the protein main chain and side chains respectively. This pattern of critical points is depicted in Figure 2.

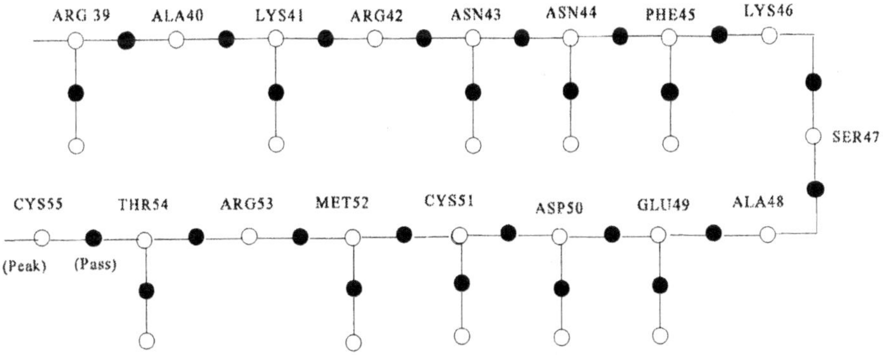

Figure 2 Planar schematization of a typical critical point motif derived from a 3Å resolution protein map

For the calculated maps of 1BP2, 1RNT and 4PTI, we found that the main branch of the longest spanning tree captures the whole amino acid sequence. Of particular importance is the observation that successive peaks in the main branch of the longest spanning tree correspond to successive residues in the sequence, and that, in general, each residue is associated with one and only one peak in the main branch of the tree. Thus at 3 Å, the electron density distribution of the protein along its main chain can be represented as a sequence of *super-atoms*, which correspond to the residues. These results are summarized in Table II. It also was found that the critical points capture accurately the conformation of the protein, with the proviso that (as mentioned before) they are located close to the centre of mass of C_αCON rather than to the C_α atoms. Figure 3 shows a superposition of a tracing through the C_αs and a tracing through the peaks in the main branch of the principal tree for 1RNT.

Table II Location and connectivity of critical points for the calculated electron density maps of 1BP2, 1RNT and 4PTI at 3 Å resolution

	1BP2	1RNT	4PTI
No. of critical points	532	365	196
Total No. of spanning trees	25	19	10
Total No. of peaks in the main branch of the longest spanning tree	134	109	60
No. of residues represented in the main branch of the longest spanning tree	123	104	58
No. of residues in the main branch of the longest spanning tree represented by one and only one peak	114	99	56

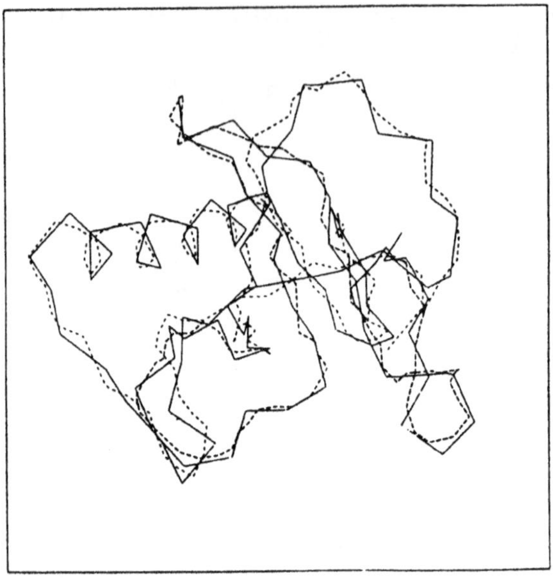

Figure 3 Perspective view depicting the superposition of the C (dark) chain and the peaks of the main branch of the longest spanning tree (light) for the protein 1RNT

Figure 4 and Table III show results from calculated and experimental maps of penicillopepsin. For the calculated map, it is found that a single tree connects 321 or 99% of the residues, For the experimental map, the chain is broken down into seven trees, which trace significant portions of the protein main chain with 278 or 86% of the residues represented in their main branches. However, we found that only 55% of the residues were captured by one and only one peak along the main branches of the trees. This is due to the fact that high density peaks are often surrounded by a constellation of lower density ones. Better results are obtained when the set of critical points is subjected to a pruning procedure before one attempts to connect them. In addition, the graph-generation algorithm did not yield a single spanning tree capturing the whole protein. Rather, the backbone of the protein mapped onto several trees of varying lengths. Improvement in the connectivity results was achieved by enforcing a peak-pass-peak motif, that is by enforcing the presence of a pass between two connected peaks. The construction of the tree(s) was carried out from the collection of peak-pass-peak building blocks, rather than the collection of single critical points.

```
Seq  1-50     A-A-S-G-V-A-T-N-T-P-  T-A-N-D-E-E-Y-I-T-P-  V-T-I-G-G-T-T-L-N-L-  N-F-D-T-G-S-A-D-L-W-  V-F-S-T-E-L-P-A-S-Q-
Theor         Tree01.......................01 • Tree01.......01 Tree01..............................
Exper         ••••• Tree01.........01 • ••••• • • ••• •  Tree02........................02 ••• Tree02.     .02 Tree02

Seq  51-100   Q-S-G-H-S-V-Y-N-P-S-  A-T-G-K-E-L-S-G-Y-T-  W-S-I-S-Y-G-D-G-S-S-  A-S-G-N-V-F-T-D-S-V-  T-V-G-G-V-T-A-H-G-Q-
Theor          ...................................................  ...  ..... ...tree01 Tree01
Exper           .02 Tree02............................... .. .02 • ••••••• Tree02 .. .. .              •

Seq  101-150  A-V-Q-A-A-Q-Q-I-S-A-  Q-F-Q-Q-D-T-N-N-D-G-  L-L-G-L-A-F-S-S-I-N-  T-V-Q-P-Q-S-Q-T-T-F-  F-D-T-V-K-S-S-L-A-Q-
Theor              . . .01 Tre01................01 Tree01    .         . 01 Tree01.. .
Exper           02 Tree.02 Tree02••••• • • ••• Tree02.  ...             02   Tree02. ..                       02 Tree03

Seq  151-200  P-L-F-A-V-A-L-K-H-Q-  Q-P-G-V-Y-D-F-G-F-I-  D-S-S-K-Y-T-G-S-L-T-  Y-T-G-V-D-N-S-Q-G-F-  W-S-F-N-V-D-S-Y-T-A-
Theor                   ............................ ... . . .            . . . . .             01 Tre01 Tree01
Exper                    ............. .. . .... .   . ..              .  ....03 Tree02

Seq  201-250  G-S-Q-S-G-D-G-F-S-G-  I-A-D-T-G-T-T-L-L-L-  L-D-D-S-V-V-S-Q-Y-Y-  S-Q-V-S-G-A-Q-Q-D-S-  N-A-G-G-Y-V-F-D-C-S-
Theor          .   . .  ...................01 Tree01 .. .   .     ...  .    . .01 Tree01
Exper          . ..  ....  ......................02 Tree04...  ... . ... 04 T04  Tree05.... .     05   Tree  06 Tree 07

Seq. 251-300  T-N-L-P-D-F-S-V-S-I-  S-G-Y-T-A-T-V-P-G-S-  L-I-N-Y-G-P-S-G-D-G-  S-T-G-L-G-G-I-Q-S-N-  S-G-I-G-F-S-I-F-G-D-
Theor          .............01 Tree01................ ..... .. .  01 Tree01 ....... ...... ..
Exper          Tree04.. 04  Tree02..............02      •••••••  •••     •   • • Tre07 Tree04

Seq. 301-323  I-F-L-K-S-Q-Y-V-V-F-  D-S-D-G-P-Q-L-G-F-A-  P-Q-A
Theor          .   .  . . ..............................  . 01
Exper           04 Tree02  .  ..................  .....  ..  02
```

Figure 4 Critical point network in experimental and calculated maps of 3APP

Table III Location and connectivity of critical points for the calculated and experimental electron density maps of 3APP at 3 Å resolution

	3APP(calc'd)	3APP(expt't)
No. of critical points	1576	1544
Total No. of spanning trees	37	223
No. of spanning trees considered as principal*	1	7
Total No. of peaks in the main branch of the principal spanning tree(s)	353	325
No. of residues represented in the main branch of the principal spanning tree(s)	321	278
No. of residues in the main branch of the principal spanning tree(s) represented by one and only one peak	267	176

Figure 5 shows the tree construction results for the antifreeze protein. In this case 60 (or 91%) of the residues are represented in the trees. We see, however, that the tracing of the chain is broken down into 10 trees. Going back to the electron density map, it is clear that these breaks are due to electron density voids in between the residues. Without further intelligence, the tree construction algorithm is unable to fill in the gaps and make the connection.

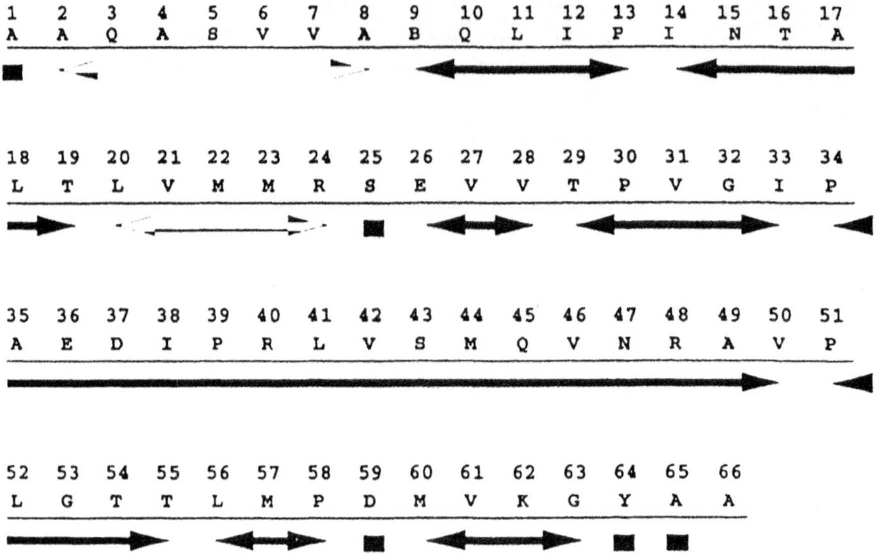

Figure 5 Critical point networks in the experimental map of antifreeze protein

At low resolution, as expected, fewer critical points are found and connectivity among sequential residues is found only in the case of helices. Connectivity of the β-sheet peaks no longer follows the chain direction. These results correspond to the observations reported by Lesk[22]. In 5 Å calculated maps, it is found that solvent regions are characterized by the presence of low density critical points, primarily passes, pales and pits while the protein regions are populated with higher density points. The topological features thus can be used for envelope definition.

Test calculations with the protein structure 1RNT at 3 Å and 5 Å suggest the presence of a hierarchy between the peaks and passes at 5 Å and those at 3 Å. As illustrated in Figure 6, individual critical points at 5 Å are found to be associated with individual or multiple points at medium resolution.

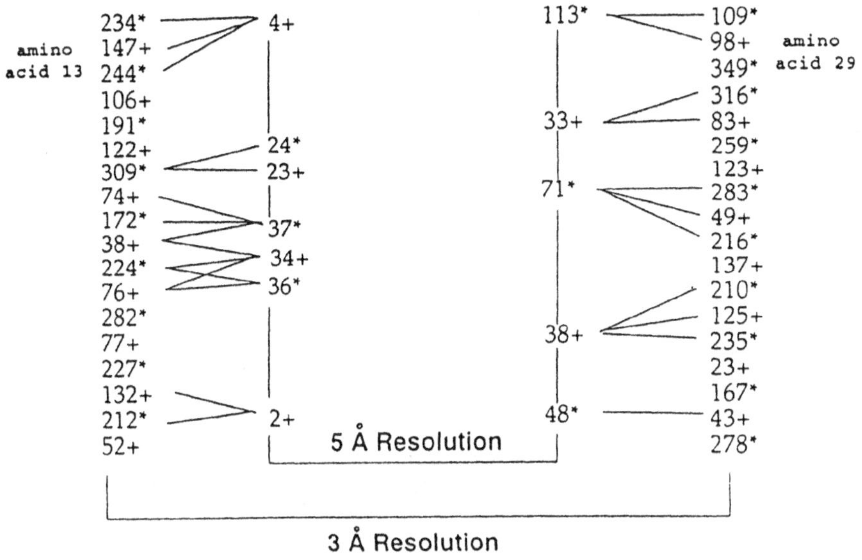

Figure 6 Relationship between critical points at 3 and 5 Å for 1RNT

3.2.2 *Recognizing secondary-structure motifs in critical point segments*

Once a critical-point graph is constructed, it can be analysed so as to classify substructures in the protein. In particular, a geometrical analysis of segments of connected critical points can be used for the recognition of secondary-structure motifs. In general, torsion angles among four sequential peak critical points and distances between the peaks at the extremities of the four-member segment are the most useful parameters for the identification of β sheets and α helices. The length of a secondary-structure segment and the existence of parallel segments with similar geometry are also useful criteria for motif recognition. Specific rules or templates to be used in recognition procedures are usually resolution dependent, since the critical points catch different atomic moieties at different resolution levels .

Since the geometrical parameters may have value ranges that match more than one type of secondary structure and since different parameters may yield conflicting results for a given segment, a method has to be used to establish a *confidence level* in the results obtained. One possible approach is through the calculation of *certainty factors* , as implemented in the diagnosis system MYCIN[23]. For the purpose of recognition, it is useful to have available for each parameter of interest frequency distributions, *f(ssm,g)* and *f(not_ssm,g)*, given a secondary-structure motif *ssm* and a geometric constraint *g*. These were computed, using 63 non-homologous protein structures retrieved from the PDB, for α helices and β sheets and for geometric constraints based on the torsion angle t_{1234}, distance d_{14} and angle a_{123}. The frequency distributions can be used to calculate measures of belief based on *f(ssm,g)* and disbelief based on *f(not_ssmn,g)*, in the assignment of a given secondary- structure motif to critical points. These in turn are used to calculate certainty factors. ranging from -100 to + 100. Figure 7 maps certainty factors for both α-helix and β-sheet hypotheses onto the sequence, as obtained for a calculated map of 4PTI. As can be seen, the approach correctly identifies helices and sheets in the structure. It should be noted that the high certainty factor associated with a helical motif in a turn region is not unexpected. Both motifs have similar geometry and are differentiated by their length.

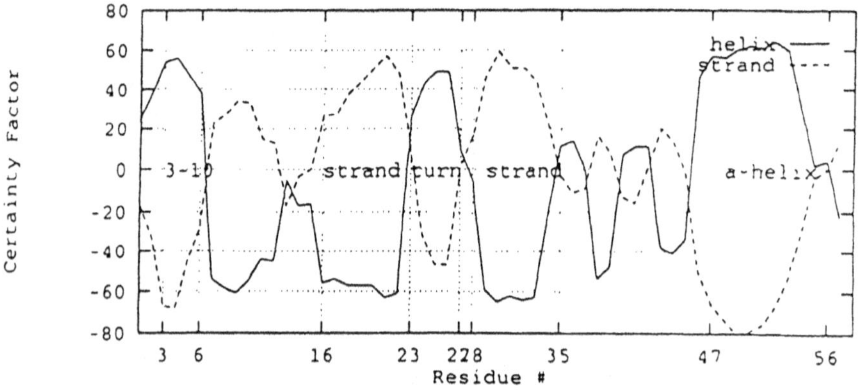

Figure 7 Certainty factors for helix and sheet motifs in 4PTI at 3 Å

A Bayesian approach has also been applied to the identification of secondary structure in critical point graphs[24]. As with the MYCIN approach, this probabilistic method was relatively successful in classifying sequences of critical points that were geometrically well resolved. For experiments that involved noisy experimental data, the accuracy of both methods decreased.

3.2.3 *Mapping Sequence Onto Structure*

Information derived from the low level analysis of the protein image can also be used to help classify peak critical points in terms of amino acid residues[25]. For each peak along the backbone of the protein we can derive an environment which characterizes properties of the critical point. An environment contains information such as peak height, distance to solvent, participation in secondary structure and volume. Using the threading approach proposed by Lathrop and Smith[26], we have applied a scoring function that provides a statistical preference of a residue for a given environment. This function is used to find an optimal threading of the protein sequence onto the structure of critical points.

3.3 HIGH-LEVEL MOLECULAR SCENE ANALYSIS

The output of medium-level scene analysis is an image partitioned into separate entities with some of these entities having been classified or identified. To achieve the aim of scene analysis, these local interpretations must be reconciled into a holistic interpretation of the image and, ultimately, an understanding of the information conveyed by the image Reconstructing the full image from its parts involves applying the rules and constraints of the domain as well as assessing consistency with previously analysed similar scenes. In doing so, different modes of reasoning may be invoked, for example case-based reasoning and induction. What is clear is that high-level scene analysis requires intelligence and relies on the ability to retrieve pertinent knowledge and combine it with current information about the scene under analysis.

A crucial step in the attempt to extract meaningful conclusions from a body of data is the process of information organization or knowledge representation. How effectively we answer questions of interest depend, in a large part, on how well the knowledge representation scheme can make explicit salient features of the information. This is clearly illustrated by the different representation models routinely used in protein crystallography to extract meaningful information about the protein under study. The results of the structure determination process are commonly stored using a descriptive numerical model to represent atomic identification, coordinates and thermal vibration parameters. While this representation model is ideal for performing numerical calculations, it is not suitable for reasoning about important features of the protein, such as shape and spatial relationships. Indeed, alternative representations are used, for example topological diagrams to depict adjacency, connectively and spatial relationships of secondary structure segments and a variety of visual representations suitable for display on graphics stations.

To address this issue, we have designed a knowledge representation scheme[27] that consists of three interrelated representations: a descriptive representation, which serves as a knowledge base of molecular scenes and two depictive representations, which are used to carry out the visual and spatial analysis of an image. The knowledge base is being implemented as a frame system[28] and is organized according to both the structural and conceptual hierarchies of molecular structures. As shown in Figure 8, the structural hierarchy separates the informaton according to the building block model of moelcular structures while the conceptual hierarchy organizes information according to the classes, sub-classes and instances of the structural building blocks. To capture spatial features, we use a representation model that is based on a spatial description logic (SDL)[29] specifically tailored for structured objects. SDL extends traditional description logics[30] by adding the notion of a symbolic image. This image preserves structural motifs, which may be at different levels of the protein structural hierarchy, and their spatial relationships. The model allows representation of the conceptual hierarchy through a subsumption hierarchy and of the structual hierarchy though embedding. A similarity-based structured concept formation tool (IMEM)[31] has been designed and implemented for classifying SDL images. The visual component of our representation model scheme is based on critical point mapping, which has already been described in section 3.1.1.

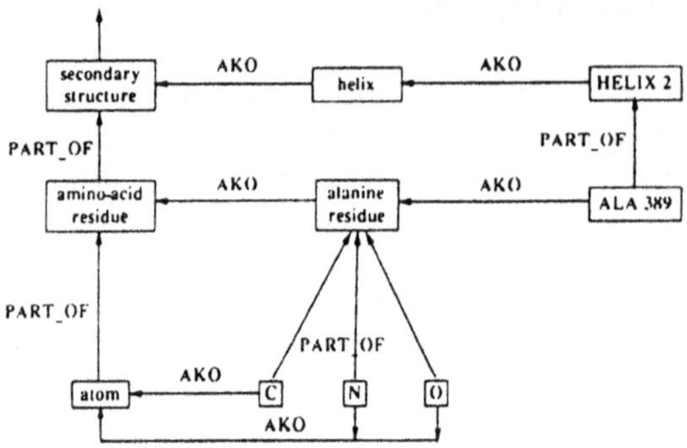

Figure 8 Structural and conceptual hierarchies in the molecular knowledge base

Once the tools for knowledge representation and reasoning have been assembled, strategies must be designed to combine and reconcile the partial interpretations these tools may yield so as to reconstruct the full image and achieve a coherent global interpretation. The approach we are implementing for high-level molecular scene analysis is based on the FORR (FOr the Right Reasons) Architecture)[32]. This architecture is well suited to our own problem because it recognizes and accommodates different types of knowledge as well as different learning techniques. An important component of the FORR architecture is the use of advisors. These advisors have highly specialized and specific knowledge. Their task is to provide advice (through comments) on possible decisions based solely on their narrow view of the problem. The global perspective is then achieved by combining the advisors' comments and reconciling possible conflicting advices. Advisors tailored to the molecular scene domain, for example advisors that comment on disulfide bridge, helix, threading, are currently under constructions as well as the underlying infrastructure needed to combine their input.

Our work on high-level molecular scene analysis is still at a preliminary stage and what was presented in this section were two major issues faced at this level and the approaches we are using to resolve them. Perhaps what emerges most clearly from the discussion is the importance of knowledge and experience. To be effective and efficient, programs for automated map interpretation will require the capacity to learn and improve their performance. The construction of a knowledge base will allow new knowledge to be retained. The FORR architecture, on the other hand, provides the advisors opportunities for training and practice so as to learn and improve their performance.

4. Concluding Remarks

Re-phrasing protein structure determination within a scene analysis framework offers the opportunity to fully exploit the accumulating knowledge on crystal and molecular structures and move to a knowledge-based approach to structure determination. In the process of implementing such an approach several important issues will have to be addressed. In particular, one will have to consider carefully the trade-off between speed and intelligence. When dealing with a broad search space that may contain many local minima, as is often the case at the beginning of a structure determination exercise, one may favour coarser image interpretations for the sake of gaining speed. Adding intelligence may be delayed until the number of possible search paths has decreased to a reasonable level. Also, as in any knowledge-based approach, one will have to strive to achieve the delicate balance set by Barrow and Tenenbaum[33]so as to avoid "oversight" (not seeing things that are really present), while not suffering from "hallucination" (seeing things that are not present at all..

5. Acknowledgements

Financial assistance from the Natural Sciences and Engineering Research Council of Canada in the form of operating and collaborative research grants is gratefully acknowledged.

6. References

1. Miller, R., DeTitta, G.T., Jones, R., Langs, D.A., Weeks, C. M. and Hauptman, H.A. (1993) On the application of the minimal principle to solve unknown structures, *Science* **259**, 1430-1433; Miller, R. and Weeks, C.M. (1998) Shake-and-Bake: Applications and advances, in S. Fortier (ed.), *Direct Methods for Solving Macromolecular Structures*, Kluwer Academic Publishers, Dordecht.

2. Sheldrick, G. (1998) SHELX applications to macromolecules, in S. Fortier (ed.), *Direct Methods for Solving Macromolecular Structures*, Kluwer Academic Publishers, Dordecht.

3. Fischler, M.A. and Firschein, O. (1987) *Intelligence. The Eye, the Brain, and the Computer,* Addison-Wesley, Menlo Park ; Marr, D. (1982) *Vision*, W.H. Freeman and Company, New York; Duda, R.O. and Hart, P.E. (1973) *Pattern Recognition and Scene Analysis*, John Wiley & Sons, New York.

4. Everitt, B.S. (1993) *Cluster Analysis*, 3rd edition, Edward Arnold, London.

5. Dean, P.M. (1995) *Molecular Similarity in Drug Design*, Blackie Academic & Professional, Glasgow.

6. Jones, T.A. (1985) Interactive computer graphics: FRODO, in H.W. Wyckoff, C.H.W. Hirs and S.N. Timasheff (eds.) *Methods of Enzymology*, **115**, Academic Press, New York, pp. 157-171.

7. Jones, T.A., Bergdoll, M. and Kjeldgaard, M. (1990) O: a macromolecule modeling environment, in C.E. Bugg and S.E. Ealick (eds.) *Crystallographic and modelling methods in molecular design,* Springer-Verlag, New York, pp. 189-199.

8. Bernstein, F.C., Koetzle, T.F., Williams, G.J.B., Meyer, E.F.,Jr., Brice, M.D., Rodgers, J.R., Kennard, O., Shimanouchi, T.S. and Tasumi, M. (1977) The protein data bank: a computer-based archival file for macromolecular structures, *J. Mol. Biol.* **112**, 535-542; Abola, E.E., Bernstein, F.C., Bryant, S.H., Koetzle, T.F. and Weng, J. (1987) Protein data bank, in F. H. Allen, G. Bergerhoff, and R. Sievers (eds.), *Crystallographic Databases-Information Content, Software Systems, Scientific Applications*, Data Commission of the International Union of Crystallography, Bonn/Cambridge/Chester, pp. 107-132.

9. James, M.N.G. and Sielecki, A.R. (1983) Structure and refinement of penicillopepsin, *J. Mol. Biol.* **163**, 299-361.

10. Jia, Z., DeLucca, C.I., Chao H. and Davies, P.L. (1996) Structural basis for the binding of a globular antifreeze protein to ice, *Nature* **384,** 285-288.

11. Collaborative Computational Project, Number 4. (1994) The CCP4 suite: programs for protein crystallography, *Acta Cryst.* **D50**, 760-763

12. Cowtan, K. (1998) Introduction to density modification, in S. Fortier (ed.), *Direct Methods for Solving Macromolecular Structures*, Kluwer Academic Publishers, Dordecht.

13. Bader, R.W. (1990) *Atoms in Molecules - A Quantum Theory*, Clarendon Press, Oxford ; Smith, V.H. Jr., Price, P.F. and Absar. I. (1977) Representation of the electron density and its topographical features, *Isr. J. Chem.* **16**, 187- 197.

14. Johnson, C.K. (1976) Interpolation polynomials, minimal spanning trees and ridge-line analysis in density map interpretation, *Proc. Am. Crystallogr. Assoc. Meet.*, Evanston, IL, USA, Abstract B1; Johnson, C.K. (1977) Peaks, passes, pales and pits: a tour through the critical points of interest in density maps, *Proc. Am. Crystallogr. Assoc. Meet.*, Asilomar, CA, USA, Abstract JQ6

15. Johnson, C.K. (1977) *ORCRIT. The Oak Ridge Critical Point Network Program*, Chemistry Division, Oak Ridge National Laboratory, USA.

16. Leherte, L., Fortier, S., Glasgow, J. and Allen, F.H. (1994) Molecular scene analysis: a topological approach to the automated interpretation of protein electron density maps, *Acta Cryst.* **D50**, 155-166.

17. Swanson, S.M. (1994) Core tracing depicting connections between features in electron density, *Acta Cryst.* **D50**, 695-708.

18. Hilditch, C.J. (1969) Linear skeletons from square cupboards, *Mach. Intell.* **4**, 403-420.

19. Greer, J. (1974) Three-dimensional pattern recognition: an approach to automated interpretation of electron density maps of proteins, *J. Mol. Biol.* **82**, 279-301 .

20. Marr, D. and Nishihara, H.K. (1978) Representation and recognition of the spatial organization of three-dimensional shapes, *Proc. R. Soc. London Ser. B.* **200**, 269-294.

21. Wilson, R.J. (1985) *Introduction to Graph Theory*, 3rd edition, Longman Scientific and Technical, New York.

22. Lesk, A.M. (1991) *Protein Architecture*, D. Rickwood and B.D. Hames (eds.), IRL Press, Oxford.

23. Shortliffe, E.H (1976) *Computer-based medical consultations: MYCIN*, Elsevier, New York.

24. Leherte, L., Glasgow, J.I., Fortier, S., Baxter, K. and Steeg, E. (1997) Analysis of three-dimensional protein images, *Journal of AI Research (JAIR)*, in press.

25. Baxter, K., Steeg, E., Lathrop, R., Glasgow, J, and Fortier, S. (1996) From electron density and sequence to structure: integrating protein image analysis and threading for structure determination, in *Proceedings of the 4th International Conference on Intelligent Systems for Molecular Biology*, AAAI/MIT Press, Menlo Park.

26. Lathrop, R. and Smith, T. (1996) Global optimum protein threading with gapped alignment and empirical pair score functions, *Journal of Molecular Biology* **255(4)**, 641-665.

27. Glasgow, J., Fortier, S., Conklin, D., Allen, F.H. and Leherte, L. (1995) Knowledge representation tools for molecular scene analysis, *Proceedings of the 28th Hawaii intl. conf. on system sciences,* Hawaii.

28. Minsky, M. (1975) A framework for representing knowledge, in P.H. Winston (ed.) *The psychology of computer vision*, McGraw Hill, New York, pp. 211-277.

29. Conklin, D. and Glasgow, J. (1992) Spatial analogy and subsumption, in D. Sleeman and P. Edwards (eds.), *Machine Learning: Proceedings of the 9th International Conference ML(92),* Morgan Kaufmann, New York, pp. 111-116.

30. Nebel, B. (1990) *Reasoning and revision in hybrid representation systems*, Springer-Verlag, Berlin.

31. Conklin, D., Fortier, S., Glasgow, J.I. and Allen, F.H. (1996) Conformational analysis form crystallographic data using conceptual clustering, *Acta Cryst.* **B52**, 535-549.

32. Epstein, S.L. (1994) For the right reasons: the FORR architecture for learning in a skill domain, *Cognitive Science* **18,** 479-511.

33. Barrow, H.G. and Tenenbaum, J.M. (1981) Computational vision, Proc. IEEE, **69**, 572-595.

MEMBRANE PROTEINS SOLVED BY ELECTRON MICROSCOPY AND ELECTRON DIFFRACTION

A. HOLZENBURG
School of Biochemistry and Molecular Biology
School of Biology
University of Leeds, Leeds LS2 9JT, UK

1. Prologue

I have been appointed rather late to give a lecture at this school and for this reason, as all members in the audience will have been aware, there were no lecture notes for my contribution. I tried to compensate for this shortcoming by offering a special tutorial, and a small crowd has actually taken me up on this offer. From these and any subsequent discussions I have learnt a lot for myself and the open, friendly and enthusiastic attitude of most of the students made this school a unique experience.

The original title of this lecture was scheduled as "Membrane proteins solved by electron microscopy and electron diffraction". However, since this is a school aimed at disseminating knowledge of practical approaches as well as the theory behind it, it seems more appropriate to view this contribution under the heading "How to solve membrane proteins ...".

S. Fortier (ed.), Direct Methods for Solving Macromolecular Structures, 361–380.

2. Introduction, overview and some clarifying remarks

While soluble proteins, domains thereof or peptides are ideally suited for 3-D crystallisation and X-ray analysis to a resolution at which individual atoms can be visualised, the situation changes abruptly when one deals with membrane proteins which, by their very nature, prefer a 'two-dimensional environment' and are therefore very difficult to crystallise in three dimensions. For this reason viable alternatives are needed and the most obvious alternative is to grow two-dimensional (2-D) crystals (crystals that are only one unit cell thick). These 2-D crystals can be readily analysed in the electron microscope and are the prerequisite for electron crystallographic studies. At this point we should differentiate between the two main classes of membrane proteins: integral membrane proteins and membrane-associated proteins. Both require a '2-D matrix' in order to fulfill their functional role in a physiological setting, and this matrix is the lipid bilayer. Integral membrane proteins will want to span the entire membrane in order to be active while for membrane-associated proteins the contact with the lipid surface is crucial. As with 3-D crystals, it is the quality of the 2-D crystals that ultimately determines the resolution and several 2-D crystallisation strategies have been developed. For integral membrane proteins, a possible way forward is via reconstitution into 2-D protein (and lipid) crystals: The idea is to mix detergent-protein and detergent-lipid mixed micelles at a suitable lipid-protein and lipid-detergent ratio, pH, ionic milieu and temperature and then deplete the resulting ternary micelles (detergent-lipid-protein) of the detergent component (e.g. by dialysis) in order to induce the formation of 2-D crystals of the protein, ideally within lipid sheets rather than vesicles [1]. Other approaches to 2-D crystals of integral membrane proteins include in situ methods, batch methods etc. and they have been thoroughly reviewed in the literature [2]. The 2-D crystallisation of membrane-associated proteins can be very effectively carried out on planar lipid films [3]: Take a small trough filled with buffer containing the protein to be crystallised and add a dilution lipid and a ligand lipid (a lipid to which the protein to be crystallised has a very high affinity) dissolved in e.g. chloroform/hexane. The lipids will then orientate themselves at the air-water interface, the protein will recognise the ligand lipid, form complexes with it and crystals will eventually form via lateral diffusion and self-organisation of the lipid-protein complexes.

For electron microscopical observation, 2-D crystalline samples can either be embedded in negative stain (e.g. uranyl acetate) for low to intermediate resolution studies or in vitreous water (no ice crystals are present that could damage the protein), tannin or glucose. The latter three embedding media preserve the structure of proteins to high resolution [4]. In keeping with this it is crucial to observe the specimens and record electron micrographs at low doses (< 1000 e^-/nm^2) and low temperature (< -150 °C; "cryo" studies). The electron micrographs are then digitised (unless one has access to on-line image acquisition facilities via e.g. a slow-scan CCD camera), Fourier transforms are calculated, phase and amplitude information is extracted, data sets from different crystals are merged, and finally, via Fourier synthesis, a projection map is calculated. Tilting the specimen relative to the incident beam will give access to 3-D structural information.

At this point, we have not yet talked about electron diffraction and have only dealt with images. With images it is important to remember that the Fourier transform of an image provides both, amplitude and phase information. The latter is related to the phase structure of the object and should not be confused with the propagating wave front phases which are lost in the image. When dealing with images, there is no phase problem as far as the crystallographic structure factor phases are concerned. Just to keep things clear in our minds: when speaking in the following about amplitude and phase information, we can say that the phases are concerned with the positions of centres of mass and the amplitudes describe the volumes of these centres of mass.

In order to be able to record electron diffraction patterns with sufficient signal-to-noise ratios, quite large (in the region of a few 1000 unit cells) and well-ordered 2-D crystals are required. In other words, the biggest challenge (= substitute for the old fashioned term 'problem') is to grow 2-D crystals of sufficient quality, that is size and order. Once electron diffraction patterns have been recorded, one would use the amplitude information from these patterns and combine it with the phase information as obtained from the Fourier transforms of the images in order to calculate the structure. We know that images provide us with both, amplitude and phase information, so why would one want to do that ? This is because the amplitude information from the images is modulated by an 'electronoptical shortcoming' called the contrast transfer function and suffers from an exponential fall-off with increasing resolution so that the amplitude information cannot be taken as read but requires tricky correction and scaling procedures. Furthermore, the amplitude information from images tends to be quite scattered. For these reasons it is very advantageous to record electron diffraction patterns. More specifically, the usage of electron diffraction amplitudes bestows several benefits: there are no disturbing effects due to specimen drift (which can be severely limiting) since the diffraction pattern is translationally invariant, and electron diffraction amplitudes are not affected by the contrast transfer function. These positive effects together with the fact that the accumulated electron dose required for recording electron diffraction patterns is below the dose required for images, it is not astonishing that electron diffraction patterns contain information to higher resolution than the Fourier transform of a corresponding image. However, in contrast to the 'poorer' image amplitudes, image phases can be easily corrected for the contrast transfer function, are very reliable (less scatter when compared to the image amplitudes) and readily extendable using the now well established maximum entropy methodology [5]. In other words, image phases and electron diffraction amplitudes constitute an ideal combination. In biological sciences, it was this combination of information that originally bore the name "electron crystallography". Robert M Glaeser has, in 1985, defined electron crystallography as ..." any use of electron diffraction, electron scattering or direct imaging in which the analysis of data is essentially crystallographic in its execution or in its objectives", and I think this is a very useful definition.

Following on from what has been said above, it should be evident that the situation in electron crystallography is quite different from X-ray crystallography where structures are either solved to high resolution or not solved at all. With electrons, one can solve a structure over a wide resolution range: With regards to membrane proteins, high resolution information (< 0.8 nm) has been obtained for e.g. the well known bacteriorhodopsin [6], LHC-II [7] and porin [8] to name but a few. At intermediate (1.5

to 0.8 nm) or low resolution (above 1.5 nm, typically 2-3 nm) structures have been obtained for a great number of membrane proteins and it should be stressed that establishing structure-function relationships does not always depend on structural details at atomic resolution. This is especially true with big proteins and protein complexes. In the following, I am going to use three examples to illustrate these points.

3. The Examples

3.1. PHOTOSYSTEM II

The first example is concerned with photosystem II (PSII) from higher plants. PSII is a protein-pigment complex located in the thylakoid membrane of plants and cyanobacteria (for an overview see e.g. Nicholson *et al.* [9]). Within chloroplasts, PSII occupies primarily those parts of the thylakoid membrane system known as grana. Upon illumination, it provides the plant with chemical potential and the environment with oxygen allowing us to adopt an aerobic lifestyle. PSII is a transmembrane complex, it is big (approaching 1 MDa), and it is extremely labile, i.e. it truly is an X-ray crystallographer's nightmare. Out of the more than 20 polypeptides, the major ones are

- the reaction centre heterodimer (D1/D2) + Cyt b-559 (= reaction centre),
- the extrinsic oxygen evolution-enhancing subunits of 33, 23 and 16 kDa (OEE-33, OEE-23, OEE-16) constituting the oxygen evolution complex (OEC), and
- the core chlorophyll-binding proteins of 47 and 43 kDa (CP47, CP43).

All of the above polyeptides are part of the so-called PSII core. The peripheral light-harvesting antennae comprise a number of polypeptides of 25 to 30 kDa and complement the core light-harvesting proteins CP47 and CP43 in their function to trap light energy as efficiently as possible and to transfer it to the reaction centres. Between 8 and 20 copies of these peripheral light-harvesting polypeptides are associated with

PSII and the major light-harvesting antenna polypeptide is called light-harvesting complex II (LHC-II, see 3.2.).

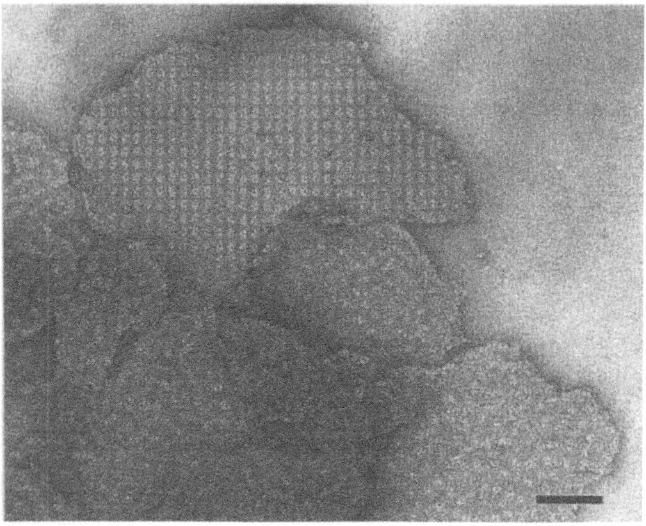

Figure 1. Electron micrograph of a negatively stained (protein appears white) *in situ* 2-D crystal of PSII (a x b = 17.7 x 20.1 nm, $\gamma = 91°$, *p*1) surrounded by membranes containing non-crystalline PSII. The scale bar corresponds to 100 nm.

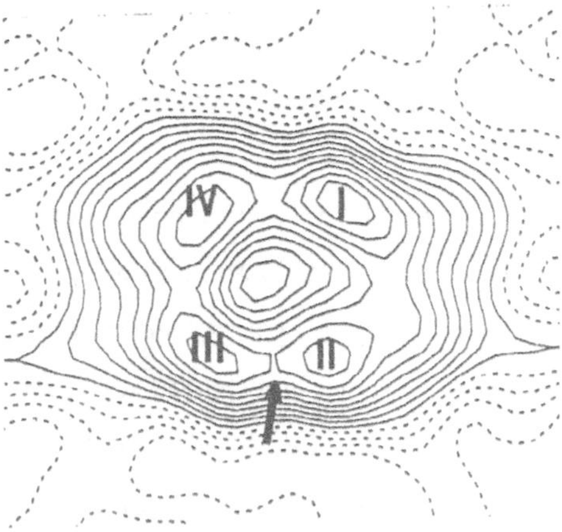

Figure 2. Fourier projection map of native PSII revealing four domains (the location of domain V is indicated by an arrow) and a central protein deficit. One unit cell is shown.

In order to investigate the structure of native PSII, one has to analyse the complex whilst in its native membrane environment. Electron microscopical analysis of negatively stained *in situ* 2-D crystals (Figure 1) resulted in a projection map at low resolution (3 nm) showing the complex to consist of four major domains surrounding a central stain dip (Figure 2).

By tilting the specimen relative to the incident electron beam, as mentioned earlier, one can gain access to 3-D structural information. However, the impracticality to tilt a specimen by as much as 90° means that not all of the 3-D information is retrieved (missing cone problem) and as a result of this the resolution along z is worse compared to the resolution in the a,b plane. By calculating a 'wedge diagram' (extent of data plotted as radius in the x^*y^* plane *versus* z^*; cf. Henderson et al. [6]) one can determine as to whether the information is complete within a given 'wedge'. With this being the case and employing a maximum tilt angle of 60° (87% of the Fourier space has been sampled with % completeness = sin 60 x 100), the resolution along z is worse by a factor of about 1.3.

Looking at a plot (e.g. like the one in Holzenburg *et al.*[10]) showing the modulation of the amplitudes and phases along z^* for each reflection (h,k), one can easily see that the phase information is more reliable than the amplitude information. Slices through the 3-D structure reveal 4 major domains and a central deficit on one side (known as the lumenal side) of the complex while the other side (referred to as stromal side) is mainly characterised by two domains forming a central ridge to the structure (Figure 3e and f). The overall 3-D structure of PSII is highly anisotropic going from top to bottom, left to right or back to front. It also protrudes much more from the lumenal side of the membrane than from the stromal side with which it is flush aiding its organisation within the chloroplast membrane system (Figure 4). Estimating the mass of the complex using the enclosed volume gives a number that agrees well with the unit cell containing one PSII complex, i.e. PSII operates as a monomer *in vivo* and not as an oligomer.

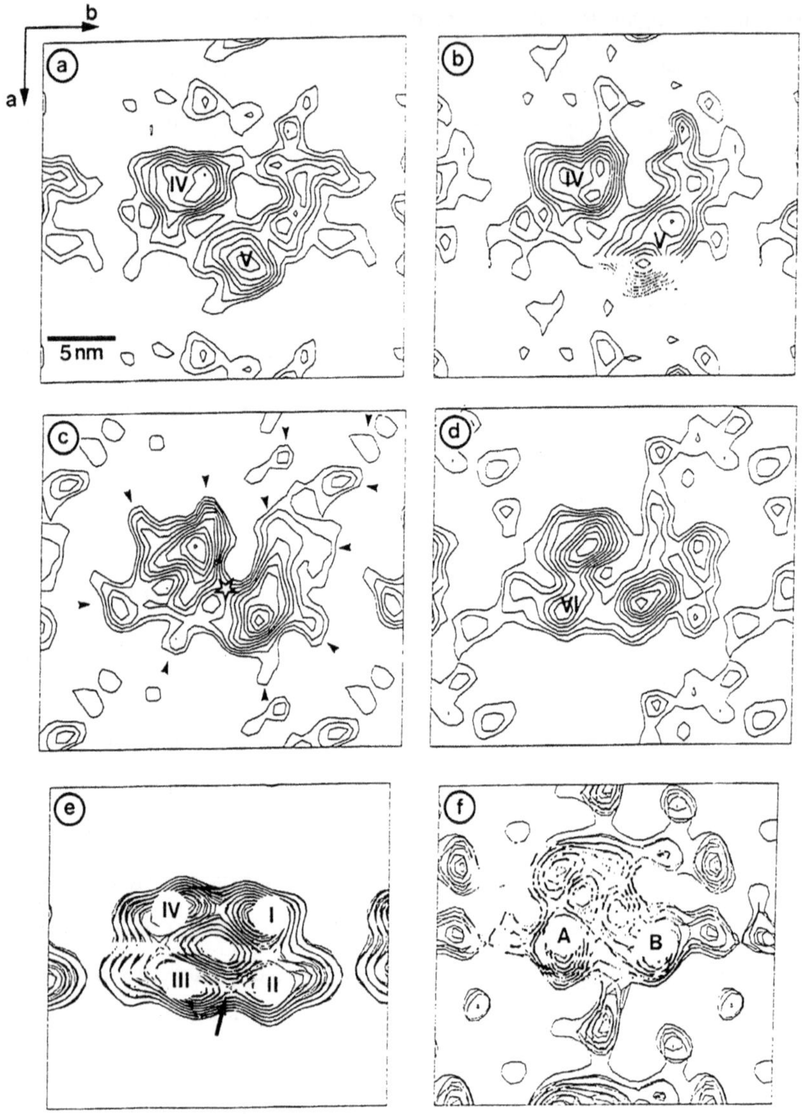

Figure 3. Slices through the 3-D structure of PSII before (e,f; native) and after (a-d) Tris-treatment. The lumenal side of the complex (a,b) is dominated by two large domains IV and V. At the centre of the membrane (c), the star marks a point of pseudo-twofold symmetry. Towards the stromal side (d), another domain (VI) is apparent between domains IV and V. As yet unidentified peripheral features are marked by arrowheads (see [11]). In (e) and (f), three consecutive lumenal and stromal slices, respectively, have been superimposed. The labels in (e) correspond to those in Figure 2, and in (f), the two domains forming a central ridge are labelled (A and B). Reproduced from Ford *et al.* [11].

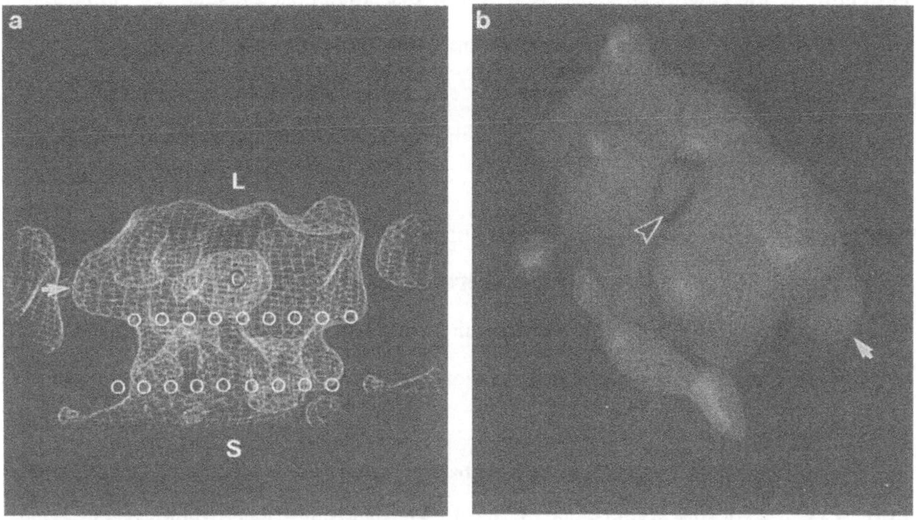

Figure 4. 3-D chicken wire (a) and surface relief (b) presentations of native PSII. The arrows point at identical features in (a) and (b) to facilitate comparison. The intramolecular cavity is highlighted by an arrowhead in (b) and labelled "C" in (a). In (a), lumenal (L) and stromal (S) faces as well as the estimated position of the thylakoid membrane (open circles) are indicated.

Furthermore, the central protein deficit is now discernable as a distinct intramolecular cavity (Figure 4), a biologically extremely important structural feature as we shall see. From the electron microscopical point of view, another observation deserves attention. It is interesting to note that the features in the projection map are identical to those on the lumenal side of the 3-D map. This is due to the fact that these 2-D crystal containing membranes have a tendency to adsorb lumenal face down onto the support film. Together with preferential stain trapping between the lumenal face and the support film this causes a non-uniform stain distribution along *z* with a very strong bias towards lumenal features and very little information on the stromal side when analysed in projection only.

In order to be able to relate some of the observed domains to specific polypeptides of PSII, 2-D crystals were subjected to simple Tris-washing, which removes the three extrinsic OEE-polypeptides, and, after biochemical and functional characterisation, were again analysed in three dimensions [11]. As a result of the OEC removal, the stain could reach the interior parts of the complex much better, the crystals obtained were better ordered and adsorbed both ways (lumenal side up and down) to the support film. These factors allowed the retrieval of a combined data set showing an unbiased and more detailed 3-D structure with a maximum resolution of 1.8 nm. As a result, the 3-D map after Tris-washing revealed a further domain (V) between domains II and III and a domain VI located between IV and V within the boundaries of the membrane (Figure 3). By calculating a 3-D Fourier vector difference map (native minus Tris-washed) and displaying this difference map within the native 3-D map, the location of the OEC could be determined (Figure 5). The data revealed that (i) there is only one copy of each OEE-polypeptide per PSII and (ii) all of them are located on the lumenal side of the complex where they coincide with domains I, II and III and help forming the intramolecular cavity. Washing the OEC off resulted not only in an opening of this cavity but also in a drastic decrease in the oxygen evolution activity, so that a true structure-function relationship could be established. For many years, scientists have proposed a special environment required for oxygen evolution (after all, this process takes place at a redox potential of around +1V): an intramolecular cavity is a perfect solution [11,12].

Following on from this, the question arises which OEE-polypeptide corresponds to which domain. This was solved by treating the 2-D crystals with different concentrations of NaCl inducing a differential removal of the OEE-polypeptides [12]. Since they are all located on the lumenal side and since preferential staining can reveal information specific to the lumenal side (see above), an assignment

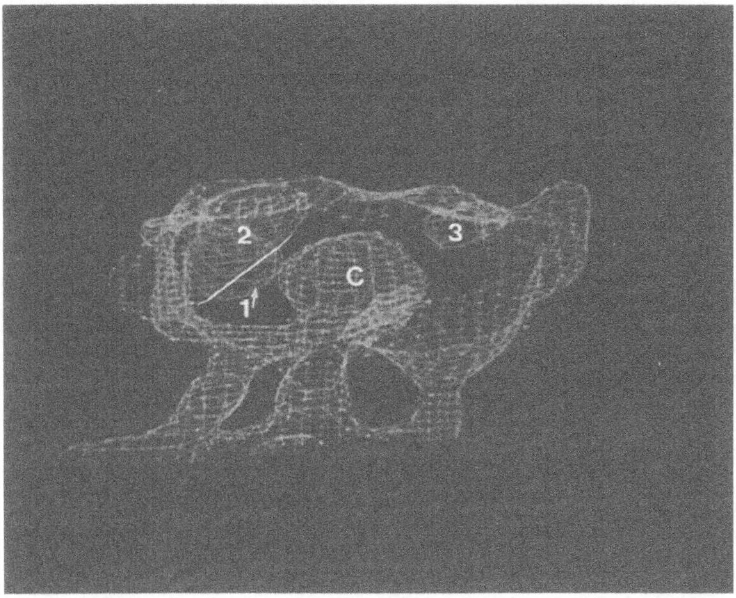

Figure 5. 3-D Fourier vector difference map (labelled 1-3) displayed inside the native PSII 3-D map (C = cavity). After differential removal of the OEE-polypeptides [12], 1 could be assigned to OEE-33, 2 to OEE-23 and possibly another 10-kDa extrinsic polypeptide, and 3 to OEE-16. Modified from Ford *et al.* [11].

can be introduced without the need for further 3-D analysis [12]. By comparing the different data sets it became clear that domain III houses OEE-16, II houses OEE-23, and I OEE-33. Due to the relatively large volume in the 3-D difference map asssigned to OEE-23 (Figure 5) and the finding that a further 10-kDa polypeptide is required for the binding of OEE-23 to the lumenal face of the complex, it seems very probable that

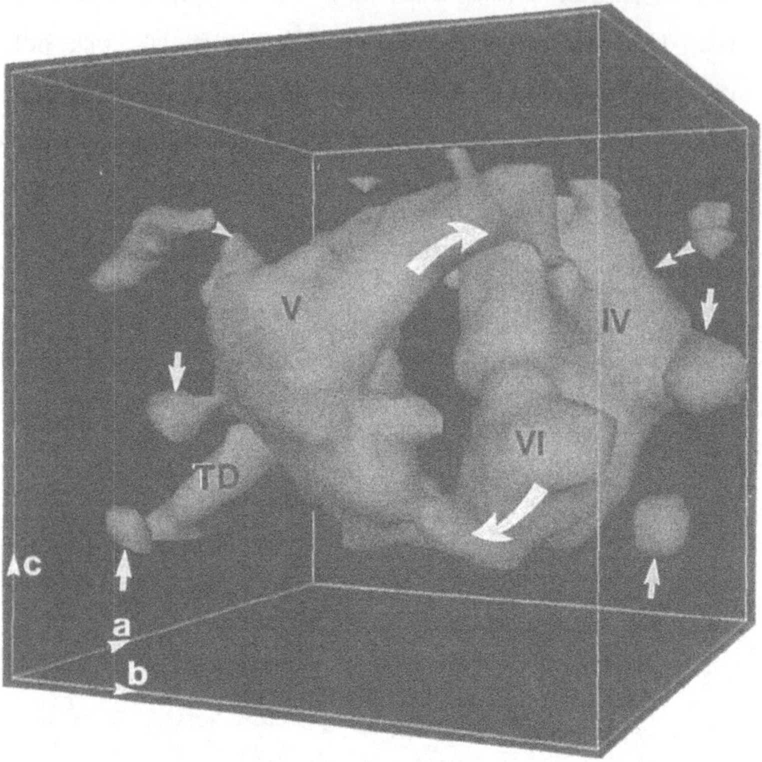

Figure 6. Surface rendered 3-D map of Tris-treated PSII (viewed side-on) revealing domains IV, V, and VI, spherical densities delineating the putative lipid bilayer (arrows), and the embracing architecture of the major domains (curved arrows). The arrowheads point at small unidentified lumenal domains. Reproduced from Ford *et al.* [11].

both the10-kDa polypeptide as well as OEE-23 are part of domain II. Additional information on the identity of domains IV and V was gained from cross-linking and other biochemical and biophysical studies suggesting, in conjunction with the 3-D data after Tris-washing [11], that domain IV corresponds to CP47. This would leave only domain V as possible candidate for CP43 and domain VI, which is sandwiched between domains IV and V, is therefore very likely to house the reaction centre (Figure 6).

The described experiments show how one can tackle the structure of a large and fragile membrane protein complex to a resolution that is not high but nevertheless biologically meaningful. Striving for higher resolution data for the native complex, 2-D crystals of PSII have been embedded in vitreous water and data were recorded under cryo-conditions (see 2.). The crystals, at that point in time, were not large enough to record electron diffraction patterns, so all the information had to come from images necessitating the recording and evaluation of thousands of electron micrographs prior to the averaging over 53 small but well-ordered crystalline areas [13]. Averaging was carried out in reciprocal space using the lattice vectors to facilitate the alignment. The feasibility of this approach has so far only been demonstrated by Perkins *et al.* [14] for purple membrane preparations where the results could be validated by comparing them to a known data set. With PSII, the approach was re-assessed by checking the self-consistency of the final structure factors that allowed the calculation of a projection map to 1.3 nm resolution from data sets that, when judged individually, showed strong reflections only up to about 5 nm resolution [13]. It is therefore important to realise that an effective merging can be carried out as long as there are enough reflections to unambiguously determine the lattice parameters making sure that the crystalline areas to be merged are isomorphous. The general strategy is summarised in Figure 7. The cryo-electron crystallographic data to 1.3 nm resolution represent the highest resolution structure so far for the native complex and confirm earlier estimates of the molecular mass as well as the assignments of three central domains to CP47, CP43 and the reaction centre heterodimer D1/D2 plus cytochrome b-559 (Figure 8). The data also proved valuable for discussing the evolution of reaction centres from anoxygenic to oxygenic photosynthesis [13]. Even though the map shows details to 1.3 nm resolution, the light-harvesting antenna polypeptides (LHCs), which are expected to be located around the core of PSII, are not well defined and their likely positions in the map are characterised by low density features only. This could be due to the LHCs being disordered in the native membrane. However, the major light-harvesting antenna polypeptide LHC-II can be isolated and undergoes oligomerisation into trimeric complexes which lend themselves for the outstanding electron crystallographic studies carried out by da Da Neng Wang, Werner Kühlbrandt and Yoshinori Fujiyoshi [7].

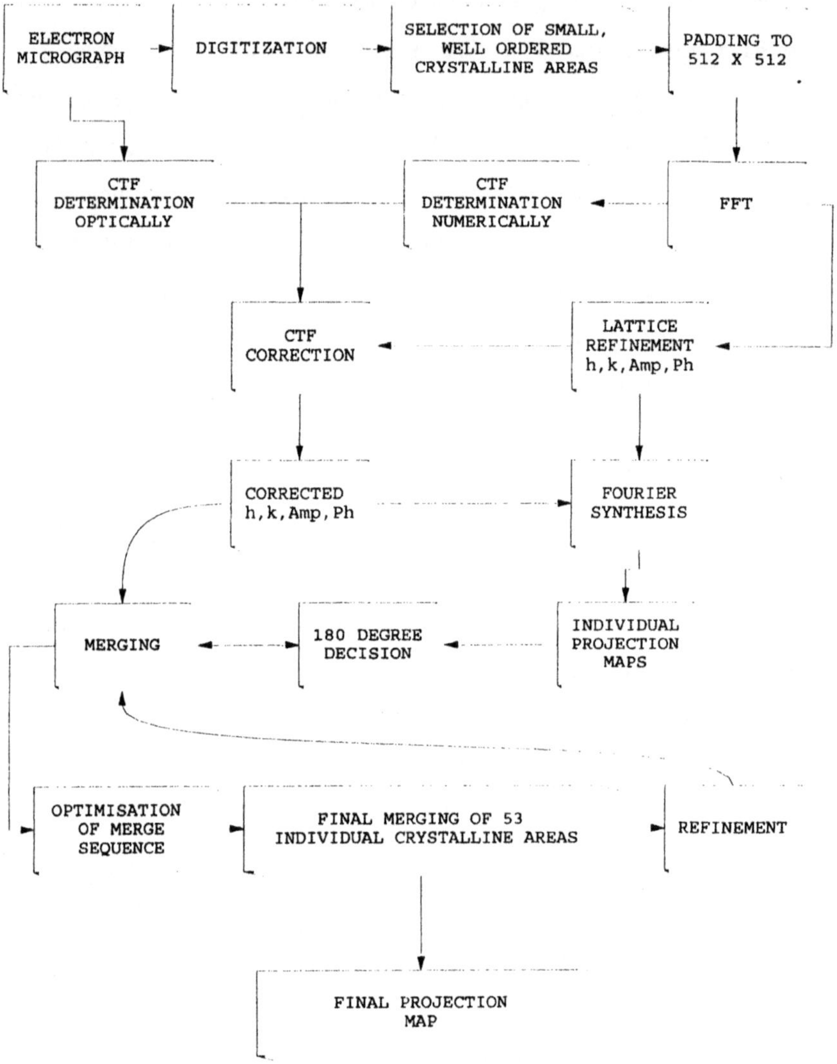

Figure 7. Flow chart describing the stages included in the averaging of the PSII image cryo-data in reciprocal space. The 180° decision relates to the application of appropriate rotation functions for crystals that are rotated with respect to each other (for more details, see Stoylova *et al.* [13]).

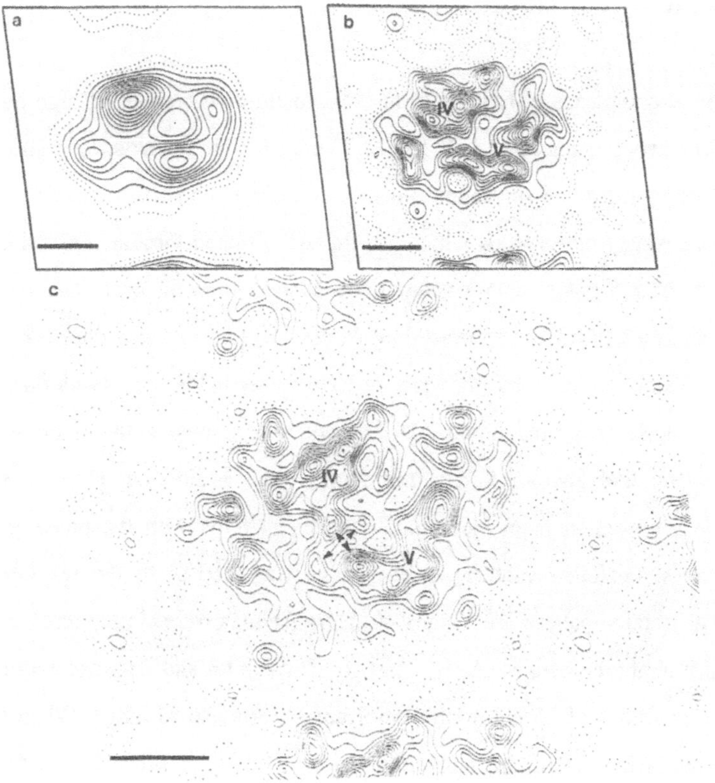

Figure 8. Fourier projection maps of native PSII in the frozen-hydrated state (unstained) after amplitude and phase correction calculated to (a) 5 nm, (b) 2.5 nm, and (c) 1.3 nm resolution. In (c), the central regions within domain VI are highlighted by arrowheads. Other domains are labelled as above. Dashed contour lines are drawn at 1σ and 2σ, and the first solid contour line at 3σ. The scale bar corresponds to 5 nm.

3.2. LHC-II

The trimeric complexes of LHC-II form 2-D protein-lipid crystals of high quality. Due to the plane group being $p321$ ($a = b = 12.95$ nm), fusion of smaller crystals into larger ones did not constitute a stumbling block since there is no need to consider any of the problems arising from crystals which are flipped (rotated through 180° along an axis parallel to the a,b plane) and/or rotated with respect to each other. The crystals were initially analysed in different embedding media at liquid nitrogen temperature [4] and electron diffraction patterns showing reflections to 0.346 nm resolution could be recorded. The first high-resolution 3-D data sets were calculated to 0.6 nm resolution in the a,b plane and about 0.8 nm resolution in the z direction [15] by combining amplitude information from electron diffraction patterns with the phase information from images, i.e. the resolution in the image was inferior to the resolution in the electron diffraction pattern. Secondary structure features were clearly resolved showing three transmembrane helices (A, B, C), two of which (A and B) cross over each other [15]. The positions of 15 chlorophyll molecules were also assigned. When looking at the modulation of amplitudes and phases along z^* [15], it becomes clear how much more reliable the electron diffraction amplitude information is versus the amplitude information from images [e.g. 10]. However,

- at high tilt, electron diffraction maxima can be considerably displaced from their expected positions due to the curvature of the Ewald sphere [16]. This effect is reduced at higher accelerating voltages (i.e. shorter wavelength).
- It is also crucial that the specimens are absolutely flat. Is this not the case, then one can observe a blurring of the reflections at high angles [16].

Finally, analysing the crystals at liquid helium temperature using Yoshinori Fujiyoshi's 400-kV electron cryomicroscope equipped with an extremely stable top entry stage, resulted in Fourier transforms of images which extended to a resolution comparable to that of electron diffraction data (!) [7]. The lower temperature in conjunction with the special stage results in less specimen drift and allows to record images at higher electron doses. Furthermore, the higher accelerating voltage improves coherence, decreases specimen damage (also contrast but that is compensated for by the higher affordable doses), Ewald sphere curvature and the attenuation of the envelope of the contrast transfer function at higher resolution. The resolution of the final 3-D map was 0.34 nm in the *a,b* plane and better than 0.49 nm orthogonal to it, which allowed, the polypeptide chain to be traced, individual amino acids as well as pigment molecules to be identified, and a detailed energy transfer mechanism to be proposed [7].

3.3. HUMAN BLOOD COAGULATION FACTOR IX

The third and last example is concerned with a low resolution study of the blood coagulation factor IX (FIX). Using the lipid layer technique mentioned earlier, FIX was 2-D crystallised under quasi-physiological conditions, bound to the surface of a lipid layer. The 2-D crystals were negatively stained, images were recorded in the electron microscope and a projection map was calculated to 3 nm resolution [17]. The unit cell ($a = b = 14.7$ nm, $\gamma = 97.1°$, *p*2) contained two molecules related by a point of twofold rotational symmetry and each molecule consists of a large flat domain and a small stalk-like domain attached to the large domain off-centre (Figure 9). Because one can deduce from the stain distribution that the small domain is pointing towards the lipid layer, the orientation of the factor relative to the membrane plane can be determined. A piece of information that cannot be obtained by any other method.

By combining this low resolution information with the known X-ray data [18] one can assign the large domain to the catalytic domain and the second epidermal growth factor-like (EGF-2) peptide and the stalk-like domain to the N-terminal γ-carboxyglutamic acid-rich domain (Gla domain; responsible for membrane binding) and EGF-1 (Figure 9). This way a complete picture of the *in vivo* situation is obtained. A very good example of how synergistic a combined X-ray/electron crystallographic approach can be. But this is not to say that the lipid layer technique should only be considered for low resolution studies. Quite the opposite is true, since this approach has in a number of cases led to very high quality 2-D crystals that diffract electrons to a resolution of better than 0.4 nm [e.g. 3].

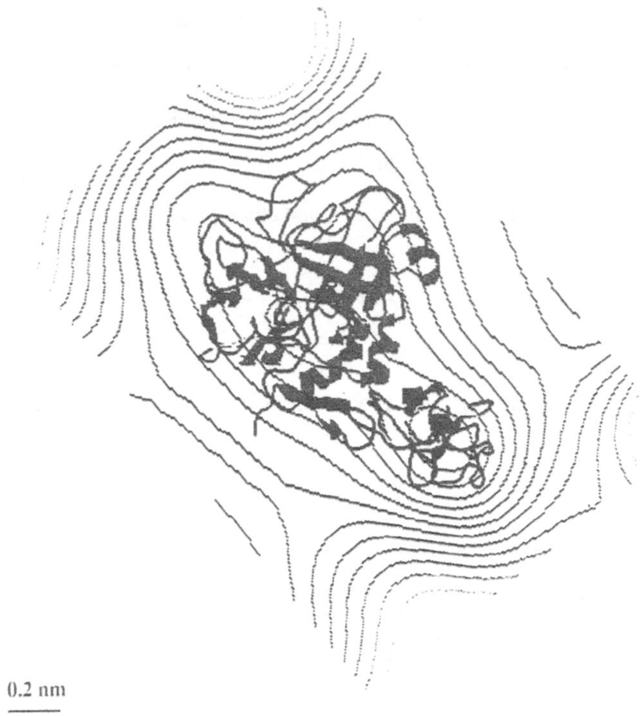

Figure 9. Projection map of lipid-bound FIX with the X-ray structure [18] superimposed (light grey = catalytic domain and EGF-2, dark grey = EGF-1 and Gla domain).

4. Epilogue

Since its begginings in the 1930s, electron micrsopy has come a long way. Over the last years, advances in biological electron crystallography (specifically, the engineering of better microscopes, improvements of specimen preparation and image processing techniques) have made huge leaps forward. While it is to be hoped that the field will continue to flourish, the biggest gains could perhaps be encountered, if the X-ray and electron communities worked together more closely, underpinned by a sound basis of mutual understanding and appreciation.

Acknowledgements

I like to thank Dr. Robert C. Ford and his coworkers at UMIST as well as my past and current laboratory members, particularly Dr. Svetla Stoylova, for sustaining the *tour de force* on PSII. I am also grateful to Dr. Stoylova for providing Figures 7 and 9. The financial support by the UK Biotechnology and Biological Sciences Research Council, the Academic Development Fund of the University of Leeds, the Leeds Centre for Molecular Recognition in Biologial Sciences, NATO, the Nuffield Foundation, and the Royal Society is greatly acknowledged.

References

1. Holzenburg,A. (1995) Electron microscopical analysis of ion channels. In: Ion channels - a practical approach (ed. R.H. Ashley) 269-290, IRL Press, Oxford.
2. Jap,B.K., Zulauf,M., Scheybani,T., Hefti,A., Baumeister,W., Aebi,U. and Engel,A. (1992) 2-D crystallisation: from art to science. *Ultramicroscopy* **46**, 45-84.
3. Brisson,A., Olofsson,A., Ringler,P., Schmutz,M. and Stoylova,S. (1994) Two-dimensional crystallisation of proteins on planar lipid films and structure determination by electron crystallography. *Biol. Cell* **80**, 221-229.
4. Wang,D.N. and Kühlbrandt,W. (1991) High-resolution electron crystallography of light-harvesting chlorophyll *a/b*-protein complex in three different media. *J. Mol. Biol.* **217**, 691-699.
5. Gilmore,C.J., Nicholson,W.V. and Dorset,DL. (1996) Direct methods in protein electron crystallography: the *ab initio* structure determination of two membrane protein structures in projection using maximum entropy and likelihood. *Acta Cryst. A* **52**, 937-946.
6. Henderson,R., Baldwin,J.M., Ceska,T.A., Zemlin,F., Beckmann,E. and Downing, K.H. (1990) Model for the structure of bacteriorhodopsin based on high-resolution electron cryo-microscopy. *J. Mol. Biol.* **213**, 899-929.
7. Kühlbrandt,W., Wang,D.N. and Fujiyoshi,Y. (1994) Atomic model of plant light-harvesting complex by electron crystallography. *Nature* **367**, 614-621.
8. Walian,P.J. and Jap,B.K. (1990) 3-Dimensional electron diffraction of Phoe porin to 2.8 Å resolution. *J. Mol. Biol.* **215**, 429-438.

9. Nicholson,W.V., Ford,R.C. and Holzenburg,A. (1996b) A current assessment of photosystem II structure. *Bioscience Reports* **16**, 159-187.
10. Holzenburg,A., Bewley,M.C., Wilson,F.H., Nicholson,W.V. and Ford,R.C. (1993) Three dimensional structure of photosystem II *Nature* **363**, 470-472.
11. Ford,R.C., Rosenberg,M.F., Shepherd,F.H., McPhie,P. and Holzenburg,A. (1995) Photosystem II 3-D structure and the role of the extrinsic subunits in photosynthetic oxygen evolution. *Micron* **26**, 133-140.
12. Holzenburg,A., Flint,T.D., Shepherd,F.H. and Ford, R.C. (1996) Photosystem II: Mapping the locations of the oxygen Evolution-enhancing subunits by electron microscopy. *Micron* **26**, 121-127.
13. Stoylova,S., Flint,T.D., Ford,R.C. and Holzenburg,A. (1997) Projection structure of photosystem II *in vivo* determined by cryo-electron crystallography. *Micron*, in press.
14. Perkins,G.A., Downing,K.H. and Glaeser,R.M. (1995) Crystallographic extraction and averaging of data from small image areas. *Ultramicroscopy* **60**, 283-294.
15. Kühlbrandt,W.and Wang,D.N. (1991) Three-dimensional structure of plant light-harvesting complex determined by electron crystallography. *Nature* **350**, 130-134.
16. Baldwin,J.and Henderson,R. (1984) Measurement and evaluation of electron diffraction patterns from two-dimensional crystals. *Ultramicroscopy* **14**, 319-336.
17. Stoylova,S., Gray,E., Barrowcliffe,T.W., Kemball-Cook,G. and Holzenburg,A. (1997) Electron crystallographical study of lipid-bound human factor IX. *Thromb. Haemost.* **Suppl.** 590
18. Brandstetter,H., Bauer,M., Huber,R., Lollar,P. and Bode, W. (1995) X-ray structure of clotting factor IXa - active site and module structure related to Xase activity and hemophilia B. *Proc. Natl. Acad. Sci. U.S.A.* **92**, 9796-9800.

THE PHASE PROBLEM: A PROBLEM IN CONSTRAINED GLOBAL OPTIMIZATION

HERBERT A. HAUPTMAN
Hauptman-Woodward Medical Research Institute, Inc.
73 High Street, Buffalo, New York 14203-1196

Abstract

It is now almost 200 years since Gauss, a teenager at the time, formulated his famous principle of least-squares and used it to determine, for the first time, the orbit of one of the asteroids, a problem which had defeated astronomers for years. When applied to the crystallographic phase problem, least-squares leads directly to the formulation of the minimal principle, which effectively replaces the phase problem by one of constrained global minimization. Shake-and-Bake, the computer software package which implements this formulation of the phase problem, provides a completely automatic solution of this problem. Shake-and-Bake requires that diffraction intensities to a resolution of 1.2Å, at least, be available. Structures having as many as 600 independent non-hydrogen atoms have been routinely solved in this way; the ultimate potential of the method is still not known.

When single-wavelength anomalous scattering (SAS) diffraction data are available, the phase problem may again be formulated as a problem in global optimization. Although the objective function has a myriad of local maxima, its global maximum is readily accessible and easily identified by virtue of its isolation. The ability to determine the global maximum of the objective function represents the latest and most successful attempt to go directly from the known probabilistic estimates of the three-phase structure invariants to the values of the individual phases. The relationship between the maxima of the objective function and the solutions of the newly formulated system of SAS tangent equations plays a key role in this development.

1. Introduction

The techniques of modern probability theory lead to the joint probability distributions of arbitrary collections of diffraction intensities and their corresponding phases. These distributions constitute the foundation on which direct methods are based. They have provided the unifying thread from the beginning, *ca* 1950, until the present time. They have led, in particular to the (first) minimal principle [1-3] which has found expression in the ***Shake-and-Bake*** formalism [4,5], a computer program which provides a completely automatic solution to the phase problem, *ab initio*, provided that diffraction data to at least 1.2Å are available. Our experience shows that structures having as many as 600 independent non-H atoms are routinely accessible to this approach and suggests that its ultimate potential is greater still.

S. Fortier (ed.), Direct Methods for Solving Macromolecular Structures, 381–388.

It should perhaps be pointed out that the minimal principle, the theoretical basis of *Shake-and-Bake*, replaces the phase problem by one of *constrained* global minimization, in sharp contrast to an earlier formulation by Debaerdemaeker & Woolfson [6], in which the (unconstrained) global minimum was sought. The distinction is crucial, not only on the theoretical level but in the applications as well: Non only does the constrained global minimum yield accurate values of the individual phases *ab initio* but this formulation suggests how the minimum is to be reached (*via* the *Shake-and-Bake* program) and how to identify it. The failure to impose the constraints, as is done in the earlier work, greatly limits the usefulness of that approach, especially since it is not at all clear how to reach or identify the unconstrained global minimum.

One naturally anticipates that, with the availability of single-wavelength anomalous-scattering data, the ability to determine phases *ab initio* will be strengthened. This expectation is in fact realized here. Specifically, an SAS maximal principle is formulated which, even though unconstrained, nevertheless strengthens the earlier minimal principle by incorporating SAS estimates of the cosines and sines of the three-phase structure invariants. The initial applications show that, in this way, the phase problem is solvable, *ab initio*, even for macromolecules, when SAS diffraction data alone are available at a resolution of about 2.5Å.

This work represents the latest attempt to go directly from known estimates of the three-phase structure invariants to the values of the individual phases (*cf.*, for example, (Han *et al.*, [7] and Hauptman and Han [8]). However, instead of attempting to solve by least-squares a redundant system of linear equations, as was done in the earlier work, the formulation presented here transforms the problem into one of global optimization, a problem with a surprisingly easy solution. Furthermore, as shown by the initial application, briefly described here, the results derived here represent a substantial improvement over the earlier work.

Although other phasing methods employing SAS data have been devised (*e.g.* Fan *et al.* [9]), with the exception of the two references cited in the previous paragraph, none of these exploits the information contained in the SAS estimates of the three-phase structure invariants. Thus, this work marks a radical departure from the earlier approaches.

Once again, the probabilistic theory of the (three-phase) structure invariants, initiated in the SAS case in 1982 [10], plays the central role. It should perhaps be stressed at the outset that, owing now to the breakdown of Friedel's law and contrary to all earlier belief, unique values for all the structure invariants in the whole interval from 0 to 2π are determined since the enantiomorph is fixed by the observed magnitudes $|E|$. It is believed that the ability to fix the enantiomorph *ab initio* accounts for the unexpected result described here.

The approach adopted here is similar to that used in the derivation of the minimal principle but is suitably modified in order to take into account the availability of the SAS diffraction data. Not only is one led in this way to the SAS maximal principle but an important connection with the SAS tangent formula, the analogue of the traditional tangent formula, is established. Two remarkable properties of the SAS maximal function emerge: (a) the easy accessibility and ready identification of its global maximum and (b) the isolated character of this maximum.

2. The Non-SAS Case

2.1. THE NATURE OF THE CONSTRAINTS

For a structure consisting of N identical atoms in the unit cell the normalized structure factor E_H is defined by

$$E_H = \frac{1}{N^{1/2}} \sum_{j=1}^{N} \exp(2\pi i H \cdot r_j). \tag{2.1}$$

Since the number of equations (2.1) exceeds by far the number of unknown atomic position vectors r_j, elimination of the r_j's leads to a system of equations among the normalized structure factors E_H alone:

$$F(E) = 0, \tag{2.2}$$

each function F of which may be written as a function of known magnitudes |E| and unknown phases ϕ:

$$F(E) \to F(|E|;\phi) = 0. \tag{2.3}$$

The system of equations (2.3) leads directly to a system of identities among the phases:

$$F(|E|;\phi) \to G(\phi/|E|) = 0 \tag{2.4}$$

where the G's are functions of the phases ϕ that depend upon the known parameters |E|. One infers then that the phases are, of necessity, constrained to satisfy the system of identities (2.4).

2.2. THE PROBABILISTIC BACKGROUND

If one assumes that the atomic position vectors r_j are the primitive random variables, uniformly and independently distributed in the unit cell, then the normalized structure factors E, as functions of the primitive random variables r_j, Eq. (2.1), are themselves random variables. Since the magnitudes |E| are obtainable from the diffraction experiment, the phases ϕ may be regarded as random variables, the conditional probability distribution of any collection of which, assuming as given appropriate magnitudes |E|, may be found by standard techniques. Thus the conditional probability distribution of the triplet

$$\phi_{\mathbf{HK}} = \phi_{\mathbf{H}} + \phi_{\mathbf{K}} + \phi_{-\mathbf{H}-\mathbf{K}}, \tag{2.5}$$

given the three magnitudes

$$|E_{\mathbf{H}}|, \ |E_{\mathbf{K}}|, \ |E_{\mathbf{H+K}}|, \tag{2.6}$$

is known to be

$$P(\Phi/\kappa_{HK}) = \frac{1}{2\pi I_0(\kappa_{HK})} \exp(\kappa_{HK} \cos\Phi_{HK}), \tag{2.7}$$

where Φ represents the triplet ϕ_{HK} (Eq. (2.5)), κ_{HK} is given by

$$\kappa_{HK} = \frac{2}{N^{1/2}} |E_H E_K E_{H+K}|, \tag{2.8}$$

and I_0 is the Modified Bessel Function. From Eq. (2.7) one finds the conditional expected value of cos ϕ_{HK} to be

$$\mathcal{E}(\cos\phi_{HK}) = \frac{I_1(\kappa_{HK})}{I_0(\kappa_{HK})}, \tag{2.9}$$

where I_1 is the Modified Bessel Function, and infers that the larger the value of the parameter κ_{HK} the smaller is the variance of the cosine.

2.3. THE MINIMAL PRINCIPLE

In view of § 2.2 one defines the minimal function, a function of the phases ϕ, by means of

$$m(\phi) = \frac{\sum_{H,K} \kappa_{HK} \left\{ \cos\phi_{HK} - \frac{I_1(\kappa_{HK})}{I_0(\kappa_{HK})} \right\}^2}{\sum_{H,K} \kappa_{HK}} \tag{2.10}$$

and infers that the global minimum of m(ϕ), where the phases ϕ are constrained to satisfy the system of identities (2.4), yields the true values of the phases for some choice of origin and enantiomorph (the minimal principle). In this way the phase problem is formulated as one of *constrained* global minimization, with emphasis on the word *constrained*: the *unconstrained* global minimum of m(ϕ) does not solve the phase problem; the *constrained* global minimum does. The reader is referred to DeTitta *et al.* [3], Weeks *et al.* [4], and Miller *et al.* [5] for further details, in particular how the computer program *Shake-and-Bake* converges to the constrained global minimum of m(ϕ). (See also Miller & Weeks, these Proceedings.)

3. The SAS Case

Once again the probabilistic theory of the (three-phase) structure invariants, initiated in the SAS case in 1982 [10], plays the central role. It should perhaps be stressed at the outset that, owing now to the breakdown of Friedel's Law and contrary to all earlier belief, unique values for all the structure invariants in the whole interval from 0 to 2π are determined since the enantiomorph is fixed by the observed magnitudes |E|. It is believed that the ability to fix the enantiomorph, *ab initio*, accounts for the unexpected result described here (§3.5).

The approach adopted here is similar to that used in the derivation of the minimal principle but is suitably modified in order to take into account the availability of the SAS diffraction data. Not only is one led in this way to the SAS maximal principle, but an important connection with the SAS tangent formula, the analogue of the traditional tangent formula, is established. Two remarkable properties of the SAS maximal function emerge: (a) the easy accessibility and ready identification of its global maximum and (b) the isolated character of this maximum.

3.1. THE PROBABILISTIC BACKGROUND

With the assumption that SAS diffraction data are available, the conditional probability distribution $P(\phi)$ of the triplet

$$\phi_{\mathbf{HK}} = \phi_{\mathbf{H}} + \phi_{\mathbf{K}} + \phi_{-\mathbf{H}-\mathbf{K}}, \tag{3.1}$$

given the six magnitudes

$$|E_{\mathbf{H}}|,\ |E_{-\mathbf{H}}|,\ |E_{\mathbf{K}}|,\ |E_{-\mathbf{K}}|,\ |E_{\mathbf{H}+\mathbf{K}}|,\ |E_{-\mathbf{H}-\mathbf{K}}|, \tag{3.2}$$

is known to be [10],

$$P(\Phi) = \left[2\pi I_0\left(A_{\mathbf{HK}}\right)\right]^{-1} \exp\left\{A_{\mathbf{HK}} \cos\left(\Phi - \omega_{\mathbf{HK}}\right)\right\} \tag{3.3}$$

where Φ represents the triplet $\phi_{\mathbf{HK}}$ (Eq. (3.1)), I_0 is the Modified Bessel Function and $A_{\mathbf{HK}}$ and $\omega_{\mathbf{HK}}$ are expressed in terms of the six magnitudes (3.2) and the (presumed known) complex-valued atomic scattering factors $\mathfrak{f}$. Hence $A_{\mathbf{HK}}(>0)$ and $\omega_{\mathbf{HK}}$ are here assumed to be known for every pair $(\mathbf{H},\mathbf{K})$. Note that, owing to the breakdown of Friedel's Law, the six magnitudes (3.2) are, in general, distinct.

In view of Eq. (3.3), the most probable value of $\phi_{\mathbf{HK}}$ is $\omega_{\mathbf{HK}}$, and the larger the value of $A_{\mathbf{HK}}$ the better is this estimate of $\phi_{\mathbf{HK}}$:

$$\phi_{\mathbf{HK}} = \phi_{\mathbf{H}} + \phi_{\mathbf{K}} + \phi_{-\mathbf{H}-\mathbf{K}} \approx \omega_{\mathbf{HK}}. \tag{3.4}$$

3.2. THE SAS MAXIMAL PRINCIPLE

In exact analogy with the derivation of the minimal principle in the non-SAS case, one now defines the SAS maximal function $M(\phi)$ by means of

$$M(\phi) = \frac{\sum\limits_{\mathbf{H},\mathbf{K}} A_{\mathbf{HK}} \cos\left(\phi_{\mathbf{H}} + \phi_{\mathbf{K}} + \phi_{-\mathbf{H}-\mathbf{K}} - \omega_{\mathbf{HK}}\right)}{\sum\limits_{\mathbf{H},\mathbf{K}} A_{\mathbf{HK}}} \tag{3.5}$$

and infers that the global maximum of $M(\phi)$ yields the true values of the phases for some choice of origin (the SAS maximal principle). There remains the problem of finding the global maximum of $M(\phi)$, a problem presumed to be difficult by virtue of the existence of a myriad of local maxima of $M(\phi)$. The solution, however, turns out to be unexpectedly straightforward. How the problem is solved via the system of SAS tangent equations is described next [11].

3.3. THE SYSTEM OF SAS TANGENT EQUATIONS

Eq. (3.4) implies the SAS tangent formula: For each fixed value of the reciprocal lattice vector H

$$\tan \phi_{\mathbf{H}} = \frac{\sum_{\mathbf{K}} A_{\mathbf{HK}} \sin(\omega_{\mathbf{HK}} - \phi_{\mathbf{K}} - \phi_{-\mathbf{H}-\mathbf{K}})}{\sum_{\mathbf{K}} A_{\mathbf{HK}} \cos(\omega_{\mathbf{HK}} - \phi_{\mathbf{K}} - \phi_{-\mathbf{H}-\mathbf{K}})} \tag{3.6}$$

where the sign of sin $\phi_{\mathbf{H}}$ is given by the numerator of (3.6) and the sign of cos $\phi_{\mathbf{H}}$ by the denominator. Thus the tangent formula determines a unique value for $\phi_{\mathbf{H}}$ when the values of all other phases are assumed to be known.

Although (3.6) differs from the standard tangent formula only in the presence of the non-zero estimates $\omega_{\mathbf{HK}}$ of the three-phase structure invariants $\phi_{\mathbf{HK}}$, as well as the completely different set of weights $A_{\mathbf{HK}}$, these differences are of fundamental importance in the applications. As described in the sequel, the present formulation solves the phase problem for macromolecules in the SAS case with diffraction data to 2.5Å resolution; not surprisingly, the same claim cannot be made for the standard tangent formula when used in the same way.

3.4. THE MAXIMAL PROPERTY OF THE SAS TANGENT FORMULA

Fix the reciprocal lattice vector **H**. Assume that the values of all phases other than $\phi_{\mathbf{H}}$ are specified arbitrarily. Then the maximal function $M(\phi)$ becomes a function, $M(\phi_{\mathbf{H}}/\phi)$, of the single phase $\phi_{\mathbf{H}}$. It is then readily shown that, as a function of $\phi_{\mathbf{H}}$, $M(\phi_{\mathbf{H}}/\phi)$ has a unique maximum in the whole interval $(0, 2\pi)$ and the value of $\phi_{\mathbf{H}}$, which maximizes $M(\phi_{\mathbf{H}}/\phi)$ is given by the SAS tangent formula (3.6).

3.5. HOW TO FIND SOLUTIONS OF THE SYSTEM OF SAS TANGENT EQUATIONS

Specify arbitrarily initial values for all the phases ϕ. Fix **H**. Calculate a new value for the phase $\phi_{\mathbf{H}}$ by means of the SAS tangent formula (3.6), in this way, in view of § 3.4, increasing the initial value of the maximal function $M(\phi)$. Fix $\mathbf{H}' \neq \mathbf{H}$. Calculate a new value for $\phi_{\mathbf{H}'}$, using Eq. (3.6), the new value for $\phi_{\mathbf{H}}$, and initial values for the remaining phases, thus increasing still further the value of $M(\phi)$. Continue in this way to obtain new values for all the phases, thus completing the first iteration and, in the process, continuously increasing the value of $M(\phi)$. Complete as many iterations as necessary in order to secure convergence. Convergence is assured since the iterative process yields a monotonically increasing sequence of numbers, the values of $M(\phi)$, bounded above by unity. Evidently the process leads to a solution of the system of tangent equations (3.4) and, at the same time, a local maximum of the maximal function $M(\phi)$.

The applications show that only rarely does convergence require more than thirty iterations, and usually fewer than twenty suffice.

3.6. THE TALLEST PEAK, TOTALLY ISOLATED, SMOOTH AND RIPPLE FREE, RESTS ON THE BROADEST BASE

The process described in §3.5 always leads to a local maximum of $M(\phi)$, the number of which is legion. It is natural therefore to ask: Will the process ever lead to the global maximum? Preliminary calculations, based on three structures ranging in complexity from 1000 to 4000 independent atoms and in resolution from 3.0 to 2.5Å, show unequivocally that the answer is yes, frequently! Quite unexpectedly, the success rate is high, usually in the range of 10 to 15%.

A remarkable additional feature of the SAS maximal function is the total isolation of its global maximum. Thus, while the values of the local maxima are continuously distributed in a rather narrow range, the value of the global maximum exceeds these by far. This remarkable property of $M(\phi)$ not only makes it easy to identify its global maximum and the associated sets of values for the phases, but no doubt accounts as well for the unexpectedly large circle of convergence surrounding the global maximum.

3.7. THE SAS CORRESPONDENCE PRINCIPLE

It is clear from §§3.4 and 3.5 that there corresponds to every solution of the system of SAS tangent equations (3.6) a local maximum of $M(\phi)$, and conversely (the SAS correspondence principle).

3.8. THE LINEAR CONGRUENCE CONNECTION

The problem of going from the estimated values $\omega_{\mathbf{HK}}$ of the three-phase structure invariants ϕ_{HK} (2.5) to the values of the individual phases ϕ may be formulated as the problem of solving the redundant system of linear congruences

$$\phi_{\mathbf{H}} + \phi_{\mathbf{K}} + \phi_{-\mathbf{H}-\mathbf{K}} \equiv \omega_{\mathbf{HK}} (\text{modulo } 2\pi) \tag{3.7}$$

each with weight $A_{\mathbf{HK}}$. This was the point of view adopted in the earlier work of Han *et al.* [7] and Hauptman and Han [8] in which the system (3.7) was transformed into a redundant system of linear equations

$$\phi_{\mathbf{H}} + \phi_{\mathbf{K}} + \phi_{-\mathbf{H}-\mathbf{K}} = \omega_{\mathbf{HK}} + 2\pi n_{\mathbf{HK}} \tag{3.8}$$

and the attempt was made, with limited success, to determine the integers $n_{\mathbf{HK}}$ in such a way as to make the system (3.8) self consistent. The resulting redundant system of linear equations (3.8) was then solved by least-squares.

The SAS maximal principle may thus be re-interpreted as yielding the solution of the redundant system of linear congruences (3.7).

3.9. THE INITIAL APPLICATION

The method described here was applied, with experimentally determined diffraction data, to the *ab initio* solution of the phase problem for the platinum derivative of the previously known macromomycin structure [12] consisting of approximately 750 protein atoms and 150 solvent molecules and crystallizing in the space group $P2_1$. With diffraction data to 2.5Å resolution, 150,000 three-phase structure invariants with largest A values were estimated. These involved 2710 phases whose values were to be determined.

One hundred solutions of the system of SAS tangent equations (3.6) were obtained using initial values of the phases chosen at random. Each of these trials converged to solution in five to eight cycles. Of the 100 trials, 17 yielded the same global maximum of the SAS maximal function $M(\phi)$ (3.5). Since the macromomycin structure had been previously determined, it was possible to calculate the average initial phase error using the known phases from the refined structure. This turned out to be 49° for all 2710 phases. It should be stressed that this solution of the phase problem for macromomycin was strictly *ab initio* in the sense that the only information needed were the observed SAS diffraction intensities at 2.5Å resolution; and the resulting map was interpretable.

With error-free diffraction intensities, the same calculation, again using SAS estimates for 150,000 three-phase structure invariants, yielded the values of 2120 phases with an initial average phase error of 30°.

Research supported by the National Institutes of Health Grant No. GM46733.

4. References

1. Hauptman, H. (1991) A minimal principle in the phase problem, in *Crystallographic Computing 5 from Chemistry to Biology*; Proceedings of the International School of Crystallographic Computing", Bischenberg, France (1990); D. Moras, A.D. Podjarny & J.C. Thierry (eds.), pp. 324-332. IUCr Oxford University Press.
2. Hauptman, H., Velmurugan, D. and Fusen, H. (1991) The minimal principle solves some crystal structures", in *Direct Methods of Solving Crystal Structures* (H. Schenk, Ed.), Proceedings of the International School of Crystallography, Erice, Trapani, Italy, April (1990), Plenum Publ., New York, New York, pp. 403-406.
3. DeTitta, G., Weeks, C., Thuman, P., Miller, R. and Hauptman, H. (1994) Structure solution by minimal function phase refinement and Fourier filtering. I. Theoretical basis, *Acta Cryst.* **A50**, 203-210.
4. Weeks, C.M., DeTitta, G.T., Hauptman, H.A., Thuman, P., and Miller, R. (1994) Structure solution by minimal function phase refinement and Fourier filtering. II. Implementation and applications, *Acta Cryst.* **A50**, 210-220.
5. Miller, R. Gallo, S.M., Khalak, H.G. and Weeks, C.M. (1994) *SnB*: Crystal structure determination *via Shake-and-Bake*, *J. Appl. Cryst.* **27**, 613-621.
6. Debaerdemaeker, T. and Woolfson, M.M. (1983) On the application of phase relationships to complex structures. XXII. Techniques for random phase refinement, *Acta Cryst.* **A39**, 193-196.
7. Han, F., DeTitta, G. and Hauptman, H. (1991) TELS: Least-Squares solution of the structure invariant equations, *Acta Cryst.* **A47**, 484-490.
8. Hauptman, H. and Han, F. (1993) Phasing macromolecular structures *via* structure invariant algebra, *Acta Cryst.* **D49**, 3-8.
9. Fan. H., Hao, Q., Gu, Y., Qian, J. and Zheng, C. (1990) Combining direct methods with isomorphous replacement or anomalous scattering data. VII. *Ab initio* phasing of one-wavelength anomalous scattering data from a small protein, *Acta Cryst.* **A46**, 935-939.
10. Hauptman, H. (1982) On integrating the techniques of direct methods with anomalous dispersion, *Acta Cryst.* **A38**, 632-641.
11. Hauptman, H. (1996) The SAS maximal principle: A new approach to the phase problem, *Acta Cryst.* **A52**, 490-496.
12. Van Roey, P. and Beerman, T.A. (1989) Crystal structure analysis of auromomycin apoprotein (macromomycin) shows importance of protein sidechains to chromophore binding selectivity, *Proc. Natl. Acad. Sci.-USA* **86**, 6587-6591.

SHAKE-AND-BAKE: APPLICATIONS AND ADVANCES

RUSS MILLER[1,2] and CHARLES M. WEEKS[1]
[1]*Hauptman-Woodward Med. Res. Inst., Inc., 73 High St., Buffalo, NY 14203 USA*
[2]*SUNY-Buffalo, Dept. of Comp. Sci., Buffalo, NY 14260 USA*

Abstract

Shake-and-Bake is a direct-methods procedure which has provided *ab initio* solutions for protein structures containing as many as 600 independent non-H atoms, provided that good-quality diffraction data are available to 1.1Å resolution. This algorithm extends the range of conventional direct methods by repetitively, unconditionally, and automatically alternating reciprocal-space phase refinement with filtering in real space to impose constraints. Choosing good values for parameters, such as the number of peaks used for density modification, selecting the optimum phase-refinement method, and performing an adequate number of refinement cycles, are critical for successful large-molecule applications. Large structures lacking atoms of any element heavier than oxygen require different parameter settings. In this paper, it is shown that with high probability a solution can be obtained for any one of six small protein structures within one day on a single SGI R10000 processor.

1. Introduction

The majority of small-molecule organic crystal structures having fewer than 100 independent non-H atoms are solved using reciprocal-space direct methods. These methods are based on the fact that there exist linear combinations of phases (the structure invariants), whose values, in principle, depend only on the magnitudes of the normalized structure factors. The conditional probability distributions of the structure invariants permit individual invariant values to be estimated as first proposed by Hauptman and Karle [1]. Conventional direct methods begin to fail in the 100-200 atom range because the accuracy of the underlying probabilistic phase relationships decreases as the size of the structure increases. Fortunately, however, the size of molecular structures amenable to phasing by direct methods can be increased significantly if these methods are augmented by the imposition of physically-meaningful constraints in real space.

Real-space phase-improvement methods, commonly known as density-modification methods, are widely used in macromolecular crystallography [2]. The potential for real-space constraints to improve phases in the context of small-molecule direct methods was recognized by Jerome Karle [3], who found that even a relatively small, chemically-sensible fragment extracted from an electron density map by manual interpretation could be parlayed into a complete solution by transformation back to reciprocal space and then performing additional iterations of phase refinement.

S. Fortier (ed.), Direct Methods for Solving Macromolecular Structures, 389–400.

The tremendous increases in computer speed in recent years have made it feasible to consider a 'multisolution' approach [4] in which each trial structure is repeatedly cycled back-and-forth between real and reciprocal space, while performing optimization alternately in each space. This compute-intensive process, which requires the use of two Fourier transforms during each cycle, forms the basis of the synergistic *Shake* (phase refinement) and *Bake* (density modification) procedure in which the power of reciprocal-space phase refinement is automatically augmented by filtering to impose the phase constraints implicit in real space [5-7]. A detailed description of *Shake-and-Bake* can be found in two recent reviews [8,9].

1.1. RECIPROCAL-SPACE PHASE REFINEMENT

Multisolution direct-method procedures require sets of starting phases which are subjected to refinement. In recent years, random number generators have been used to assign initial phase values [10,11]. In the *Shake-and-Bake* procedure, phases are assigned initial values by first generating trial structures consisting of randomly positioned atoms (thereby imposing an atomicity constraint from the outset) and then computing structure factors.

The tangent formula of Karle & Hauptman [12] provides a simple, but highly effective, means for extracting phase values from the three-phase or triplet invariants. Phase expansion and/or refinement in reciprocal space can be accomplished through successive applications of this relationship which, in either its original or a weighted form [13], is at the heart of widely-used conventional multisolution phasing programs, including MULTAN [14] and SHELXS-86 [15].

The *Shake-and-Bake* algorithm permits alternative optimization strategies as well as the tangent formula during the phase refinement step. In particular, one powerful technique is to reduce the value of an objective function, such as the minimal function $R(\phi)$ [16-18]. Minimal functions and the minimum principle are discussed in detail in another paper (H. Hauptman) in these proceedings. $R(\phi)$ is a measure of the mean-square difference between the values of the triplet invariants calculated using a trial set of phases and their expected values as predicted by the conditional probability distribution. The minimal function is expected to have a constrained global minimum when the phases are equal to their correct values for some choice of origin and enantiomorph. $R(\phi)$ can also be written to include contributions from higher-order (quartet) invariants, but this option has not been shown to be computationally efficient. Experimentation has thus far confirmed that when the minimal function is used actively in the phasing process, and solutions are produced, the final trial structure corresponding to the smallest value of $R(\phi)$ is a solution. Therefore, $R(\phi)$ is also an extremely useful figure-of-merit for selecting those trials which have converged to solution.

Parameter shift [19] is a seemingly simple search technique that has proven to be quite powerful as an optimization method when used to reduce the value of the minimal function, provided that appropriate choices of parameter values are made. The phases are ranked in decreasing order with respect to the values of the associated normalized structure-factor magnitudes. Each phase, in turn, is allowed to take on two or more alternative values, and the value of $R(\phi)$ corresponding to each is computed. The phase value yielding the minimum for $R(\phi)$ is taken as the refined value and immediately replaces the old value. Based on experimentation with known structures, it has been determined that, in terms of running time and percentage of trial structures that produce a solution, an excellent strategy is to perform a maximum of two 90° phase shifts and perform three complete iterations through the phase set [7]. Under these conditions, the notation

PS(90°,2,3) will be used. When parameter-shift phase refinement is applied in centrosymmetric space groups, only a single shift of 180° is required for each phase. In acentric space groups, phases having restricted values may either be constrained to these values (restricted PS) or treated as general phases (unrestricted PS). In either case, space-group restrictions are reapplied during structure-factor calculation.

It is interesting to note that, although tangent-formula phase refinement alone can produce solutions under favorable circumstances, parameter-shift refinement has not been used successfully apart from the dual-space *Shake-and-Bake* procedure. Other optimization methods have also been tried in the context of *Shake-and-Bake*. Parameter shift has some features in common with simulated annealing [20] but has fewer parameters and lacks the hill-climbing properties of the latter. Although simulated annealing was found to have potentially twice the success rate of parameter shift when applied to certain small-molecule test cases providing that good choices were made for all parameter values, it proved difficult to select the right values in advance [21]. Genetic algorithms were found to be inefficient as an optimization method in *Shake-and-Bake* [22].

1.2. REAL-SPACE CONSTRAINTS

In the applications reported to date, automatic real-space electron-density map interpretation in the *Shake-and-Bake* procedure consists of selecting an appropriate number of the largest peaks (equal to or less than the expected number of atoms in the structure) to be used as an updated trial structure without regard to chemical constraints other than a minimum allowed distance. These peaks are then regarded as atoms, and a structure-factor calculation imposes the atomicity constraint. If markedly unequal atoms are known to be present, appropriate numbers of peaks (atoms) can be weighted by the proper atomic numbers during transformation back to reciprocal space. Thus, *a priori* knowledge concerning the chemical composition of the crystal is utilized, but no knowledge of constitution is required or used during peak selection. It is useful to think of peak picking in this context as simply an extreme form of density modification appropriate when atomic-resolution data are available. The entire dual-space refinement procedure is repeated for an appropriate number of cycles which have been determined empirically by experimentation with known datasets [7].

2. *SnB*: An Implementation of *Shake-and-Bake*

The *Shake-and-Bake* procedure has been implemented in an efficient and easy-to-use program, *SnB* [23]. Pertinent information concerning *SnB*, including the complete *User's Manual*, may be accessed from the home page on the World Wide Web at http://www.hwi.buffalo.edu/SnB. Stand-alone UNIX executables for SGI, SUN, IBM, and DEC alpha workstations as well as PC/Linux versions may be downloaded without cost to academic users. *SnB* is also available on the Cray T3D and Cray C90 at the Pittsburgh Supercomputing Center, the TMC CM-5 at NCSA, and the IBM SP2 at the Cornell Theory Center.

2.1 THE DEVELOPMENT OF *SnB*

The development of *SnB* began in earnest in 1992 after the decision was made to create an easy-to-use program that could be widely distributed to the crystallographic community. The program was designed using a rapid-prototyping model [24]. Existing crystallographic routines for phase refinement, Fourier summation, peak picking, and structure-factor calculation (written in FORTRAN) were updated and modified so that they would blend together and keep all of the required information in memory instead of continually writing intermediate results to files. A user-friendly, ASCII-based front end was designed in C. C is a standard language on all UNIX systems and was chosen over FORTRAN for the front end primarily because of its enhanced string handling and system interface capabilities. C was also chosen as the core language for the back end on account of its ability to handle dynamic memory allocation and system integration. The critical crystallographic routines remained in FORTRAN.

The front end was designed in a highly modular way, allowing for migration of the code to a wide variety of sequential, parallel, and distributed computing environments. The choice of a basic ASCII interface was made at a time when a GUI interface would have been much more difficult to design and much less portable. In addition, efforts were made to bring the code into POSIX compliance so that porting the code from one platform to another would require a minimum of effort. After a number of Beta-tests were performed and the program refined, *SnB* vers. 1.0 was publicly released in 1994. Program maintenance, parameter tuning, and optimizations of the code continued from 1994-1996, and the web site was established from which the current version and *User's Guide* can be downloaded.

2.2 THE DEVELOPMENT OF *SnB* VERS. 2.0

By 1996, it became apparent that the use of *SnB* was slowly but steadily increasing, primarily for solving difficult atomic-resolution crystal structures that had resisted solution by more conventional means. When considering options for dealing with some of the computational inefficiencies in *SnB* vers. 1, it was determined that many of the gross inefficiencies resulted from the rapid-prototyping effort. It appeared that the best solution for overcoming these inefficiencies and allowing for future innovations was to re-implement the system from the ground up. During the past year, the core crystallographic routines were redesigned and rewritten by CMW. These routines will serve as the foundation for a new implementation of *SnB*, which will be released as *SnB* vers. 2.0 in the later part of 1997. Thus, the code and major data structures for *SnB* vers. 2, which is being demonstrated at this conference for the first time, are much more tightly integrated than the original version of the system. This results in a program with significantly improved performance, which is critical if *SnB* is to be routinely applied in reasonable computing environments to structures in the 1000 non-H atom range.

Additional crystallographic options have been incorporated into the new system, including the following:

- Normalized structure-factor magnitudes can now be calculated within *SnB* itself.
- An inverse Fourier transform has been added, which the user can now choose to invoke as an alternative to the structure-factor routine that was previously the only option for proceeding from real to reciprocal space.

- A variant of low-density elimination [25] is now available as a density modification option, in addition to the peak-picking routine used in the original version.
- The Fourier-refinement module can now work with reflections in addition to those used in the invariant-based phase refinement. This is expected to result in cleaner final maps.
- Options for handling atoms that fall on space-group-specific special positions have been added.

While optimizations, enhancements, and additional routines continue to be added, an effort has just been initiated to design a new user-friendly interface for *SnB* vers. 2.0. When considering the plethora of new environments that are available, including Windows-based Intel-based PCs, it became clear that using C as the language to redesign the front-end system might not be the best choice. After much discussion, it was decided that a combination of JAVA and HTML will be utilized to provide a front-end that will be highly portable among machines that implement a mature browser, such as Netscape or Internet Explorer.

2.3 *SnB* AND PARALLEL COMPUTERS

Parallel computers have played a major role in the development of *Shake-and-Bake* and *SnB*. Parallel computers, including the Intel iPSC Hypercube and Thinking Machines Corporation CM-2, were used from the beginning of this research effort since the existing departmental VAX machines and Sun workstations simply did not have the cycles necessary to support this effort. Once the *Shake-and-Bake* procedure was established, machines such as the Thinking Machines Corporation CM-5 and the Cray T3D were used in order to secure enough cycles so as to establish a basic proof of concept and begin optimizing the procedure. Once the procedure was reasonably efficient, it was ported to traditional workstations, including those manufactured by Sun and SGI. It was on these workstations that the development effort continued. Recently, *SnB* vers. 1.5 (the latest public release of *SnB*) has been retrofitted for parallel machines, including the HP/Convex Exemplar, SGI Origin, Cray T3E, and IBM SP2, in order to perform production runs, typically on very large structures.

3. Progressing Towards Macromolecules

In Table 1, a chronological listing of important *SnB* applications is given. The first significant application of *SnB* to an unknown structure was in 1992, with the solution of two polymorphic forms of a cycloheptapeptide, Ternatin I & II [6], with 110 and 105 non-H atoms in the asymmetric unit, respectively. The data for these structures were collected in 1982 and 1990, respectively, and a significant effort had been made over the years to solve them with traditional tangent-based direct methods as well as molecular replacement. All such attempts had failed. Ternatin was finally solved by *SnB* in approximately 70 minutes on a Thinking Machines Corporation CM-5, with a final postmortem analysis showing a nearly 1% success rate.

The *SnB* development team at the Hauptman-Woodward Institute then proceeded to apply *Shake-and-Bake* successfully to three previously known small proteins (crambin [26], gramicidin A [27], and rubredoxin [28]), all of which are much larger than structures routinely amenable to direct methods in the past. In fact, previous attempts to solve crambin *ab initio* using pure conventional tangent-formula-based direct methods had been unsuccessful [29]. Next, the first two successful applications of *Shake-and-Bake* to previously unknown structures of significant size (the 40-residue Er-1 pheromone [33] and the 12-residue tetrameric designer peptide alpha-1 [34]) came out of David Eisenberg's lab at UCLA. Both investigations were performed without the direct involvement of the *SnB* authors. Other significant new structures included two conotoxins with resolutions in the 1.1-1.2Å range solved by Jennifer Martin's group at the University of Queensland [35,36]. The 64-residue scorpion toxin (Tox II) [37], the largest successful *Shake-and-Bake* application so far, had been previously solved, but the number of residues and amino-acid sequence were deliberately withheld from the *SnB* team. The only information supplied was that the protein was composed of approximately 500 atoms and contained four disulfide bonds.

The final two entries in Table 1 are previously unknown structures solved in 1996. Vancomycin is a glycopeptide antibiotic that is regarded as the last line of defense against bacteria that have become resistant to antibiotics such as penicillin. This structure was solved simultaneously and independently in two laboratories [38,39] using different programs that implement the *Shake-and-Bake* paradigm, namely *SnB* [23] and Sheldrick and Gould's new variant of the *Shake-and-Bake* procedure which employs peaklist optimization [40]. Finally, the 289-atom proprietary structure crystallizing in space group I4 was solved by the authors after an *SnB* user requested assistance. This structure could only be solved by vers. 1.5 after a patch was inserted to eliminate all peaks occurring on special positions. Once this correction was made, the I4 structure exhibited an astonishingly high frequency of successful trials for a structure of its size (see next section). Interested individuals are referred to the *SnB* web site for an up-to-date listing of interesting, non-proprietary applications.

TABLE 1. Chronology of interesting *SnB* applications

Structure	Year	Space Group	Resolution	Non-H Atoms		'Heavy' Atoms
				Compound	Total	
Ternatin [6]	1992	$P2_12_12_1$	~0.9Å	104	105,110	None
Riboflavin Tetrabutyrate [30]	1993	P1	N.A.	188	188	None
Crambin [31]	1993	$P2_1$	0.83	327	~400	S6
Gramicidin A [32]	1993	$P2_12_12_1$	0.86	272	317	None
Rubredoxin [32]	1993	$P2_1$	1.0	395	497	FeS6
Er-1 Pheromone [33]	1994	C2	1.0	303	328	S7
Alpha-1 Peptide [34]	1994	P1	0.92	N.A.	471	Cℓ1
α-Conotoxin PnIA [35]	1995	$P2_1$	1.1	110	122	S4
α-Conotoxin GI [36]	1995	$P2_1$	1.2	97	118	S4
Scorpion Toxin II [37]	1995	$P2_12_12_1$	0.96	508	624	S8
Vancomycin [38]	1996	$P4_32_12$	0.9	202	258	Cℓ8
I4 Peptide [41]	1996	I4	1.1	248	289	S10

4. Improving Performance

The user of *SnB* vers. 1 only needs to supply (i) basic crystal data including space group, cell constants, and the contents of the asymmetric unit and (ii) an input reflection file consisting of *h, k, l* and the normalized structure-factor magnitudes, |E|. Cost-effective default values are provided for the control parameters (displayed following each query) based on extensive experimentation with several known test structures [7,42,43]. These parameters are summarized in Table 2. Several depend on structure size and can be expressed as a function of n, the number of independent non-H atoms in the asymmetric unit. In the following discussion, the term 'success rate' refers to the percentage of trial structures which converge to solution, and 'efficiency' or 'cost-effectiveness' refers to the number of solutions per unit of time (*e.g.*, hours or days) using a particular computing platform. *SnB* solutions for known test structures are automatically confirmed on the basis of mean phase error or the cosine figure-of-merit [31, eq. 5]. In all cases the minimal function, $R(\phi)$, also served to identify solutions unambiguously.

TABLE 2 Default parameter values for the *SnB* structure-determination procedure. The quantity n is the number of non-H atoms in the asymmetric unit.

Parameter	Default
Invariant generation:	
Number of phases	$10n$
Number of triples	$100n$
Number of negative quartets	0
Starting atoms/random trial	min (n,100)
Number of *SnB* cycles:	
Parameter-shift (PS) refinement	$n/2$
Tangent-formula refinement	$n/4$
PS phase refinement:	
Size of phase shift	90°
Max. number of shifts	2
Number of iterations	3
Exploit restricted phases?	No
Number of peaks to select	$[0.8n,n]$
Exploit heavy atoms?	Yes
Number E-Fourier steps	2

A few comments concerning Table 2 and the operation of *SnB* vers. 1 are in order. In practice, it is not necessary to use more than 100 randomly positioned atoms as an initial trial structure for the first cycle. During subsequent cycles, choosing n peaks to recycle through the procedure is satisfactory for smaller structures. However, for large structures containing a significant number of atoms with low occupancy or high thermal motion, trial structures composed of less than n peaks (*e.g.*, $0.8*n$) give better performance. The geometry of trials that are solutions can be improved by E-Fourier recycling [44], and the user can select the number of such Fourier refinement cycles (*i.e.*, *SnB* cycles with no phase refinement) and the number of peaks to be used in each cycle. It is often useful to build, over the course of several cycles, from the number of peaks used during the *Shake-and-Bake* stage to the approximate total number of atoms expected in the structure. When atoms with atomic numbers greater than 10 are present, the user has the option of weighting the appropriate number of largest peaks in the structure-factor calculations. Unequal weighting has resulted in accelerated convergence to solution in cases where a small number of sulfur, iron, or chlorine atoms is present.

The relative efficiency of tangent-formula and parameter-shift phase refinement in *Shake-and-Bake* has been compared for several datasets [42,43]. In the case of tangent refinement, the minimal function is also computed, but used only as a figure-of-merit. Regardless of which refinement method is used, optimization proceeds most rapidly when there is immediate feedback of each refined phase value. In general, the tangent formula solves small structures (<100 atoms) more cost-effectively, but the relative efficiency of parameter shift increases quickly with increasing structure size. In addition, tangent-formula cost-effectiveness is highly dependent on the number of phase-refinement iterations (*i.e.*, the number of passes through the list of phases) per complete *Shake-and-Bake* cycle, and the number of iterations per cycle must be chosen judiciously if high efficiency is, in fact, to be achieved. This is especially true for structures in space group P1 where it is never advisable to perform more than one iteration of tangent refinement per cycle. Parameter-shift efficiency does not exhibit such strong dependence on the number of refinement iterations, and it is advised that three iterations be performed. Given a fixed number of machine cycles, it is important to consider the trade-off between the number of trial structures processed and the number of cycles processed per trial structure. In the case of small molecules, *n*/4 cycles are adequate for tangent refinement, and *n*/2 cycles are advised for parameter shift.

Using the data for six proteins, a study was recently undertaken with *SnB* vers. 2 to determine whether the default recommendations associated with *SnB* vers. 1 are also wise choices for attempting to solve large molecules with *SnB* vers. 2. For purposes of this experiment, *n* is taken to be the number of atoms in the asymmetric unit for the protein molecule itself. Initial experiments with crambin confirmed that certain recommendations such as the atom:phase:invariant ratio, the number of parameter-shift iterations, and unequal peak weighting in the presence of heavier atoms, appear to be valid for larger structures. However, these initial experiments suggested that further investigations were in order. The results of these investigations are presented in Tables 3-5.

Based on speculation that something interesting might happen if the number of peaks used in the structure-factor calculation was reduced substantially (especially if heavier atoms were present), the number of recycled peaks was varied over a broad range of values. The results for the six protein structures are given in Table 3. In all but one case, the optimum number of peaks to use appears to be significantly less than half the number of atoms in the protein molecule itself. The exception to this rule is gramicidin A, the only one of the six structures containing solely first row elements. The variation in success rate is especially significant for scorpion toxin II, the largest of the test structures. The original *SnB* vers. 1 attempt at solving the structure consisted of processing trial structures for 255 (or *n*/2) cycles of unrestricted parameter shift (maximum of two 90° phase shifts; three iterations) with 400 peaks selected for use in the structure-factor calculation. This yielded one solution out of 1619 trials. In the present study, approximately 18,000 such trials were generated, and the lowest phase errors obtained after 255 cycles were 58, 74, and 79°, one of which was detectable based on the minimal function. These three trials were then subjected to further refinement and found to converge to solution (mean phase error less than 40°) after 270, 295, and 320 cycles, respectively. Approximately 3300 400-peak trials were then refined for 500 cycles. Three of these trials converged to solution, but only after more than 480 *SnB* cycles. Therefore, one may conclude that the original *SnB* solution of Tox II was indeed serendipitous! In contrast, the 200-peak trials are much more likely to lead to solution (Table 3).

Both restricted and unrestricted parameter-shift phase refinement are compared to tangent refinement in Table 4. It can be seen that, in all six cases, unrestricted parameter shift is superior to restricted parameter shift, and both forms of parameter shift produce more solutions than the tangent formula. Again, the results for gramicidin A stand out. No tangent-based solutions were obtained after 275 cycles, in contrast to the 17 solutions obtained with unrestricted parameter shift. In fact, all 2000 gramicidin A trials in this experiment were subjected to a total of 500 refinement cycles, and only one tangent solution was obtained (at about cycle 350). After 500 cycles, there were 53 unrestricted parameter-shift solutions.

TABLE 3. Influence of the number of peaks used in the structure-factor calculation on the success rate of *Shake-and-Bake* with unrestricted parameter-shift (90°,2,3) phase refinement.

Protein	Trials	Cycles	Phases	Peaks Used							
				25	50	100	150	200	250	300	400
Vancomycin	1000	200	2000	0.8%	0.4%	0.6%	0.3%	0.2%	0.2%	---	---
I4 Peptide	1000	250	1900	44.0	53.0	52.0	50.0	45.0	42.0	---	---
Gramicidin A	5000	275	3000	---	0	0.3	0.5	1.1	0.9	0.7%	---
Crambin	5000	200	3000	2.6	3.3	3.7	3.6	2.9	2.9	2.7	---
Rubredoxin	1000	400	4000	5.0	5.7	6.2	6.0	5.4	5.1	3.9	3.4%
Toxin II	>2000	500	5000	---	---	1.0	---	1.4	---	0.4	0.1

TABLE 4. Success rates for different phase-refinement options. The number of peaks selected for each structure was chosen based on the optimum values reported in Table 3.

Protein	Trials	Cycles (~n)	Phases	Peaks	Parameter-Shift (90°,2,3) Unrestricted	Parameter-Shift (90°,2,3) Restricted	Tangent Formula
Vancomycin	2500	200	2000	100	0.6%	0.5%	0.4%
I4 Peptide	1000	250	1900	50	53.0	44.0	25.0
Gramicidin A	2000	275	3000	200	0.9	0.4	0
Crambin	2000	300	3000	100	4.8	3.7	2.2
Rubredoxin	1000	400	4000	150	6.0	5.2	4.0
Toxin II	>2000	500	5000	200	1.4	1.0	0.6

As shown by the results presented in Table 5, gramicidin A continues to behave differently when the question of the optimum number of refinement cycles is considered. It is most efficient to perform $n/2$ or fewer cycles for four of these six proteins. The exceptions are gramicidin and Tox II, the largest structure, for which $3n/4$ cycles is at least as good as $n/2$ cycles. Therefore, it may be advisable to perform at least $3n/4$ cycles for larger structures so as to insure that solutions are not missed altogether. The efficiency of solution for gramicidin A is highest at 500 cycles (ca. $1.75n$), the maximum tested. The suggestion is that large structures lacking any atoms heavier than oxygen are especially likely to require many cycles for convergence to solution to be achieved. This also appears to be true for the smaller, but difficult, structure ternatin ($n \cong 100$) for which oxygen is also the heaviest element.

TABLE 5. Success rates as a function of the number of unrestricted parameter-shift (90°,2,3) refinement cycles. The numbers of cycles are expressed in terms of n, the number of independent non-H atoms in the protein. The numbers of trials, phases, and Fourier peaks used for each structure are given in Table 3.

Protein	SnB Refinement Cycles Performed					
	$0.25n$	$0.5n$	$0.75n$	n	$1.25n$	$1.5n$
Vancomycin	0.1%	0.4%	0.4%	0.6%	0.7%	---
I4 Peptide	27.0	40.0	48.0	53.0	---	---
Gramicidin A	0	0.4	0.6	0.9	1.2	2.0%
Crambin	3.1	4.1	4.6	4.8	---	---
Rubredoxin	4.6	5.5	5.9	6.0	---	---
Toxin II	0.05	0.5	1.0	1.4	---	---

SnB throughput for both versions, in terms of trials per day, for the six proteins is summarized in Table 6. Assuming that *n*/2 cycles are performed and an optimum number of peaks selected in each cycle, it can be determined from the expected success rates and trial turnover that there is a high probability that a solution will be found in a single day for all structures except Tox II using an SGI R10000 processor. If more cycles (~*n*) are performed for each trial processed, there is a strong likelihood that even Tox II will be solved on a single workstation in a single day.

TABLE 6. *SnB* success rate and throughput (given as trials per day) using unrestricted parameter-shift (90°,2,3) phase refinement, the optimum number of peaks (highest success rate) as given in Table 3, and *n*/2 cycles. The numbers of trials per day are for an SGI R10000 processor.

Structure	Phases	Cycles	Peaks	Success	Trials per Day	
		(*n*/2)		Rate	Vers. 1	Vers. 2
Vancomycin	2000	100	100	0.4%	382	391
I4 Peptide	1900	125	50	40.0	540	274
Gramicidin A	3000	135	200	0.4	258	572
Crambin	3000	150	100	4.1	443	1029
Rubredoxin	4000	200	150	5.5	226	294
Toxin II	5000	250	200	0.5	73	109

5. Concluding Remarks

The ultimate potential of the *Shake-and-Bake* approach to the *ab initio* structure determination of macromolecules is unknown. Based on the study of six proteins reported here, it is possible, however, to suggest guidelines that maximize the probability of obtaining a solution for a structure containing several hundred atoms. If several atoms of moderately 'heavy' elements (S, Cl, Fe) are present, it may be best to select a number of peaks equivalent to about 40% of the non-H protein atoms in the asymmetric unit. If no such elements are present, the 80% rule suggested for *SnB* vers. 1 is a better choice. It is always best to use parameter shift rather than the tangent formula as a refinement method, especially if only carbon, nitrogen, and oxygen are present. Furthermore, the parameter-shift refinement should always be unrestricted. If the structure is very large (more than 500 non-H atoms) or contains only the lighter elements, it appears to be prudent to perform at least as many refinement cycles as there are independent non-H protein atoms.

6. Acknowledgments

The *Shake-and-Bake* algorithm and the *SnB* program have been made possible by the financial support of grants GM-46733 from NIH and IRI-9412415 from NSF. The authors would like to express their appreciation to Steve Gallo for all of his work in the development of *SnB* vers. 1, to Jan Pevzner for assisting with the coding of the inverse Fourier transform in *SnB* vers. 2, to Brett Miller for assistance in obtaining and tabulating some of the results from *SnB* vers. 2 presented in Section 4, and to Prof. Herbert Hauptman for his continued inspiration and support.

7. References

[1] Hauptman, H.A. and Karle, J. (1953) *Solution of the Phase Problem. I. The Centrosymmetric Crystal*, ACA Monograph No. 3, Polycrystal Book Service, Dayton, OH.

[2] Podjarny, A.D., Bhat, T.N., and Zwick, M. (1987) *Ann. Rev. Biophys. Biophys. Chem.* **16**, 351.

[3] Karle, J. (1968) Partial structural information combined with the tangent formula for noncentrosymmetric crystals, *Acta Cryst.* **B24**, 182-186.

[4] Germain, G. and Woolfson, M.M. (1968) On the application of phase relationships to complex structures, *Acta Cryst.* **B24**, 91-96.

[5] Weeks, C.M., DeTitta, G.T., Miller, R., & Hauptman, H.A. (1993) Applications of the minimal principle to peptide structures, *Acta Cryst.* **D49**, 179-181.

[6] Miller, R., DeTitta, G.T., Jones, R., Langs, D.A., Weeks, C.M., and Hauptman, H.A. (1993) On the application of the minimal principle to solve unknown structures, *Science* **259**, 1430-1433.

[7] Weeks, C.M., DeTitta, G.T., Hauptman, H.A., Thuman, P., and Miller, R., (1994) Structure solution by minimal function phase refinement and Fourier filtering: II. Implementation and applications, *Acta Cryst.* **A50**, 210-220.

[8] Weeks, C.M. and Miller, R. (1996) *SnB*: Applying *Shake-and-Bake* to proteins, in P. Bourne and K. Watenpaugh (eds.), *Proceedings of the Macromolecular Crystallography Computing School*, Bellingham, WA. An electronic version can be viewed at http://www/sdsc.edu/Xtal/IUCr/CC/School96/.

[9] Weeks, C. and Miller, R. (1997) Macromolecular Phasing by *Shake-and-Bake*, in K.S. Wilson, G. Davies, and S. Bailey (eds.), *Recent Advances in Phasing, Proceedings of the CCP4 Study Weedend*, DL-CONF-97-02, CCLRC, Daresbury Laboratory, Warrington, UK, in press.

[10] Baggio, R., Woolfson, M.M., Declercq, J.-P., and Germain, G. (1978) On the application of phase relationships to complex structures. XVI. A random approach to structure determination, *Acta Cryst.* **A34**, 883-892.

[11] Yao, J.-X. (1981) On the application of phase relationships to complex structures. XVIII. RANTAN - random MULTAN, *Acta Cryst.* **A37**, 642-664.

[12] Karle, J. and Hauptman, H. (1956) A theory of phase determination for the four types of non-centrosymmetric space groups 1P222, 2P22, $3P_12$, $3P_22$, *Acta Cryst.* **9**, 635-651.

[13] Hull, S.E. and Irwin, M.J. (1978) On the application of phase relationships to complex structures. XIV. The additional use of statistical information in tangent-formula refinement, *Acta Cryst.* **A34**, 863-870.

[14] Main, P., Fiske, S.J., Hull, S.E., Lessinger, L., Germain, G., Declercq, J.P., and Woolfson, M.M. (1980) *MULTAN80: A system of computer programs for the automatic solution of crystal structures from x-ray diffraction data* (Universities of York and Louvain).

[15] Sheldrick, G.M. (1985) *SHELXS86. Program for the solution of crystal structures* (University of Gottingen, Germany).

[16] Debaerdemaeker, T. and Woolfson, M.M. (1983) On the application of phase relationships to complex structures. XXII. Techniques for random phase refinement, *Acta Cryst.* **A39**, 193-196.

[17] Hauptman, H.A. (1991) A Minimal Principle in the Phase Problem, in D. Moras, A.D. Podnarny & J.C. Thierry (eds.), *Crystallographic Computing 5: From Chemistry to Biology*, IUCr Oxford Univ. Press, pp. 324-332.

[18] DeTitta, G.T., Weeks, C.M., Thuman, P., Miller, R. and Hauptman, H.A. (1994) Structure solution by minimal function phase refinement and Fourier filtering: Theoretical basis, *Acta Cryst.* **A50**, 203-210.

[19] Bhuiya, A.K. and Stanley, E. (1963) The refinement of atomic parameters by direct calculation of the minimum residual, *Acta Cryst.* **16**, 981-984.

[20] Kirkpatrick, S., Gelatt, C.D., and Vecchi, M.P. (1983) Optimization by simulated annealing, *Science* **220**, 671-680..

[21] Khalak, H.G., Weeks, C.M., Miller, R., and Bloebaum, C.L. (1994) Simulated annealing applied to molecular structure determination, in *Proceedings 5th AIAA/NASA/USAF/ISSMO Symposium on Multidisciplinary Analysis and Optimization*, Panama City, FL.

[22] Chang, C.-S., DeTitta, G.T., Miller, R., and Weeks, C.M., (1994) On the application of parallel genetic algorithms in x-ray crystallography, in *Proceedings of the 1994 Scalable High Performance Computing Conference*, IEEE Computer Society Press, pp. 796-802.

[23] Miller, R., Gallo, S.M., Khalak, H.G., and Weeks, C.M. (1994) *SnB*: Crystal structure determination *via Shake-and-Bake*, *J. Appl. Cryst.* **27**, 613-621.

[24] Gallo, S.M., Miller, R., and Weeks, C.M. (1996) The design of a portable scientific tool: A case study using *SnB*, in *Proceedings Supercomputing '96 Conference*. Can be viewed at http://www.supercomp.org/sc96/proceedings/.

[25] Shiono, M. and Woolfson, M.M. (1992) Direct-space methods in phase extension and phase determination. I. Low-density elimination, *Acta Cryst.* **A48**, 451-456.

[26] Hendrickson, W.A. and Teeter, M.M. (1981) Structure of the hydrophobic protein crambin determined directly from the anomalous scattering of sulfur, *Nature* **290**, 107-113.

[27] Langs, D.A. (1988) Three-dimensional structure at 0.86Å of the uncomplexed form of the transmembrane ion channel peptide gramicidin A, *Science* **241**, 188-191.

[28] Dauter, Z., Sieker, L.C., and Wilson, K.S. (1992) Refinement of rubredoxin from *Desulfovibrio vulgaris* at 1.0Å with and without restraints, *Acta Cryst.* **B48**, 42-59.

[29] Sheldrick, G.M., Dauter, Z., Wilson, K.S., Hope, H., and Sieker, L.C. (1993) The application of direct methods and Patterson interpretation to high-resolution native protein data, *Acta Cryst.* **D49**, 18-23.

[30] Ferrara, J., personal communication.

[31] Weeks, C.M., Hauptman, H.A., Smith, G.D., Blessing, R.H., Teeter, M.M., and Miller, R. (1995) Crambin: A direct solution for a 400 atom structure, *Acta Cryst.* **D51**, 33-38.

[32] Hauptman, H.A. (1995) Looking ahead, *Acta Cryst.* **B51**, 416-422.

[33] Anderson, D.H., Weiss, M.S., and Eisenberg, D. (1996) A challenging case for protein crystal structure determination: the mating pheromone Er-1 from *Euplotes raikovi*, *Acta Cryst.* **D52**, 469-480.

[34] Prive, G., Ogihara, N., Wesson, L., Cascio, D., and Eisenberg, D. (1995) A designer peptide at high resolution: *Shake-and-Bake* solution of a 400 atom structure, Abstr. W008, Proc. Am. Crystallogr. Assoc. Meeting, Montreal.

[35] Hu, S.-H., Gehrmann, J., Guddat, L.W., Alewood, P.F., Craik, D.J., and Martin, J.L. (1996) The 1.1Å crystal structure of the neuronal acetylcholine receptor antagonist, α-conotoxin PnIA from *Conus pennaceus*, *Structure* **4**, 417-423.

[36] Guddat, L.W., Martin, J.L., Shan, L., Edmundson, A.B., and Gray, W.R. (1996) Three-dimensional structure of the α-conotoxin GI at 1.2Å resolution, *Biochemistry* **35**, 11329-11335.

[37] Smith, G.D., Blessing, R.H., Ealick, S.E., Fontecilla-Camps, J.C., Hauptman, H.A., Housset, D., Langs, D.A., and Miller, R. (1996) The *ab initio* structure and refinement of a scorpion protein toxin, Abstr. E1146, IUCr Meeting, Seattle.

[38] Loll, P.J., Bevivino, A.E., Korty, B.D., and Axelsen, P.H. (1997) Simultaneous recognition of a carboxylate-containing ligand and an intramolecular surrogate ligand in the crystal structure of an asymmetric vancomycin dimer, *J. Am. Chem. Soc.* **119**, 1516-1522.

[39] Schafer, M., Schneider, T.R., and Sheldrick, G.M. (1996) Crystal structure of vancomycin, *Structure* **4**, 1509-1515.

[40] Sheldrick, G.M. and Gould, R.O. (1995) Structure solution by iterative peaklist optimization and tangent expansion in space group P1, *Acta Cryst.* **B51**, 423-431.

[41] *SnB* solution by the authors. On account of its proprietary nature no further information can be revealed concerning the structure itself.

[42] Weeks, C.M., Hauptman, H.A., Chang, C.-S., and Miller, R. (1997) Structure Determination by Shake-and-Bake with Tangent Refinement, *ACA Transactions Symposium* **30**, in press.

[43] Chang, C.-S., Weeks, C.M., Miller, R., and Hauptman, H.A. (1997) Incorporating tangent refinement in the *Shake-and-Bake* formalism, *Acta Cryst.* **A53**, in press.

[44] Sheldrick, G.M. (1985) SHELX-84, in G M. Sheldrick, C. Kruger & R. Goddard (eds.), *Crystallographic Computing 3: Data Collection, Structure Determination, Proteins, and Databases*, Clarendon Press, Oxford, pp. 184-189.

SHELX APPLICATIONS TO MACROMOLECULES

GEORGE M. SHELDRICK
University of Göttingen
Institut für Anorganische Chemie, Tammannstraße 4,
D-37077 Göttingen, Germany

Abstract

Direct methods inspired by *Shake & Bake* involving iteration between real and reciprocal space are able to solve structures with several hundred independent atoms, but still require data to *atomic resolution* (about 1.2Å). It is planned to add a new program, SHELXD, incorporating such algorithms in a future release of the SHELX system. Initial trials of the preliminary test version of SHELXD have been reasonably successful, and include the *ab initio* solution of several large 'small-molecules' that had resisted attempts using previous version of SHELXS, as well as the *ab initio* solution of three unknown (and several known) proteins from the native data alone.

1. Introduction

A feature that probably contributed significantly to the rapid acceptance of the conventional direct methods program SHELXS-86 was the E-Fourier recycling [1,2] shown in Figure 1 that was used to complete the structure obtained from direct methods. Usually a couple of cycles were sufficient. Since the E-Fourier recycling was only applied to the 'best' solution, and only E-values greater than (say) 1.2 were employed, the computing requirements were modest. Very often this procedure was able to find every atom (except perhaps disordered solvent molecules), which users found very convenient. On a few occasions the E-Fourier recycling succeeded in extracting the solution from a rather dubious set of direct methods phases, but despite this strong hint, it did not occur to the author that it could itself be effective as a 'direct method'. This required the development of the *Shake & Bake* philosophy by Weeks, Miller and Hauptman at Buffalo [3,4], which inspired much of the work reported here.

S. Fortier (ed.), Direct Methods for Solving Macromolecular Structures, 401–411.

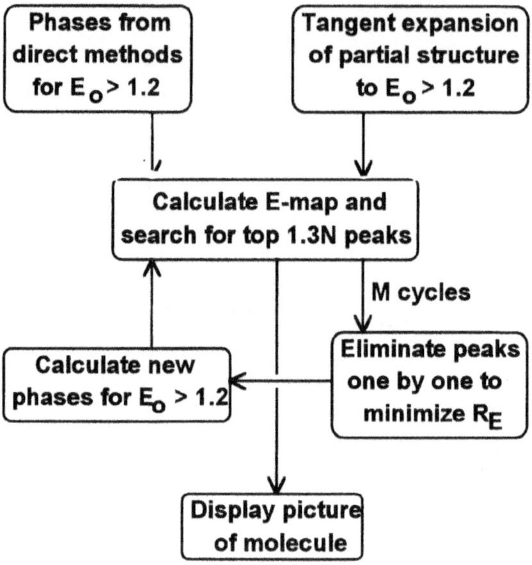

Figure 1. E-Fourier recycling as used in SHELXS-86 to improve phases from direct methods.

2. Peaklist Optimisation

Fine tuning of the *E*-Fourier recycling method since SHELXS-86 was distributed included Sigma-A weighted difference Fourier maps [5] and the use of the correlation coefficient [6] between E_c^2 and E_o^2 to decide which atoms to delete:

$$CC = \left[\sum wE_o^2E_c^2 \cdot \sum w - \sum wE_o^2 \cdot \sum wE_c^2 \right] \Big/ \left\{ \left[\sum wE_o^4 \cdot \sum w - \left(\sum wE_o^2\right)^2 \right] \cdot \left[\sum wE_c^4 \cdot \sum w - \left(\sum wE_c^2\right)^2 \right] \right\}^{1/2}$$

The correlation coefficient is more sensitive than the *R*-factor in the important early stages, and appears to give a very good indication of the true phase error (e.g. Figure 2). Tests on rubredoxin by Sheldrick and Gould [7] showed that the elimination of atoms to improve the correlation coefficient (*peaklist optimisation*) was very effective at expanding the structure from the iron and four sulfur atoms to all ca. 400 atoms, provided that the resolution was better than 1.3Å.

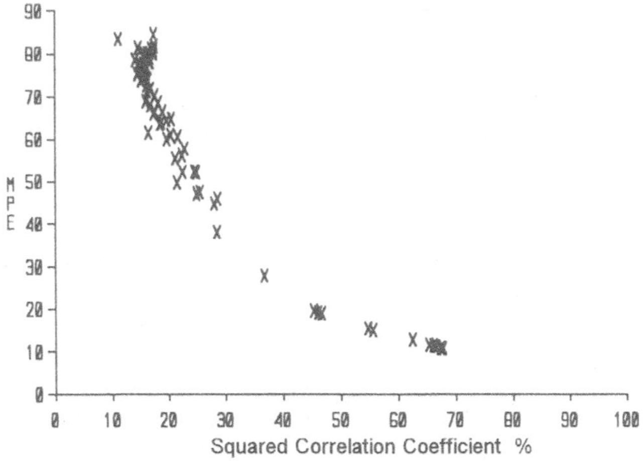

Figure 2. E-weighted mean phase error (MPE) from direct methods attempts as a function of the square of the correlation coefficient between E_c^2 and E_o^2 for crambin (0.92Å data kindly provided by Håkon Hope).

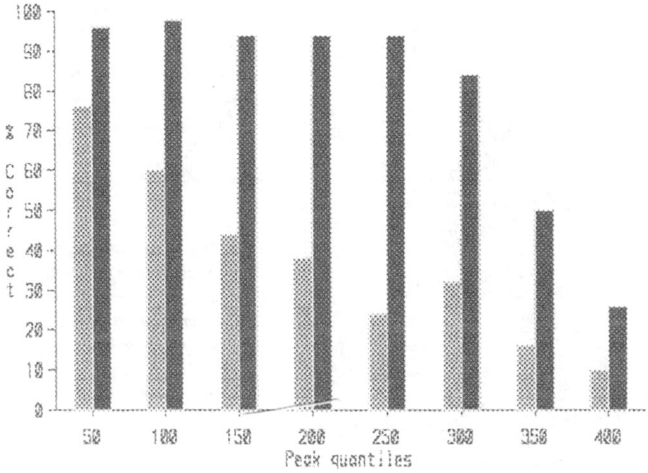

Figure 3. Histogram showing the percentage of correct peaks (within 0.3Å of the true atomic positions) before (gray) and after (black) iterative ***peaklist optimisation*** for rubredoxin with the data truncated to 1.2Å. There are 50 peaks in each quantile, so the first pair of columns are for peaks 1-50 sorted on peak height, the second pair for peaks 51-100, etc.

3. Application to the Solution of an Unknown Protein

Provided that a few heavier atoms can be located by for example Patterson interpretation, *peaklist optimisation* can be used to complete the structure and so - in the exceptionally favourable case of a small macromolecule that diffracts to atomic resolution and contains a few heavier atoms - provides a method of *ab initio* structure solution. Frazão *et al.* [8] were able to solve the structure of an unknown cytochrome c_6 in this way. The best sequence identity with a protein of known structure was only about 24%, so molecular replacement would have been difficult, but undoubtedly the structure could also have been solved - albeit at a higher cost in synchrotron beam-time - by MAD phasing. The iron and three sulfurs were located by automated Patterson interpretation [9] and the full structure was expanded from them by *peaklist optimisation* using synchrotron data collected to 1.1Å (although the 1.2 to 1.1Å shell was extremely weak) at the EMBL outstation in Hamburg. Figure 4 shows the same region of the structure at different stages of the structure determination.

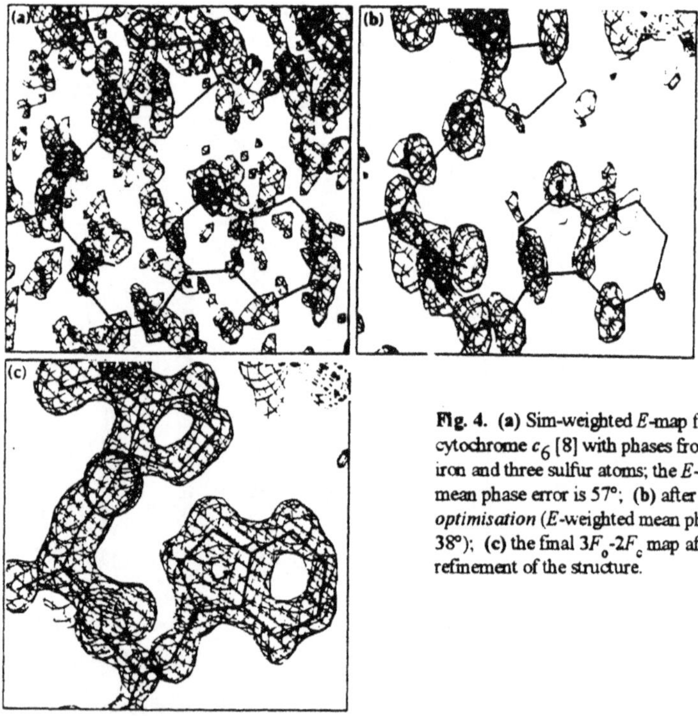

Fig. 4. **(a)** Sim-weighted E-map for a cytochrome c_6 [8] with phases from the iron and three sulfur atoms; the E-weighted mean phase error is 57°; **(b)** after *peaklist optimisation* (E-weighted mean phase error 38°); **(c)** the final $3F_o$-$2F_c$ map after refinement of the structure.

Iterative application of *peaklist optimisation* enabled about 90% of the protein atoms to be identified from the peaklist alone without the need to examine any maps; this was however required to find the remaining atoms, which had high thermal displacement parameters or were disordered.

4. A Real/Reciprocal Space Recycling Method for Structure Determination

Inspired by the *Shake & Bake* Philosophy described in the preceding lecture, Sheldrick and Gould [7] turned the *peaklist optimisation* procedure into a full *ab initio* method for structure determination by the addition of the tangent formula in the reciprocal space stage. Their algorithm (Figure 5) could start from random phases for a number of trials, or the initial phases could be generated by (a) a rotation search (to maximise $\Sigma E_c^2(E_o^2-1)$ for the largest E-values) for a known small fragment (a small piece of α-helix proved very effective) or (b) threefold Patterson superposition from vector triangles identified in the sharpened Patterson peaklist (to exploit the presence of heavier atoms such as sulfur or phosphorus). Since these two methods of generating slightly better than random starting phases are not able to position the origin of the space group, all calculations were performed on data expanded to the effective space group $P1$. Expansion to $P1$ may in any case increases the chances of this approach converging to the correct solution, but increases the computer time required.

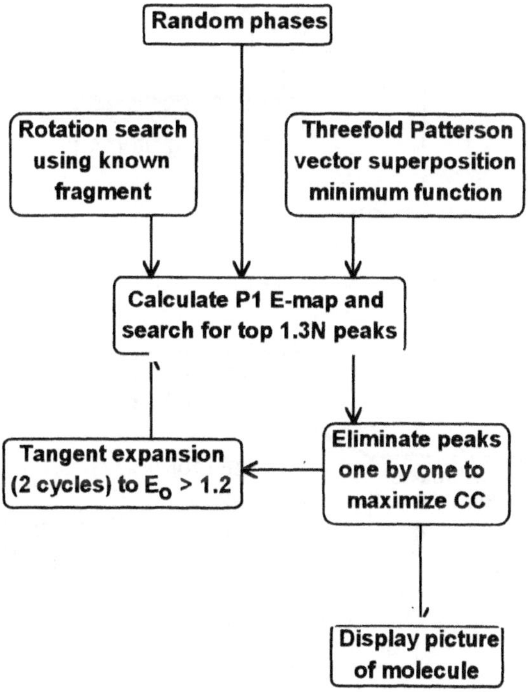

Figure 5. Real/reciprocal space recycling starting from random or almost random phases proposed by Sheldrick & Gould [7]. The correlation coefficient CC was calculated for all data expanded to the space group $P1$.

Tests showed that the *peaklist optimisation* was much more effective than simply accepting the top N peaks, but that it takes about the same CPU time as three structure factor calculations, and so is slower. Starting with slightly better than random phases from the rotation search or Patterson superposition map considerably increased the success rate. The method was successful in solving several structures with more than 200 atoms in the asymmetric unit, but proved very expensive in consumption of computer resources.

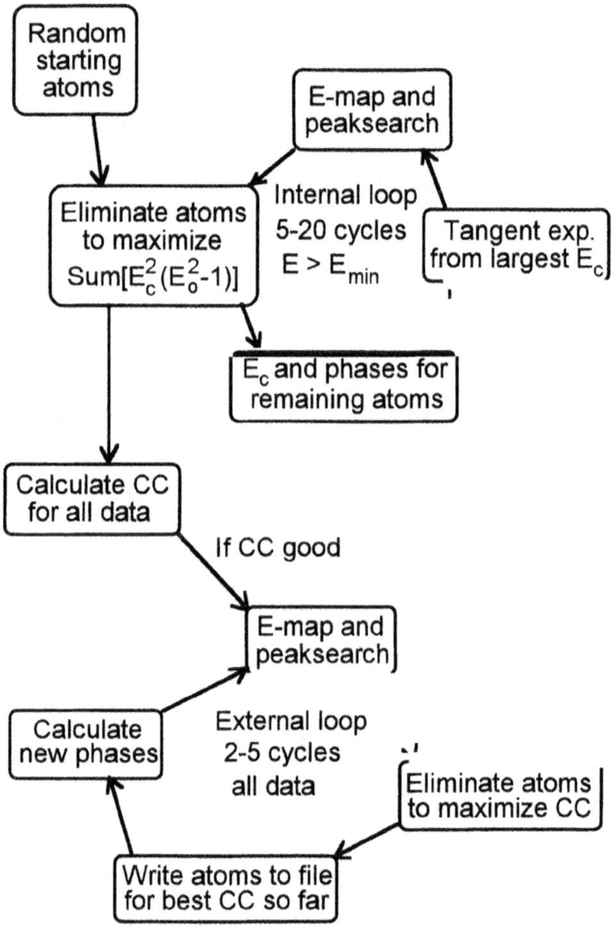

Figure 6. The ***half-baked*** approach, as incorporated in SHELXD. It is repeated indefinitely, restarting from random atoms, until interrupted! It may be performed either applying the symmetry of the space group or after expanding all data to the effective space group *P*1 (in which case the starting phases may be generated by a rotation search). Usually not more than two tangent cycles are applied per internal loop cycle. E_{min} is normally chosen to be in the range 1.2 to 1.6.

The computer time required could be reduced considerably by calculating the correlation coefficient for only the largest E-values, for which structure factors were required anyway to provide initial phases for the tangent refinement. However the correlation coefficient proved much less effective when not applied to the full range of E-values. The solution was to divide the procedure into an *internal loop*, in which a specified number of peaks were eliminated subject to the condition that $\Sigma E_c^2(E_o^2-1)$ remained as large as possible, alternating with tangent phase refinement, and an *external loop*, applied only for solutions with good values of CC (for all data), in which *peaklist optimisation* as described above was applied using all data so that the final structure was as complete as possible. The new procedure (which has somehow acquired the name *half-baked*) is illustrated in Figure 6.

In general, it appears to be computationally more efficient to expand the data to an effective space group of $P1$ for monoclinic structures; a larger percentage of trials lead to a solution, more than compensating for the increased cycle time. For higher symmetry it may be better to impose the full space group symmetry. It should also be possible to include twinning in the external loop; sometimes it is easier to guess the twin law than the space group, in which case the data could be expanded to $P1$.

The procedure described above is philosophically similar to *Shake & Bake*, but relative to *Shake & Bake* it does more of the work in real than in reciprocal space. It appears to be roughly comparable in its ability to solve difficult structures. One structure solved at about the same time by both programs, but using two different synchrotron data-sets, is vancomycin, a glycopeptide antibiotic of crucial medical importance in the struggle against the evolution of antibiotic resistant bacteria. The unexpurgated solution obtained by Schäfer, Schneider & Sheldrick [10] using the *half-baked* procedure is shown in Figure 7. The data were 99.3% complete to 1.09Å, the edge of the image plate used for synchrotron data collection. Including solvent there are 313 atoms in the asymmetric unit in $P4_32_12$. 2000 trials with 8 cycles in the internal loop gave one solution; the CC of 75.5% was well separated from the rest (the next largest CC was 57.9%, for an incorrect solution). The CPU time used corresponded to a mere 4 VAX-years.

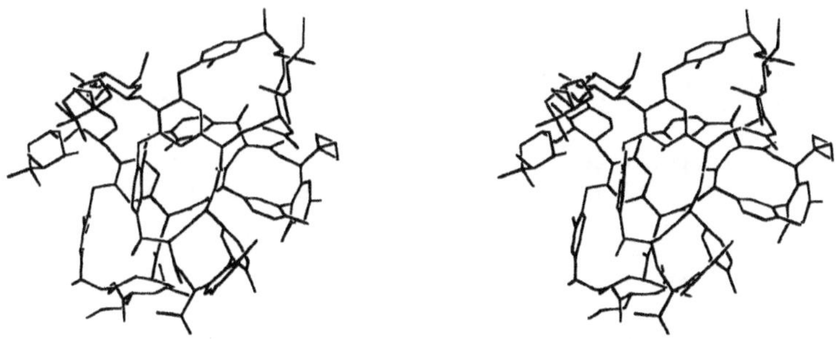

Figure 7. Stereo view of the unedited *ab initio* solution of the structure of vancomycin. The two antibiotic molecules are almost complete, and form a tight unsymmetrical dimer.

5. Exploiting the Presence of Heavier Atoms

If the structure contains *heavier* atoms such as sulfur or metal atoms, the control parameters can be set so that the *internal loop* (Figure 6) is used to locate these atoms, e.g. by starting from $2N_H$ random atoms or E-map peaks and eliminating N_H of them. Even if not all the heavy atoms are correctly identified, the external loop may well be able to expand to the complete structure. If reliable criteria can be found for recognizing whether the correct heavy atoms have been located, the process can be made very efficient; the internal loop operates on a very small number of atoms and it is also possible to use a slightly higher E-threshold for this loop, both of which reduce the CPU time required for it, and a large number of internal loop trials can be performed for each solution that is expanded to the to the complete structure in the external loop.

TABLE 1. Crambin test, internal loop searching for 3 disulfide bonds, external loop expanding to full structure. The 0.9Å low-temperature data were collected and provided by Håkon Hope.

```
1625 E-sig(E) > 1.500 used to generate 77607 unique TPR
Try 19,   CC(HA) = 19.03%,   PATFOM = 13.80

Peak      x      y      z     self cross-vectors
99.9 0.3019 0.1253 0.1020  19.2
                           15.6

96.7 0.2571 0.0783 0.1028  22.4  2.0
                           18.4 14.3

96.7 0.3914 0.1707 0.4511  13.0  8.6  9.6
                           18.2 13.8 18.4

93.7 0.4373 0.1292 0.4262  11.1  9.1 10.3  2.1
                           16.3 35.5 12.2 11.3

90.7 0.0794 0.2353 0.0483  11.5  9.4  7.9 15.5 16.9
                            0.9 17.0 15.6 10.4 12.9

85.7 0.1098 0.3147 0.0591  13.1  8.6  7.5 14.6 16.0  1.9
                           13.2 11.5 10.3 16.1  7.7  0.0

Peaklist optimization cycle 1  CC=30.05% for 41 atoms
Peaks: 99 97 97 93 92 88 15 15 -14 14 -13 -12 -12 -12 -12

Peaklist optimization cycle 2  CC=47.75% for 108 atoms
Peaks: 99 95 93 92 85 81 34 -34 34 34 34 -33 33 33 33 33

Peaklist optimization cycle 3  CC=70.62% for 240 atoms
Peaks: 99 95 91 90 80 77 36 36 36 35 35 35 34 34 34 33 33

Peaklist optimization cycle 4  CC=81.57% for 354 atoms
Peaks: 99 96 92 88 74 73 37 37 37 36 36 35 35 35 35 35 34
Fragments: 310 2 1 1 1 1 1 1 1 1 1 1 1 1 1 1 1 1 1 1 1 1
```

Table 1 illustrates the application of this strategy to the solution of crambin, a 46-residue protein containing three disulfide bridges. The data had been collected (by Håkon Hope) to a resolution limit of 0.83Å, although the outer data are not quite complete. In this case a rather rigorous figure of merit could be applied: only those 6-atom solutions from the internal loop were accepted (i.e. the structure was then completed using the external loop) when these six atoms formed three S-S bonds with no sulfur atom in common. In the *crossword table* shown above, the three disulfide bridges can be identified by their distances of about 2Å. The external loop of *peaklist optimisation* leads to the essentially full structure in 4 cycles with a convincing CC (values greater than 70% are invariably correct). A minus sign in the list of peak heights indicates that that peak was rejected in the elimination procedure. The final line shows that there is a connected fragment of 310 atoms, plus a number of well-defined water molecules that do not bond to other atoms.

Recently we were successful in solving hirustasin, an unknown 55-residue protein with five disulfide bridges, by the same technique [11]; in this case solutions with four or more disulfide bridges were expanded. In fact the program located 9 of the 10 sulfur atoms correctly. In this case it was interesting that the data had only been measured to 1.2Å and that the data in the range 1.4 to 1.2Å are very noisy. Previous attempts to solve hirustasin by molecular replacement and SIR methods had failed. The two stage approach, first trying to locate the heavier atoms, may posses a hidden advantage when the resolution barely reaches atomic resolution: in the internal loop the resolution may not be so critical because the distances between heavy atoms are greater than those between light atoms. Then in the external loop the full data are used, so a good data to parameter ratio is preserved throughout.

In the absence of disulfide bridges, the Patterson superposition function provides a useful figure of merit, as discussed in detail in a previous talk. One efficient algorithm is to expand further only those solutions that after the internal loop have either the largest CC so far (based on the full data but only for the heavy atoms) or the best sum of the Patterson superposition minimum functions for all pairs of heavy atoms. There are indications that the sum of the square roots of the minimum functions discriminates a little better. We refer to this figure of merit as PATFOM. We have been able to solve *ab initio* the structures of a number of known and unknown metalloproteins using this algorithm, the largest having 8 iron atoms and 170 residues in the asymmetric unit [12].

6. Conclusions and Future Prospects

In reciprocal space, the phase refinement algorithm enables the structure to escape from local minima in search of a global minimum, and shows some similarities to *simulated annealing*. This is achieved by the correlation between reflection phases and intensities, and the dominant role of reflections with large E-values. In real space, the powerful constraint of an atomic model (the key to the success of *Shake & Bake* ?)

provides detail that may have been lost in the reciprocal space stage, and forces convergence towards a physically reasonable solution.

To extend the method to lower resolution, density modification may not prove sufficiently incisive as a replacement for peak-picking. More promising is the method used in ARP [13,14] to fill density with atoms. Alternatively, instead of using individual atoms, typical groups of 3-5 atoms (e.g. peptide units) could be fitted to the density (given a fast computer !).

Real/reciprocal space recycling has the potential to solve small macromolecular structures, especially when heavier atoms - perhaps even just disulfide bonds - can be exploited. The major drawback is that data are required to almost atomic resolution. All the structures with more than about 300 unique non-hydrogen atoms solved so far by purely *ab initio* methods from the native data alone contained sulfur or heavier atoms, and in all cases data had been collected to 1.2Å or better (even if it wasn't there!).

I am very grateful to T. Bes, M.A. Carrondo, Z. Dauter, C. Frazão, E. Parisini, H. Hope, K. Johnsen, J. Kärcher, V. Lamzin, P. Lubini, P. Mittl, J. Priestle, M. Schäfer, T.R. Schneider, I. Uson, and K.S. Wilson for their excellent data, help and advice in the course of this work. We thank the the Eropean Commission for support: contract number ERBCHBGCT940731.

7. References

1. Sheldrick, G.M. (1982) Crystallographic algorithms for mini- and maxi-computers, in D. Sayre (ed.), *Crystallographic Computing,* Clarendon Press, Oxford, pp. 506–514.

2. Sheldrick, G.M. (1990) Phase annealing in SHELX-90: direct methods for larger structures, *Acta Cryst.* **A46**, 467–473.

3. Miller, R., DeTitta, G.T., Jones, R., Langs, D.A, Weeks, C.M., and Hauptman, H.A. (1993) On the application of the minimal principle to solve unknown structures, *Science* **259**, 1430–1433.

4. Miller, R., Gallo, S.M., Khalak, H.G., and Weeks, C.M. (1994) SnB: crystal structure determination via Shake-and-Bake, *J. Appl. Cryst.* **27**, 613–621.

5. Read, R.J. (1986) Improved Fourier coefficients for maps using phases from partial structures with errors, *Acta Cryst.* **A42**, 140–149.

6. Fujinaga, M. and Read, R.J. (1987) Experiences with a new translation-function program, *J. Appl. Cryst.* **20**, 517–521.

7. Sheldrick, G.M. & Gould, R.O. (1995) Structure solution by iterative peaklist optimization and tangent expansion in space group P1, *Acta Cryst.* **B51**, 423–431.

8. Frazão, C., Soares, C.M., Carrondo, M.A., Pohl, E., Dauter, Z., Wilson, K.S., Hervás, M., Navarro, J.A., De la Rosa, M., and Sheldrick, G.M. (1995) *Ab initio* determination of the crystal structure of cytochrome c_6 and comparison with plastocyanin, *Structure*, **3**, 1159–1169 .

9. Sheldrick, G.M., Dauter, Z., Wilson, K.S., Hope, H., and Sieker, L.C. (1993) The application of direct methods and Patterson interpretation to high-resolution native protein data. *Acta Cryst.* **D49**, 18–23.

10. Schäfer, M., Schneider, T.R., and Sheldrick, G.M. (1996) Crystal structure of vancomycin, *Structure* **4**, 1509–1515.

11. Usón, I., Mittl, P., Priestle, J.P., and Sheldrick, G.M. (1997) To be published.

12. Parisini, E., Lamzin, V.S., and Sheldrick, G.M. (1997) To be published.

13. Lamzin, V.S. and Wilson. K.S. (1993) Automatic refinement of protein models, *Acta Cryst.* **D49**, 129–147.

14. Perrakis, A.,Sixma, T.A., Wilson, K.S., and Lamzin, V.S. (1997) The world according to wARP: improvement and extension of crystallographic phases, CCP4 Meeting, York, January 1997.

AB INITIO DETERMINATION OF A LOW RESOLUTION MOLECULAR ENVELOPE BY DIRECT METHODS

Charles W. Carter, Jr.
Department of Biochemistry and Biophysics,
CB #7260 University of North Carolina at Chapel Hill,
Chapel Hill, NC 27599-7260

There is considerable variety in the methods by which low resolution phasing has been attempted (Podjarny & Urzhumtsev, 1997). These include clustering of electron density histograms (Lunin *et al.*, 1990), refinement of random positions for a small number of large scatterers (Podjarny & Urzhumtsev, 1997) or a large number of essentially point scatters (Subbiah, 1991), and experimental contrast variation (Bentley *et al.*,1984; Carter *et al.*, 1990). I will be concerned here only with the latter approach and its use in determining an envelope for the *B. stearothermophilus* tryptophanyl-tRNA synthetase (TrpRS).

Practical successes, with unsolved macromolecular structures, of the Bayesian paradigm for statistical direct methods developed by Bricogne (Bricogne, 1984; Bricogne, 1988; Bricogne, 1993), have resulted from applying the methods to discrete aspects of phase determination: heavy-atom parameter refinement (De La Fortelle & Bricogne, 1997), envelope determination by contrast variation (Carter et al., 1990), density modification (Carter & Xiang, 1997; Doublié *et al.*, 1994; Xiang *et al.*, 1993), and refinement (Irwin & Bricogne, 1996). Several of these applications make a forceful case that conditional probability methods, and particularly the real space formulation based on maximum entropy (Bricogne, 1984), though still in its infancy, represents a substantial improvement over traditional implementations. The application described here made use of the traditional direct methods, as implemented in the program, MITHRIL (Gilmore, 1984). It was carried out before the current implementations of entropy maximization, sampled hypothesis permutation, and scoring became available.

1. Suitability of the Monoclinic TrpRS Crystals

Monoclinic crystals of the TrpRS:tryptophan complex, space group $P2_1$, provided an excellent test case for *ab initio* phase determination based on amplitudes for the envelope transform obtained from solvent contrast variation. The asymmetric unit of these crystals contained three enzyme dimers (Mr 35,600) in the asymmetric unit and the diffraction diagram has a low resolution 3m point group (Figure 1). Three important features of

S. Fortier (ed.), Direct Methods for Solving Macromolecular Structures, 413–419.

noncrystallographic symmetry suggested its use for low-resolution phase determination. First, the low resolution, high-symmetry space group could facilitate *ab initio* phasing, by analogy to small molecule problems. Second, the apparent 32 point group was closely aligned with the crystallographic axes, suggesting that noncrystallographic 6-fold symmetry based on space group P321 might be employed, *ab initio*, to refine phases determined by tangent formula refinement. Third, the expectation that the asymmetric unit contents should be arranged as 3 dimers provided important prior chemical information that proved useful in distinguishing good phase sets.

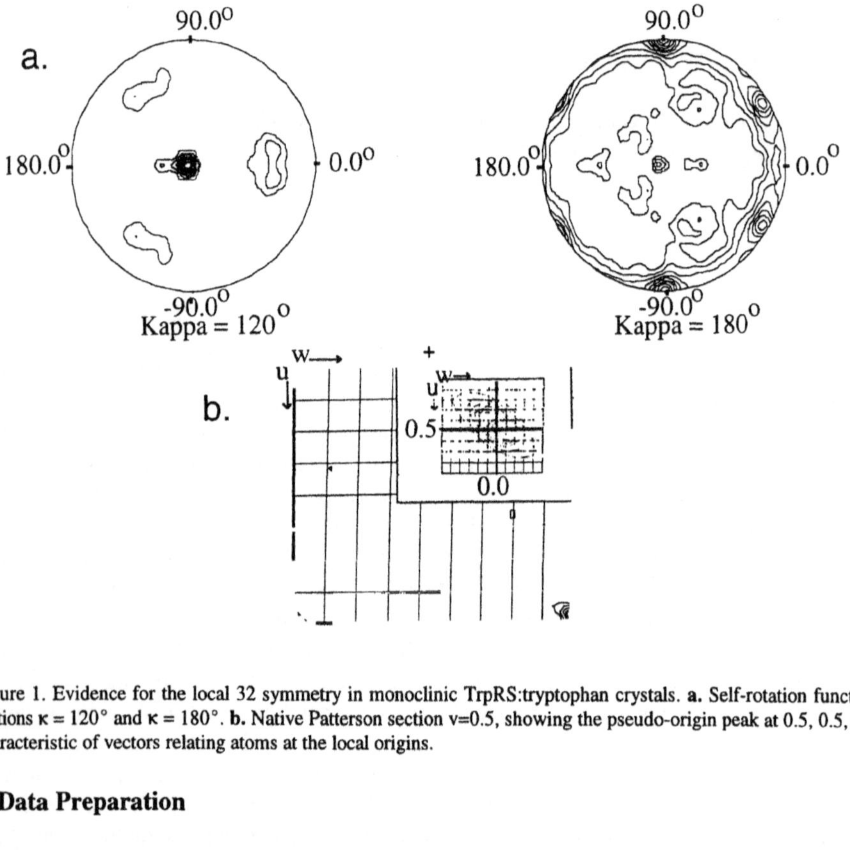

Figure 1. Evidence for the local 32 symmetry in monoclinic TrpRS:tryptophan crystals. **a.** Self-rotation function sections κ = 120° and κ = 180°. **b.** Native Patterson section v=0.5, showing the pseudo-origin peak at 0.5, 0.5, 0.0 characteristic of vectors relating atoms at the local origins.

2. Data Preparation

Amplitudes from a contrast variation series were obtained by integration in space group $P2_1$. They were reduced to P 3 m1 and amplitudes for the molecular envelope were estimated using the program, GFROMF, written for this purpose, as described in another chapter (Carter, 1998). In the absence of useful information from the Wilson plot of the data, normalization was carried out empirically, by assuming an absolute scale based on the volume and unit cell contents in the crystals, and an effective number of atoms consistent with a reasonable fraction of $|E_H| > 1.0$. No sharpening was applied to the data.

Scaling between datasets was consistent with equilibrating the mean scattering power for each data set in a peak at about 10.5 Å in the radial intensity distribution (Figure 2). This result provided a rough, but independent estimate for the effective number of scatterers in the unit cell. The effective radius of such scatterers implied a volume of ~600 Å^3 per scatterer at low resolution (Harker, 1953). Hence to the resolution of this study, there were approximately 78 scatterers per asymmetric unit, or ~13 per monomer. In turn, this value afforded an estimate for the absolute scale of the data by confirming the value we had found empirically to be useful for normalization (Carter et al., 1990).

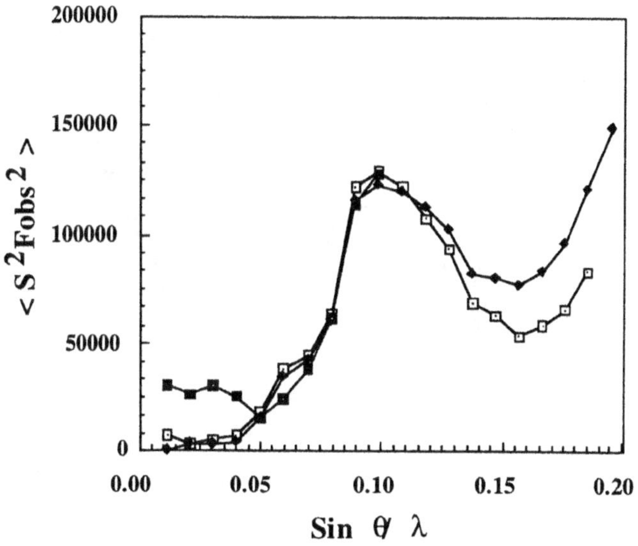

Figure 2. Scaling of contrast-variation datasets used to generate{$|G_H|$} and the estimation of $N_{effective}$. The scaled datasets correspond most closely at the peak in the radial intensity distribution, which is characteristic of scattering from group ("glob"; Harker, 1953) scatterers.

3. Phase determination

The largest 34 of these were input to the direct methods program, MITHRIL (Gilmore, 1984), using default options together with Hull-Irwin weighting. Four of 32 phase sets had reasonable figures of merit and the corresponding maps were examined. One set (that with the highest combined figure of merit) revealed a density distribution consistent with the prior prejudice that the asymmetric unit had three enzyme dimers related by the noncrystallographic three-fold axis (Figure 3a). As is often the case for direct methods solutions, this solution was essentially centrosymmetric. Phases for this map were applied in space group $P2_1$, and that map was averaged, without solvent flattening, within a simple hexagonal envelope coextensive with an asymmetric unit (Figure 3b). Averaging induced a decisive breakdown of centrosymmetry, while preserving the nature of the "dimeric"interactions in the 32 point group (Figure 3c).

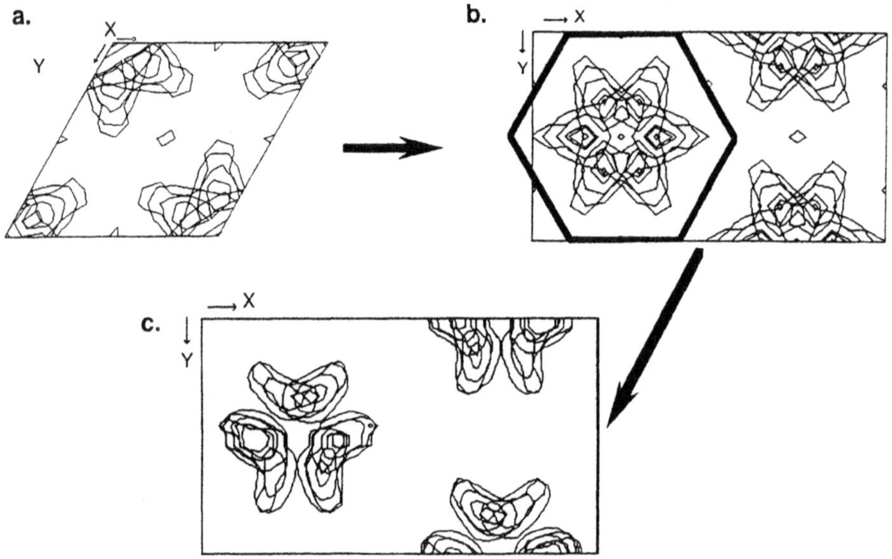

Figure 3. Evolution of the envelope density map. **a.** Map from tangent refinement of the 34-reflection phase set with the highest figure of merit in space group P 321. **b.** The same map, reindexed into space group P 21. The dark, octagonal boundary represents a section of the (right) prismic envelope used for 6-fold averaging. **c.** The 170-reflection map after 6-fold averaging. All maps have a resolution of 18 Å.

4. A Brief Post-Mortem

The resulting structure (Carter, 1990) bears a striking resemblance to the 18 Å envelope of the structure of the TrpRS-Trp-5'AMP complex (Doublié et al., 1995) (Figure 4). However, the direct phase determination was nevertheless only a partial success. Attempts to extend phases using MESF and phase permutation failed. In view of this qualification, the close resemblance of the envelope to the actual structure raises two different kinds of questions, which should be addressed in closing.

The resulting structure (Carter, 1990) bears a striking resemblance to the 18 Å envelope of the structure of the TrpRS-Trp-5'AMP complex (Doublié et al., 1995) (Figure 4). However, the direct phase determination was nevertheless only a partial success. Attempts to extend phases using MESF and phase permutation failed. The threefold axis of the monoclinic crystals has since been determined to be a 3_1 screw axis, so the structure is technically incorrect, consistent with the inability to extend phases. In view of these qualifications, the close resemblance of the envelope to the actual structure raises two different kinds of questions, which should be addressed in closing.

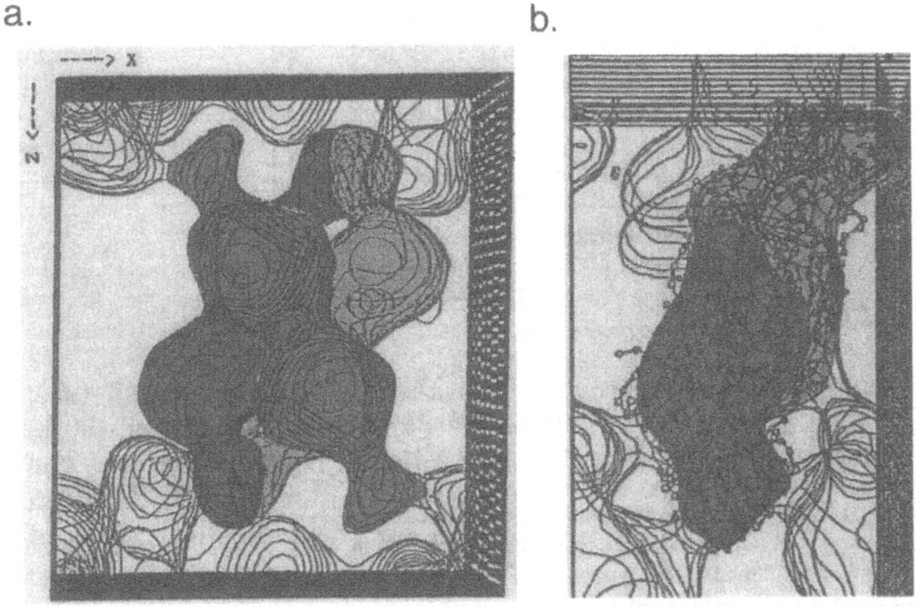

Figure 4. Comparison of the 18 Å envelopes from the *ab initio* determination described here containing three dimers (**a**) and a single dimer of the atomic resolution structure (**b**) of the TrpRS:tryptophanyl-5'AMP complex (Doublié, et al., 1995). The orientation of the two maps was selected to highlight the similarity of their shapes. Both are divided into four "domains", corresponding to the Rossmann fold (larger) and small helical (smaller) domains of the two monomers in the dimer. The characteristic "W"shape of the dimer is evident from in the darkly shaded subunit behind the others in **a**, and in **b**. The atomic structure of the molecule is superimposed on **b**.

4.1. TO WHAT EXTENT WAS THIS STRUCTURE DETERMINED DIRECTLY, *AB INITIO*?

Although several different datasets were used to assemble the amplitudes of the envelope transform, the phasing itself used only the single set of amplitudes, $\{|G_H|\}$, with no prior knowledge of phases. Phase sets were selected according to a sampling protocol and generated using the tangent formula following the default options of a standard direct methods program. Thus, conditional probability was the only source of phase information for this set of amplitudes. The resulting structure is not a macromolecular structure, but an envelope. Prior information concerning the likely quaternary structure of the enzyme was useful in recognizing the best phase set. So the phasing did follow, in spirit, the requirements for an *ab initio* structure determination.

4.2. TO WHAT EXTENT WAS THIS STRUCTURE CORRECT?

We have subsequently solved ~10 different structures at resolutions between 2.9 - 2.0 Å. We have not solved this crystal form, however. Nevertheless, it is possible to check the correctness of the solution by several comparisons with the known structures. Qualitatively, the comparison in Figure 4 is favorable. The outlines compare well with those of the envelope calculated from the known structure in another crystal form. Each shows the monomer divided into two domains, one (the Rossmann dinucleotide-binding fold) significantly larger than the other. Moreover, a molecular replacement solution for a closely-related space group (unliganded TrpRS; Ilyin unpublished) had rotational angles that correspond to the orientations in the envelope density map.

Despite these successes, there are also now signs that the success was incomplete. The gradient of the envelope boundary is very steep close to the 2-fold axis, very flat in the second domain. Thus, it was easy to generate models with different volumes by small shifts in the contours. However, the volume of the published envelope is wrong. We cannot fit a TrpRS molecule into the envelope.

A more important problem relates to the suitability of this phase set for phase extension and extrapolation to the $\{|F_H|_{obs}\}$. Considerable effort failed to confirm this structure by extending it to higher resolution. Maximum entropy solvent flattening (Xiang *et al.*, 1993) did not work, suggesting that the structure was incompatible with the imposition of the envelope constraint; nor did phase permutation work. It was unclear at the time whether or not this failure to extrapolate came from the reduction of the $|G|$ values to noise or to other errors in the model.

4.3. PROBABLE REASON FOR FAILURE: AN INCORRECT LOW RESOLUTION SPACE GROUP

The reasons for our failure to extend this structure have begun to become clear from structures we have now solved of the ligand-free TrpRS and a slightly different polymorph of the tryptophan complex. From these, we know that although the noncrystallographic point group in reciprocal space is indeed 32, the translational parameters in real space, however, are wrong. The low resolution cell has a 3_1 screw axis. So the structure is technically incorrect, consistent with the inability to extend phases. In retrospect, the screw axis is apparent from approximate systematic extinctions along the c* axis. Thus, we actually probably did extract meaningful information about the shape of the TrpRS molecule from the

$\{|G_H|\}$ amplitudes, but fell short of a useful direct phase determination for the crystal as a whole.

5. References

1. Bentley, G., Lewit-Bentley, A., Finch, J. T., Podjarny, A. D. & Roth, M. (1984). Low Resolution Structure of the Nucleosome Core Particle of Chromatin Analyzed by Neutron Solvent Contrast Variation. *Journal of Molecular Biology* 176, 55-75.
2. Bricogne, G. (1984). Maximum Entropy and the Foundations of Direct Methods. *Acta Crystallographica* A40, 410-445.
3. Bricogne, G. (1988). A Bayesian Statistical Theory of the Phase Problem. A Multichannel Maximum-Entropy Formalism for Constructing Generalized Joint Probability Distributions of Structure Factors. *Acta Crystallographica* A44, 517-545.
4. Bricogne, G. (1993). Direct Phase Determination by Entropy Maximization and likelihood Ranking: Status Report and Perspectives. *Acta Crystallographica* D49, 37-60.
5. Carter, C. W., Jr. (1990). Efficient Factorial Designs and the Analysis of Macromolecular Crystal Growth Conditions. *METHODS: A Companion to Methods in Enzymology* 1, 12-24.
6. Carter, C. W., Jr. (1998). The solvent contrast variation method. *in Direct Methods for Solving Macromolecular structures* (Fortier, S., ed.), pp. In Press. Kluwer Academic Publishing, Dordrecht.
7. Carter, C. W., Jr., Crumley, K. V., Coleman, D. E., Hage, F. & Bricogne, G. (1990). Direct Phase Determination for the Molecular Envelope of *Bacillus stearothermophilus* Tryptophanyl-tRNA Synthetase by X-ray Contrast Variation. *Acta Crystallographica* A46, 57-68.
8. Carter, C. W., Jr. & Xiang, S. (1997). Phase Improvement Using Conditional Probability Methods: Maximum Entropy Solvent Flattening and Phase Permutation. *Methods in Enzymology* 277, In press.
9. De La Fortelle, E. & Bricogne, G. (1997). Maximum-Likelihood Heavy-Atom Parameter Refinement for Multiple Isomorphous Replacement and Multiwavelength Anomalous Diffraction Methods. *Methods in Enzymology* 276, 472-494.
10. Doublié, S., Bricogne, G., Gilmore, C. J. & Carter, C. W., Jr. (1995). Tryptophanyl-tRNA synthetase crystal structure reveals an unexpected homology to Tyrosyl-tRNA synthetase. *Structure* 3, 17-31.
11. Doublié, S., Xiang, S., Gilmore, C. J., Bricogne, G. & Carter, C. W. J.(1994). Overcoming Non-Isomorphism by Phase Permutation and Likelihood scoring: Solution of *Bacillus stearothermophilus* Tryptophanyl-tRNA Synthetase. *Acta Crystallographica* A50, 164-182.
12. Gilmore, C. J. (1984). MITHRIL: a computer program for Direct *Methods. Journal of Applied Crystallography* 17,42-46.
13. Harker, D. (1953). The Meaning of the Average $|F^2|$ for Large Values of the Interplanar Spacing. *Acta Crystallographica* 6, 731-736.
14. Irwin, J. & Bricogne, G. (1996). A test of maximum-likelihood refinement of macromolecular structures with Buster and TNT. *Acta Crystallographica* A 525 suppl., C-86.
15. Lunin, V. Y., Urzhumtsev, A. G. & Skovoroda, T. P. (1990). Direct Low-Resolution Phasing from Electron-Density Histograms in Protein Crystallography. *Acta Crystallographica* A46, 540-544.
16. Podjarny, A. D. & Urzhumtsev, A. G. (1997). Low Resolution Phasing. *Methods in Enzymology* 276, 641-658.
17. Subbiah, S. (1991). Low-Resolution Real Space Envelopes: An Approach to The *Ab Initio* Macromolecular Phase Problem. *Science* 252, 128-133.
18. Xiang, S., Carter, C. W., Jr., Bricogne, G. & Gilmore, C. J. (1993). Entropy Maximization Constrained by Solvent Flatness: a New Method for Macromolecular Phase Extension and Map Improvement. *Acta Crystallographica* D49, 193-212.

MODEL BASED VERY LOW RESOLUTION PHASING

A.PODJARNY#, A.URZHUMTSEV#* AND V.LUNIN*
#UPR de Biologie Structurale, IGBMC , 67404 Illkirch , France
**IMPB RAN, Puschino, Moscow region, 142292, Russia*

1. Introduction.

This paper deals with phasing macromolecular data at very low resolution. The definition of «low resolution» depends on the object being studied; in what follows the resolution below 6-8Å will be referred to as «low resolution» and the one below 20-25Å as «very low resolution» (VLR). These two limits define the resolution zone where the contribution of the bulk solvent is strong and uncorrelated to that from the macromolecule itself. At higher resolution the contribution is negligible, and at lower resolution it is strong but roughly proportional to the one of the macromolecule [1]. Measuring the very low resolution X-ray data is technically difficult, and many research groups do not collect them. However, they carry information that can be useful. This paper discusses their importance for improving the molecular images and proposes methodologies for obtaining the VLR phases.

2. Do very low resolution data have any information ?

The inclusion of VLR reflections can substantially modify a macromolecular image. This was shown by Podjarny et al. [4] in the case of tRNA, where a phase extension procedure for phasing the VLR data drastically improved the image. For calculated data [5] it was clearly shown that the exclusion of only 1% of the data (29 reflections out of 2500) completely destroys the molecular image at 6Å resolution. In the case of SIR phase errors, the molecular envelope keeps its position but the electron density peaks are shifted. In the case of missing VLR reflections the effect is inverted: the envelope is lost but the peaks are at their places [5], as the exclusion of VLR terms causes large scale modulations of the density in the unit cell. The fact that the peaks are at the right place has important consequences. Firstly, when a map is calculated at high resolution, its peaks have a high contrast and such density modulation does not «hide» them completely. This has allowed crystallographers to ignore VLR data for a long time. Secondly, it gives a possibility of automatically determining the molecular envelope from such maps. The knowledge of the envelope can be used to improve the molecular image. The phases of its structure factors can be used as a good approximation to the phase values of VLR reflections. If their amplitudes are available, simple adding them to the Fourier calculation can completely change the map [4,5]. Calculated amplitudes can be used to give corresponding weights for the Fourier coefficients through the comparison with the experimental ones.

S. Fortier (ed.), Direct Methods for Solving Macromolecular Structures, 421–431.

3. How to use the information from very low resolution data ?

Therefore, VLR data do carry important information, first of all, on the shape of molecule. Such an information can be used in different cases [6], for example:
- in density modification procedures for the image improvement;
- in molecular replacement if the *internal* differences between two molecules are large;
- if diffraction data are not available at higher resolution;
- in the case of very large molecular complexes, like the ribosome; etc.

If VLR amplitudes have been measured, the determination of their phases by isomorphous replacement is difficult while not impossible [6]. In the case of viruses where practically all VLR reflections are centrosymmetric, a good approximation can be done by calculating the structure factors in a spherical shell. In the general case, a searching procedure based on some *a priori* knowledge of the density can be applied to find these phases. In what follows, we will describe two model-based approaches for obtaining VLR phases. The first one is an *ab-initio* method, which is based on the construction of many Few Atoms Models (FAM). The second one is the use of the molecular replacement at very low resolution using a known envelope of the molecule.

4. The FAM method for VLR phasing.

The Few-Atoms-Models method [7] is an *ab initio* phasing technique for very low-resolution diffraction data. It generates phase sets from a large number of Few Atoms Models and filters by an amplitude correlation criteria. The basic observation underlying the method is that these filtered phase sets form clusters, and one of them is close to the right solution (Figure1).

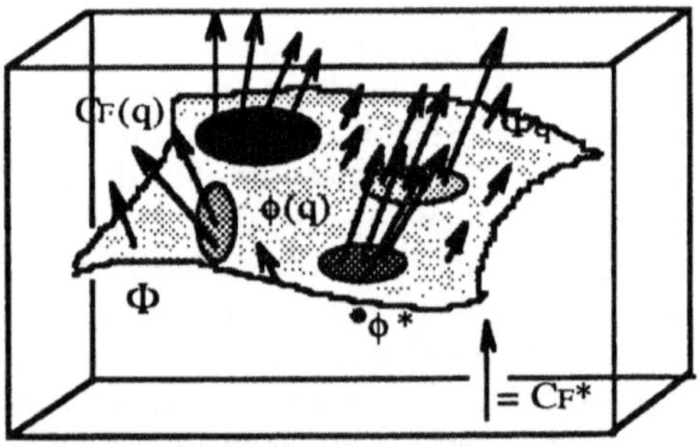

Figure 1. Schematic presentation of the distribution of the phase sets in the FAM method. Every phase set is presented by a point in the phase space, Φ. Φ_q is a subspace of phase sets, φ(q), calculated from FAM models. The length of an arrow is proportional to the corresponding amplitude correlation, CF(q). The variants with CF>CF* are forming few clusters, one of them is close to the correct solution, φ* (thick point).

The FAM method consists of the following main steps :

a) Model generation: A very large number (millions) of simple models are generated. Each of them consists of the same small number (usually less than 10) of large gaussian spheres. The coordinates of these spheres are generated randomly and no conditions are imposed to prevent overlap.

b) Model selection in reciprocal space: A structure factor set at very low resolution is calculated from each model. If the correlation between the calculated and observed amplitudes, defined for example by

$$Corr(F^{obs}, F^{calc}) = \Sigma_s(F^{obs}(s) \cdot F^{calc}(s)) \;/\; [(\Sigma_s(F^{obs}(s))^2) \cdot \Sigma_s(F^{calc}(s))^2]^{1/2} \quad (1)$$

is below a given threshold the model is rejected. The threshold is chosen such that the number of accepted models is adequate to sample the space; test calculations place this number around a few hundred. The uncentered correlation (1) is useful for maximizing the agreement between the strongest amplitudes at very low resolution.

c) Merging the model phase sets : The phase sets calculated from the selected models are then merged through a clustering procedure. This procedure is based on the definition of a distance in a multidimensional "phase space". All the phases of each set are considered as a single point in this phase space, where the dimensionality is the number of phases and the coordinates of this point are the values of the phases. The distance D_{jk} between points j and k is determined in terms of the corresponding densities ρ_j and ρ_k as

$$D_{jk} = \kappa * [\int(\rho_j(r)-\rho_k(r))^2 d^3r]^{1/2} \quad (2)$$

where $\kappa = \kappa_j = [\int(\rho_j(r))^2 d^3r]^{-1/2}$ is the same for all models. Here ρ_j stands for an electron density distribution calculated with the Fourier coefficients $\{ F^{obs}(s) \exp(i\phi_j^{calc}(s)) \}$, with the phases from the j-th model. This distance (2) is related to the usual map correlation coefficient C_{jk} as

$$C_{jk} = \kappa^2 * [\int(\rho_j(r)\rho_k(r))\, d^3r] = 1 - (D_{jk}^2/2) \quad (3)$$

The process of clustering [8] is then done as a function of a parameter δ. Initially, δ=0 and each point is an individual cluster. δ is then continuously increased, until it reaches the value of the smallest Djk; then points j and k are merged into and replaced by a new point. As δ increases, the different points are merged, until finally they converge to a single one. The tree can then be read starting from the single unique cluster at the top and going down to the smaller ones; each cluster division represents a "level". Inside each cluster, the phase $\phi(s)$ and its weight w(s) can be obtained by averaging all the original phases from the starting models, as follows:

$$w(s) \cdot \exp(i\,\phi(s)) = \Sigma_j \exp(i\,\phi_j^{calc}(s)) / N \quad (4)$$

where the index j goes over all the N points in the cluster.

Step (c) is applied as the points in phase space selected in (b) are generally observed to group in a small number of clusters. It is expected that one of the resulting phase sets corresponds to the correct solution, and that it can be identified by other criteria like molecular packing or symmetry considerations. Note that in the FAM solution the choice of enantiomer is arbitrary.

5. Application of the FAM method to the tRNA synthetase complex

5.1. GENERATION OF THE FAM MODELS.

The algorithm described above was tested using the neutron diffraction data set from the tRNA^{Asp}- AspRS [9] complex in the cubic form (space group I432, a=354 Å). These tests were conducted at 50Å (31 reflections), where the experimental data are practically complete. Different numbers of pseudo-atoms were tried, ranging from 1 to 10. To measure the effectiveness of the FAM technique in a test case where the answer is known [10], the correlation Cor(ϕ,ϕ^{true}) for the different phase sets was calculated after every step. The FAM distributions follow the random one for values of Cor(ϕ,ϕ^{true}) < 0.45, but for the 1- and 2-atoms cases become significantly larger when Cor(ϕ,ϕ^{true}) > 0.45 (see Fig. 2). Note that this effect diminishes as the number of atoms increases; it is largest for 1-atom FAMs, remains important for 2-atoms FAMs, but almost disappears for 10-atoms FAMs, as in this latter case the number of variables (30 atomic co-ordinates) approaches the number of phases (31 phases). This shows that the filtering power of this first stage of FAM, as compared with just random phase sets, depends on its ability to generate good phase sets with fewer degrees of freedom.

5.2. MODEL SELECTION BY AMPLITUDE CORRELATION.

The filtering of the phase sets from the FAM modelisation is further enhanced by choosing the solutions with a large correlation in F. This is shown in Fig. 2 for the 2-atoms FAM selection, with cut-offs at 0.74 (corresponding to 1σ deviation from the mean amplitude correlation) and 0.81 (corresponding to 2σ).

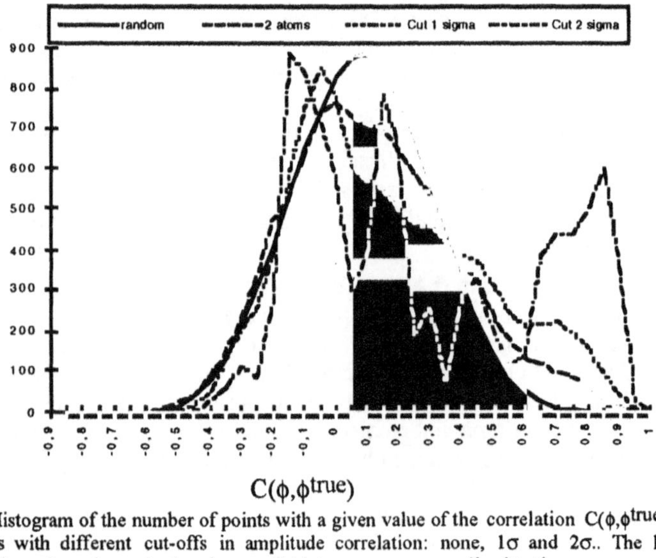

Figure 2 Histogram of the number of points with a given value of the correlation C(ϕ,ϕ^{true}) for 2-atoms FAMs with different cut-offs in amplitude correlation: none, 1σ and 2σ.. The histogram obtained with random phases is also shown; all histograms are normalized to the same area.

This last curve, corresponding to 1179 possible points out of 500,000 generated ones, shows peaks for high values of $C(\phi,\phi^{true})$ indicating the creation of a densely populated zone of phase sets near the correct one.

5.3. CLUSTERING.

In the general case when the correct solution is not known, the identification of the dense zones after filtering is done by the clusterisation procedure described above, which leads to a "cluster tree". Fig.3a presents the results of the application of this procedure to the data described in 5.2. There are three main clusters, noted as C11, C12 and C13.

a) b)

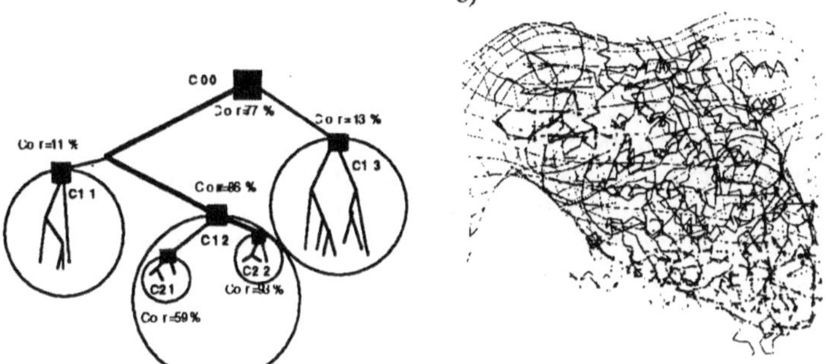

Figure 3.
a) Labeled scheme of the cluster tree, indicating the main nodes and corresponding $C(\phi,\phi^{true})$ values
b)Overlap of the 50Å-resolution synthesis calculated with FAM phases of cluster C22 with the atomic model of the AspRS complex in the cubic cell. The shape of the density follows the model closely. The additional densities correspond to symmetry related molecules (not shown)

Investigation of these points allows the clear choice of one of them (C12), based on the criterion that the core of the density should not be on symmetry elements (here - on the dyad axis x=y, z=0). C12 is the largest cluster, and it can be further divided in two points, C21 and C22 ; again, one point is clearly better than the other by the same criterion. In this way, the correct pathway could be followed without any comparison with the correct density. When the process was finished, the synthesis corresponding to the "cluster" phase sets were compared with the correct one at 50 Å resolution. By going down two levels of the cluster tree, the correlation with the correct synthesis increases from 77% to 93%; however, going one level further does not increase this value anymore. Fig. 3b shows the 50 Å resolution map calculated with the phases and figures of merit from the correct cluster superposed with the correct model. This figure shows that the predicted envelope follows closely the borders of the model, and has therefore useful phasing information.

6. Application of the FAM method to the T50S ribosomal particle

The FAM method was applied to experimental diffraction data of T50S crystals, space group $P4_12_12$ (or $P4_32_12$), a=b=498 Å, c=198 Å [11,12]. The different parameters (number of atoms, number of trials, etc) had been previously optimized in tests with simulated ribosome data [7]. The observed amplitudes were used for d>60Å. The space group used in the calculations was $P4_12_12$; the results are also valid for $P4_32_12$ by changing z into -z. One million models of 5-gaussian spheres ($B=34000Å^2$) were generated in step (a) of which 560 variants with amplitude correlation (1) higher than 86% were chosen in step (b).

The cluster tree showed clearly separated subsets. A path to the solution was established up to the 5th level, choosing the correct cluster at each level by the criterion that the density on crystallographic dyads should be weak. This criterion was chosen since the particles do not form dimers and should not have extensive interfaces. This criterion was quite clear for levels 1-4; it became less so at level 5, as most of the models with density on top of the dyads were already ruled out. Therefore, the choice was no longer clear and the process was finished at this level.

Averaging phase sets inside the cluster according to formula (4) weights down the higher resolution reflections because the phases are more dispersed; the mean weight decreases as a function of resolution. For example, a threshold in weight of 0.3 leads to an effective resolution around 110Å. This limit could not be improved just by changing the parameters (e.g., the number of pseudo-atoms). The strong reflections at the lowest resolution dominated the correlation of the observed and the calculated amplitudes at step (b), leading in all cases to similar results.

To solve this problem and to increase the resolution of the image, the FAM method was further developed and re-run. To increase the sensitivity to weaker reflections, the uncentered amplitude correlations (1) were replaced by the more commonly used centered ones,

$$\mathrm{Corr}(F^{obs},F^{calc}) = \\ [\Sigma_s(F^{obs}(s)-\langle F^{obs}\rangle)(F^{calc}(s)-\langle F^{calc}\rangle)]/[(\Sigma_s(F^{obs}(s)-\langle F^{obs}\rangle)^2)(\Sigma_s(F^{calc}(s)-\langle F^{calc}\rangle)^2)]^{1/2} \quad (5)$$

calculated around the mean values $\langle F^{obs}\rangle$ and $\langle F^{calc}\rangle$ of $\{F^{obs}(s)\}$ and $\{F^{calc}(s)\}$, respectively. Three million 10-atoms models with $B=10000Å^2$ were generated (step (a)) inside a sphere of 110Å radius centered in the middle of the envelope obtained previously thus covering it completely. In step (b) different filtering criteria were applied simultaneously in different resolution ranges, as follows:

1) 60Å<d<110Å (69 reflections) : centered amplitude correlation (5) larger than 0.45 (corresponding mean value in the original run was 0.23 for the selected models).
2) 110Å<d (20 reflections): map correlation (3) larger than 0.7 with the previous map.

These filters selected 950 variants and a cluster tree was built. The tree showed several major clusters, but only one corresponding map had no noise density on the crystallographic dyad (x=y, z=0) and improved envelope definition. The effective resolution for this synthesis is about 80Å (mean weight > 0.3), thus extending the phase information from the earlier limit of 110 Å. Note that the corresponding envelope (Fig. 4) is not spherical and its peaks and valleys resemble the shape of the ribosome particle as seen by electron microscopy [13].

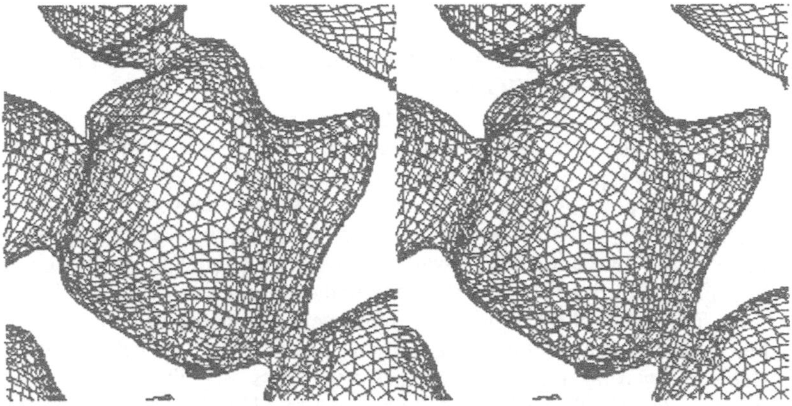

Figure 4. Stereo view (in the space group $P4_32_12$) of FAM density after the final run, showing the topological features of the T50S particle. The density level is chosen to include 30% of the unit cell volume inside the envelope.

7. Molecular replacement at very-low resolution

Doing molecular replacement ([14] and references therein) at very low resolution poses two particular problems:

1) The very low resolution reflections have a very strong contribution from the disordered solvent.
2) The available searching object is not a molecular model but an envelope, e.g., from electron microscopy reconstruction.

Urzhumtsev & Podjarny [15] investigated these problems using model and experimental data. They showed that at resolutions lower than approximately 20-30 Å the contribution of the disordered solvent to the observed structure factors becomes proportional to the contribution of the molecule [1], and that a simple envelope model can take the solvent contribution into account. Therefore, at this resolution range, molecular replacement searches can be carried out with either atomic models or molecular envelopes. In this work a protocol for low resolution molecular replacement was developed and tested [15]. At this low resolution the signal peak can be both broad and displaced from the correct one. Therefore, the peaks of the rotation function are supplemented by their closest neighbors with a

15-20° scanning step. For the translation step, the usual functions were replaced by the calculation of Corr(F^{obs},F^{calc}) (formula (5)) as a function of the translation vector. Also, all points of translation searches with correlation value above a certain level (and not only the positions of the peaks) are kept for the final analysis by rigid body refinement.

It should be noted that the problems outlined above are not faced during usual molecular replacement applications as the very low resolution data are seldom measured, however some of proposed measures can be possible useful for difficult cases at usual resolution.

7.1 MOLECULAR REPLACEMENT SEARCH USING EXPERIMENTAL RIBOSOME DIFFRACTION DATA

The low-resolution molecular replacement protocol was applied to the T50S diffraction data [11]. The searching object was an image of this particle as reconstructed at 28 Å from the electron microscopy views of a tilt series of the crystalline arrays [13]. The program AMoRe [16] was used for the rotation function, the translation search was done with an amplitude correlation function developed independently, and the rigid body refinement was done using AMoRe again. The result of the rigid body refinement was confirmed by a complete 6-dimensional search in a large "box" around the indicated orientation and position, which also generated mean values and dispersions (σ). This work followed the protocols developed with simulated ribosome data [15].

Table 1. Peaks of the low-resolution molecular replacement procedure (after the refinement step) with correlation value higher than 3σ for at least one of the resolution ranges studied. The intermolecular distance is measured between the centers of mass of neighboring molecules. Larger values are preferred; a value lower than 100 Å implies strong overlap. Peaks A,C,D and E are close to each other through symmetry or enantiomer operations. Peak C is the best; it has the strongest value in both resolution ranges, an acceptable intermolecular distance, and the second highest value of the rotation function.

peak name	height at 90Å-∞ (in σ)	height at 90-150Å (in σ)	space group	intermol. distance, (in Å)	Rotation Position	Rotation peak height
A	3.0	2.5	$P4_12_12$	132	1	41.8
B	2.9	3.1	$P4_12_12$	76	1	41.8
C	3.7	3.3	$P4_32_12$	142	2	39.1
D	3.5	3.0	$P4_12_12$	120	5	37.0
E	3.1	2.9	$P4_12_12$	134	5	37.0
F	3.0	3.0	$P4_32_12$	132	6	35.7
G	2.9	3.0	$P4_12_12$	63	7	31.6

In order to minimize the possible discrepancies between the electron microscopy envelope and the crystallographic image, the molecular replacement searches were carried out at the lowest possible resolution where the rotation function could yet show a signal. Such resolution was estimated as 90Å. Two resolution ranges (90Å-∞ and 90-150Å) were used for the translation and refinement steps. In contrast to FAM, in the molecular replacement method the model fixes the enantiomer, and therefore both space groups need to be explored. The results of these searches are shown in Table 1; there are clear indications for the choice of peak C using the criteria of amplitude correlation and intermolecular distance. The resulting packing is very similar (Fig.5) to the one obtained by the FAM method, where no assumptions were made either about the number of molecules or their shape.

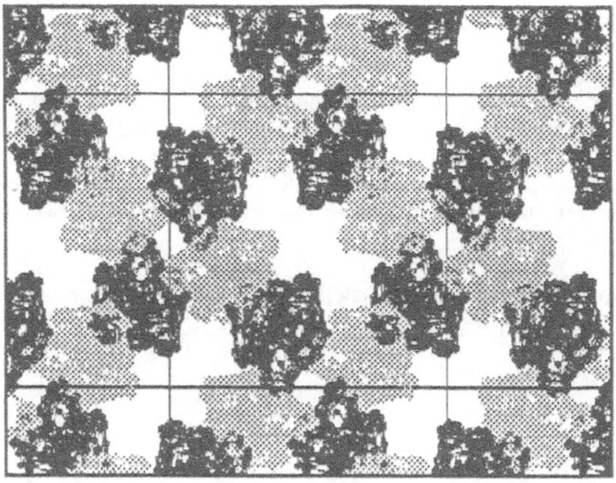

Figure 5. C-axis projection of the symmetry related particle envelopes, placed by molecular replacement, showing their packing in a full unit cell.

8. Conclusions

Two low resolution phasing techniques which explore either the whole phase space or some specific subspace have been analyzed. In the FAM case, the variants with best values of the search criterion are grouped in a small number of clusters which can be easily identified. One of these clusters is usually very close to the correct solution of the phase problem while others can be very far from it. It is important to note that the phase set with the best value of the criterion does not necessarily belong to this correct cluster. This observation explains, in particular, the problems with searches selecting a *single* solution. As it was observed, the character of this distribution does not depend on the particular information and criterion used. For example, it can be noted in addition that the LAPS method developed by Volkmann [17] based on the maximum likelihood criterion found the solution for the T50S case also through a cluster oriented search. In the molecular replacement case, the programs have been adapted to the particular problems of searching in the VLR range. In this case also, several criteria are necessary to find the right solution.

All these observations indicate that at the very low resolution end the available information and search criteria are *weak* in the sense that in general they cannot indicate unambiguously the correct solution; additional information is necessary. At higher resolution, the same information and criteria, e.g., an atomic model and the amplitude correlation, can be *strong* enough to indicate a single solution. The particular low resolution cases where the information is very accurate and the same criteria can unambiguously identify the right solution remain the exception rather than the rule.

Acknowledgments.

The data of T50S (X-ray diffraction amplitudes and electron microscopy diffraction envelope) were provided by A.Yonath and co-workers as a part of an ongoing collaboration; we thank them for making these data available. The neutron data for the $tRNA^{Asp}$- AspRS complex were measured by M. Roth, A. Lewitt-Bentley and D. Moras; we thank them for making these data available. The low resolution molecular replacement algorithms were developed in collaboration with J.Navaza. We also thank Z. Berkovitch-Yellin, P.Dumas, D.Moras, B.Rees, M. Roth, F. Schlünzen, J.C. Thierry and N.Volkmann for useful discussions and their continuing interest and encouragement. This work was supported by a CNRS-RAS collaboration. AGU and ADP were supported by the Centre National de la Recherche Scientifique (CNRS) through the UPR 9004, by the Institut National de la Santé et de la Recherche Médicale, by the Centre Hospitalier Universitaire Régional. VYL was supported by RFBR grant 97-04-48319.

References

1. Urzhumtsev, A.G. and Podjarny, A.D. (1995) On the problem of solvent modeling in macromolecular crystals using diffraction data: 1. The low resolution range, *Joint CCP4 and ESF-EACBM Newsletter on Protein Crystallography*, ***32***, 12-16.
2. Lunin, V.Yu. (1991) Use of the electron-density-syntheses histograms for solving of the phase problem in protein crystallography, Dr.Sci.Thesis, Institute of Crystallography, RAS, Moscow.
3. Urzhumtsev, A.G., Lunin, V.Yu. and Luzyanina, T.B. (1989) Bounding a molecule in a noisy synthesis, *Acta Cryst.*, **A45**, 34-39.
4. Podjarny, A.D., Schevitz, R.W. and Sigler, P. (1981) Phasing low resolution macromolecular structure factors by matricial direct methods, *Acta Cryst.*, **A37**, 662-668.
5. Urzhumtsev, A.G. (1991) Low-resolution phases: their influence on SIR-syntheses and retrieval with double-step-filtration, *Acta Cryst.*, **A47**, 794-801.
6. Podjarny, A.D. and Urzhumtsev, A.G. (1997) Low resolution phasing, in C.W.Carter, Jr., and R.M.Sweet, (eds), *Methods in Enzymology* **276A**, Academic Press, San Diego.,. pp. 641-658.
7. Lunin, V.Yu., Lunina, N.L., Petrova, T.E., Vernoslova, E.A., Urzhumtsev, A.G. and Podjarny, A.D. (1995) On the ab initio solution of the phase problem for macromolecules at very low resolution: the Few Atoms Model method, *Acta Cryst.*, **D51**, 896-903.
8. Lunin, V.Yu., Urzhumtsev, A.G. and Skovoroda, T.A. (1990) Direct low-resolution phasing from electron-density histograms in protein crystallography, *Acta Cryst.*, **A46**, 540-544.
9. Podjarny, A.D., Rees, B., Thierry, J.-C., Cavarelli, J., Jesior, J.C., Roth, M., Lewitt-Bentley, A., Kahn, R., Lorber, B., Ebel, J.-P., Giegé, R. and Moras, D. (1987) Yeast $tRNA^{Asp}$-aspartyl-tRNA synthetase complex : low resolution crystal studies, *J.Biomol.Struct.Dynam.*, **5**, 187-198.
10. Urzhumtsev, A.G.,.Podjarny, A.D. and Navaza, J. (1994) On the solution of the molecular replacement problem at very low resolution, *Joint CCP4 and ESF-EACBM Newsletter on Protein Crystallography*, **30**, 29-36.
11. Volkmann, N., Hottenträger, S., Hansen, H.A.S., Zaytzev-Bashan, A., Sharon, R., Yonath, A. and Wittmann, H.G. (1990) Characterization and preliminary crystallographic studies on large ribosomal subunits from *Thermus thermophilus*, *J.Mol.Biol.*, **216**, 239-241.

12. Urzhumtsev, A.G., Vernoslova, E.A. and Podjarny, A.D. (1996) Approaches to very low resolution phasing of the ribosome 50S particle from *Thermus thermophilus* by the Few-Atoms-Models and Molecular Replacement methods, *Acta Cryst.*, **D52**, 1092-1097.
13. Yonath, A., Leonard, K.R. and Wittmann, H.G. (1987) A tunnel in the large ribosomal subunit revealed by three-dimensional image reconstruction, *Science*, **236**, 813-816
14. Rossmann, M.G. (1990) The molecular replacement method, *Acta Cryst*, **A46**, 73-82.
15. Urzhumtsev, A.G., Podjarny, A.D. (1995) On the solution of the molecular replacement problem at very low resolution: application to large complexes, *Acta Cryst.*, **D51**, 888-895.
16. Navaza J. (1994) AMoRe : an automated package for molecular replacement , *Acta Cryst.*, **A50** 157-163.
17. Volkmann, N., Schlunzen, F., Urzhumtsev, A.G., Vernoslova, E.A., Podjarny, A.D., Roth, M., Pebay-Peyroula, E., Berkovitch-Yellin, Z., Zaytsev-Bashan, A. and Yonath, A. (1995).On ab-initio phasing of ribosomal particles at very low resolution, *Joint CCP4 and ESF-EACBM Newsletter on Protein Crystallography*, **32**, 23-32.

USE OF NON-CRYSTALLOGRAPHIC SYMMETRY FOR *AB INITIO* PHASING OF VIRUS STRUCTURES

MICHAEL S. CHAPMAN*[†], ERIC BLANC*, JOHN E. JOHNSON[∞], ROBERT MCKENNA[‡], SANJEEV MUNSHI[§], MICHAEL G. ROSSMANN[¶] & JUN TSAO[%]

* *Department of Chemistry & Institute of Molecular Biophysics, Florida State University, Tallahassee, FL 32306-3015, USA.*

[∞] *Department of Molecular Biology, The Scripps Research Institute, 10550 N. Torrey Pines Rd., MB-13, La Jolla, CA 92037, USA.*

[‡] *Department of Biological Sciences, University of Warwick, Coventry, CV4 7AL, UK.*

[§] *WP44-B122, Merck & Co. Inc., West Point, PA 19486, USA.*

[¶] *Purdue University, Department of Biological Sciences, West Lafayette, IN 47907, USA*

[%] *Center for Macromolecular Crystallography, 260 BHS THT 79, University of Alabama, UAB Station, Birmingham, AL 35294, USA*

Abstract: Several virus structures have now been determined through *ab initio* phasing at very low (~20Å) resolution followed by extension to high resolution using their non-crystallographic symmetry (NCS). The methods are described, and *post-mortem* investigations of phase determination are analyzed, particularly for relevance to the more challenging general case of *ab initio* determination of protein structure.

1. Scope

A generous definition of "*ab initio* phasing" can include molecular replacement, MAD methods *etc.* (see elsewhere, this volume). Virus structures are among the few large structures solved with phases calculated from single monochromatic data sets without an atomic model. Here, "*ab initio*" will be used only when there is no prior source of phases. "*Ab initio*" has been used to describe the calculation of **all** phases from scratch, or the calculation of high resolution phases from low resolution phases determined by

---

[†] To whom correspondence should be addressed.

S. Fortier (ed.), Direct Methods for Solving Macromolecular Structures, 433–442.

other methods [1]. Here, to distinguish the two, the latter will be termed "phase extension" which is discussed more fully in the accompanying article [2].

2. Process and Theory, an Introduction

Initial phasing is somewhat unique, but phase improvement and extension by NCS is a specific form of density modification. Similar iterative techniques are used as shown in Figure 1 of the accompanying article [Chapman, 1997 #1737]. For viruses, the equivalence of reciprocal- and real-space procedures is especially instructive. Phases can be considered calculated from a weighted vector sum of neighboring structure factors (see accompanying article [3] and references therein).

3. Viruses as a special case

Although it is hoped that similar techniques may one day be routinely applicable to protein structure determination, there are several reasons why it was with virus structures that *ab initio* techniques bore some of their first fruits:

1. *High order non-crystallographic symmetry*: The three successful high resolution *ab initio* determinations used N = 60-fold NCS. Phasing power is thought to be proportional to $\sqrt{N}$ [4]. 60-fold averaging can be forgiving of poor *ab initio* phases and has "recovered" from mean phase errors of 86.3° (3.7° from random) [5].

2. *Well-defined point-group operations*: With icosahedral 532 point group symmetry there are only a few discretely different relative orientations of symmetry axes [6]. Up to 60 NCS operators are specified with just 3 orientational and 3 positional parameters, the number usually required for each operator. The NCS parameters can be determined more precisely using methods such as the locked rotation and translation functions [7, 8]. As phase extension proceeds, NCS parameters can be refined by methods that are easier to apply with high symmetry – optimization of the agreement between electron density at equivalent points [9].

3. *Large unit cells, and closely-spaced reciprocal lattices*: The dependence of the refined phase on neighboring structure factors decreases rapidly (see accompanying article [3]). With small unit cell:molecular volume, reciprocal lattice points are coarsely spaced, few neighboring reflections have significant weight, and the new phase, being mostly dependent on its old value, improves little. With large unit cell:molecular volume, other reflections have significant weight, there is more interdependence and potential for the phase to improve. Interdependence is also greater with more reflections that are neighbors by NCS transformation, *i.e.* both large unit cells and high-order NCS contribute to phasing power for viruses.

4. *Phasing models can be built with few parameters and more confidence*: This results from both simple low resolution structure and the large number of low

resolution reflections. For a virus with a 350 Å unit cell, a 30 Å model can be refined against and phase 6,600 reflections. For a protein with a ~ 50 Å unit cell, the molecular shape is less predictable, and only 19 reflections are available to determine model parameters. Then there are fewer starting phases for extension. Therefore, it may be necessary to have a more sophisticated higher resolution model to start phasing, with the additional challenges that this brings.

Thus, viruses are particularly well suited to NCS constraints, and this has facilitated the application of *ab initio* methods that are still far from routine for large proteins. *Post mortem* analysis of *ab initio* phasing is facilitated by the availability of accurate (model-independent) final phases that can be used to evaluate progress. Maps calculated with 60-fold NCS refined phases are of quality similar to the best protein maps, obtained with especially careful anomalous dispersion techniques [10, 11]. Coming from some of the few successful, albeit special-case applications, analyses offer insights perhaps helpful to more challenging *ab initio* protein structure determinations.

4. Successes of various initial phasing models

4.1. SUB-MOLECULAR RESOLUTION

4.1.1. Southern Bean Mosaic Virus (SBMV)

The forebear of high resolution studies was the 22.5 Å phase determination of SBMV [12]. Between 60 and 35 Å, spherically averaged structure amplitudes agreed well with those of a solid sphere with diameter of 281 Å, the nearest neighbor distance for packed spheres in the crystal. Centric phases were calculated from the model, refined using NCS and extended to 22.5 Å where the agreement broke down between experimental structure amplitudes and those calculated by back transformation of the map. The limitation to 22.5 Å was blamed on coincidence of the 532 icosahedral symmetry and special points of 23 crystallographic symmetry, and the resulting failure to break the centric nature of phases. In retrospect, with only 10-fold non-crystallographic symmetry, SBMV would be expected to be one of the more difficult phase determinations.

4.1.2. Polyoma virus

A more sophisticated, non-centric starting model was built by decorating a sphere with cylinders, consistent with electron microscopic images [13]. Maps calculated by phase extension to the 22.5 Å resolution limit were consistent with later high resolution structure determination [14].

4.1.3. Non-parametric models

Most progress has involved simple parametric models fit to the diffraction data. An alternative approach, requiring fewer *a priori* assumptions, has roots in methods of low resolution phase calculation that use small numbers of large spheres [15, 16] or "condensing gasses" of large numbers of small spheres [17]. A lattice covers a unique region of the icosahedral volume. By comparison of calculated and observed structure amplitudes, it is determined which lattice points are occupied by *pseudo*-atoms [18]. Genetic algorithms are used to find good models among the many possible. For the best, occupancies of the pseudo-atoms are refined, then calculated phases are improved by iterative NCS averaging (§2). In 50% of trials, phases were successfully determined for simulated error-free 24 Å poliovirus data. It is not known whether the methods will be applicable to real structures, but the simulations provided insights into phase determination which will be discussed later.

4.2. TOWARDS ATOMIC RESOLUTION

4.2.1. Prelude of things to come: bacteriophage MS2 et al.

With viruses, NCS was used first to improve inaccurate phases obtained by isomorphous replacement. Then it was used to improve molecular replacement phases calculated from structures that were first closely related, but later decreasingly so, as confidence improved in NCS phase refinement. In what was then an extreme case, the structure of the bacteriophage MS2 was determined starting with 13 Å phases calculated from a completely unrelated model, SBMV. Extension to 3.4 Å led to an uninterpretable map, but the phases were good enough to determine the binding sites of heavy metals. This overcame the formidable problem of determining > 90 heavy atom sites per crystallographic asymmetric unit. Isomorphous replacement phases were extended successfully to 3.3 Å resolution. This initial phase determination suggested that it might be possible to start at very low resolution, that a very crude initial model might suffice, and gave valuable insights into the phasing process (see later).

4.2.2. Canine parvovirus (CPV)

The initial phasing model for CPV was similar to that used for SBMV, except that a spherical shell allowed the nucleic acid and protein to have different uniform densities. CPV showed the critical importance of precise phasing models. Phases for a model can be calculated by back-transformation of an electron density map. However, an analytical calculation of structure factors is useful for parameter refinement [5]:

$$\vec{F}_h \propto \sum_n^N \Gamma\{R_i, R_o, \rho_{NA}\} \exp\{2\pi j \vec{h} \cdot \vec{S}_n\}; \qquad \Gamma = G\{R_o\} - \left(\frac{R_i}{R_o}\right)^3 (1-\rho_{NA}) G\{R_i\} \qquad (1)$$

$$G = 3[\sin X - 2X \cos X]/X^3; \qquad X = 2\pi|h|R \qquad (2)$$

where R_o, R_i are the outer and inner shell radii, ρ_{NA} is the density of the nucleic acid (relative to the protein), $j = \sqrt{-1}$, h is the reflection index, and S_n is the position of the nth virus in the unit cell. By comparing calculated and observed structure amplitudes it is possible to determine R_o, R_i, ρ_{NA} and S_n by systematic search or least-squares refinement. Refinement is facilitated by analytical partial derivatives [5].

How are the phases calculated? The crystal is the convolution of point scatterers (exponential term) and spherical shells (Γ term). In most cases, N is small (1 or 2 per asymmetric unit) and the exponential generates only centric terms. Γ, the analytical expression for the Fourier transform of a spherical shell, also has only centric terms. Its effect on the phase of the point scatterer term is either to change it by 180° ($\Gamma < 0$) or to leave it unchanged ($\Gamma > 0$).

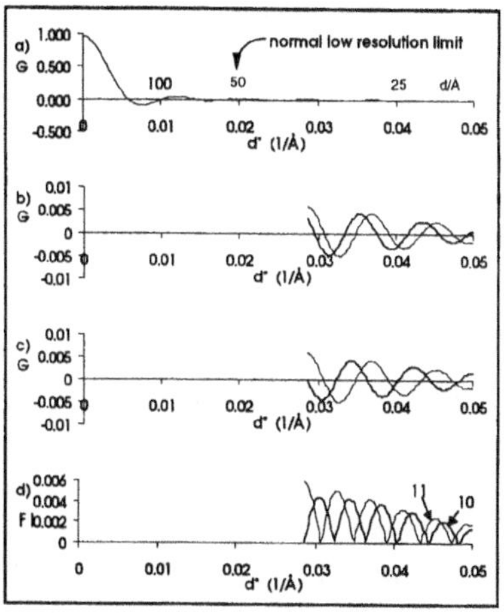

Figure 1: Ambiguities of phase determination. (a) The Fourier transform of a solid sphere with radius of ~ 125 Å. In determining parameters for a phasing model, its agreement with experimental structure amplitudes is maximized. Data are normally only available at resolutions where non-spherical components of the virus gradually dominate those of a perfect sphere. (b) G-functions of "a" in thick line and with radius changed by 4% in thin line. Within the 35 to 20 Å resolution range, ~½ the calculated phases are incorrect (G > 0 instead of G < 0 et vice versa). (c) With 8% error in radius, most phases are the Babinet opposites. Providing enough phases are mutually consistent, extension might yield a Babinet solution which can be corrected before an atomic model is built. Thus, this error is not as serious as in (b). (d) Panels a-c show phased (signed) G-functions, but it is only magnitudes that are observable. Within the available narrow resolution window, it is possible to nearly superimpose the G-functions of discretely different models. Different peaks superimpose at 22 Å, showing that complete low resolution data could resolve the ambiguity. With this single parameter solid sphere, the superimposition in (d) is not good. When additional degrees of freedom are added by fitting inner radius, density levels and position, a nearly exact superimposition can be obtained over a finite resolution range.

Figure 1 illustrates how discretely different models may have similar fit to the diffraction data. Multiple optima are obvious in systematic searches [5]. The wrong choice can lead to an incorrect low resolution physical model. However, with extension to ~ 3 Å, Babinet inversion can be detected and corrected prior to building an atomic model. More damaging to phasing prospects are parameter sets between the optima. Simulations showed that these generate mixtures of correct phases and Babinet opposites from which convergence is not possible [19]. The tests showed that spherical shell radii need to be within 3% of one of the models corresponding to a correct or Babinet phase solution. Systematic fine search and least-squares refinement can meet these stringent criteria [5], but crystal packing calculations and electron microscopy might not. Lack of required precision is one of several possible explanations for the initial difficulties of several phase determinations (see below).

De novo phase determination for CPV was not completely successful due to failure to refine the point symmetry location as the resolution increased. This was only realized with analysis of partial isomorphous derivative data [9]. As with MS2, the heavy atom sites had negative peaks, indicating that the extended phases were Babinet-inverted. The 532 symmetry location was refined to maximize the heavy atom peaks. Extension was restarted with isomorphous replacement phases. Retrospective analysis showed that the *ab initio* phasing failed due to a ~ 2 Å error in its position. The 2 Å precision achieved at low resolution was not sufficient for extension beyond 9 Å. Average phase error of 41° quickly dropped to < 10° upon correcting the position, demonstrating that with greater experience, complete *ab initio* determination would have been possible [9].

4.2.3. Bacteriophage ϕX174

Phasing of ϕX174 was initiated prior to completion of CPV. Spherically symmetric models with unrefined radii from packing calculations failed, as did early attempts with phases derived from electron microscopic (EM) images [20]. Success came with an unrelated atomic model fit into the EM envelope. Retrospective analysis showed that phases could have been extended from the EM image had the point symmetry location been accurate. Why did the atomic model work, but not the EM image directly? Perhaps the higher resolution structure factors from an atomic model led to a more precise virus position. Experience with CPV also suggests that phasing could fail if, due to experimental error, the size of the EM image was not within the 3% required precision. It might be necessary to refine against the x-ray data, the EM magnification or the contrast transfer function [21] that can change the size of a near-spherical image.

4.2.4. Nudaurelia capensis ω virus (NωV)

Once again, several early attempts with spherical shell models and EM density failed. Following careful reprocessing of the low resolution data, and fitting parameters of a hollow shell model by correlation with diffraction data, it was possible to extend phases from 18 Å to 10 Å [22, 23]. Maps calculated by extension to higher resolution were uninterpretable. Success came by building an atomic model into the low resolution map (using homology to distant relatives) from which phases were calculated. These were improved by averaging and the process was iterated, extending to higher resolution.

5. Phase extension, the process

Comments are restricted to differences with methods used for proteins [3].

Envelope: The symmetry of a virus normally belongs to a closed point group. The envelope needs only to distinguish different assemblies and not individual proteins. Relatively crude envelopes often suffice at the start, even polyhedra defined by bisecting normal planes to vectors between neighboring viral centers. Spheres, or a combination of spheres and planes are often used [9, 24], but always set to err on the side of generous envelopes. When surface features are prominent, structure determinations have benefited from more detailed envelopes [20]. As phase recombination following 30+-fold NCS prevails over minor masking errors, envelopes can be improved during refinement by comparing the electron density at specific points with those of points to which it might be related by NCS [25]. Automatic solvent definition [26] can also be used. At very low resolution, the scattering of viral nucleic acid is significant, and it is better to flatten this to an average value that is independent from that chosen for the solvent [5]. At high resolution, scattering from the disordered nucleic acid can be considered to be the same as from the external solvent.

Filling missing observations: Even with data collected from hundreds of crystals, virus data sets rarely approach completion. It has been argued that with appropriate processing of accurate data and phases, only $1/60^{th}$ of a data set for an assembly with 60-fold NCS is required to generate a map equivalent in quality to that of a complete data set without NCS [4, 27]. Thus the Herculean effort of completing a virus data set is rarely undertaken. The potential consequences of missing data are dire. A reflection does not contribute equally to all points of a map related by NCS. Thus a map calculated with partial data should not be expected to have the exact NCS that will be constrained. In reciprocal space, this is equivalent to calculating phases from neighboring structure factors, some of which have been incorrectly reset to zero. This shows a potential future benefit of reciprocal space implementation, in which the missing reflections could be ignored. In real-space implementations, it is now routine to complete the data by calculation of missing observations by back-transformation of the current map. There has not been a systematic study of how weighting of the filled reflections relative to observed reflections might be optimized to improve convergence.

Related are the effects of resolution truncation. Series termination errors, greatest near the viral surface, need not be icosahedrally symmetric. Simulations showed benefit in leaving a surface margin unaveraged [18], but series termination is usually less problematic with real data.

Extension: Especially without prior phases, resolution can only be extended in small steps (see accompanying article [3]). Progress is usually followed with graphs of the correlation coefficient between observed and back-transformed structure magnitudes plotted against resolution. A gradual decrease might indicate imprecise point symmetry, or poor envelope definition. Sudden changes might indicate the presence of multiple solutions (see below) and/or that more gradual extension might be required.

Programs: As few virus structures are solved, some prefer programs such as *RAVE* [28] that have been tested extensively in many protein structure determinations. These are not well optimized for problems as large as viruses. Increases of speed have been achieved with programs exploiting highly parallel computer architecture [29], or by loading the entire map onto cheap physical memory that has recently become available (EB & MSC, unpublished).

6. *Post mortem* analysis of phase determination

Histograms comparing final and intermediate phases have been revealing (Figure 2). Analysis of MS2 [30] showed that the initial (unsuccessful) refinement converged towards the Babinet solution, but some phases were close to the enantiomer or enantiomeric Babinet solution. Application of 532 symmetry is consistent with any of these solutions, but their co-existence was unexpected, because of their incompatible electron densities (that result in uninterpretable maps). The multiple solutions were thought linked to prior observations of correlation coefficient (above) oscillating with resolution, indicating variable quality of phase determination. Experience of several structure determinations suggested that reciprocal space was being partitioned with regions converging upon different solutions. As each phase depends only on neighboring structure factors, phases for each region could become consistent with each other and the NCS, but inconsistent with other regions of reciprocal space. Overall statistics might indicate little, because it would only be on the boundaries of regions where two solutions might "fight" each other and structure factors might be inconsistent with each other.

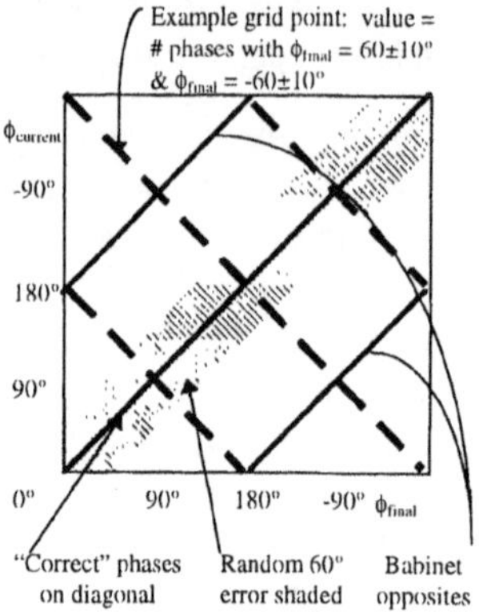

Figure 2: Calculation of histograms for analysis of phases [30] and expected positions of peaks. Solutions of the correct hand are shown in solid line, and enantiomeric solutions are dashed.

The sensitivity of *ab initio* phasing to starting model can now be rationalized on the basis that for success, the vast majority of phases must belong to one consistent set.

As with MS2, competing solutions appeared limiting in the uninterpretable initial maps of NωV [23]. By contrast, the final *ab initio*-derived phases of φX174 and CPV showed only modest random error about a single Babinet solution [9, 20]. However, intermediate phases of φX174 showed multiple solutions, and also repeated switching of large numbers of phases between the correct, Babinet and enantiomeric solutions during phase refinement [20]. This might have been due to slight (~ 1Å) adjustments of the symmetry location, but is not known for certain. It should be emphasized that although the *ab initio* determinations have been more exhaustively analyzed, phases for some other viruses have been difficult to refine [31-33]. Perhaps similar problems occur with phases originally derived from isomorphous molecular replacement.

With the ability to compare to accurate final phases, had we chanced upon a general limitation of symmetry averaging? Did competing internally consistent regional phase sets limit phase refinement of proteins? The structure determination of human rhinovirus 50 (HRV50) (Blanc *et al.*, in progress) presented an opportunity for an intermediate test of this hypothesis. Phases were determined not *ab initio*, but by molecular replacement from the closely related 3.5 Å HRV1A structure [34]. They were refined and extended to 2.4 Å resolution using 15-fold symmetry averaging. However, the map was disappointing, looking more like 4 Å resolution than 2 Å. Following model-building and real-space refinement [35], phases were calculated from the atomic model and re-extended from 4 Å resolution. Although refinement continues, the resulting improved map enabled the building of a structure with R_{free} = 25% using all data to 2.0 Å. Unexpectedly, comparison of the two phase sets showed no evidence of multiple phase sets -- just random phase error about a single solution. Furthermore, searches through reciprocal space revealed that there was no significant correlation in the phase error of neighboring reflections.

The conclusion is anecdotal, because only one of the detailed analyses did not involve *ab initio* determination. There is a preliminary suggestion that low resolution *ab initio* phase determinations are more susceptible to competing solutions than higher resolution phase determinations. At both high and low resolution, different phase solutions may be consistent with the symmetry averaging. It may be that at high resolution, the effect of a poor model is near-random phase, whereas at low resolution, it can lead to large regions of reciprocal space with phase solutions that are internally consistent, consistent with the NCS, but compete with different solutions in other regions. It appears that competing solutions can persist through much phase refinement.

7. Acknowledgments & References

The structure determinations were supported by the National Institutes of Health (MGR; JEJ), the National Science Foundation (MGR) and the Lucille P. Markey Charitable Trust (MGR). Recent work was supported by the National Science Foundation (MSC).

1 Rossmann, M. G. (1995) *Ab initio* phase determination and phase extension using non-crystallographic symmetry, *Current Opinion in Structural Biology* **5**, 650-655

2 Chapman, M. S., Blanc, E., McKenna, R., Munshi, S., Rossmann, M. G. and Tsao, J. (1997) Use of non-crystallographic symmetry for *ab initio* phasing of virus structures, in Fortier, S. (ed.), *NATO Advanced Study Institute on Direct Methods for Solving Macromolecular Structures*, Kluwer Academic Publishers, Dordrecht, The Netherlands

3 Chapman, M. S. (1997) Introduction to the use of non-crystallographic symmetry in phasing, in Fortier, S. (ed.), *NATO Advanced Study Institute on Direct Methods for Solving Macromolecular Structures*, Kluwer Academic Publishers, Dordrecht, The Netherlands

4 Arnold, E. and Rossmann, M. G. (1986) Effect of errors, redundancy, and solvent content in the molecular replacement procedure for the structure determination of biological macromolecules , *Proc. Natl. Acad. Sci. USA* **83** , 5489-93

5 Chapman, M. S., Tsao, J. and Rossmann, M. G. (1992) *Ab initio* Phase Determination for Spherical Viruses: Parameter Determination for Spherical Shell Models, *Acta Crystallographica* **A48**, 301-312

6 Caspar, D. L. D. and Klug, A. (1962) Physical principles in the constructuion of regular viruses, *Cold Spring Harbor Symposium in Quantitative Biology* **27**, 1-24

7 Tong, L. and Rossmann, M. G. (1990) The Locked Rotation Function, *Acta Crystallographica* **A46**, 783-792

8 Tong, L. (1996) The Locked Translation Function and Other Applications of a Patterson Correlation Function, *Acta Crystallographica* **A52**, 476-79

9 Tsao, J., Chapman, M. S., Wu, H., Agbandje, M., Keller, W. and Rossmann, M. G. (1992) Structure Determination of Monoclinic Canine Parvovirus, *Acta Crystallographica* **B48**, 75-88

10 Burling, F. T., Weis, W. I., Flaherty, K. M. and Brünger, A. (1996) Direct Observation of Protein Solvation and Discrete Disorder Using Experimental Crystallographic Phases, *Science* **271**, 72-77

11 Zhou, G., Wang, J., Blanc, E. and Chapman, M. S. (1997) Determination of the Relative Precision of Atoms in a Macromolecular Structure, *Acta Crystallographica* ***submitted***

12 Johnson, J. E., Akimoto, T., Suck, D., Rayment, I. and Rossmann, M. G. (1976) The Structure of Southern Bean Mosaic Virus at 22.5 Å Resolution, *Virology* **75**, 394-400

13 Rayment, I., Baker, T. S., Caspar, D. L. D. and Murakami, W. T. (1982) Polyoma virus capsid structure at 22.5 Å resoution , *Nature* **295** , 110-15

14 Stehl, T., Yan, Y. and Harrison, S. C. (1994) The structure of murine polyomavirus complexed with an oligosaccharide receptor fragment, *Nature* **369**, 160-3

15 Lunin, V. Y., Lunina, N. L., Petrova, T. E., Vernoslova, E. A., Urzhumtsev, D. A. G. and Podjarny, A. (1995) On the *ab initio* solution of the phase problem for macromolecules at very low resolution: the few atoms model method, *Acta Crystallographica* **D51**, 896-903

16 Podjarny, A. (1997) Model based approaches at very low resolution, in Fortier, S. (ed.), *NATO Advanced Study Institute on Direct Methods for Solving Macromolecular Structures,* Kluwer Academic Publishers, Dordrecht, The Netherlands

17 Subiah, S. (1991) Low-resolution Real-Space Envelopes: An Approach to the Ab Initio Macromolecular Phase Problem, *Science* **252**, 128-133

18 Miller, S. T., Hogle, J. M. and Filman, D. J. (1996) A Genetic Algorithm for the *Ab Initio* Phasing of Icosahedral Viruses, *Acta Crystallographica* **D52**, 235-251

19 Tsao, J., Chapman, M. S. and Rossmann, M. G. (1992) *Ab initio* Phase Determination for Viruses with High Symmetry: A Feasibility Study., *Acta Crystallographica* **A48**, 293-301

20 McKenna, R., Xia, D., Willingmann, P., Ilag, L. and Rossmann, M. G. (1992) Structure Determination of Bacteriophage ϕX174, *Acta Cryst.* **B48**, 499-511

21 Toyoshima, C. and Unwin, N. (1988) Contrast transfer of frozen hydrated specimins: Determination from pairs of defocussed images, *Ultramicroscopy* **25**, 279-92

22 Munshi, S., Liljas, L., Cavarelli, J., Bomu, W., McKinney, B., Reddy, V. and Johnson, J. E. (1996) The 2.8 Å structure of a T=4 animal virus and its implications for membrane translocation of RNA, *Journal of Molecular Biology* **261**, 1-10

23 Munshi, S., Liljas, L. and Johnson, J. E. (1997) Joys and Traps of Phase Extension: Structure Determination of *Nudaurelia capensis ω Virus*, *Acta Crystallographica* **in preparation**

24 Rossmann, M. G., Arnold, E., Erickson, J. W., Frankenberger, E. A., Griffith, J. P., Hecht, H., Johnson, J. E., Kamer, G., Luo, M., Mosser, A., Rueckert, R., Sherry, B. and Vriend, G. (1985) Structure of human common cold virus and functional relationship to other picornaviruses. , *Nature* **317** , 145-153

25 Rossmann, M. G., McKenna, R., Tong, L., Xia, D., Dai, J.-B., Wu, H., Choi, H.-K. and Lynch, R. E. (1992) Molecular Replacement Real-Space Averaging, *J. Appl. Cryst.* **25**, 166-80

26 Wang, B. C. (1985) Resolution of the Phase Ambiguity in Macromolecular Crystallography , *Methods in Enzymology* **115** , 90-112

27 Chapman, M. S., Minor, I., Rossmann, M. G., Diana, G. D. and Andries, K. (1991) Human rhinovirus 14 complexed with antiviral compound R 61837, *Journal of Molecular Biology* **217**, 455-63

28 Jones, T. A. (1992) a, yaap, asap, @£? A set of averaging programs, in Dodson, E., Gover, S. and Wolf, W. (ed.), *Proceedings of the CCP4 Study Weekend,*, Science and Engineering Research Council, pp. 91-105

29 Cornea-Hasegan, M. A., Zhang, Z., Lynch, R. E., Marinescu, D. C., Hadfield, A., Muckelbauer, J. K., Munshi, S. and Rossmann, M. G. (1996) Phase refinement and extension by means of non-crystallographic symmetry averaging using parallel computers, *Acta Crystallographica* **D51**, 749-59

30 Valegard, K., Liljas, L., Fridborg, K. and Unge, T. (1991) Structure Determination of the Bacteriophage MS2, *Acta Crystallographica* **B47**, 949-960

31 Acharya, R., Fry, E., Stuart, D., Fox, G., Rowlands, D. and Brown, F. (1989) The three-dimensional structure of foot-and-mouth disease virus at 2.9 Å resolution, *Nature* **337**, 709-716

32 Larson, S. B., Koszelak, S., Day, J., Greenwood, A., Dodds, J. A. and McPherson, A. (1993) Three-dimensional Structure of Satellite Tobacco Mosaic Virus at 2.9 Å resoluion., *Journal of Molecular Biology* **231**, 375-91

33 Ban, N. and McPherson, A. (1995) The structure of satellite panicum mosaic virus at 1.9 Å resolution, *Nature Structural Biology* **2**, 882-890

34 Kim, S., Smith, T. J., Chapman, M. S., Rossmann, M. G., Pevear, D. C., Dutko, F. J., Felock, P. J., Diana, G. D. and McKinlay, M. A. (1989) Crystal Structure of Human Rhinovirus Serotype 1A (HRV1A), *J. Mol. Biol.* **210**, 91-111

35 Chapman, M. S. (1995) Restrained Real-Space Macromolecular Atomic Refinement using a New Resolution-Dependent Electron Density Function, *Acta Crystallographica* **A51**, 69-80

HALF TITLE PAGE
4. LATEST DEVELOPMENTS

Direct Methods, and the use of Synchrotron Radiation for Macromolecular Crystallography

Robert M. Sweet[†], John M. Skinner[†], and Thomas C. Terwilliger[‡]

[†] *Biology Department, Brookhaven National Laboratory*
[‡] *Structural Biology, Los Alamos National Laboratory*

1. Introduction

There is a nice match between some of the capabilities of direct phasing methods (DM) and the sort of problem that can be approached at a synchrotron-radiation (SR) source. We will begin by describing some of these matches, will discuss the ways that the SR community has developed methods to collect high enough quality data for DM to work, and will discuss ways of treating the data to extract the best possible signal.

2. The harmony between direct methods and the synchrotron

What are the possible uses of direct methods in connection with diffraction data measured at a SR x-ray source? Consider the situations where there is a good match of capabilities and needs (+) or a bad match (-).

Ab Initio Structures

- \+ One needs to take only one data set at the resolution limit
- \+ Often this approach is necessary for very dense crystals into which heavy atoms cannot be soaked.
- \+ Direct methods benefits from the very low resolution data that can be measured easily with synchrotron radiation.
- \- On a certain scale, direct methods still require a lot of computing time
- \- There are (still) resolution limits to the structures that can be solved – right now the limit seems to be roughly 1.3Å.

Use of direct methods to resolve phase ambiguity with single-wavelength anomalous data

- \- Probably the method is useless if synchrotron radiation is available – one would simply measure MAD data with that heavy-atom derivative.

Multi-wavelength Anomalous Diffraction (MAD) structures

- \+ Direct methods already are useful and will find increasing use in the solving of the positions of many (5-100) anomalous scatters for structures to be determined at modest resolution, for example in the use of seleno-methionine as a heavy atom.

S. Fortier (ed.), Direct Methods for Solving Macromolecular Structures, 445–450.

- \+ We may be able to introduce anomalous scatterers to the solvent to produce a solvent-volume mask of the molecule, the phases from which can be extended by DM
- \- Probably the direct methods are useless for "Super-MAD" (use of a few *very* anomalous scatterers such as lanthanides). This is because the small number of scatterers perhaps can be solved by inspection of Patterson-function maps. But (+) note that Maximum-Likelihood methods, the child of Direct Methods, is useful for the precise refinement of heavy-atom parameters.

3. How do we take the data?

The answer is, *very carefully!* To demonstrate some of the tools and techniques that are used in the accurate measurement of diffraction data, we'll focus on the sort of experiment that is probably the most difficult: determination of a structure by the MAD method when one has a large number of modest size anomalous scatters in a large molecule. The signals that must be measured in this experiment are the real ($\Delta f'$) and imaginary ($\Delta f''$) components of the scattering, shown schematically in Figure 1. Typically, these signals represent only a few percent of the total scattering from the crystal. Also typically, the anomalous and dispersive signals are estimated from small differences among measurements – Friedel (or Bijvoet) mates, the same reflections measured at different x-ray wavelengths.

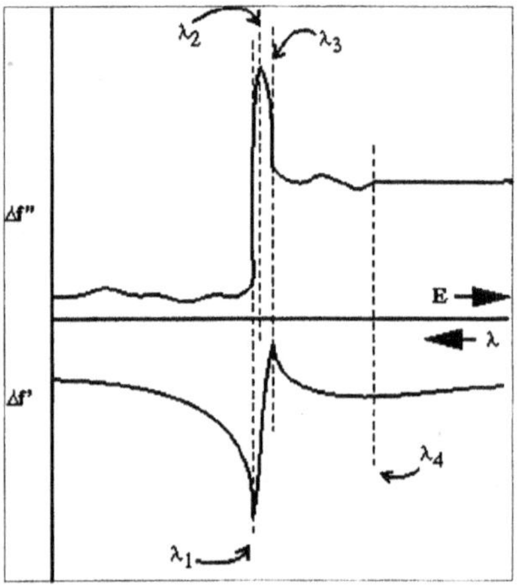

Figure 1: Simulated imaginary ($\Delta f''$) and real ($\Delta f'$) curves for an anomalous scatterer.

Wayne Hendrickson and his colleagues defined the standard [1] in their early experiments. Because in the MAD method one wants to take data in a way to optimise the measurement of small differences, one tries to measure reflections that are to be compared (a) close together in time, (b) with a similar geometry, perhaps employing the "Friedel Flip" to acquire precise Friedel mates (F_{hkl} and F_{-h-k-l}), or (c) on the same image, in this case making use of mirror symmetry in the diffraction pattern to measure Bijvoet pairs.

Working to produce protocols for operation of beamlines at the National Synchrotron Light Source at Brookhaven National Laboratory, we have attempted to make measurements of this sort more nearly automatic and certainly increasingly reliable. This has required the close integration of available hardware and software to give the complete system. The system is based (at the time of this writing) on a Charge-Coupled Device-based (CCD) detection system that was produced for us by Walter Phillips and Martin Stanton [2]

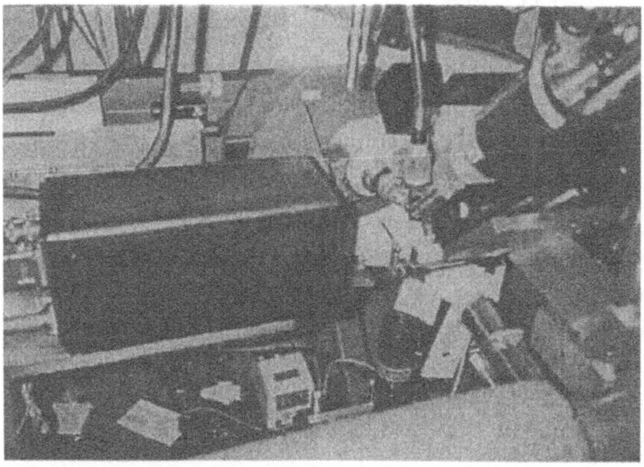

Figure 2: The Brandeis CCD-based detector mounted on the Nonius diffractometer at beamline X12-C at the NSLS.

of Brandeis University under a grant from the National Science Foundation. The detector rides on the θ arm of a Nonius κ-axis goniometer (Figure 2). It is under the control of a Graphical User Interface (GUI) that we have prepared to control the program MADNES [3] for diffractometer control and ACE [4] for beamline control. This entire software system is described in some detail elsewhere [5]. The GUI, which has been refined continually over several years, provides an easy-to-use system that integrates the hardware and all of the software, with many processes running independently on at least three different computers, under a unified control. The software provides automatic execution of many important functions. For example, routine realignment of the diffractometer to the synchrotron's x-ray beam requires the pressing of just one (GUI) button. The changing of x-ray wavelengths during MAD data collection is performed automatically according to a script written by the operator, and each change is checked by the measurement of an x-ray absorption spectrum from the heavy atoms in the crystal, and a numerical analysis of this spectrum to provide a precise positioning of the monochromator. (An example of a view of the control-computer screen during this process is shown in Figure 3.) The software automatically logs a record of the data taken, and writes files that record all of the available parameters describing the *way* the data are being taken – file names, x-ray wavelength, detector position, *etc.*

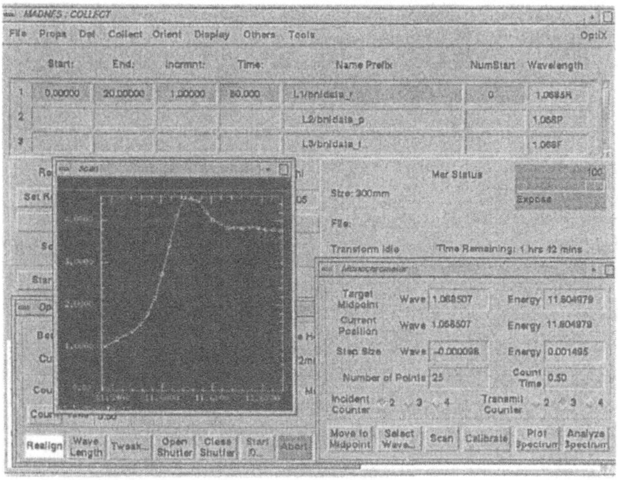

Figure 3: The **COLLECT** screen in MARMAD showing an absorption spectrum that has been fitted to select the precise wavelength for MAD data collection.

An additional important consideration is to define the very best protocol for the treatment of data in order to extract the signal from which one will determine the positions of the heavy atoms. Several different "derived" values have been used in recent times for the solving of the heavy-atom positions, whether Patterson-function or Direct Methods are used. One can use the anomalous ΔF values from the data taken at a wavelength chosen to give the maximum possible signal (λ_2 in Figure 1). One might find the maximum dispersive signal: $F(\lambda_1) - F(\lambda_3)$ or $F(\lambda_1) - F(\lambda_4)$. The parameter F_A defined by Hendrickson [1] is intended for this purpose, but often one finds (especially with difficult structures where the signal is weak) that there are numerous values that are estimated to be higher than they actually are. Terwilliger has defined a method (program MADMRG [6]) by which individually all of the anomalous and then all of the dispersive signals measured at several wavelengths can be combined to produce averaged signals, normalized to a single arbitrary wavelength. Finally, Terwilliger has defined an algorithm (program MADBST [7]) to produce a Bayesian estimate of F_A.

Using data given to us before publication by S. Doublié, from a sub-structure with 15 Se atoms in a structure of 108,000 Da [8], we have applied several different treatments to assess the strength of the signal from the MAD data. The test was to calculate an origin-removed Patterson function from the derived amplitudes, and then to calculate a correlation coefficient between that function and one *calculated* from the refined positions and

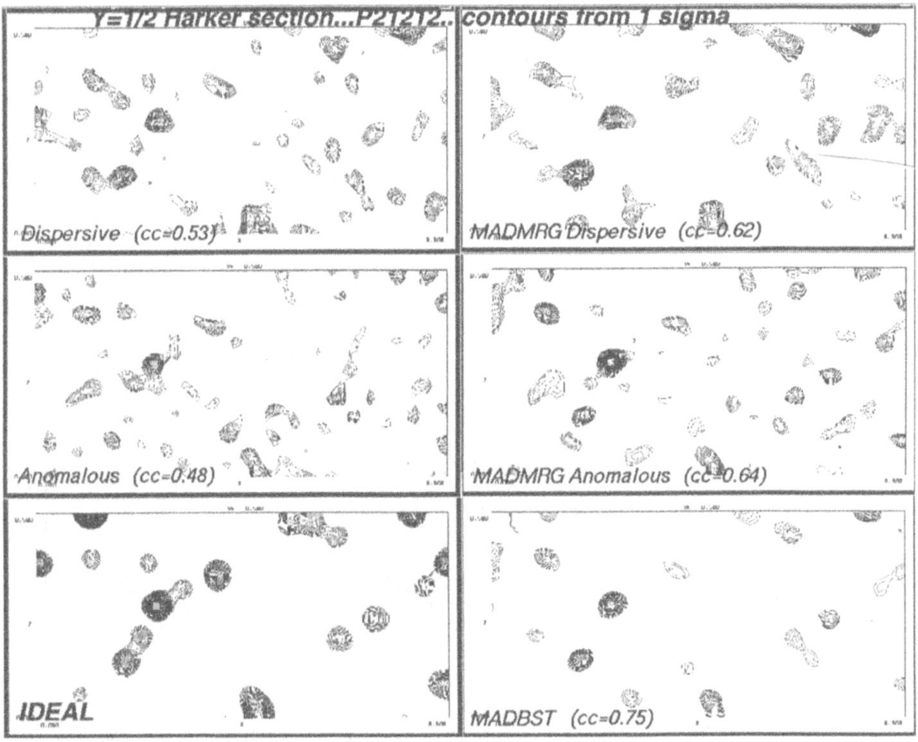

Figure 4: Section from a Patterson function calculated from S. Doublié's structure after different treatments of the data.

occupancies of the heavy atoms.

The results of this test are shown in Figure 4 where one can see the V = ½ section from Patterson functions calculated with Dr. Doublié's data, treated in different ways. One can see a progressive improvement between the model Patterson function (IDEAL) and the derived function as one progresses from use of simple differences (Anomalous is from λ_2 and Dispersive is λ_4 - λ_1) to averaged signals (from MADMRG) and Bayesian-corrected signals (from MADBST). Considering that the structure was solved (at 2.5Å resolution) with the use of the simple Dispersive differences (15 Se atoms were found with program SHELXS [this volume]), one might expect that many atoms will be found in the future with the use of F_A values from MADBST.

4. Conclusions

The synchrotron community and the direct-methods community already collaborate to assist experimenters in the rapid determination of new structures. Clear examples are in the *ab initio* determination of high-resolution structures and the location of anomalous scatterers for MAD phasing. There is more that we can do together to provide a virtual pipeline from the diffractometer to the electron-density map. MAD is an important key, and the principal burden will be to organize software to provide a seamless flow of data from the measuring programs to data-reduction, to scaling, *etc.* Other places where DM may prove to be valuable is in determination of phases for molecular envelopes that can be derived from MAD treatment of heavy atoms in the interstitial solvent of a crystal, and in the completion and extension of phases that might be measured by the Three-Beam method.

This work was supported by the United States Department of Energy Offices of Health and Environmental Research and of Basic Energy Sciences, and by the National Science Foundation. Figures are reproduced here with permission of the copyright-owners.

5. References

1. Hendrickson, W.A. and Ogata, C.M. (1997) "Phase determination from Multiwavelength Anomalous Diffraction Measurements." in Carter, C.W., Jr., and Sweet, R.M. *eds.* **Methods in Enzymology 276**, Macromolecular Crystallography, Part A. pp 494 - 523.

2. Stanton, M., Phillips, W.C. and O'Mara, D. (1994) "CCD-based detector for x-ray crystallography", *Proc. Soc. Photo-Opt. Instr. Eng.* 2278 16-20.

3. Messerschmidt, A. and Pflugrath, J.W. (1987) *J. Appl. Cryst.* **20,** 306.

4. Feng, S. K., Siddons, D.P., and Berman, L. (1994) *Nuclear Instruments and Methods in Physics Research* A **347** 603-606.

5. Skinner, J.M., LaBarca, R.S., and Sweet, R.M. (1996) *Nuclear Instrumentation and Methods in Physics Research A* **383**, 627 - 630.; Sweet, R.M. and Skinner, J.M. (1997). Proceedings of the IUCr Macromolecular Crystallography Computing School P.E. Bourne and K. Watenpaugh (Eds.) SDSC Inc. San Diego CA (in the press). — http://www.sdsc.edu/Xtal/IUCr/CC/School96/bs/bs.htm; Skinner, J.M., and Sweet, R.M. (1997) Integrated software for a macromolecular crystallography synchrotron beamline, *submitted.*

6. Terwilliger, T.C., "Multiwavelength Anomalous Diffraction Phasing of Macromolecular Structures: Analysis of MAD Data as Single Isomorphous Replacement with Anomalous Scattering Data using the MADMRG program." in Carter, C.W., Jr., and Sweet, R.M. *eds.* **Methods in Enzymology 276**, Macromolecular Crystallography, Part A. pp 530 - 537.

7. Terwilliger, T. C (1994) *Acta Cryst.* **50**: 11-16.

8. Doublié, S. (1997) personal communication.

THE LIKELIHOOD BASED CHOICE OF PRIORS IN STATISTICAL APPROACHES TO THE PHASE PROBLEM

V.Y.LUNIN
Institute of Mathematical Problems of Biology, RAS
Pushchino, Moscow Reg., 142292 Russia (lunin@impb.serpukhov.su)

1. Introduction

The methodology of employing statistical approaches to the phase problem is to first consider all hypothetically possible structures, to then find some typical relationships between phases and to finally postulate then that most frequent relationships are valid for the studied structure too. To put this idea into mathematical form it is necessary to introduce first an ensemble of possible structures, i.e. to point out which structures are considered as possible and to define for every possible structure its probability. Usually the probabilities are defined in terms of prior distributions of atomic positions in the unit cell. These priors reflect a preliminary knowledge and influence significantly the final conclusions. Thus the choice of the most suitable prior is an important step in the application of statistical approaches.

One of the widespread statistical recipes is to use the prior which results in the maximal likelihood (ML) value, defined in the crystallographic framework as the probability to obtain the set of experimentally determined magnitude values, provided atomic coordinates are distributed with the test prior. As all statistical methods ML-principle never assures the "true answer" when applied to an isolated object, but reveals its good properties in the systematic use only. We discuss below some examples of the use of this principle at different stages of phase problem solution.

2. How is it possible to calculate the likelihood value?

The customary way is to derive first the joint probability distribution for the set of both magnitudes and phases of structure factors and to integrate it then with respect to phases, supposing magnitudes are equal to the observed values. Nevertheless, this integration generally can not be performed in closed form. A possible escape is to use a diagonal approximation [4,8,2], which is equal to the additional hypothesis that structure factors may be considered as independent random variables.

An alternative way is the use of the concept of generalized likelihood (GL) [7], which is defined as the probability to get a set of magnitudes close enough rather than

S. Fortier (ed.), Direct Methods for Solving Macromolecular Structures, 451–454.

equal to the observed values. In this approach we can use computer-based simulating procedures to calculated GL values.

3. The estimates of errors in phases calculated from preliminary models.

When a preliminary atomic model for the studied structure is present it is reasonable to consider atomic coordinates as random variables centered around the model atom positions. To define the prior more properly it is necessary to specify the expected deviation of model coordinates from the true ones. Depending on this estimate of the model quality, different conclusions can be made about the expected values of errors in phases calculated from the preliminary model. To find the most appropriate hypothesis for the model quality the ML-principle coupled with the diagonal approximation were applied [4,7,8]. Fig. 1 shows results of the use of this approach in tests when the true phases were known. It was found [4] that ML-estimates of phase errors were correct when obtained for unrefined models, but became overestimated when used for refined structures. The exclusion of test set reflections from the refinement [3] and the use of these reflections only for the calculation of the likelihood value (Maximal Marginal Likelihood-estimates) allow [5,8] to overcome this problem (Fig.1b).

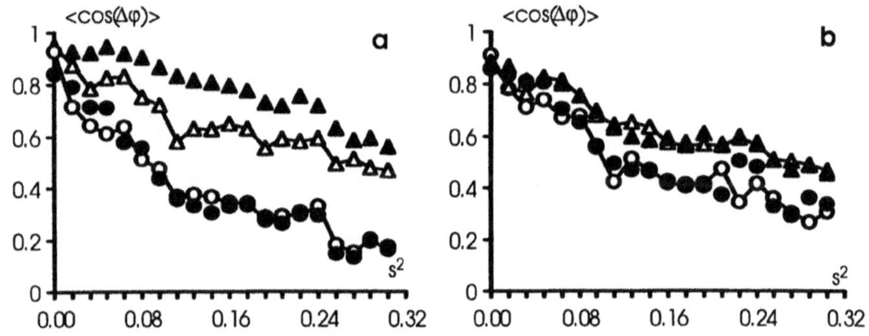

Fig. 1. Estimated (closed) and true (open) phase errors: a) ML-estimates; an atomic model with random coordinate errors before (circles) and after (triangles) reciprocal space refinement; b) MML-estimates; the same model after refinements without (circles) and with (triangles) the use of stereochemical restraints.

4. The use of a generalized likelihood for the choice of priors.

The recently suggested FAM-procedure for low-resolution phasing [6] consists of: a) generating a large number of pseudo-atomic models; b) selection of ones resulted in a good correlation between calculated and observed magnitudes; c) grouping the selected sets of calculated phases into a small number of clusters (Fig. 2.) and d) calculating of the centroid phases for every cluster. The output of the procedure suggests a small number of possible solutions of low-resolution phase problem. A GL-based test [7] consists of generating a number of random models into the corresponding molecular region masks. It is expected that the largest number of models with high correlation of calculated and observed magnitudes will be obtained when using the true mask. Fig. 3 shows that this expectation is satisfied to some extent only. The GL-test provides the true answer when applied to a small number of top-level clusters, but fails when used to detect fine features.

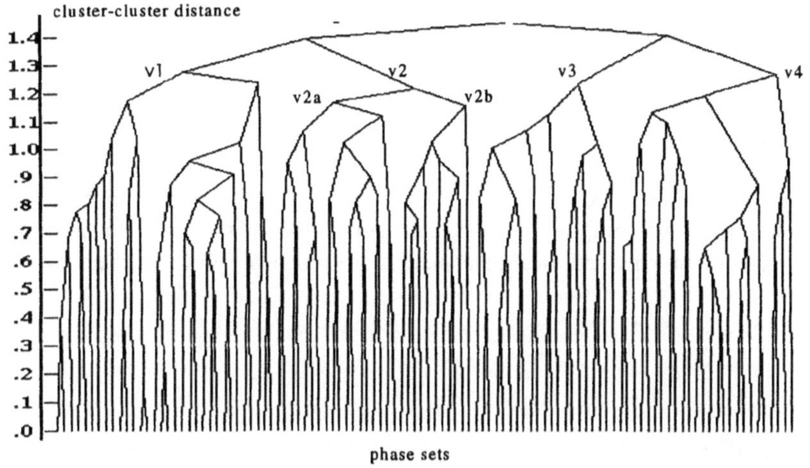

Fig. 2. The clusterization tree shows the order in which phase sets are joining together forming clusters. 100 phase sets from 1000 ones calculated from 2-atoms random models were analyzed.

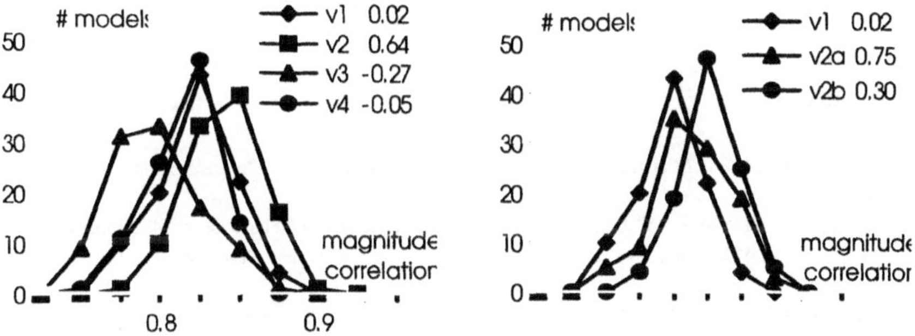

Fig. 3. Distribution of the values of the correlation between calculated and observed magnitudes when generating 100-atoms random models into different masks. The figures in the legend show for every cluster the correlation of the centroid phases with the true ones.

5. Likelihood-gradient approximation for the prior.

When starting from the uniform prior, the guaranteed way to increase likelihood value is to take as the first approximation of the prior the value of the gradient of the likelihood with respect to the prior. (To be more precise, it is necessary to consider here a conditional likelihood, provided that the phases fixing origin/enantiomorph are specified and to take linear combination of the constant prior and the conditional likelihood gradient.). The map of the prior may be used to detect favorable and unfavorable regions for atoms in the unit cell. A special technique of explicit representation of the saddle point approximation for j.p.d. [1] was developed to calculate this gradient approximation. In our tests the model atoms were found to be in "favorable' regions and there were no atoms in "unfavorable" ones. Nevertheless there existed "favorable" regions which did not contain model atoms.

6. Conclusions.

The three considered examples of the use of ML-principle of the prior choice coupled with different ways of the likelihood calculation revealed that this principle may be used to some extent only and must be applied in the practical work with a necessary caution.

Acknowledgments

The author thanks N.Lunina, T.Skovoroda, T.Petrova, A.Urzhumtsev and A.Podjarny for their contributions at the different stages of the work.

7. References

1. Bricogne, G. (1984) Maximum Entropy and Foundation of Direct Methods, *Acta Cryst.* **A40**, 410-445.
2. Bricogne, G. and Gilmor, C.J. (1990) A multisolutional Method of Phase Determination by Combined Maximization of Entropy and Likelihood. I. Theory, Algorithms and Strategy, *Acta Cryst.*, **A46**, 284-297.
3. Brünger, A.T. (1992) Free R value: a novel statistical quantity assessing the accuracy of crystal structures, *Nature (London)* **354**, 472-474.
4. Lunin, V.Yu. and Urzhumtsev, A.G. (1984) Improvement of Protein Phases by Coarse Model Modification, *Acta Cryst.* **A40**, 269-277.
5. Lunin, V.Yu; and Skovoroda, T.P. (1995) R-free Likelihood Based Estimates of Errors for Phases Calculated from Atomic Models, *Acta Cryst.* **A51**, 880-887.
6. Lunin, V.Yu., Lunina, N.L., Petrova, T.E., Vernoslova, E.A., Urzhumtsev, A.G., and Podjarny, A.D. (1995) On the ab-initio Solution of the Phase Problem for Macromolecules at Very Low Resolution: the Few Atoms Model Method, *Acta Cryst.* **D51**, 896-903.
7. Lunin, V.Yu., Lunina, N.L., Petrova, T.E., Urzhumtsev, A.G. and Podjarny A.D. (1997) On the *Ab Initio* Solution of the phase Problem for Macromolecules at Very Low resolution. II. Generalized Likelihood Based Approach to the Cluster Discrimination, submitted to *Acta Cryst.* **D**.
8. Read, R. (1986) Improved Fourier Coefficients for Maps Using Phases from Partial Structures with Errors, *Acta Cryst.* **A42**, 140-149.

DEVELOPMENTS IN MAXIMUM ENTROPY AND LIKELIHOOD

C.J.Gilmore and W.V. Nicholson
Department of Chemistry
University of Glasgow
Glasgow G12 8QQ, Scotland, UK.

1. Introduction

The purpose of this chapter is to outline some recent developments in the use of the ME method in macromolecular electron diffraction. It includes the following:

- The *ab initio* solution of halorhodopsin from two-dimensional data at 6Å resolution. The use of envelopes is discussed.
- Phase extension from 15Å for the blood porin data, also in two dimensions.
- Attempts to solve the bacteriorhodopsin structure *ab initio* from 3-D data including the use of the DM program for phase refinement.

2. Halorhodopsin

2.1 *AB INITIO* STUDIES

Electron diffraction amplitudes and electron micrograph derived crystallographic phases from halorhodopsin to 6 Å resolution have been reported by Havelka, Henderson, Heymann, and Oesterhelt [1]. Within the 6 Å resolution limit, there are 76 unique reflections in projection. The data were normalised with an imposed overall temperature factor of zero. Although the data have a nominal 6Å resolution, there are only 16 reflections with a resolution better than 8Å and these are all weak with a maximum U-magnitude of 0.012, which corresponds to an E magnitude of <0.5. The effective resolution of the data from the viewpoint of direct methods is, therefore, around 10Å. The plane group is centric: p4gm.

The origin was left undefined, and 8 reflections with a maximum resolution of 15Å were given permuted phases. This generated 256 nodes. Analysis of the associated LLGs using the t-test gave unambiguous and correct single phase indications for six of these reflections all of which were structure seminvariants. In addition the phase of reflection 6, $\phi(6)$, was indicated to be $\pi+\varphi(7)$. Fixing the origin by setting $\varphi(6)=\pi$ to match the correct structure resulted in the definition of $\varphi(7)$ also. (This is not 'cheating': we are ensuring that there is a match between the origin of our maps and that of the

S. Fortier (ed.), Direct Methods for Solving Macromolecular Structures, 455–462.

published structure in order to facilitate comparisons.) At this point eight reflections were uniquely phased, and the map correlation coefficient at this point was 0.97. These phases were used to define the root node of a new phasing tree. A further 8 reflections in the resolution range 6-10Å were then permuted. The lowest U-magnitude in this set was 0.006. This corresponds to an E of 0.24, which is a magnitude normally inaccessible to routine direct methods. From log-likelihood gain (LLG) analysis all eight reflections were unambiguously phased and of these only one was wrong.

At this point there was sufficient detail in the centroid map to make use of the $P(\delta q)$ function [2-5]. $P(\delta q)$ can be used with any maximum entropy map, $q^{ME}(\boldsymbol{x})$. A set of reflections is chosen using the standard criteria, and their phases permuted. Each phase permutation gives rise to a difference map $\delta q(\boldsymbol{x})$ which is a Fourier synthesis using the coefficients $U_{\mathbf{h}}^{obs} - U_{\mathbf{h}}^{ME}$ where $U_{\mathbf{h}}^{ME}$ is the Fourier coefficient of $q^{ME}(\boldsymbol{x})$. $P(\delta q)$ is calculated for each phase permutation from:

$$P(\delta q) \propto -\int_V \frac{\delta q^2(\mathbf{x})}{q^{ME}(\mathbf{x})} d^3\mathbf{x} \tag{1}$$

A minimum value of $P(\delta q)$ is expected for the correct phase set. This technique has several useful features:

(i) Each phase permutation requires only one Fourier synthesis and a map division and so is very fast, and large numbers of phase combinations can be generated.

(ii) It acts as a pre-filter for entropy maximisation; the reflections to be permuted are first subjected to this filter and only those with a certain minimum $P(\delta q)$ (typically 10-25%) are passed on for entropy maximisation.

(iii) The likelihood function uses only moduli, but $P(\delta q)$ incorporates phases into the calculation. It therefore acts as a useful tool in exploring structure-factor space from the current node.

The technique preferentially selects those phases that build density where it is already well defined. This is very appropriate in this situation where we have a clear envelope. It does, however, need to be used with care. In particular, $q^{ME}(\boldsymbol{x})$ must have developed sufficient detail. Those $q^{ME}(\boldsymbol{x})$ maps based on a very small basis sets or utilising only small U-magnitudes may have insufficient contrast for its successful use. In this case one obtains a set of $P(\delta q)$ values that are virtually constant. It is also advantageous to choose reflections for which there is a small but finite extrapolated magnitude from the current $q^{ME}(\boldsymbol{x})$.

A set of 12 reflections in the resolution range 6-15Å were selected, giving rise to 4096 phase sets, each of which was filtered *via* $P(\delta q)$. The 100 phase sets with the lowest values of $P(\delta q)$ were kept, and passed on to entropy maximisation. The best eight nodes based on LLG estimates were kept. The centroid map corresponding the best of these is shown in Figure 1(a). The true map is shown in Figure 1(b). The basis set reflections have a mean phase error of 9° for 28 reflections and a map correlation coefficient of 0.90.

These results are broadly similar to those derived by Dorset [6] using the Sayre equation.

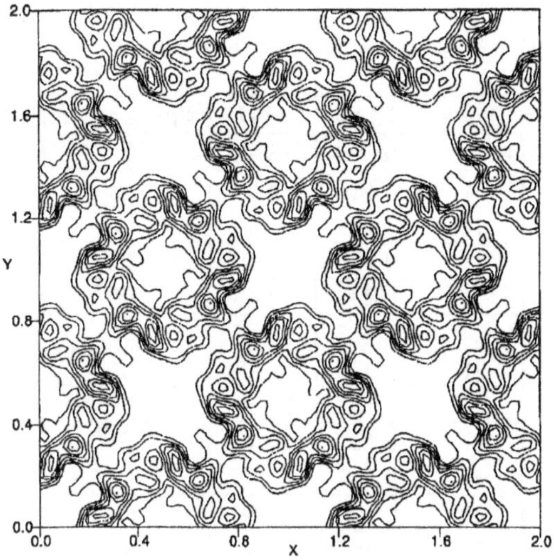

(a)

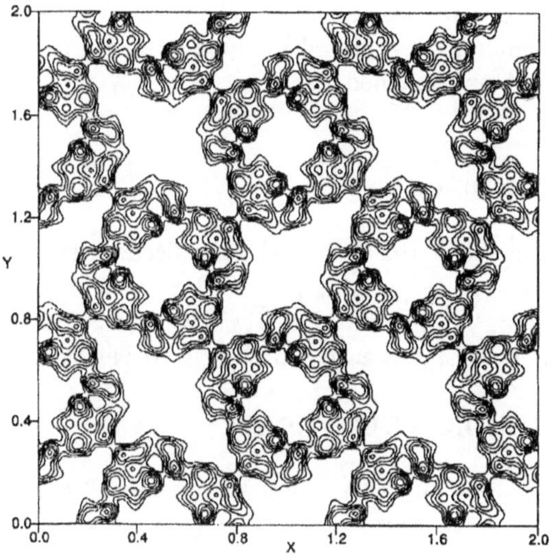

(b)

Figure 1. (a) The best map obtained for halorhodopsin (b) the true map derived from 76 image phases.

2.2 USING ENVELOPES

It is a relatively straightforward process to use envelopes in an active way in the ME method. To demonstrate this, a basis set of 17 reflections at 15Å resolution was used to define a root node. Seven reflections at 10Å resolution were then given permuted phases and the nodes thus generated subjected to entropy maximisation with the envelope from the root node used as a constraint. The analysis using *NS+LLG* is shown in Table 1. It can be seen that all the selected phase sets have low errors.

Node	LLG	NS+L	Phase error
4	0.158	83.94	7
8	0.186	84.02	16
12	0.145	85.41	15
16	0.173	85.59	25
20	0.098	84.43	15
24	0.115	84.82	24
28	0.090	84.15	23
32	0.104	84.67	33
36	0.129	84.18	0
40	0.207	84.15	9
48	0.110	83.89	18
52	0.204	83.59	7

Table 1. The best nodes for halorhodopsin when the envelope is used in the ME equations as a non-uniform prior. N is the number of atoms in the unit cell, S is the node entropy and L the log-likelihood gain.

3. Studies on an Aqua-porin.

This is a data set comprising 192 2-dimensional reflections at a resolution of *ca.* 6Å from Bing Jap [7]. The plane group is p4gm. Nine reflections at 15Å were used to define a root node using image derived phases. The object of this calculation was to examine the feasibility of phase extension to 6Å using ME methods.

3.1 DEFINING THE UNIT CELL CONTENTS

How many atoms are there in a 6Å projection? It is very difficult to define, but it is essential to have some concept of this figure. Direct methods are not, in general, too critical of the exact number of atoms in the cell, but one cannot be too liberal in this matter. We used likelihood to help define the number: different numbers of C atoms were used in the normalisation, definition of the root node at 15Å and subsequent entropy maximisation. The LLG extrapolating to 10Å was computed for each of the eight cell contents. Figure 2 shows a plot of LLG against cell contents. A clear maximum around 264 atoms is shown and this figure was used in subsequent calculations.

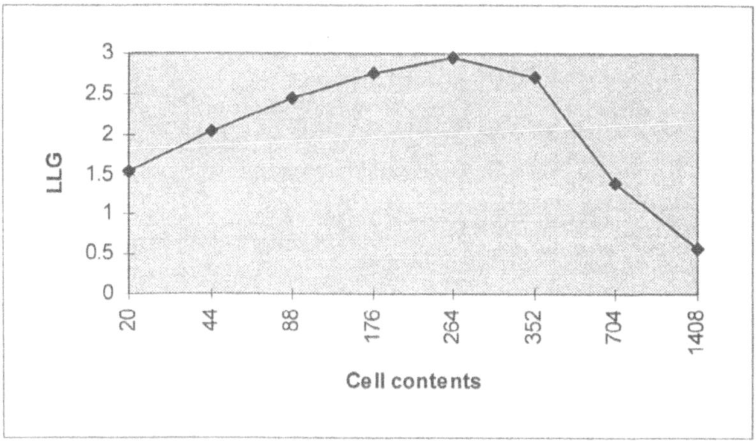

Figure 2. The plot of LLG against cell contents (defined as the number of C atoms in the unit cell).

3.2 PHASE EXTENSION

Attempts to extend phases directly from 15 to 6Å were a disaster with all the LLGs negative, and uninterpretable maps. Extension to 10Å was, however, straightforward. Seven reflections at 10Å were given permuted phases generating a 129 node phasing tree (including the root node). The results of the analysis are shown in Table 2. The map corresponding to node 72 is shown in Figure 3(a) with the true map in Figure 3(b). The map correlation coefficient is 0.90.

Node	LLG	Entropy	Score	Phase error
72	2.21	-1.30	17.45	9
37	1.46	-0.85	14.45	34
35	1.85	-1.16	12.81	46
53	1.44	-1.11	12.80	43
126	1.49	-1.21	11.78	33
128	1.17	-1.03	10.44	21
61	1.69	-1.03	10.05	55
103	1.24	-1.15	9.87	27

Table 2. The best phase sets for aqua-porin.

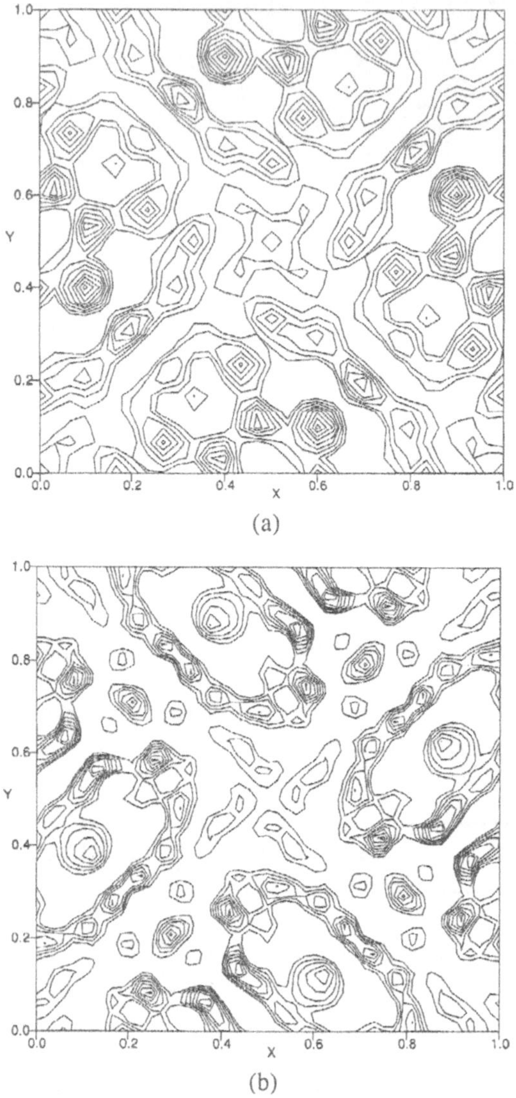

Figure 3. (a) The best map at 10Å for the aqua-porin (b) the map computed from the image phases.

4. Attempts to Solve *Bacteriorhodopsin* ab initio using 3D Data

No one has yet solved a protein data set ab initio using 3D data: it is a very difficult challenge indeed. We present here our attempts to do so using the 3D data from bacteriorhodopsin [8]. The space group is P3 which is non centrosymmetric as well as having no centric projections.

The data were normalised and a rootless single level tree was constructed by permuting the phases of 12 reflections. In a full factorial design this would generate 16,777,216 nodes! However the use of the Golay code as a source of experimental design [9] reduces this to 4096 (which still takes several days on a network of UNIX workstations!). The nodes were analysed as usual. Since there is a floating origin, map correlation software provided by Lunin and Lunina [10] was used to generate optimum shifts to match the true structure with the centroid maps from the best nodes. Table 3 summarises the data for these nodes.

Node	Phase error before DM	map cc before DM	Phase error after DM	map cc after DM
129	83	0.16	68	0.21
192	40	0.67	46	0.53
616	93	0.04	81	0.13
1531	60	0.47	48	0.53
1553	71	0.34	50	0.55
3354	75	0.18	69	0.26
3700	44	0.60	57	0.37
3709	65	0.37	60	0.37

Table 3. Phase sets and their corresponding map correlation coefficients before and after the application of the DM program.

It can be seen that there is often a dramatic improvement especially with some very poor phase sets, but that the better sets acquire worse phase errors. However, the overall trend is encouraging and the method needs to be further examined. The structure is far from being solved at this stage.

Acknowledgements

Grateful acknowledgements are due to Doug Dorset and Bing Jap, and we thank V. Lunin and L. Lunina for their map correlation software, and Adrian Lapthorn for help with the DM program. This research was supported by grants from the Human Frontier Science Project and the EPSRC.

References

1. Havelka, W.A., Henderson, R., Heymann, J.A.W. and Oesterhelt, D. (1993) Projection structure of halorhodopsin from Halobacterium halobium at 6Å resolution obtained by electron cryo-microscopy. *J.Mol.Biol*, **234**, 837-846
2. Bricogne, G. and Gilmore, C.J. (1990) A multisolution method of phase determination by combined maximisation of entropy and likelihood. I. Theory, algorithms and strategy, *Acta Cryst.* **A46,** 284-297.
3. Gilmore, C.J., Bricogne, G. and Bannister, C. (1990) A multisolution method of phase determination by combined maximisation of entropy and likelihood. II Application to small molecules, *Acta Cryst.* **A46,** 297-308.
4. Henderson, K and Gilmore, C.J. (1989) The challenge of X-ray and neutron powder diffraction in J. Skilling (ed.), *Maximum Entropy and Bayesian Methods* Kluwer Academic Publishers, Amsterdam pp. 233-239.
5. Gilmore, C.J., Nicholson, W.V., and Dorset, D.L. (1996) Direct methods in protein electron crystallography: the ab initio structure determination of two membrane protein structures in projection using maximum entropy and likelihood, *Acta Cryst.* **A52,** 937-946.
6. Dorset, D.L. (1995) Direct structure determination by electron crystallography: protein data sets. *Micron,* **6**, 511-520.
7. Jap, B.K. and Li, H.L. (1995) Structure of the osmo-regulated H20-channel, aqua-chip, in projection at 35Å resolution, *J.Mol.Biol.* **251**, 413-420.
8. Henderson, R., Baldwin, J. M., Downing, K. H., Lepault, J. and Zemlin, F. (1986) Structure of purple membrane from *Halobacterium halobium*: recording, measurement and evaluation of electron micrographs at 3.5Å resolution, *Ultramicroscopy.* **19**, 147-178.
9. Bricogne, G. (1993) Direct phase determination by entropy maximisation and likelihood ranking: status report and perspectives, *Acta Cryst.* **D49**, 37-60.
10. Lunin, V.Y. and Lunina, N.L. (1996) The map correlation coefficients for optimally superimposed maps, *Acta Cryst.* **A52**, 365--368.

EXTENDING THE RESOLVING POWER OF *SHAKE-AND-BAKE*

CHARLES M. WEEKS[1], RUSS MILLER[1,2] and HERBERT A. HAUPTMAN[1]
[1]*Hauptman-Woodward Med. Research Inst.itute, Inc., 73 High Street, Buffalo, NY 14203 USA*
[2]*SUNY-Buffalo, Dept. of Comp. Sci., Buffalo, NY 14260 USA*

Abstract

Direct methods are based on the fact that suitable conditional probability distributions of structure invariants permit their values to be estimated. As with traditional multisolution direct-method protocols, the dual-space *Shake-and-Bake* procedure typically fails at resolutions below 1.1-1.2Å, although solutions are obtainable at 1.4Å in favorable cases. It is the nonavailability of sufficiently accurate invariant estimates, and not the peak-picking algorithm employed in the real-space density-modification step, that appears to impose the initial limitation on the resolving power of *Shake-and-Bake*. The feasibility of using three-beam diffraction to provide a limited number of sufficiently accurate estimates and a formalism for incorporating this information into the *Shake-and-Bake* procedure are discussed. The same formalism will accommodate SIR and SAS information as well.

1. *Shake-and-Bake* at 1.1-1.4Å

Shake-and-Bake is a state-of-the-art algorithm for the direct phasing of single-crystal diffraction data. This procedure extends the range of successful applications of conventional direct methods from approximately 100 independent non-H atoms to small macromolecules containing over 600 atoms. The improvement made possible by *Shake-and-Bake* depends largely on unconditional cyclic alternation of reciprocal-space phase refinement with filtering in real space to impose physically-meaningful constraints. The theoretical background of *Shake-and-Bake*, its implementation in the computer programs *SnB* [1] and SHELX[15], and details regarding successful applications to very high resolution (1.0Å or better) protein data are described elsewhere in these Proceedings [2,3].

Table 1 contains a summary of the ranges of resolution for which solutions have been obtained for a group of protein test data sets truncated at various resolutions. All of these structures are solvable at 1.1Å. In fact, the I4 peptide and two smaller conotoxin structures [10,11] were originally solved using the *SnB* program with 1.1 or 1.2Å data. As resolution decreases, success rates (percentages of trial structures going to solution) also decrease and, ultimately, no solutions are found.

Success rates may be increased, or the range of resolutions over which solutions are obtainable may be extended, by making suitable adjustments to the values of certain control parameters. In particular, increasing the ratio of triplet invariants to phases or

S. Fortier (ed.), Direct Methods for Solving Macromolecular Structures, 463–468.

increasing the number of *SnB* cycles are especially beneficial as the resolution decreases to 1.2Å or lower. It is interesting to note that, using 500 *SnB* cycles with truncated 1.1Å toxin II data and a *phase/invariant ratio (P/I)* of 1:10, no solutions were found in 2000 trials. However, when the ratio was increased to 1:20, the success rate was 1 in 200 trials. Similar results were obtained at 1.1Å using the crambin data for which the 300-cycle success rates were 0.2 and 1.1% for P/I ratios of 1:10 and 1:30, respectively. Using the full (0.86Å) gramicidin A data set, the 275-cycle success rate (P/I ratio = 1:10) was 1.1%, but the *earliest* solution (2000 trials) occurs at cycle 327 using the 1.1Å truncated data and a P/I ratio of 1:25. It should be noted that, in extensive studies with the complete crambin (0.83Å) and gramicidin A (0.86Å) data sets, the optimum P/I ratios, from the perspective of either success rate or computational efficiency, were 1:10 for crambin and 1:10-1:20 for gramicidin A.

At present, the available evidence concerning the relationship (if any) between resolution and the number of peaks selected during the density modification step is conflicting. It has been shown that, using the full data sets, the optimum number of peaks recycled as atoms for crambin and toxin II are 100 and 200, respectively [3]. Crambin results at 1.1Å (1.1% success when 100 peaks were selected *vs.* 0 solutions out of 5000 trials when 200 peaks were selected) were consistent with the observations for the full data. On the other hand, there are 0 solutions out of 2500 toxin II trials when 200 peaks are selected at 1.2Å, but there are 5 solutions out of 750 trials when only 100 peaks were selected (1000 cycles; P/I ratio 1:80).

It is interesting to note that the two test structures for which solutions have not been achieved at a resolution lower than 1.1Å are gramicidin A, which has no atoms heavier than oxygen, and rubredoxin, which has a cluster of 'heavier' atoms. In contrast, the two structures (vancomycin, I4 peptide) with the most well-separated heavier atoms (Cl, S) can be solved with truncated 1.3 or 1.4Å data. It seems likely that the presence of several heavier atoms facilitates solvability at lower-than-normal resolution. At 1.4Å, there is some overlap of minimal function values for solutions and nonsolutions such that the histogram does not have its typical biomodal shape, but the best values still earmark solutions. At 1.5Å, there are no solutions for any of the test structures. Some vancomycin and I4 peptide trials do have mean phase errors in the 50-60° range, but these trials are not identifiable on the basis of any known figure of merit.

TABLE 1. Successful *SnB* applications have been made in the indicated resolution ranges. The resolution range defines the resolution of the full data set and the lowest resolution of the corresponding truncated data set for which *SnB* solutions have been obtained.

Protein	Space Group	Protein Atoms	Total Atoms	'Heavy' Atoms	Resolution Range
Vancomycin [4]	$P4_32_12$	202	258	Cl8	0.9-1.4Å
I4 Peptide [5]	I4	248	289	S10	1.1-1.4
Gramicidin A [6]	$P2_12_12_1$	272	317	---	0.86-1.1
Crambin [7]	$P2_1$	327	~400	S6	0.83-1.2
Rubredoxin [8]	$P2_1$	395	497	FeS6	1.0-1.1
Scorpion Toxin II [9]	$P2_12_12_1$	508	624	S8	0.96-1.2

2. 'Unified' *Shake-and-Bake*

The phase-refinement segment of the *Shake-and-Bake* cycle utilizes either the tangent formula (Table 2, Eq. 4) or a parameter-shift routine targeted at reducing the value of the minimal function (Table 2, Eq. 3) so as to improve the phase values. The minimal function is a measure of the mean-square difference between the triplet cosine-invariant values and their expected values as predicted by the ratio of modified Bessel functions, $I_1(A)/I_0(A)$. In an attempt to determine which part(s) of the *Shake-and-Bake* procedure are responsible for failure at 1.5Å, the expected cosine-invariant values for crambin were replaced by their actual known values. Approximately 25% of the random trials then converged to solution in fewer than 100 cycles with peak picking used as the density-modification technique. In fact, this result held even when the crambin data set was truncated to 2.0Å. Thus, it seems likely that it is primarily the availability of sufficiently accurate invariant estimates for reciprocal-space phase refinement, and not the procedures employed in the real-space density-modification step, that impose the first limitation on the resolving power of *Shake-and-Bake.*

These observations suggest that the key to extending the range of *Shake-and-Bake* to 1.5Å or lower resolution lies in providing more accurate invariant estimates. As shown in Table 2 (Eq. 6-9), the mathematics of the phase-refinement process may be modified to include the case where information is available about individual invariant values and not just their cosines. This provides for a *unified* treatment of any information about invariant values regardless of the source of that information. Several years ago, it was demonstrated by Hauptman that single isomorphous replacement [12] and single anomalous scattering [13] data can be integrated with direct methods to estimate invariant values with appropriate weights. Recent esperimental work with lysozyme crystals has shown that it is possible to use three-beam diffraction to measure the values of a carefully selected set of triplet invariants with a mean phase error around 20° [14]. As suggested by Table 3, information from SIR and SAS experiments could be combined with multiple-beam measurements using the unified *Shake-and-Bake* formalism.

The data for crambin truncated to 1.5Å resolution were used to test the efficacy of the unified tangtent formula (Table 2, Eq. 8), and the results of these tests are presented in Figure 1 and Table 4. Solutions were conclusively identified on the basis of mean phase error but, in all cases, the unified tangent figure of merit (Table 2, Eq. 9) was diagnostic and perfectly correlated with mean phase error.

In the first experiment (Figure 1a), the 1000 reflections with the largest |E| values were used to generate the 10,000 triplet invariants with the largest A values (998 reflections actually occurred in these invariants). Several sets of values were computed for these invariants. The first set was error-free, and eight additional sets were created with random errors of 10, 20, ..., 80°. *Shake-and-Bake,* with 100 peaks being picked in the density-modification step, was then applied to a sample of 1000 random trials using each set of invariants. The effects of pure phase refinement (a: 100 phase-refinement iterations or passes through the phase list for 1 *SnB* cycle) were compared to 100 iterations of refinement spread over 10 cycles (b: 10 iterations, 10 cycles) and then over 100 cycles (c: 1 iteration, 100 cycles) consisting of equal amounts of 'shaking' and 'baking'. In all cases, the success rate was in the range 11-16% when the initial mean invariant error was $\leq 30°$. For all amounts of initial error, the results were best when phase refinement and peak picking were alternated. The superiority of method (c) became more pronounced as the amount of initial invariant error increased. Refinement schemes a, b, and c produced

0, 1, and 13 solutions, respectively, when the initial invariant error was 60°, and scheme c had 1 solution starting with 70° invariant error if more than 100 cycles were performed.

TABLE 2. The mathematics of 'shaking' (phase refinement).

	Conventional *Shake-and-Bake*		Unified *Shake-and-Bake*	
Invariants (mod 2π):	$\Phi_{HK} = \phi_H + \phi_K + \phi_{-H-K} \approx 0$	(1)	$\Phi_{HK} = \phi_H + \phi_K + \phi_{-H-K} \approx \omega_{HK}$	(6)
Weights:	$W_{HK} = A_{HK} = \frac{2}{N^{½}}\lvert E_H E_K E_{-H-K}\rvert$	(2)	W_{HK} varies	
Minimal Function:	$R(\phi) = \frac{1}{\sum_{H,K} W_{HK}} \sum_{H,K} W_{HK}\left(\cos\Phi_{HK} - \frac{I_1(W_{HK})}{I_0(W_{HK})}\right)^2$	(3)	$R(\phi) = \frac{1}{2\sum_{H,K} W_{HK}} \sum_{H,K} W_{HK}\left\{\left[\cos(\Phi_{HK}) - \cos(\omega_{HK})\right]^2 + \left[\sin(\Phi_{HK}) - \sin(\omega_{HK})\right]^2\right\}$	(7)
Tangent Formula:	$\tan\phi_H = \frac{-\sum_K \lvert E_K E_{-H-K}\rvert \sin(\phi_K + \phi_{-H-K})}{\sum_K \lvert E_K E_{-H-K}\rvert \cos(\phi_K + \phi_{-H-K})}$	(4)	$\tan\phi_H = \frac{\sum_K W_{HK}\sin(\omega_{HK} - \phi_K - \phi_{-H-K})}{\sum_K W_{HK}\cos(\omega_{HK} - \phi_K - \phi_{-H-K})}$	(8)
Tangent Figure-of-Merit:	$m(\phi) = 1 - \left[\frac{1}{\sum_{H,K} W_{HK}} \sum_{H,K} W_{HK}\cos(\Phi_{HK})\right]$	(5)	$m(\phi) = 1 - \left[\frac{1}{\sum_{H,K} W_{HK}} \sum_{H,K} W_{HK}\cos(\Phi_{HK} - \omega_{HK})\right]$	(9)

TABLE 3. Invariant information and its role in unified *Shake-and-Bake*.

Invariant Type	Φ_{HK}	$\text{Cos}\Phi_{HK}$	$\text{Sin}\Phi_{HK}$	W_{HK}
Conventional	0	$I_1(A)/I_0(A)$	-----	A_{HK}
SIR[12]	$0\,(A_{SIR} > 0)$; $\pi\,(A_{SIR} < 0)$	$I_1(A)/I_0(A)$	-----	$\lvert A_{SIR}\rvert$
SAS[13]	ω_{HK}	$\cos\omega_{HK}$	$\sin\omega_{HK}$	A_{SAS}
3-Beam Measurement	ω_{HK}	$\cos\omega_{HK}$	$\sin\omega_{HK}$	$\frac{I_1(W_{HK})}{I_0(W_{HK})} = \cos(\Phi_{HK}\text{error})$

TABLE 4. Solutions per 1000 trials for 1.5Å crambin data after one iteration of unified tangent formula phase refinement for different numbers of cycles. The experiment involved 971 phases and 5000 triplet invariants, and 100 peaks were selected in each cycle.

	Initial Random Invariant Error			
Cycles	0°	20°	40°	60°
1-20	140	90	22	0
21-40	19	34	33	0
41-60	3	7	24	1
61-80	0	2	21	0
81-100	0	3	24	2
101-600	---	---	---	22

In a second, similar experiment (Figure 1b), the number of invariants was reduced with concomitant reduction in the number of accessible phases. Solutions were obtained in all cases tested when the initial invariant error was 50° or less. Table 4 shows when solutions occurred as a function of cycle for various amounts of initial error. More cycles are required to achieve solution as the initial error increases, and it is clear that more than 100 cycles should be performed to insure solution when the starting invariant error is 60°.

3. Concluding Remarks

A study of six protein data sets has demonstrated that it is routinely possible to solve structures with 1.1Å data using the *Shake-and-Bake* technique. To insure solvability at 1.1Å, it is advisable to use an increased number of invariants or perform more cycles than would be necessary if 1.0Å or higher resolution data were available. This is especially

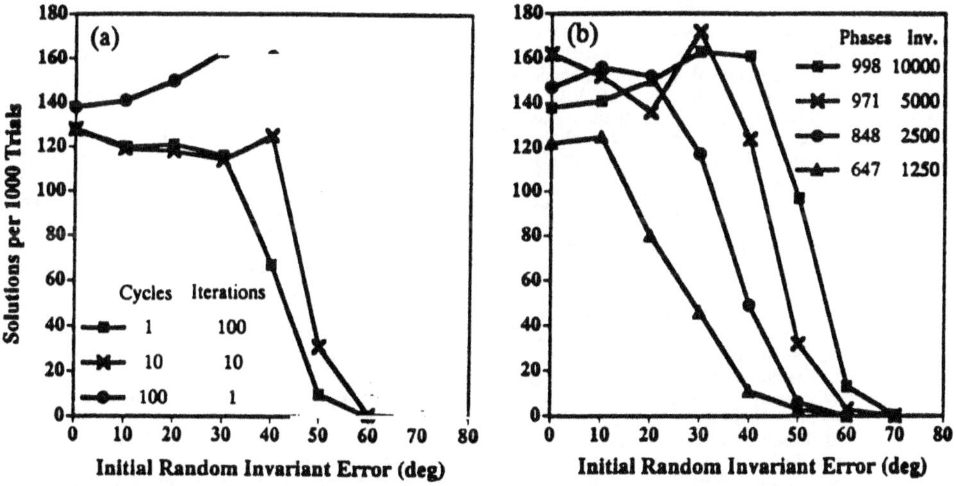

Figure 1. Application of the unified tangent formula to 1.5Å data for crambin. (a) Results of varying the number of phase refinement iterations and *Shake-and-Bake* cycles (998 phases, 10000 triplet invariants, and 100 peaks selected in each cycle). (b) Results of varying the number of triplet invariants used (100 cycles, 1 phase refinement iteration, and 100 peaks).

true if the resolution is only 1.2Å. Some structures, especially those containing several heavier atoms (e.g., S, Cl), can be solved at a resolution as low as 1.4Å.

Successful application of *Shake-and-Bake* to 1.5Å or lower resolution data appears to require that supplemental information such as SIR, SAS, or 3-beam diffraction be used to estimate the values of at least some invariants with increased accuracy. Low phase/invariant ratios are possible provided that invariant error is also low. Random mean invariant errors as large as 60° can be tolerated provided that a dual-space phase improvement scheme is utilized.

4. Acknowledgments

The *Shake-and-Bake* algorithm and the *SnB* program have been made possible by the financial support of grants GM-46733 from NIH and IRI-9412415 from NSF.

5. References

1. Miller, R., Gallo, S.M., Khalak, H.G., and Weeks, C.M. (1994) *SnB*: Crystal structure determination *via Shake-and-Bake*, *J. Appl. Cryst.* **27**, 613-621.
2. Hauptman, H.A. (1997) The phase problem: A problem in constrained global optimization, in S. Fortier (ed.), *Direct Methods for Solving Macromolecular Structures*, Kluwer Academic Publishers, Dordrecht, The Netherlands, 25th Course, Intl. School of Crystallography, NATO ASI Series, Erice, Sicily, May (1997).
3. Miller, R. and Weeks, C.M. (1997) *Shake-and-Bake*: Applications and advances, in S. Fortier (ed.), *Direct Methods for Solving Macromolecular Structures*, Kluwer Academic Publishers, Dordrecht, The Netherlands, 25th Course, Intl. School of Crystallography, NATO ASI Series, Erice, Sicily, May (1997).
4. Loll, P.J., Bevivino, A.E., Korty, B.D., and Axelsen, P.H. (1997) Simultaneous recognition of a carboxylate-containing ligand and an intramolecular surrogate ligand in the crystal structure of an asymmetric vancomycin dimer, *J. Am. Chem. Soc.* **119**, 1516-1522.
5. *SnB* solution by the authors. On account of its proprietary nature no further information can be revealed concerning the structure itself.
6. Langs, D.A. (1988) Three-dimensional structure at 0.86Å of the uncomplexed form of the transmembrane ion channel peptide gramicidin A, *Science* **241**, 188-191.
7. Hendrickson, W.A. and Teeter, M.M. (1981) Structure of the hydrophobic protein crambin determined directly from the anomalous scattering of sulfur, *Nature* **290**, 107-113.
8. Dauter, Z., Sieker, L.C., and Wilson, K.S. (1992) Refinement of rubredoxin from *Desulfovibrio vulgaris* at 1.0Å with and without restraints, *Acta Cryst.* **B48**, 42-59.
9. Smith, G.D., Blessing, R.H., Ealick, S.E., Fontecilla-Camps, J.C., Hauptman, H.A., Housset, D., Langs, D.A., and Miller, R. (1997) The *ab initio* structure determination and refinement of a scorpion protein toxin, Acta Cryst., in press.
10. Hu, S.-H., Gehrmann, J., Guddat, L.W., Alewood, P.F., Craik, D.J., and Martin, J.L. (1996) The 1.1Å crystal structure of the neuronal acetylcholine receptor antagonist, α-conotoxin PnIA from *Conus pennaceus*, *Structure* **4**, 417-423.
11. Guddat, L.W., Martin, J.L., Shan, L., Edmundson, A.B., and Gray, W.R. (1996) Three-dimensional structure of the α-conotoxin GI at 1.2Å resolution, *Biochemistry* **35**, 11329-11335.
12. Hauptman, H. (1982) On integrating the techniques of direct methods and isomorphous replacement. I. The theoretical basis, *Acta Cryst.* **A38**, 289-294.
13. Hauptman, H. (1982) On integrating the techniques of direct methods with anomalous dispersion. I. The theoretical basis, *Acta Cryst.* **A38**, 632-641.
14. Weckert, E., Schwegle, W., and Hummer, K. (1993) Direct phasing of macromolecular structures by three-beam diffraction, *Proc. R. Soc. Lond. A*, **442**, 33-46.
15. Sheldrick, G. (1997) SHELX Applications to Macromolecules, in S. Fortier (ed.), *Direct Methods for Solving Macromolecular Structures*, Kluwer Academic Publishers, Dordrecht, The Netherlands, 25th Course, Intl. School of Crystallography, NATO ASI Series, Erice, Sicily, May (1997).

REINVESTIGATION OF THE USE OF PATTERSON MAPS TO EXTRAPOLATE DATA TO HIGHER RESOLUTION

Extrapolation of X-ray Data to Higher Resolution

David A. Langs
Hauptman-Woodward Medical Research Institute, Inc.
73 High Street, Buffalo, NY 14203 USA

1. Introduction

The idea of using suitably modified Patterson maps for the purpose of extrapolating X-ray diffraction amplitudes to higher resolution is not new. It was first proposed by Karle and Hauptman [1] in 1964 as a means of refining E-values to obtain values which were more consistent with the point atom approximation upon which direct methods probability relationships are based. What K&H proposed is that one could compute a sharpened, origin-removed Patterson function using $|E|^2-1$ as coefficients, and then modify the density to remove unexpected negative regions, as well as peaks which represented vectors which would place molecules closer than a van der Waals contact to proper symmetry elements such as rotation axes, inversion centers and mirror planes. When one back transformed the modified map, one not only obtained refined estimates for the observed $|E|^2-1$ amplitudes, but also obtained values for terms which were not included in the original map. Each cycle of refinement included calculation, modification, and back transformation of the map, and renormalization of the extrapolated $\langle||E|^2-1|\rangle$ values to fit the expected distribution, and insure that negative $|E|^2$ values were reset to zero as their smallest possible value. One must also eliminate reflections corresponding to space group extinctions, as the modified Patterson will not respect this constraint. Only one subsequent publication [2] appears to have used this Patterson refinement scheme to improve the `accuracy' of the measured E-values to solve a structure.

The K&H calculations used CuKα data of the $P2_12_12_1$ structure of arginine dihydrate, which was the first noncentric structure ever solved by tangent formula methods [3]. K&H were able to extrapolate the 1406 measured data in five refinement cycles, each time increasing the resolution limit to include an additional 20 % of data until 2688 terms were obtained. They noted that the Patterson map had better peak resolution in agreement with the solved structure, and also that the extrapolated data could be used in a passive manner to obtain improved estimates of the triples cosine invariants within the observed sphere by means of the B3,0 formula [4], but no details of this analysis were given. Also no quantitative figures were given as to how well the extrapolated data agreed with the expected values based on the refined structure, or whether these extrapolated terms could

S. Fortier (ed.), Direct Methods for Solving Macromolecular Structures, 469–473.

be actively used to strengthen the phasing process.

This presentation will reinvestigate both if the $|E|^2$-1 sharpening is optimal, and whether a more effective density modification scheme than simply zeroing the negative Patterson density can be found. One can assess the effectiveness of various alterations of the extrapolation procedure by computing the correlation coefficient (CC) between extrapolated amplitudes and their `true' values computed from the low resolution refined structure. Calculations were performed for a half dozen CuKα data sets for structures varying in size from about 100 to 624 non-Hydrogen atoms. Attempts were made to extend the resolution of the CuKα data sets to ~ 0.5Å resolution. The CC agreements were computed for 30 successive resolution shells of data having roughly the same number of reflections. The first 6 shells of data comprised the original set of measured data. Details of these analyses are described in a manuscript [5] that was submitted to Acta Cryst. Section A on the occasion of Herbert Hauptman's 80th birthday on February 14, 1997.

2. **Analysis**

Sophisticated methods to derive better E-values that fit the anisotropy, variance and higher moments of the temperature factor distribution surprisingly did not improve the Patterson extrapolation results. In fact if we compare $|F|^2$, $|F/\langle fj \rangle|^2$, and $|F/\langle fj \rangle \exp(Beff\ s^2)|^2$ as Patterson coefficients, successive sharpening by introducing <fj> the average atomic form factor and Beff the effective temperature of the data, does significantly improve the results of the extrapolation process. In all cases tested it was found that the best effective temperature for the Patterson method is in the vicinity of Bmin, the lowest B-values for the ordered mainchain atoms of the structure. Bmin is usually 2 to 5 units less than the Wilson Biso for structures that scatter to CuKα resolution. Biso is in turn slightly less than the effective anisotropic B-value which leads to the greatest direct methods significance [6]. Recall that Sheldrick's preference for $|E^3F|^{1/2}$ as Patterson coefficients also emphasizes the need not to oversharpen the signal in the data. In the paragraphs that follow, when referring to an $|E|^2$ Patterson, we will consider a sharpened, origin preserved Patterson map based on optimizing the value of Beff in the vicinity of that known for the refined structure.

Patterson density modification schemes which simply zeroed the density below some fixed threshold were compared. If the threshold was too low, only a small handful of pixels having the most negative values were effected and the extrapolated values of the observed |E|'s were not significantly perturbed, and volunteer values for data outside the observed set were not significantly non-zero. The $|E|^2$ values for the observed data were essentially all positive. As this threshold was elevated to the zero value suggested by K&H, the power for extrapolating reliable values for the larger $|E|^2$ terms outside the data set increased, but the percentage of negative extrapolated $|E|^2$ terms within the observed set also increased. Further raising this threshold improved our ability to estimate the larger extrapolated $|E|^2$ terms better, but at a cost of having poorer estimates of the weaker terms, and a larger percentage of negative $|E|^2$ values within the observed data set. If the threshold was raised too high, to include too few of the strongest Patterson vectors, the procedure would fail to reliably indicate even our strongest extrapolated data.

A sliding threshold procedure was adopted in an attempt to both maximize efforts to initially estimate the strongest extrapolated terms with greater reliability and reduce the percentage of negative $|E|^2$ terms within the observed data set. If the most negative feature in the initial Patterson map was *e.g.* -N units, it was found to be most effective if the first density modification threshold was set at +2N units, and then this threshold value was gradually lowered by 3N/ncycle units over the next ncycles of refinement until it finally reached the original value of -N units. Not only were the CC agreements best using this protocol, the percentage of negative $|E|^2$ within the observed data set approached zero, provided that they were reset to their observed values for all cycles of the refinement except the very last in which they were allowed to assume their refined value.

3. **Results**

Using this protocol it was possible to extend the 0.86Å observed data set for the 317 atom $P2_12_12_1$ structure of gramicidin A [7] to ~0.5Å resolution with CC's in the range of 0.39 to 0.71 units over the extrapolated $\sin\theta/\lambda$ range from 0.86 to 0.50Å. In comparison the K&H procedure produced a slightly better CC of 0.41 for the first unobserved shell closest to the measured data, but these values gradually fell to -.03 units over the remaining 23 resolution shells out to 0.50Å.

This new procedure worked nearly as well for the other smaller test structures, but failed for a 0.96Å resolution data set for the 624 atom scorpion tox-II structure [8] which could be extended with a CC accuracy no greater than 0.20 for the first lowest resolution unobserved shell, and then rapidly degraded for higher resolution shells. This failure could be attributed to the fact that unlike the other data sets tested, 10% of the data within the observed resolution range for the tox-II structure were not recorded. The Patterson procedure was able to restore the lowest 3.0Å resolution shell of 211 unmeasured data with a CC of 0.70, but further extrapolation to recover other unobserved terms within the observed sphere of resolution was not as favorable unless these 211 terms were included as known in the `observed' Patterson map of the next round of refinement cycles.

Trials to extend data from lower initial resolution data sets, say 1.5Å, were considerably less successful, and need not be discussed at this time. The CC comparison between most CuKα resolution Patterson maps and those computed from 0.5Å error free data were extremely high, typically in the range of 0.98 to 0.99 units. This suggested that one might simply use the original Patterson function histogram as a means of improving the results at the end of the Patterson map refinement cycles. When this was done, however, the CC agreement seriously deteriorated. Although imposing the proper 0.5Å histogram will, say, in some small way improve our results, we will be obligated to have some reliable `library' of histograms to call on for various diverse macromolecular types, if we are to benefit from this procedure.

CC agreements in the range of 0.4 to 0.7 can be obtained for extrapolated data for structures having several hundreds of non-hydrogen atoms provided that the observed data are recorded to CuKα resolution, and that the observed data shell is essentially complete. But can these extrapolated E-values be exploited for phase determination in an active manner? To answer this question 96624 triples having A $\geq$ 0.4 were generated from the 3000 strongest E-values for the extended gramicidin A data set. These included 1519 |E|'s from the observed set and an additional 1481 extrapolated |E| terms. The triples were broken down into four groups, i.e. those triples including either 0, 1, 2 or 3 extrapolated terms. A comparison was made between $\langle \text{Cos}\Phi\text{true} \rangle$ the A-weighted averages of the true values of the cosine invariants and $\langle \epsilon(\text{Cos}\Phi) \rangle$ their associated expected values based on the $I_1(A)/I_0(A)$ estimate. The best agreement is, as expected, for the 40546 triples in group 0, where the ratio between $\langle \text{Cos}\Phi\text{true} \rangle$ and $\langle \epsilon(\text{Cos}\Phi) \rangle$ is 0.95. These ratios for groups 1, 2 and 3 are progressly worse with values of 0.74, 0.58 and 0.31 for groups of 26146, 27790 and 2142 triples, respectively.

The 2142 triples in group 3 can be rejected on the basis that the 0.31 ratio has degraded by nearly one half compared to the 0.58 value for group 2. The tangent formula solution for the original observed data set of gramicidin phased 1500 E-values through 17500 triples which had A $\geq$ 0.5. The actual solution phased only 1320 E's with a mean phase error of 39 degrees. Increasing the number of E's or the number of triples by lowering the A-value to less than 0.5 leads to significantly worse results.

Using the extended data and the top 3000 E-values and 94 thousand triples from groups 0, 1 and 2; it was possible to achieve a solution for 2600 E-values which had a significantly lower mean phase error of 35 degrees. If we inspect the accuracy of the 1477 observed E's we note a mean phase error which is appreciably better at 28 degrees. Thus it would appear that the inclusion of extrapolated E-values in the phasing process can lead to phase determinations which include a larger number of reflections with a smaller phase error as compared to the observed data. The 1477 term map disclosed 119 correct atom sites in the top 150 peaks as compared to only 105 atoms for the 1320 phases in the earlier refinement. A problem exists in that although the 2600 terms in this later refinement has a smaller phase error, only 86 correct atom sites appear in the top 150 map peaks! Numerous spurious peaks appear around the stronger atom sites in the map much like diffraction ripples due to series termination error. In order to correct this situation one could dampen the contribution of the higher order terms to the map or include a larger fraction of higher order terms into the phasing process. Common sense suggests using 7 times as many high resolution terms as compared to measured data, but in doing so it will be necessary to lower the E-threshold of the extrapolated data to a range where the average amplitude error is higher.

4. **Acknowledgement**

This work was generously supported by NIH grant GM-46733.

5. References

1. Karle, J. & Hauptman, H. (1964) Positivity, Point Atoms, and Pattersons, *Acta Cryst.* **17**, 392-396.

2. Seeman, N. C., Rosenberg, J. M., Suddath, F. L., Kim, J. J. P. & Rich, A. (1976) RNA Double-helical Fragments at Atomic Resolution : Crystal and Molecular Structure of Sodium Adenylyl-3',5'-Uridine Hexahydrate, *J. Mol. Biol.* **104**, 109-144.

3. Karle, I. L. & Karle, J. (1964) An Application of the Symbolic Addition Method to the Structure of L-Arginine Dihydrate, *Acta Cryst.* **17**, 835-841.

4. Karle, J. & Hauptman, H. (1958) Phase Determination form New Joint Probability Distributions : Space Group P1, *Acta Cryst.* **11**, 264-269.

5. Langs, D. A. (1997) Reinvestigation of the Use of Patterson Maps to Extrapolate Data to Higher Resolution, *Acta Cryst.* **A53**, in press.

6. Blessing, R. H., Guo, D. Y. & Langs, D. A. (1996) Statistical Expectation Value of the Debye-Waller Factor and |E(hkl)| Values for Macromolecular Crystals, *Acta Cryst.* **D52**, 257-266.

7. Langs, D. A. (1988) Three-Dimensional Structure at 0.86Å Resolution of the Uncomplexed Form of the Transmembrane Ion Channel Peptide Gramicidin A, *Science* **241**, 188-191.

8. Fontecilla-Camps, J. C., Habersetzer-Rochat, C. & Rochat, H. (1988) Orthorhombic crystals and three-dimensional structure of the potent toxin II from the scorpion *Androctonus australis* Hector, *Proc. Natl. Acad. Sci. USA* **85**, 7443-7447.

INTENSITY OVERSAMPLING FROM CRYSTALLOGRAPHIC SAMPLING *VIA* DISCRETE HILBERT TRANSFORMS

A.F.MISHNEV
Latvian Institute of Organic Synthesis
21 Aizkraukles St., LV 1006, Riga, Latvia

Recently the role of in-between structure factors (having half-integral Miller indices) in application to the crystallographic phase problem has been investigated [1,2]. It has been shown that when the causal Fourier transform condition (Fourier transform vanishes for negative arguments) is satisfied for the autocorrelation function (ACF), discrete Hilbert transforms (DHT) applied to intensity function provide intensity oversampling at in-between points as follows

$$I(h'+1/2,\ k'+1/2,\ L) = -\,(1/\pi^2)\sum_{h,k} I(h,k,L)/[(h'-h+1/2)(k'-k+1/2)]\,. \tag{1a}$$

$$I(h'+1/2,\ K,\ l+1/2) = -\,(1/\pi^2)\sum_{h,l} I(h,K,l)/[(h'-h+1/2)(l'-l+1/2)]\,. \tag{1b}$$

$$I(H,\ k'+1/2,\ l+1/2) = -\,(1/\pi^2)\sum_{k,l} I(H,k,l)/[(k'-k+1/2)(l'-l+1/2)]\,. \tag{1c}$$

It should be noted that the causality condition is realistic only for some specific atomic arrangements. Therefore in the present paper we derive DHT for a general electron density distribution. This problem is more complicated because the autocorrelation function (ACF) has spread to both positive and negative arguments (u,v,w) and occupies eight unit cells. In general case the continuous 3-D intensity is related to the ACF by the Fourier transform (FT)

$$I(s_1,s_2,s_3) = \int_{-a}^{a}\int_{-b}^{b}\int_{-c}^{c} A(u,v,w)\exp[2\pi i(us_1+vs_2+ws_3)]dV. \tag{2}$$

Using the centrosymmetric property of the ACF the integral may be rearranged in the following way

$$I(s_1,s_2,s_3) = 2\int_0^a\int_0^b\int_0^c A(\mathbf{u})\cos(2\pi\mathbf{us})dV + 2\int_0^a\int_0^b\int_{-c}^0 A(\mathbf{u})\cos(2\pi\mathbf{us})dV$$

$$+\ 2\int_0^a\int_{-b}^0\int_0^c A(\mathbf{u})\cos(2\pi\mathbf{us})dV + 2\int_{-a}^0\int_0^b\int_0^c A(\mathbf{u})\cos(2\pi\mathbf{us})dV$$

S. Fortier (ed.), Direct Methods for Solving Macromolecular Structures, 475–478.

$$= 2\,FT[A_1] + 2\,FT[A_2] + 2\,FT[A_3] + 2\,FT[A_4]$$

$$= 2\,I_1(s_1,s_2,s_3) + 2\,I_2(s_1,s_2,s_3) + 2\,I_3(s_1,s_2,s_3) + 2\,I_4(s_1,s_2,s_3).$$

Thus the total intensity is a sum of four terms related to the contributions from the ACF density in the corresponding unit cells. It is clear that one can immediately write three DHT equations of (1a-c) type for $I_1(s_1,s_2,s_3)$

$$I_1(h'+1/2,\, k'+1/2,\, L) = -\,(1/\pi^2) \sum_{h,k} I_1(h,k,L)/[(h'-h+1/2)(k'-k+1/2)]\,. \qquad (3a)$$

$$I_1(h'+1/2,\, K,\, l+1/2) = -\,(1/\pi^2) \sum_{h,l} I_1(h,K,l)/[(h'-h+1/2)(l'-l+1/2)]\,. \qquad (3b)$$

$$I_1(H,\, k'+1/2,\, l+1/2) = -\,(1/\pi^2) \sum_{k,l} I_1(H,k,l)/[(k'-k+1/2)(l'-l+1/2)]\,. \qquad (3c)$$

Now let $I'_2(s_1,s_2,s_3)$ be the intensity from the $A_2(u,v,w)$ shifted by translation c to the first unit cell

$$I'_2(s_1,s_2,s_3) = FT[A_2(u,v,w+1)].$$

Then by definition $I'_2(s_1,s_2,s_3)$ must satisfy the DHT. The FT of the shifted function is related to the original intensity $I_2(s_1,s_2,s_3)$ by the phase factor

$$I'_2(s_1,s_2,s_3) = \exp(2\pi i\, s_3)\, I_2(s_1,s_2,s_3),$$

which is equal to 1 for s_3 integral and -1 for s_3 half-integral. This leads to the following DHT expressions for $I_2(s_1,s_2,s_3)$

$$-\,I_2(h'+1/2,\, k'+1/2,\, L) = -\,(1/\pi^2) \sum_{h,k} I_2(h,k,L)/[(h'-h+1/2)(k'-k+1/2)] \qquad (4a)$$

$$-\,I_2(h'+1/2,\, K,\, l+1/2) = -\,(1/\pi^2) \sum_{h,l} I_2(h,K,l)/[(h'-h+1/2)(l'-l+1/2)] \qquad (4b)$$

$$-\,I_2(H,\, k'+1/2,\, l+1/2) = -\,(1/\pi^2) \sum_{k,l} I_2(H,k,l)/[(k'-k+1/2)(l'-l+1/2)]\,. \qquad (4c)$$

By shifting $A_3(u,v,w)$ and $A_4(u,v,w)$ to the first unit cell (translations $\boldsymbol{b}$ and $\boldsymbol{a}$ respectively) one obtains the corresponding expressions for $I_3(s_1,s_2,s_3)$ and $I_4(s_1,s_2,s_3)$ After adding up the equations for I_1, I_2, I_3 and I_4 we obtain the following final equations

$$I_1(h'+1/2,\, k'+1/2,\, L) - I_2(h'+1/2,\, k'+1/2,\, L) - I_3(h'+1/2,\, k'+1/2,\, L) \qquad (5a)$$
$$-\,I_4(h'+1/2,\, k'+1/2,\, L) = -\,(1/\pi^2) \sum_{h,k} I(h,k,L)/[(h'-h+1/2)(k'-k+1/2)]\,.$$

$$I_1(h'+1/2,\, K,\, l'+1/2) - I_2(h'+1/2,\, K,\, l'+1/2) - I_3(h'+1/2,\, K,\, l'+1/2) \qquad (5b)$$
$$-\,I_4(h'+1/2,\, K,\, l'+1/2) = -\,(1/\pi^2) \sum_{h,l} I(h,K,l)/[(h'-h+1/2)(l'-l+1/2)]\,,$$

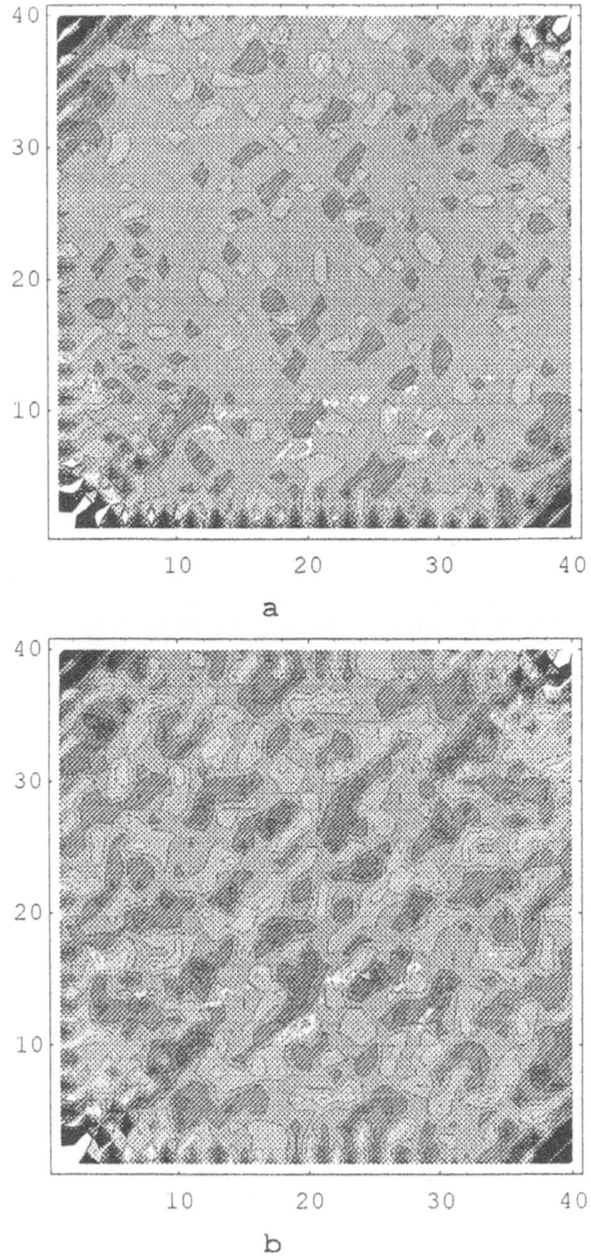

Figure 1. (a) Section of the Patterson map for a small protein, (b) the same section of the ACF map calculated using half-integral Fourier coefficients (5a-c)

$$I_1(H, k'+1/2, l'+1/2) - I_2(H, k'+1/2, l'+1/2) - I_3(H, k'+1/2, l'+1/2) \qquad (5c)$$

$$- I_4(H, k'+1/2, l'+1/2) = - (1/\pi^2) \sum_{k,l} I(H,k,l)/[(k'-k+1/2)(l'-l+1/2)] .$$

The important property of relations (5a-c) is that the left-hand sides are dependent on the phases of normal structure factors while the right-hand sides can be calculated from the observed intensities. It can be also seen that the left-hand sides of (5a-c) are the half-integral Fourier coefficients of the map, which is a superposition of four shifted ACF copies with the origins at (0, 0, 0), (0, 0, c), (0, b, 0) and (a, 0, 0). Such a map (Fig.1b), as compared with the Patterson (which is a superposition of eight shifted ACF copies, Fig.1a), can be expected to display more structural information and, in fact, provides a two-fold deconvolution of the Patterson function. It is less subjected to the peak superposition inherent to the standard Patterson, has more pronounced maxima and therefore its use may have advantage over the Patterson in molecular replacement method, superposition techniques and location of heavy atoms. Besides that the half-integral intensities can be included in the structure refinement procedures.

References

1. Mishnev, A.F. (1993) Discrete Hilbert transforms in crystallography, *Acta Cryst.*, **A49**, 159-161.
2. Mishnev, A.F. (1996) Discrete Hilbert transforms and the phase problem in crystallography, *Acta Cryst.*, **A52**, 629-633.

APPLICATIONS OF DIRECT METHODS WITH SINGLE ISOMORPHOUS REPLACEMENT OR ONE-WAVELENGTH ANOMALOUS SCATTERING DATA

H. F. Fan [a], Q. Hao [b, c, a], I. Harvey [b, c], S. S. Hasnain [c], Y. D. Liu [a], Y. X. Gu [a], C. D. Zheng [a] and H. Ke [d],

[a] *Institute of Physics, Chinese Academy of Sciences, Beijing 100080, P.R. China*

[b] *School of Applied Sciences, De Montfort University, Leicester LE1 9BH, England*

[c] *CCLRC Daresbury Laboratory, Warrington WA4 4AD, England*

[d] *Department of Biochemistry & Biophysics, University of North Carolina, Chapel Hill, NC 27599, USA*

1. Introduction

Recent developments of our study have been concentrated on the following topics: 1) the accuracy of our Cochran-distribution-based probability formula: it has been shown that the results of the formula are much more accurate than what people would usually predict; 2) an iterative procedure of combining direct methods with the solvent flattening technique: test on the experimental single isomorphous replacement (SIR) data of the known protein RNase Sa showed considerable improvement to the previous procedure; 3) direct-method phasing of the one-wavelength anomalous scattering (OAS) data of an unknown protein: the result leads to a traceable Fourier map and thus to the solution of the structure.

2. On the accuracy of the Cochran-distribution-based probability formula

Equation (1) is the probability formula we used to break the SIR or OAS phase ambiguity.

$$P_+(\Delta\varphi_h) = \frac{1}{2}+\frac{1}{2}\tanh\left\{\sin\left(\left|\Delta\varphi_h\right|\right)\left[\sum_{H'} m_{h'}m_{h-h'}\kappa_{h,h'}\sin(\Phi_3' + \Delta\varphi_{h',best} + \Delta\varphi_{h-h',best}) + \chi\sin\delta_h\right]\right\} \quad (1)$$

For details of the formula the reader is referred to reference [1]. This formula is based on the Cochran distribution [2]

S. Fortier (ed.), Direct Methods for Solving Macromolecular Structures, 479–485.

$$P(\phi_3) = \frac{1}{2\pi I_0(\kappa_{h,h'})} \exp(\kappa_{h,h'} \cos\phi_3) \tag{2}$$

For equal-atom structures,

$$\kappa_{h,h'} = 2N^{-1/2}\left|E_{-h}E_{h'}E_{h-h'}\right| \tag{3}$$

where N is the number of atoms in the unit cell. The reliability of phase estimation based on (2) depends on the 'sharpness', and thus the quantity $\kappa_{h,h'}$ of the distribution. For protein structures since N is large, the value of $\kappa_{h,h'}$ will be small. This might affect the accuracy of the phasing results. A comparison of the accuracy of estimating three-phase structure invariants by different probability formulas was reported [3]. Part of the results are shown in columns 1 to 3 of Table 1. As is seen the accuracy of estimation by the Cochran distribution is the lowest. This is true for equation (2). However the situation may be quite different for equation (1). Firstly, in equation (1) large number of three-phase structure invariants can be used jointly to predict a single phase leading usually to a much accurate estimation. Secondly, the task of equation (1) is just to make choice between the two possible signs of $\Delta\varphi$, while that of equation (2) is to predict a value in the range of 0 to 2π, the latter is surely much more complicated and inaccurate. Finally, the figure of merit m_h in equation (1) is very efficient in minimising the error of estimation. For a comparison we used a set of calculated OAS data of the protein aPP at 2Å resolution. The condition is the same as that of reference [3] for aPP. Individual phases were derived by making use of equation (1). Errors of the three-phase structure invariants calculated accordingly were than cumulated as shown in columns 4 and 5 of Table 1.

Table 1. Cumulative statistics of the three-phase structure invariants for the protein aPP

Cochran 250 largest E's		PS (true) 250 largest E's		SDs (true) 250 largest E's		Equation (1) 250 largest E's		Equation (1) total 2108 E's	
NTR	ERR	NTR	ERR	NTR	ERR	NTR	ERR	NTR	ERR
315	140	255	52	197	11	600	21	80,000	37
906	141	1021	57	1049	12	1000	25	400,000	55
3750	153	3750	98	3750	52	2288 (total)	32	807,262	71

NTR — number of triplet relations (three-phase structure invariants);
ERR — average error of three-phase structure invariants in mc (1000 mc = 2π rad)

$$\text{ERR} = \langle|\phi_{3\,true} - \phi_{3\,est}|\rangle$$

We concluded that the accuracy of estimation by equation (1) is far better than what people would usually predict. This explains why equation (1) has been so successful in dealing with the experimental OAS and SIR data of protein crystals.

3. An iterative procedure of combining direct methods with the solvent flattening technique

The major experimental error in OAS or SIR data comes from the measurement of the Bijvoet differences or the structure-factor-magnitude differences between a pair of isomorphous crystals. A considerable number of the observed differences will be incorrect not only in their magnitudes but also in their signs, which can in some circumstances have a very large effect on the estimate of the phase angle. When direct methods are used in such a case, large errors, usually not randomly distributed, will then be transferred to the solvent flattening and affect significantly the result. A new procedure [4] has been proposed for dealing with this difficulty. The procedure differs from the previous one [5] in that the direct method not only provides input phases to, but also accepts feed back phases from solvent flattening, thus forming an iterative process for breaking the ambiguities and refining the values of phases. The flow-chart is shown in *figure* 1. The new procedure was tested with the experimental SIR data of the known structure ribonuclease (RNase) Sa. For the strongest 1000 of the total 7264 reflections, the mean Fobs-weighted phase error is respectively 7.5 and 9.4 degrees lower than that of the previous procedure and that of the solvent flattening alone. Some portions of the resultant Fourier map are shown in *figure* 2 in comparison with that obtained from the previous procedure.

4. Solving an unknown protein by direct-method phasing of the OAS data

Rusticyanin is a copper-containing electron transfer protein with a very high redox potential. No similar structures are available to enable a molecular replacement solution. The crystallisation of the protein was carried out at Daresbury Laboratory. The crystals belong to the space group $P2_1$ with unit cell dimensions: a=32.43, b=60.68, c=38.01Å and β=107.82°. There is one molecule (about 1300 non-H atoms including one Cu atom) in the asymmetric unit. X-ray diffraction data were collected on station 9.5 at the Daresbury Synchrotron Radiation Source. The wavelength of 1.3765Å near the Cu K-edge was chosen to maximise the value of $\Delta f''$ (=3.87). Only one crystal was used in the experiment and it was aligned carefully so that Friedel pairs were collected on the same image plate.

The copper atom was located by the conventional direct-method program *SAPI*91 [6] using the magnitudes of anomalous differences. The direct method [7 - 9] was then used to break the phase ambiguity intrinsic to OAS data. The noise level of the direct-method map was reduced by a minimum function of the direct-method map and a Ps-function map [10]. Further improvement was obtained by using the electron density modification program DM in CCP4 suite. The resultant electron density map (*figure* 3) was clearly traceable and the backbone of the structure was then built by using the modelling program O [11]. This work was carried out at De Montfort University and Daresbury Laboratory (Harvey, Hao, Ingledew and Hasnain, unpublished results).

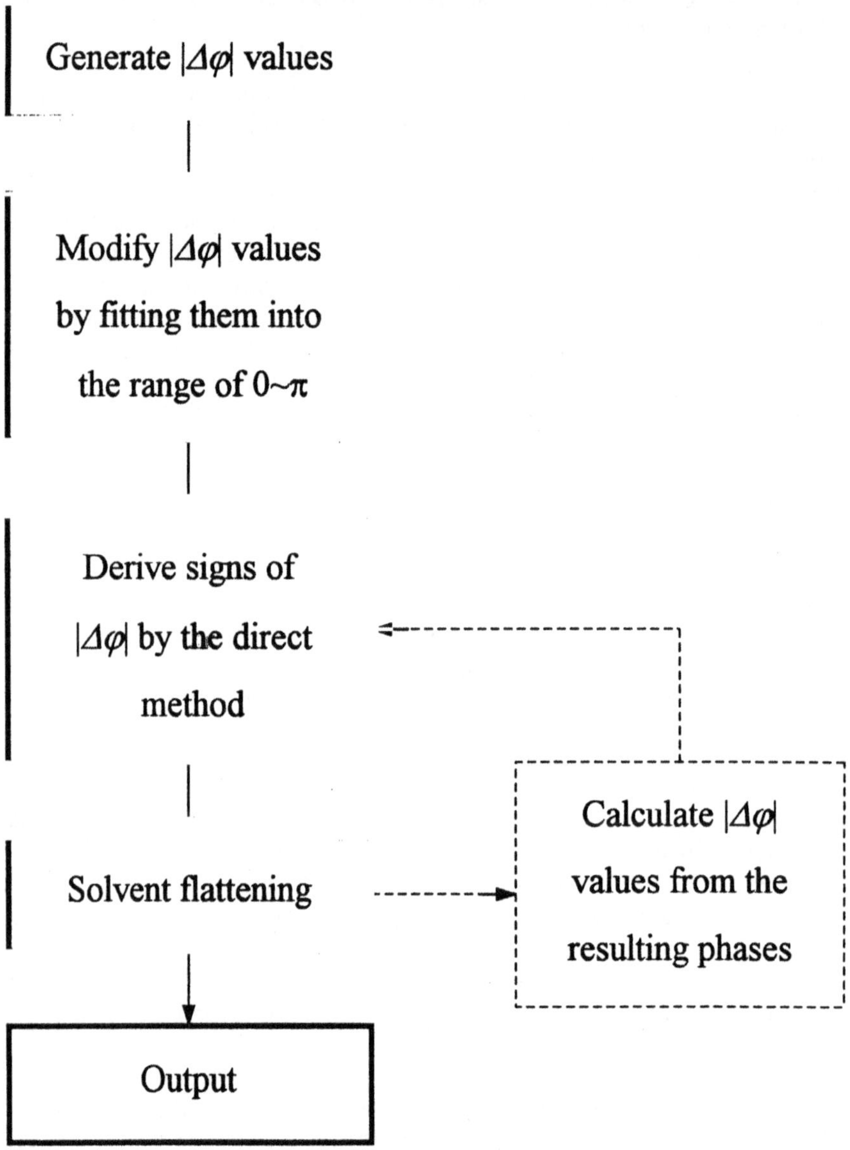

Figure 1. Flowchart of the iterative "direct method + solvent flattening". The portion shown with dashed lines is new to the previous simple "direct method + solvent flattening".

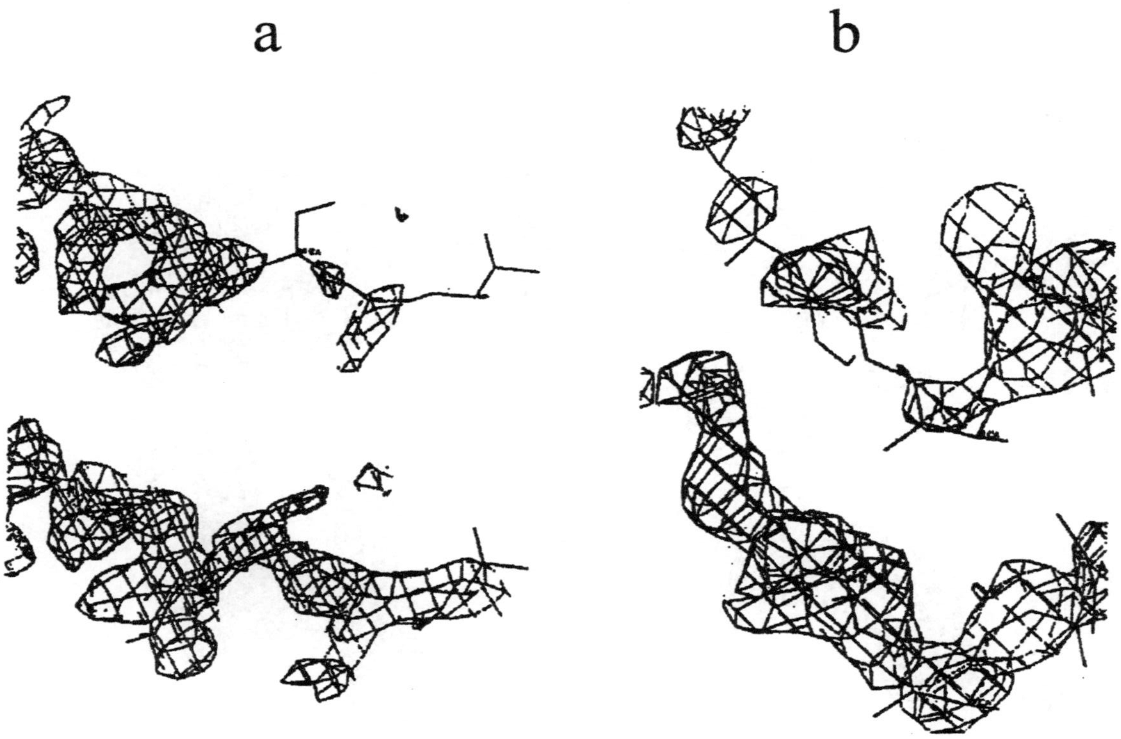

Figure 2. Comparison of Fourier maps on two different portions (a) and (b). Maps from the simple "direct method + solvent flattening" are shown on the upper part, while the corresponding maps from the iterative "direct method + solvent flattening" are shown on the lower part. All maps are contoured at 1σ.

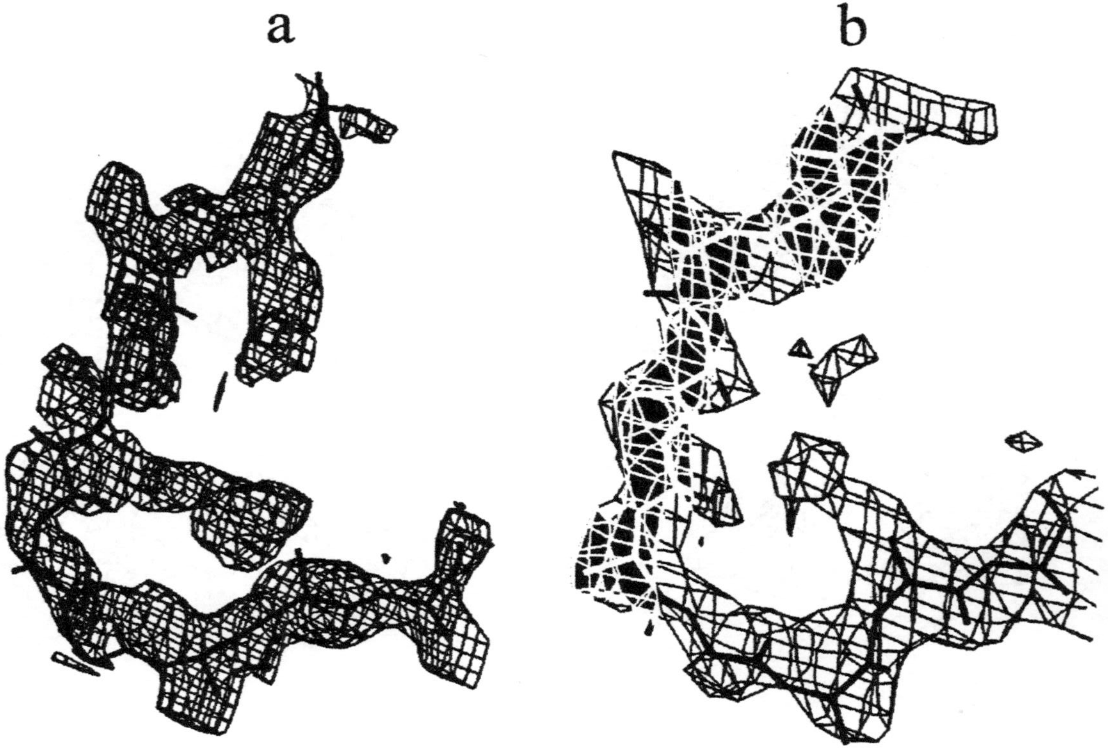

Figure 3. A section (residues 126-135) of the rusticyanin electron density map: (a) after a minimum function of direct method and Ps-function maps + density modification; (b) after one cycle of simulated annealing refinement.

The same protein was also studied independently at Cornell University using the multiwavelength anomalous diffraction (MAD) method. They have reported the structure in late 1996 [12]. However till now neither atomic positions nor diffraction phases have been released. Besides, our direct-method phases were derived before the Cornell group's publication of the structure. Hence it is fair to say, the structure was solved independently by direct phasing of the OAS data.

References

1. Fan, H. F. and Gu, Y. X. (1985) Combining direct methods with isomorphous replacement or anomalous scattering data III. The incorporation of partial structure information, *Acta Cryst.* **A41**, 280-284. See also Fan, H. F. (1997) Direct methods of solving phase ambiguities in macro-molecular structures, in this *Proceedings.*
2. Cochran, W. (1955) Relations between the phase of structure factors, *Acta Cryst.* **8**, 473-478.
3. Kyriakidis, C. E., Peschar, R. and Schenk, H. (1993) On the direct methods phase information from differences between isomorphous structure factors, *Acta Cryst.* **A49**, 557-569.
4. Gu, Y. X., Zheng, C. D., Zhao, Y. D., Ke, H. and Fan, H. F. (1997) A New Procedure of Combining Direct Methods with Solvent Flattening — Dealing with the Phase Ambiguity in Protein Crystallography, *Acta Cryst.* **D53** (submitted).
5. Zheng, X. F., Zheng, C. D., Gu, Y. X., Mo, Y. D., Fan, H.F. and Hao, Q. (1997) Use of single isomorphous replacement data of proteins — resolving the phase ambiguity and a new procedure for phase extension, *Acta Cryst.* **D53**, 49-55.
6. Fan, H. F., Yao, J. X., Zheng, C. D., Gu, Y. X. and Qian, J. Z. (1991) *SAPI* 91: a computer program for automatic solution of crystal structures from X-ray diffraction data, Institute of Physics, Chinese Academy of Sciences, Beijing 100080, P. R. China.
7. Fan, H. F. and Gu Y. X. (1985) Combining direct methods with isomorphous replacement or anomalous scattering data III. The incorporation of partial structure information, *Acta Cryst.* **A41**, 280-284.
8. Fan H. F., Hao, Q., Gu, Y. X., Qian, J. Z., Zheng, C. D. and Ke, H. (1990) Combining direct methods with isomorphous replacement or anomalous scattering data VII. Ab initio phasing of the OAS data from a small protein, *Acta Cryst.* **A46**, 935-939.
9. Hao, Q., Gu, Y. X., Zheng, C. D. and Fan, H. F. (1996) OASIS: a computer program for breaking the phase ambiguity in OAS or SIR protein data, School of Applied Sciences, De Montfort University, Leicester LE1 9BH, England and Institute of Physics, Chinese Academy of Sciences, Beijing 100080, P. R. China.
10. Hao, Q. and Woolfson, M.M. (1989) Application of the Ps-function method to macromolecular structure determination, *Acta Cryst.* **A45**, 794-797.
11. Jones, T. A., Zou, J. Y., Cowan, S. W. and Kjeldgaard, M. (1991) Improved methods for building protein models in electron density maps and the location of errors in these models, *Acta Cryst.* A47, 110-119.
12. Walter, R.L., Ealick, S.E., Friedman, A.M., Blake, R.C., Proctor, P. and Stoham, M. (1996) Multiple wavelength anomalous diffraction (MAD) crystal structure of rusticyanin: a highly oxidising cupredoxin with extreme acid stability, *J. Mol. Biol.*, **263**, 730-751.

DIRECT METHODS AND MOLECULAR REPLACEMENT TECHNIQUES: THE TRANSLATION PROBLEM

C. GIACOVAZZO[+], L. MANNA[+] & D. SILIQI[*]
[+]*Istituto di Ricerca per lo Sviluppo di Metodologie Cristallografiche CNR, c/o Dipartimento Geomineralogico, Campus Universitario, Via Orabona 4, 70125 Bari, Italy.*
[*]*Department of Inorganic Chemistry, Tirana University, Tirana, Albania*

1. Introduction

In a pioneering paper Main [1] showed that different types of prior information can be integrated into the probabilistic methods for the estimation of triplet invariants. Besides the case in which the minimal information is available (atomicity and positivity of the electron density), he considered three types of prior knowledge:
a) groups of atoms in random position and random orientation (the prior information is contained in the molecular geometry).
b) correct orientation of one or more groups of atoms randomly positioned (the crystallographer has to translate them in the correct sites).
c) knowledge of a partial structure (molecular fragment well oriented and well positioned). This case has been revisited by Giacovazzo [2] who provided a more efficient solution to the problem.
Main's formulation has been of large usefulness in direct methods applications to small molecules, but no attempt was made (at least to the knowledge of the authors) to apply it to macromolecules. In fact, the cases a) and b) are strictly related to molecular replacement techniques (see [3], and literature therein quoted): the first should help to rotate in the correct position a molecular model randomly oriented, the second to translate a well oriented molecular model to the correct site.
Quite recently the problem has been revisited by Giacovazzo, Manna, Siliqi, Bolognesi & Rizzi [4]: they considered the efficiency of the Main formula in macromolecular crystallography and applied it to face the translation problem. We will discuss here the modifications afforded to Main original formula and the experimental results so obtained. In a subsequent step Giacovazzo, Manna & Siliqi [5] showed that also quartet invariant estimates can benefit from the prior information at the point a), and argued about the combined use of triplets and quartet invariants in direct procedures. Once a macromolecule has been located, electron density modification procedures can be applied to improve and refine phases.
All the above aspects will be presented and discussed in this paper.

S. Fortier (ed.), Direct Methods for Solving Macromolecular Structures, 487–497.

2. Symbols and abbreviations

$F_{\mathbf{h}}$: structure factor with vectorial index **h**

$\phi_{\mathbf{h}}$: phase of $F_{\mathbf{h}}$

$f_j(\mathbf{h})$: scattering factor of the j-th atom

$\mathbf{C}_s \equiv (\mathbf{R}_s, \mathbf{T}_s)$: s-th symmetry operator. $\mathbf{R}_s$ is the rotational part, $\mathbf{T}_s$ the translational part.

m : order of the point group of the space group

N : number of atoms in the unit cell

N_f : number of molecular fragments (symmetry independent) with unknown position and fixed orientation.

n_i : number of atoms in the i-th molecular fragment

t_q : number of atoms (symmetry independent) whose positions are completely unknown.

q : number of atoms (symmetry equivalent included) whose positions are completely unknown

$\sum_q(\mathbf{h}) = \sum_{j=1}^{q} f_j^2(\mathbf{h})$: scattering power of the q atoms with completely unknown position.

$$\sum_N(\mathbf{h}) = \sum_{j=1}^{N} f_j^2(\mathbf{h})$$

$$\sum_{3q}(\mathbf{h}_1,\mathbf{h}_2,\mathbf{h}_3) = \sum_{j=1}^{q} f_j(\mathbf{h}_1) f_j(\mathbf{h}_2) f_j(\mathbf{h}_3)$$

$$\sum_{3N}(\mathbf{h}_1,\mathbf{h}_2,\mathbf{h}_3) = \sum_{j=1}^{N} f_j(\mathbf{h}_1) f_j(\mathbf{h}_2) f_j(\mathbf{h}_3)$$

$$\sum_{4q}(\mathbf{h}_1,\mathbf{h}_2,\mathbf{h}_3) = \sum_{j=1}^{q} f_j(\mathbf{h}_1) f_j(\mathbf{h}_2) f_j(\mathbf{h}_3) f_j(\mathbf{h}_4)$$

$$\sum_{4N}(\mathbf{h}_1,\mathbf{h}_2,\mathbf{h}_3) = \sum_{j=1}^{N} \mathrm{f}_j(\mathbf{h}_1) f_j(\mathbf{h}_2) f_j(\mathbf{h}_3) f_j(\mathbf{h}_4)$$

$\sigma_i = \sum_{j=1}^{N} Z_j^i$, where Z_j is the atomic number of the j-th atom

$\varepsilon_{\mathbf{h}}$: Wilson's factor responsible for the enhancement or reduction of the intensity of certain subsets of reflections due to particular symmetry elements.

$\Phi_3 = \phi_{\mathbf{h}_1} + \phi_{\mathbf{h}_2} + \phi_{\mathbf{h}_3}$ with $\mathbf{h}_1 + \mathbf{h}_2 + \mathbf{h}_3 = 0$

$\Phi_4 = \phi_{\mathbf{h}_1} + \phi_{\mathbf{h}_2} + \phi_{\mathbf{h}_3} + \phi_{\mathbf{h}_4}$ with $\mathbf{h}_1 + \mathbf{h}_2 + \mathbf{h}_3 + \mathbf{h}_4 = 0$

$I_0(x)$: modified Bessel function of order zero

3. The probabilistic problem

In macromolecular crystallography one has often to translate a model molecule, similar to that under study, which has been already oriented by some rotation function [6], [7], [8], [9] and [10]. Once the translation vector has been found, a refinement process starts to modify the electron density of the translated model into the electron density of the protein under study. The model molecule may be weakly correlated with the protein: in this case the correct orientation may be roughly accomplished, and consequently, the correct translation vector is unlikely found. From the point of view of direct methods the problem may be stated as follows: triplet invariant estimates are modified by making use of prior information on the orientation of the model molecule. If the new estimates are sufficiently accurate, a phasing procedure should automatically translate the model molecule in the correct position. This time the translation vector is not provided, nevertheless the electron density should be well positioned. Again a refinement process should start to modify the electron density of the model into that of the molecule under study. It may be worthwhile noting that traditional translation procedures exploit the inter-molecular vectors while direct methods use as prior information the intra-molecular vectors of the model structure only. Thus the two approaches are in some way complementary. In our probabilistic approach we will consider the model molecule as part of the molecule under study (i. e., we will neglect electron density differences between the model molecule and the corresponding part of the protein). Accordingly we will divide the protein (in agreement with Main's approach) into two parts: the first includes the N_f molecular fragments with known orientation, and their symmetry equivalents. The generic j-th atom belonging to the i-th fragment has trial positional vector $\mathbf{u}_j$, and τ_i is the shift to be applied to the i-th fragment to translate atoms to the correct positions

$$\mathbf{u}_j + \tau_i, \quad \text{for } j = 1,......n_i \;\; , \; i = 1,.....,N_f$$

The second part of the structure includes q atoms whose positions $\mathbf{r}_j$ are completely unknown. Then the structure factor may be written as

$$F_{\mathbf{h}} = \sum_{i=1}^{N_f} g_i(\mathbf{h}) + F_{q\mathbf{h}} = F_{p\mathbf{h}} + F_{q\mathbf{h}}$$

where

$$g_i(\mathbf{h}) = \sum_{s=1}^{m}\sum_{j=1}^{n_i} f_j(\mathbf{h}) \exp\left[2\pi i \mathbf{h} \mathbf{C}_s(\mathbf{u}_j + \tau_i)\right]$$

$$= \sum_{s=1}^{m} g_{is}(\mathbf{h}) \exp(2\pi i \mathbf{h} \mathbf{R}_s \tau_i),$$

$$g_{is}(\mathbf{h}) = \sum_{j=1}^{n_i} f_j(\mathbf{h}) \exp(2\pi i \mathbf{h} \mathbf{C}_s \mathbf{u}_j),$$

$$F_{q\mathbf{h}}(\mathbf{h}) = \sum_{s=1}^{m}\sum_{j=1}^{t_q} f_j(\mathbf{h})\exp(2\pi i\mathbf{h}\mathbf{C}_s\mathbf{r}_j).$$

We note: a) the primitive random variables in our probabilistic approach are the N_f shifts τ_i and the t_q atomic positions $\mathbf{r}_j$. Accordingly, the use of prior information reduces the number of primitive random variables from $(n_i \cdot N_f + t_q)$ random positions in absence of prior information to $t = N_f + t_q$. b) For the sake of simplicity the mathematical model does not allow for atoms to occupy special positions. This is not critical in most of the cases; c) The factors $g_{is}(\mathbf{h})$ do not depend on τ_i and may be calculated on the basis of prior information for any i and s. d) Our primitive random variables are considered statistically independent of each other, and are randomly distributed in the unit cell. Forbidden domains (eventually generated by the symmetry elements) are not taken into consideration. Structure factors may be normalized by observing [1] that

$$\left\langle |F_{\mathbf{h}}|^2 \right\rangle_M = \varepsilon_{\mathbf{h}}\left\{\sum_{s=1}^{m}\sum_{i=1}^{N_f}|g_{is}(\mathbf{h})|^2 + \sum\nolimits_q(\mathbf{h})\right\}:$$

then

$$E_{M\mathbf{h}} = F_{\mathbf{h}} \Big/ \left\langle |F_{\mathbf{h}}|^2 \right\rangle_M^{1/2}.$$

Besides $E_{M\mathbf{h}}$, we will also use the pseudo-normalized structure factor $E_{W\mathbf{h}}$ obtained by neglecting prior information on the model molecule:

$$E_{W\mathbf{h}} = F_{\mathbf{h}} \Big/ \left\langle |F_{\mathbf{h}}|^2 \right\rangle_W^{1/2}$$

where

$$\left\langle |F_{\mathbf{h}}|^2 \right\rangle_W = \sum\nolimits_N(\mathbf{h}).$$

4. The Main's triplet formula

The conditional probability distribution function of the triplet phase Φ_3 was stated by Main [1]:

$$P(\Phi_3) \approx [2\pi I_o(Q_3)]^{-1}\exp[Q_3\cos(\Phi_3 - \Theta_3)], \tag{1}$$

where

$$Q_3 = 2\left|E_{M\mathbf{h}_1}E_{M\mathbf{h}_2}E_{M\mathbf{h}_3}\right|\left[\frac{Q_3'^2+Q_3''^2}{\left\langle\left|F_{\mathbf{h}_1}\right|^2\right\rangle_M\left\langle\left|F_{\mathbf{h}_2}\right|^2\right\rangle_M\left\langle\left|F_{\mathbf{h}_3}\right|^2\right\rangle_M}\right]^{1/2},$$

$$Q_3' = \Re\left\{\sum_{i=1}^{N_f}\sum_{s=1}^{m} g_{is}(\mathbf{h}_1)g_{is}(\mathbf{h}_2)g_{is}(\mathbf{h}_3) + \sum\nolimits_{3q}(\mathbf{h}_1,\mathbf{h}_2,\mathbf{h}_3)\right\},$$

$$Q_3'' = \Im\left\{\sum_{i=1}^{N_f}\sum_{s=1}^{m} g_{is}(\mathbf{h}_1)g_{is}(\mathbf{h}_2)g_{is}(\mathbf{h}_3)\right\},$$

$$g_{is}(\mathbf{h}) = \sum_{j=1}^{n_i} f_j(\mathbf{h})\exp\left(2\pi i\mathbf{h}\mathbf{C}_s\mathbf{u}_j\right)$$

$$\tan\Theta_3 = Q_3''/Q_3'.$$

$\Re\{.....\}$ and $\Im\{......\}$ stand for real and imaginary part of $\{.....\}$ respectively, $\mathbf{u}_j$ is the trial position of the j-th atom (belonging to the oriented fragment).

5. The test structures

For the first practical applications Giacovazzo, Manna, Siliqi, Bolognesi & Rizzi used some structures for the solution of which molecular replacement techniques were essential. In order to manage real cases, the same model used for the original crystal structure solution was employed: it was rotated in correct position by application of some rotation function, while the translation search was performed by direct methods. The test structures are characterized below.

M-FABP (recombinant human-muscle fatty-acid binding protein) [11]. The model of adipocyte lipid binding protein (A-LBP), obtained at 2.5 Å resolution, was used as search model for molecular replacement; it shares 64% amino acid identities with M-FABP.

LPH (Lucina pectinata Hemoglobin I) [12] The molecular model of A. limacina myoglobin, was used as search model whose amino acid sequence shares 25% identities with LPH.

STM (Sea turtle myoglobin) [13] [14]. Sperm whale myoglobin was employed as search model, using data from 15.0 to 4.0 Å resolution range both for rotational and translational searches. Sperm whale myoglobin shares 63% amino acid identities with STM.

XSD (Xenophous leavis superoxide dismutase) [15]. Cu, Co bovine SOD has been used as a search model for the structure determination by means of molecular replacement. The amino acid sequence homology between the search model and XSD is ~ 50%.

6. The phasing procedure

The first applications showed that Main's formula is not powerful enough to face the lack of correlation between the model and the protein structure under study. The turning-point was the observation that reflections for which

$$\Delta E = |E_W| - |E_M| > 0 \quad (2)$$

are estimated by direct phasing procedures more accurately than reflections for which $\Delta E < 0$. Accordingly, Main's formula (1) overestimates the triplet reliability when the reflections constituting the triplet do not satisfy (2). As a consequence the practice of ordering the reflections in order of $|E_M|$ was abandoned. $|E_W|$ was used as ordering criterion. Accordingly, it was decided to express Main's reliability parameter for the triplet phases as

$$Q_3 = 2|\mathrm{E}_{\mathbf{Wh}_1}\mathrm{E}_{\mathbf{Wh}_2}\mathrm{E}_{\mathbf{Wh}_3}|\frac{\left[\Sigma_N(\mathbf{h}_1)\Sigma_N(\mathbf{h}_2)\Sigma_N(\mathbf{h}_3)(Q'^2+Q''^2)\right]^{1/2}}{\left\langle|\mathrm{F}_{\mathbf{h}_1}|^2\right\rangle_M\left\langle|\mathrm{F}_{\mathbf{h}_2}|^2\right\rangle_M\left\langle|\mathrm{F}_{\mathbf{h}_3}|^2\right\rangle_M}. \quad [3]$$

Further modifications to Main's formula were also necessary. We describe them as part of the following phasing procedure which has been implemented to solve the translation problem:

1) Observed structure factors are scaled according to standard Wilson method.
2) Structure factors corresponding to the correctly oriented model are calculated, by using the B value obtained from the observed magnitudes at the step 1. Both $|E_W|$ and $|E_M|$ are calculated.
3) A threshold is fixed for the resolution. It depends mostly on the similarity between model molecule and protein (the use of high resolution data is advisable only in case of high similarity). As a default, the resolution threshold of 3 Å is fixed.
4) Reflections are ordered according to ΔE: e.g., the first in the ordered set is the reflection with the largest positive value of ΔE, the last in the set is that with the largest negative value of ΔE.
5) A threshold TRΔE is fixed: reflections with $\Delta E >$ TRΔE are selected, and, among them, NLAR reflections with $|E_W| >$ TRE_W are used for the triplet invariant search. The threshold values are chosen so as to select a sufficiently large number of reflections among which reliable triplets could be found. The most striking feature of the process is that even reflections with small values of E_W are expected to give rise to reliable triplets provided $\Delta E > 0$.
6) Reflections for which $|E_W|$ and $|E_M|$ are simultaneously very weak are selected for the psizero figure of merit.
7) A multisolution approach is started *via* random phases. The triplets are estimated *via* the new Von Mises formula

$$P(\Phi) \approx [2\pi I_o(G)]^{-1}\exp[G\cos(\Phi_3 - \Theta_3)], \quad (4)$$

where

$$G = p_1 G',$$

$$G' = 2\left|E_{W\mathbf{h}_1}E_{W\mathbf{h}_2}E_{W\mathbf{h}_3}\right|\left[\frac{Q_3'^2+Q_3''^2}{\left\langle\left|F_{\mathbf{h}_1}\right|^2\right\rangle_M\left\langle\left|F_{\mathbf{h}_2}\right|^2\right\rangle_M\left\langle\left|F_{\mathbf{h}_3}\right|^2\right\rangle_M}\right]^{1/2} . \quad (5)$$

p_1 is a weighting factor which limits the range of G to the interval (0,6), to avoid that triplets with too high values of the reliability factor G' could dominate the phasing process. Q_3', Q_3'' and Θ_3 are defined as in (1). The reliability coefficient G' may be obtained from Q_3 by replacing in (3)

$$\left\{\sum_N(\mathbf{h}_1)\sum_N(\mathbf{h}_2)\sum_N(\mathbf{h}_3)\Big/\left\langle\left|F_{\mathbf{h}_1}\right|^2\right\rangle_M\left\langle\left|F_{\mathbf{h}_2}\right|^2\right\rangle_M\left\langle\left|F_{\mathbf{h}_3}\right|^2\right\rangle_M\right\}^{1/2} \quad (6)$$

by a unitary factor. Actually (6) is often far from unity, so that (1) and (4) work quite differently. The smaller (6) is with respect to unity, the larger the underestimate of the triplet provided by (1). On the contrary, if

$$\varepsilon\sum_N(\mathbf{h}_i) > \left\langle\left|F_{\mathbf{h}_i}\right|^2\right\rangle_M \qquad \text{for } i = 1, 2, 3,$$

assuming (6) equals to unity reduces the reliability of the triplets for which $|E_{Wi}| < |E_{Mi}|$. (4) has no theoretical basis: it has been suggested by our numerous applications and provides the probabilistic background for allowing the figures of merit (FOM's) to effectively work.

8) The solution selected among the various trials by suitable figures of merit is used as a seed for phasing the remaining reflections. Since reflections with negative ΔE values are involved in the calculations a modified form of G' will be used:

$$G' = 2p_2\left|E_{W\mathbf{h}_1}E_{W\mathbf{h}_2}E_{W\mathbf{h}_3}\right|\left[\frac{Q_3'^2+Q_3''^2}{\left\langle\left|F_{\mathbf{h}_1}\right|^2\right\rangle_M\left\langle\left|F_{\mathbf{h}_2}\right|^2\right\rangle_M\left\langle\left|F_{\mathbf{h}_3}\right|^2\right\rangle_M}\right]^{1/2}$$

p_2 is a weight factor which assumes different values according to the number of negative ΔE values. I.e., p_2 is 1 if all the three reflections contributing to the triplets have $\Delta E > 0$; $p_2 = 0.8$, 0.6 or 0.4 when only one ΔE, two ΔE or three ΔE are negative. The figures of merit usually employed for finding the correct solution among the various trials have been suitably modified [4]. The reader is referred to [4] for the details.

7. The first applications

We describe, for each test structure, the results obtained *via* our phasing procedure. In order to check if the translation problem was correctly solved we calculated the correlation factor CORR between the electron density ρ calculated from the assigned phases, and the "true" map ρ_{true} (corresponding to the published phases):

$$\mathrm{CORR} = \frac{\langle \rho\rho_{true}\rangle - \langle\rho\rangle\langle\rho_{true}\rangle}{\left(\langle\rho^2\rangle - \langle\rho\rangle^2\right)^{1/2}\left(\langle\rho^2_{true}\rangle - \langle\rho_{true}\rangle^2\right)^{1/2}}.$$

200 trials per structure are always performed, starting from random sets of phases. The results may be summarized as follows:

M-FABP - 3 Å resolution data are used: the highest value of CFOM (=0.393) recognizes the correct solution. NLAR=641 reflections are first phased, with a mean phase error of 38°. The phase expansion process leads to 7560 phased reflections with mean phase error of 72° .The value of CORR is 0.45. The electron density map calculated with all the phased reflections is of good overall quality, allowing straightforward model building interpretation throughout the polypeptide chain (see Fig. 1a).

LPH - No solution is found with 3 Å resolution data. A 4 Å cut-off is then used: the highest value of CFOM (=0.558) singles out the correct solution. NLAR=357 reflections are first phased, with a mean error of 49°. The phase expansion process leads to a total of 7085 phased reflections with mean phase error of 80°. CORR is equal to 0.29. Inspection of the electron density map, calculated with the 7085 extended phases and observed structure factors, shows regions which can be easily interpreted in terms of an atomic model compatible with the previously determined structure of the protein. In particular, regions of immediate interpretability are the α-helical segments surrounding the heme group, which is well defined (see Fig. 1b). On the other hand, less clear electron density is obtained for regions of the protein structure further away from the heme, for which in the absence of additional information, an atomic model cannot be fitted.

STM - 3 Å resolution data are used: the highest value of CFOM (=0.437) singles out the correct solution. NLAR=621 are first phased with a mean phase error of 20°. The phase expansion process leads to a total of 9753 reflections with mean phase error of 53°. CORR= 0.73. The corresponding phases allowed to calculate an electron density map which could be easily interpreted in terms of the final molecular model of the protein, the quality of the electron density being constant throughout the asymmetric unit (see Fig. 1c).

XSD - 3 Å resolution data are used: the highest value of CFOM (=0.290) singles out the correct solution. NLAR=680 reflections are first phased with mean phase error of 27°. The phase expansion process leads to a total of 19016 reflections with mean phase error of 66°: CORR= 0.57. Fig 1d shows a region comprising the Cu^{++} and Zn^{++} ions in the enzyme active site, calculated with the 19016 phased reflections. The dimeric enzyme model fits very nicely the electron density throughout the asymmetric unit, and model building, in the absence of a molecular model, could have been easily achieved.

The above tests show that the new formula (4) is able to face most of the problems usually met in macromolecular crystallography.

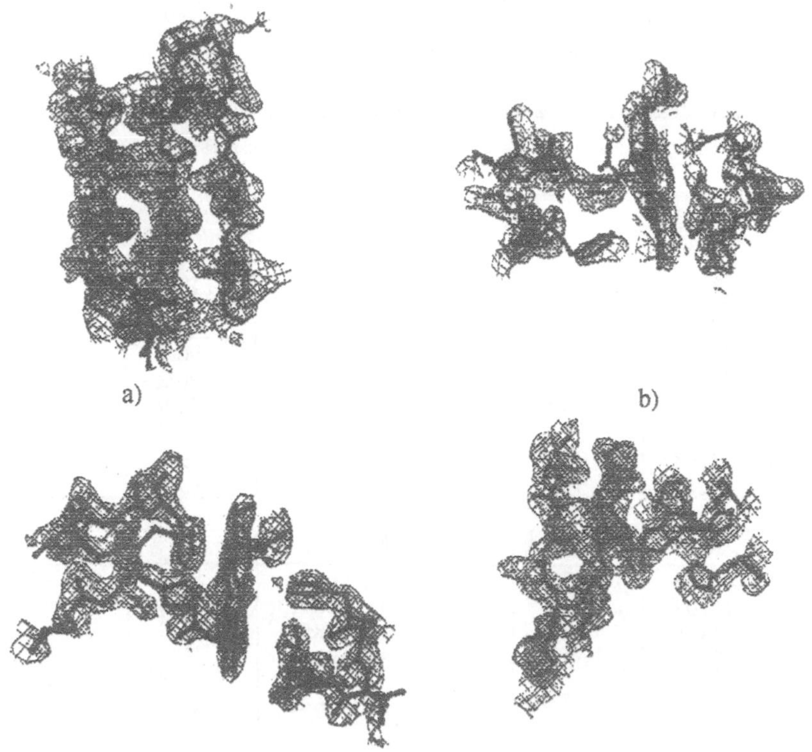

a) b) c) d)

Fig.1 - Regions of the electron density maps overlapped with the atomic models of the proteins: a) M-FABP; b) LPH; c) STM; d) XSD

8. Post-mortem analysis

Which is the largest obstacle to the solution of the translation problem? In Table 1 we show some statistical calculations for the triplet invariants estimated via (4) for M-FABP and LPH. In a) we use the translated protein molecule as the model (unitary correlation), in b) and c) we use the real model molecule (experimental data). It is quite clear that the lack of correlation is the main factor responsible for the lack of reliability of the formula. The striking point is that even large percentages of wrongly estimated triplets do not hinder the correct translation of the model structure.

9. About the role of the quartet invariants

The conditional distribution of the quartet phase Φ_4, given the orientation of one or more fragments, may be approximated by the von Mises distribution [5]

$$P(\Phi_4|R_1,R_2,...,R_7) \cong [2\pi I_0(G_4)]^{-1} \exp[G_4 \cos(\Phi_4 - \xi_4)]$$

For shortness the expressions of G_4 and ξ_4 are not given.

Table 1 - Statistical calculations for triplet invariants estimated *via* (4) for M-FABP and LPH. NR is the number of triplets with Q_3 > ARG, $\langle|\Delta\Phi_3|\rangle$ is the average phase error, % is the percentage of triplets with $|\Delta\Phi_3| > \pi/2$: a) the protein it self has been used as model molecule; b) the real model molecule has been used (up to experimental resolution)

a)

	M-FABP			LPH		
ARG	NR	%	<|ΔΦ|>	NR	%	<|ΔΦ|>
0.0	43480	77.6	57	19288	96.1	27
0.4	42683	78.0	56	19273	96.2	27
2.0	24281	86.6	45	18745	96.9	26
3.2	9936	92.6	36	17451	97.9	24
4.4	2968	94.1	32	13909	99.1	21
5.5	645	93.2	32	5116	99.7	18

b)

ARG	NR	%	<|ΔΦ|>	NR	%	<|ΔΦ|>
0.0	47078	55.4	84	20125	51.5	88
0.4	44704	55.6	84	20071	51.5	88
2.0	11431	58.8	80	17851	51.9	88
3.2	2321	62.9	75	13243	52.2	88
4.4	343	72.0	67	57321	52.7	87
5.5	19	84.2	64	927	54.3	85

It is well known that, in absence of prior information, the use of the so called positive quartets (those with large cross-magnitudes) in the phasing procedures using triplets is not advised. Indeed quartet information overlaps with that carried on by triplets, and the combined use may damage rather than improve the phasing process. When prior information is available the above statement has to be restated since the triplet reliability depends on ΔE rather than on $|E_M|$, therefore the complementarity of the triplet and quartet information has to be reformulated on a new basis. At the moment the active use of the quartets in the phasing procedures aiming at solving the translation problem is not documented. We can however claim that their use as figure of merit for recognizing the correct solution is rewarding [unpublished results].

10. The last steps

The direct procedure described in this paper may be slower with respect to conventional molecular replacement techniques. Furthermore it does not provide the translation vector, but only phases for the calculation of the respective electron density map. The method however has been improved under several aspects.

a) The vibrational atomic parameters of the search model may be taken into account (rather than assigning an overall vibrational parameter). This may increase the correlation between model and protein molecule, with favourable effects on the estimation of triplet invariants.

b) solvent flattening methods may be used to improve the quality of the phases provided by our direct procedure. The advantage of this last approach is that weights are associated to phases, and that makes the solvent flattening techniques more powerful.

Once a) and b) have been applied, a translation vector may be straightforwardly found a posteriori by a phased translation function [16]. In this way the crystallographer benefits by the simultaneous availability of a refined electron density map of the protein and of the translated atomic coordinates of the model molecule.

One of the authors (D. Siliqi) undertook this work with the support of the "ICTP Programme for Training and Research in Italian Laboratories, Trieste, Italy".

References

1. Main, P. (1976). In Crystallographic Computing Techniques, edited by F.R.Ahmed, pp 97-105. Copenhagen: Munksgaaard.
2. Giacovazzo, C. (1983). *Acta Cryst.* **A39**, 685-682
3. Rossmann, M.G. (1990). *Acta Cryst.* **A46**, 73-82
4. Giacovazzo, C., Manna, L. & Siliqi, D., Bolognesi, M. & Rizzi, M. (1997). *Acta Cryst.* To be submitted
5. Giacovazzo, C., Manna, L. & Siliqi, D. (1997). *Acta Cryst.* To be submitted
6. Rossmann, M. G. & Blow, D. M., (1962). *Acta Cryst.* **15**, 24-31
7. Tollin, P. (1966). *Acta Cryst.* **21**, 613-614
8. Crowter, R. A. & Blow, D. M., (1967). *Acta Cryst.* **23**, 544-548
9. Hoppe, W. (1957). *Acta Cryst.* **10**, 750-751
10. Huber, R. (1965). *Acta Cryst.* **19**, 353-356
11. Zanotti, G., Scapin, G:, Spadon, P., Veerkamp, J. H. & Sacchettini, J. C. (1992) *J. Biol. Chem.* **267**, 18541-18550.
12. Rizzi, M., Wittemberg, J. B., Coda, A., Fasano, M., Ascenzi,. P., Bolognesi, M. (1994). *J. Mol. Biol.* **244**, 86-99
13. Rizzi, M. Ascenzi, P. Coda, A., Brunori, M., Bolognesi, M. (1993). Rend. Fis. Acc. Lincei, s.9. **4**, 65-73.
14. Nardini, M., Tarricone, C., Rizzi, M., Lania, A., Desideri, G., De Sanctis, G., Coletta, M., Petruzzelli, R., Ascenzi, P., Coda, A., Bolognesi, M. (1995). *J. Mol. Biol.* **247**, 459-465
15. Djinovic Carugo, K., Collyer, C., Coda, A., Carri', M. T., Battistoni, A., Bottaro, G., Polticelli, F.,
16. Read, R. J., & Schierbeek, A. J. (1988). J. Appl. Cryst. **21**, 490-495

HOLOGRAPHIC METHODS IN X-RAY CRYSTALLOGRAPHY

JOHN R. SOMOZA[†], HANNA SZÖKE* AND ABRAHAM SZÖKE*
[†]*Dept. of Biochemistry and Biophysics, University of California, San Francisco, CA 94143-0448, USA. E-mail: somoza@msg.ucsf.edu.*
**Lawrence Livermore National Laboratory, Livermore, CA 94550, USA. E-mail: szoke1@llnl.gov.*

1. Introduction

In principle, the X-rays that are diffracted by a crystal provide enough information to determine the distribution of electron density in the crystal. However, due to fundamental limitations on how X-rays are detected, some of the information about these diffracted X-rays is lost. Specifically, the amplitudes of the diffracted X-rays are measured but not their phases. Furthermore, due to the sampling limitations imposed by the Bragg condition, the amplitudes of the diffraction data are not sufficient to uniquely define an electron density distribution. Instead there is a range of electron densities compatible with a given set of experimentally measured structure factor amplitudes.

Two types of information can be used to narrow down the possible electron density distributions. Most commonly, a small number of strongly diffracting atoms are soaked into the crystal of interest. If these atoms bind to specific parts of the protein, and if their presence does not strongly affect the conformation of the protein, the diffraction data that arise from these protein/heavy atom complexes can be used to determine the phases of the diffraction data and the electron density can be calculated.

It is also possible to reduce the range of possible solutions by selecting for solutions that satisfy the diffraction data amplitudes and that match our expectations of how the electron density should look. An obvious property that must be satisfied by the electron density is that it should be everywhere positive. Beyond this positivity constraint, the properties that must be satisfied by a correct electron density become more difficult to define, and depend on the resolution of the data used to describe the structure of interest. For example, at low resolution, proteins look like more-or-less globular blobs, with slightly more electron-rich blobs of density (the protein) surrounded by slightly less electron-rich density (solvent). At medium resolution, the electron density should look like connected tubes, and if we were to look closely at these tubes we might see evidence of the long strands and spirals of electron density that indicate the presence of canonical secondary structure. As the resolution increases, we see increasing structure in these tubes of density until eventually they are resolved into individual atoms.

Aside from these general properties of electron density, there are occasions when we have more specific information about the electron density. For example, there is often evidence of the approximate boundaries of a solvent region, or we might be aware of the existence of non-crystallographic symmetry.

S. Fortier (ed.), Direct Methods for Solving Macromolecular Structures, 499–502.

This *a priori* information about the electron density can be used to improve electron density maps or, in some cases, to solve a structure directly from the amplitudes of the diffracted X-rays. What is needed is a flexible, general way of selecting for electron density distributions that agree with the measured diffraction data *and* that meet real-space requirements.

2. Introduction to the holographic method

The holographic method is being developed as a general framework for incorporating experimental and *a priori* real-space information into the solution of a crystal structure [1,2,3,4]. The technique differs from other crystallographic methods in that it solves for electron density directly in real-space. It does not attempt to compute phases and then Fourier transform to obtain electron density. Instead, the electron density in the crystal is changed until it matches all of the available information. This is done by minimizing a cost function that takes into account differences between the observed and calculated structure factor amplitudes, as well as a variety of real-space constraints or restraints on the electron density. Calculating electron density in this way allows us to compute maps that simultaneously agree with all of the existing information.

3. A brief introduction to the theory

Assume that the electron density in the unit cell of a crystal is divided into a known part, $\rho_{known}(\mathbf{r})$, and an unknown part, $\rho_{unknown}(\mathbf{r})$, such that $\rho_{total}(\mathbf{r})=\rho_{known}(\mathbf{r})+\rho_{unknown}(\mathbf{r})$. The structure factors arising from the known and unknown parts of the electron density are denoted by $R(\mathbf{h})$ and $O(\mathbf{h})$, respectively. These structure factors are related to known and unknown parts of the electron density by the relations:

$$R(\mathbf{h}) = \int_{\text{unit cell}} \rho_{known}(\mathbf{r}) \exp(2\pi i \mathbf{h}\cdot\mathcal{F}\mathbf{r})\, d\mathbf{r} \tag{1}$$

$$O(\mathbf{h}) = \int_{\text{unit cell}} \rho_{unknown}(\mathbf{r}) \exp(2\pi i \mathbf{h}\cdot\mathcal{F}\mathbf{r})\, d\mathbf{r}. \tag{2}$$

The notation $R(\mathbf{h})$ and $O(\mathbf{h})$ is adopted from holography theory, where $R(\mathbf{h})$ and $O(\mathbf{h})$ denote the reference and object waves, respectively.

The structure factors that come from the whole crystal, $F(\mathbf{h})$, are simply the sum of the contributions from the known and unknown electron densities: $F(\mathbf{h})=R(\mathbf{h})+O(\mathbf{h})$. $|F(\mathbf{h})|^2$, which is experimentally measured, can be related to $R(\mathbf{h})$ and $O(\mathbf{h})$:

$$|F(\mathbf{h})|^2 = |R(\mathbf{h}) + O(\mathbf{h})|^2 = |R(\mathbf{h})|^2 + R(\mathbf{h})\, O^*(\mathbf{h}) + R^*(\mathbf{h})\, O(\mathbf{h}) + |O(\mathbf{h})|^2. \tag{3}$$

$|F(\mathbf{h})|^2$ and $R(\mathbf{h})$ are known, and what we are attempting to find is $O(\mathbf{h})$ and the corresponding electron density $\rho_{unknown}(\mathbf{r})$.

In the holographic method, electron density is solved for in real-space, so we need a good way of expressing $\rho_{unknown}(\mathbf{r})$. To do this we describe $\rho_{unknown}(\mathbf{r})$ as a sum of Gaussian basis functions of equal widths, centered on a grid that divides the unit cell into equal parts along the crystallographic axes. The grid points are located at positions $\mathbf{r}_p$, and each Gaussian blob (voxel) contains an unknown number of electrons, n(p). Thus, $\rho_{unknown}(\mathbf{r})$ is approximated by

$$\rho_{unknown}(\mathbf{r}) \approx (\pi\eta\Delta r^2)^{-3/2} \sum_{p=1}^{P} n(p) \exp\left[\frac{-|\mathbf{r}-\mathbf{r}_p|^2}{\eta\Delta r^2}\right], \tag{4}$$

where Δr is the mean grid spacing and η determines the width of the Gaussians relative to the grid spacing. This relation also allows us to express O(**h**) in terms of n(p):

$$O(\mathbf{h}) \approx \exp\left[-\eta(\pi\Delta r|\mathcal{F}^T\mathbf{h}|)^2\right] \sum_{p=1}^{P} n(p) \exp(2\pi i\mathbf{h}\cdot\mathcal{F}\mathbf{r}_p). \tag{5}$$

When this representation of O(**h**) is substituted into equation (3), we obtain a set of quadratic equations in the unknowns, n(p). The number of equations is equal to the number of measured structure factor amplitudes, and is usually not equal to the number of unknowns, P, which is the number of grid points in the unit cell. These equations are ill conditioned and their solution is sensitive to noise in the data. A (quasi) solution is obtained by minimizing the discrepancy, or cost function,

$$f_{eden} = \frac{1}{2}\sum_{\mathbf{h}} w'(\mathbf{h})^2 \left[|R(\mathbf{h}) + O(\mathbf{h})| - |F(\mathbf{h})|\right]^2. \tag{6}$$

The weights, $w'(\mathbf{h})^2$ are generally set to $1/\sigma(\mathbf{h})^2$, where $\sigma(\mathbf{h})$ is an experimentally measured estimate of the precision of an individual $|F(\mathbf{h})|$. The essence of equation (6) is that the electron density, represented by n(p), is modified to minimize the least-square difference between the calculated and observed structure factors amplitudes.

If there are MIR and/or MAD diffraction data available, they can be used to constrain the recovered electron density by defining cost functions similar to (6) for each of the heavy atom data sets (or for each wavelength in the case of MAD data). The contributions from the heavy atoms are incorporated into R(**h**) [4].

The incorporation of real-space constraints. Because we are solving for the electron density directly in real-space, it is relatively easy to make use of *a priori* knowledge we might have about the electron density. Information about the electron density is expressed by constraining or restraining individual voxels of electron density. For example, the positivity of the electron density is enforced by stipulating that all elements of the solution vector, n(p), be non-negative.

If there is a region of the unit cell for which we know or think we know the electron density (e.g. the solvent region), we can restrain the corresponding voxels to a "target" density, $n(p)_{target}$, by defining a cost function, f_{space},

$$f_{space} = \frac{1}{2}\lambda_{space} P \sum_{p=1}^{P} \tilde{w}_p^{\,2}\{n(p) - n(p)_{target}\}^2. \tag{7}$$

λ_{space} establishes the weight of f_{space} with respect to f_{eden}. The individual weights at each point, $\tilde{w}_p^2$, express the "strength of our belief" in the correctness of the target density at voxel p. Similarly, if there is a known non-crystallographic symmetry (ncs) relation within the asymmetric unit, the ncs-related regions can be "encouraged" to be similar by defining a restraining cost function similar to (7) [4].

In the presence of a real-space information, electron density is obtained by minimizing a cost function that is the sum of f_{eden} and f_{space}:

$$f_{total} = f_{eden} + f_{space}. \qquad (8)$$

4. Conclusions

The holographic method provides a new framework for determining electron density maps. The technique solves for electron density in real-space by minimizing a cost function that takes into account differences between the observed and calculated structure factor amplitudes, as well as a variety of other constraints or restraints on the electron density. At the moment, the following types of information can be used to restrain the recovered electron density: the positivity of the electron density, knowledge of the solvent region (or any other known electron density), the presence of non-crystallographic symmetry (although this restraint has not been well tested), and MIR and/or MAD data.

We have implemented the holographic method in the form of the program EDEN, which can be obtained by contacting any of the authors. The program has been tested using data from a number of different macromolecules [2,4]. In the worst cases, the method does as well as conventional crystallographic methods. In the best cases, the maps generated with EDEN are greatly improved over the maps produced using conventional methods.

5. References

1. Maalouf, G. J., Hoch, J. C., Stern, A. S., Szöke, H. & Szöke, A. (1993). Holographic methods in X-ray crystallography .III. 1st numerical results. *Acta Cryst.* A **49**, 866-871.

2. Somoza, J.R., Goodman, D.M., Beran, P., Szöke, H., Truckses, D.M., Kim, S.-H. & Szöke, A. (1995). Holographic methods in X-ray crystallography IV. A fast algorithm and its application to macromolecular crystallography. *Acta Cryst. A* **51**, 691-708.

3. Szöke, A. (1993). Holographic methods in X-ray crystallography .II. Detailed theory and connection to other methods of crystallography. *Acta Cryst.* A **49**, 853-866.

4. Szöke, A., Szöke, H. & Somoza, J.R. (1997). Holographic methods in X-ray crystallography V. Molecular replacement, multiple isomorphous replacement, multiple anomalous dispersion and non-crystallographic symmetry. *Acta Cryst. A.* **53**, 291-313.

A COMPARISON OF TWO ALGORITHMS FOR ELECTRON DENSITY MAP IMPROVEMENT BY INTRODUCTION OF ATOMICITY: SKELETONIZATION, AND MAP SORTING FOLLOWED BY REFINEMENT

F.M.D. VELLIEUX
Institut de Biologie Structurale J.-P. Ebel CEA CNRS
41 Avenue des Martyrs, 38027 Grenoble Cedex 01, France

1. Abstract

A comparison was made of two methods for electron density map improvement by the introduction of atomicity, namely the iterative skeletonization procedure of the CCP4 program DM (1), and the pseudo-atom introduction followed by refinement present in the program suite DEMON/ANGEL (2). Tests carried out using the 3.0 Å resolution electron density resulting from iterative 12-fold non-crystallographic symmetry averaging and solvent flattening for the *Pseudomonas aeruginosa* ornithine transcarbamoylase (3) indicate that pseudo-atom introduction followed by refinement performs much better than iterative skeletonization: with the former method, a phase improvement of 15.3^{o} is obtained with respect to the initial, density modification phases. With iterative skeletonization, a phase degradation of 0.4^{o} is obtained. Consequently, the electron density maps obtained using pseudo-atom phases or pseudo-atom phases combined with density modification phases are much easier to interpret. These tests also show that, for ornithine transcarbamoylase, where 12-fold non-crystallographic symmetry is present in the P1 crystals, G-function coupling leads to the simultaneous decrease of the conventional R-factor and of the free R-factor, a phenomenon which is not observed when non-crystallographic symmetry is absent from the crystal. The method is far less effective in such a case, and the results obtained suggest that the map sorting followed by refinement stage should be skipped to obtain interpretable electron density distributions.

2. Introduction

Once an electron density distribution has been obtained as the result of heavy atom or density modification, several algorithms have been proposed to further improve the quality of this electron density map by methods which introduce pseudo-atoms in the electron density.

S. Fortier (ed.), Direct Methods for Solving Macromolecular Structures, 503–511.

Such methods currently being used in macromolecular crystallography are the iterative skeletonization procedure of Wilson and Agard (PRISM, 4) which simplifies the electron density as ridge points and ridge lines connecting these points, the automatic refinement procedure of Lamzin and Wilson (ARP, 5), the weighted automatic refinement procedure of Perrakis *et al.* (wARP, 6), and the algorithm present in the program suite DEMON/ANGEL (2), which consists of the introduction of pseudo-atoms in an electron density map, followed by the refinement of these atomic coordinates and of the temperature factors of the pseudo-atoms (map sorting followed by refinement, or MSR). This idea is similar to that implemented in ARP, except that no additional pseudo-atom introduction, nor pseudo-atom deletion, are performed.

In this paper, a comparison is made between two of these algorithms, namely the iterative skeletonization procedure (as implemented in the CCP4 program DM, 1) and the procedure found in the program suite DEMON/ANGEL. The tests have been carried out at a medium resolution of 3.0 Å in a case where non-crystallographic symmetry (NCS) is present in the crystal, where such algorithms are known not to give impressive results, and at 2.3 Å resolution when NCS is absent. The calculations were carried out using an initial phase and Sim weight set resulting from iterative 12-fold averaging and solvent flattening for the *Pseudomonas aeruginosa* catabolic ornithine transcarbamoylase (OTCase, 3). For further analysis, the same tests were repeated for a structure devoid of non-crystallographic symmetry, chitinase from *Serratia marscescens* (7) at 2.3 Å resolution, using initial phases at 2.5 Å resolution resulting from a procedure of iterative solvent flattening.

3. Materials and methods

The iterative skeletonization procedure as implemented in the CCP4 program DM (1) relies on the skeletonization algorithm of Swanson (8). This is quite similar to the iterative skeletonization procedure of the program PRISM (4), which relies on Greer's algorithm (9).

Since the procedure found in the program suite DEMON/ANGEL was not described in our previous paper (2), it will be described in detail here:

First, the electron density map is sorted by decreasing density values. The result of this sorting is a sorted array containing the decreasing electron density values, together with the corresponding grid locations in the 3-dimensional electron density map array. This sorted array is read sequentially, a pseudo-atom corresponding to a water molecule oxygen is placed at the corresponding cartesian Å coordinates after estimating the maximum in the 3-dimensional map array by fitting a parabola along each axis. Following this, all electron density values within a radius of 1.52 Å from this newly introduced pseudo-atom are set to zero in the 3-dimensional map array (this value of 1.52 $Å^2$ corresponding to ca. the closest distance of approach of two carbon atoms bound by a single bond). The process of pseudo-atom introduction is then repeated, each time checking that the electron density values in the 3-dimensional map array are not less than or equal to 0.0 (otherwise, this pseudo-atom position is skipped), until all requested pseudo-atoms have been introduced in the electron density map. The output is a Protein Data Bank coordinate file containing atomic coordinates of water oxygen atoms.

The positions of these pseudo-atoms, together with their artificial temperature factor of 15.0 Å, are then refined by least squares (10) or maximum likelihood (11) refinement with a new crystallographic refinement program (12). The resulting phases can be used to compute a Sim or Sigmaa weighted electron density map (13-16), or they can be combined with the phases used in the initial map calculation to generate a phase-combined electron density distribution. It can be noted that a similar technique has been described 20 years ago (17), but has not found widespread use in macromolecular crystallography.

These two algorithms were applied separately to the 3.0 Å resolution electron density distribution resulting from the procedure of 12-fold averaging and solvent flattening used in the structure determination of *Pseudomonas aeruginosa* catabolic ornithine transcarbamoylase (3).

Iterative skeletonization was performed for a total of 20 cycles using all data to 3.0 Å resolution, with the free R-factor approach (18). 5% of the reflections were selected from the data for the free R-factor calculation, and these were changed at each skeletonization cycle.

With the program DARTH of the DEMON/ANGEL program suite, a total of 3273 atoms were introduced in the region of the averaged electron density corresponding to a single OTCase monomer. This corresponds to ca. the expected number of atomic positions, including water oxygen atoms, for an OTCase monomer. The electron density map used for pseudo-atom introduction had been calculated using a grid ca. 1/10th of the high resolution limit (3.0 Å). The resolution limits for refinement were 6.0 to 3.0 Å, the same as those used in the refinement of the atomic model of OTCase (3). During refinement, 12-fold non crystallographic symmetry constraints corresponding to the dodecamer found in the asymmetric unit were applied both for positional refinement and for the refinement of the individual atomic temperature factors. The crystallographic refinement program's protein_rep.param parameter file was modified to give the pseudo-atoms a Van der Waals radius of 1.5 Å. A total of 200 cycles maximum likelihood energy minimization were performed, followed by 17 cycles of individual atomic temperature factor refinement.

The same calculations were also performed on a structure devoid of non-crystallographic symmetry, a chitinase from *Serratia marscescens* (7). For skeletonization, the same procedure (20 skeletonization cycles with 5% of the reflections removed for the free R-factor calculation) as for OTCase was used, starting from 2.5 Å solvent flattening phases. For pseudo-atom refinement, all data between 8.0 and 2.3 Å were used, with the electron density distribution used to generate the pseudo-atoms being a 2.5 Å resolution Sim weighted electron density resulting from iterative solvent flattening. A total of 4500 atoms were introduced in the protein region of the initial map.

The resulting phase sets were compared to the set of phases for the refined structure of OTCase, which has an R-factor of 21.6% for all data between 6.0 and 3.0 Å (3). For further analysis, several electron density maps were calculated, and the real space correlation coefficient between these observed electron densities and a "control map", without any resolution cutoff and computed from the refined atomic coordinates, was calculated with the graphics display program O (19). The same calculations were performed for chitinase, comparing the phase set resulting from these density modification procedures to the phases from the refined molecular model (7).

4. Results

The results are summarized in tables 1, 2 and 3, and can be described as follows:
20 cycles of iterative skeletonization were performed for OTCase with the program DM (1). The resulting combined phase set was further away from the correct, refined phases, by 0.4^{o} (Table 1), which seems insignificant. However, the corresponding electron density map, calculated with weighted observed structure factor amplitudes and combined phases (Table 2) is less interpretable than the initial averaged electron density distribution, since the real space correlation coefficient (19) decreases by 0.1.
In contrast, refinement of the pseudo-atoms with the algorithm of the DEMON/ANGEL program suite (2) using maximum likelihood (11) both for positional and individual atomic temperature factor refinement proceeded well, since the free R-factor (18) decreased from 38.0 to 22.6%, and the R-factor decreased from 38.4 to 21.2%, for all data between 6.0 and 3.0 Å. This decrease of the free R-factor, computed with a subset of 5% of the available reflections, clearly indicates that the phases are improving at this medium resolution. Consequently, the resulting phase set shows an improvement of 15.3^{o} with respect to the initial density modification phases (Table 1). The resulting SIGMAA electron density distribution (15), computed with coefficients of the form 3m|Fo|-2D|Fc| for acentric reflections and 2m|Fo|-D|Fc| for centric reflections (20) gives a real space correlation coefficient equal to that of the initial density modification map. Examination of this SIGMAA map clearly indicates that it is superior to the density modification map (Figure 1c).

TABLE 1. Phase differences between the different phase sets obtained and the phases from the refined model of OTCase

density modification	48.4^{o}
skeletonization	48.8^{o}
refinement	33.1^{o}
combined phases	35.4^{o}

The phase set density modification is that obtained by the iterative procedure of 12-fold averaging and solvent flattening used in the structure determination of OTCase (3). The skeletonization phase set is that obtained by combining the skeletonization phases with the density modification phases, as done by the program DM (1). The refinement phase set is that obtained by refinement of the coordinates and individual temperature factors of the pseudo-atoms resulting from the run of the program DARTH. The combined phase set is a phase set resulting from phase combination between the initial density modification phases and the refinement phases [with the Hendrickson-Lattman coefficients corresponding to the density modification phases modified to ensure that the resulting combined phases differ equally from each individual phase set (2)].

Phase combination of these pseudo-atom phases with the initial density modification phases was also performed, ensuring that the resulting combined phases differed equally from each individual phase set used for phase combination (this was done because the pseudo-atom phases are dominant in the phase combination process, 2). The resulting phase set shows an improvement of 13.0^{o} with respect to the initial phases (Table 1). The resulting SIGMAA electron density distribution also has the same real space correlation coefficient as the initial density modification map, but its examination also indicates that it is superior (Figure 1d).

TABLE 2. Real space correlation coefficients between the artificial electron density distribution obtained from the refined atomic coordinates [calculated with the program O (19)], and several electron density distributions.

E.D. map	<CCF>
AVG	0.59
DM	0.49
SIGMAA	0.59
SIGMAA-phi comb	0.59

AVG: initial density modification map (3);
DM: electron density map computed from combined phases resulting from 20 cycles of skeletonization;
SIGMAA: electron density distribution obtained with modified SIGMAA coefficients (15; 20) and phases from the refined pseudo-atom coordinates;
SIGMAA-phi comb: same than SIGMAA, except that the phases used result from phase combination between the initial density modification phases and the pseudo-atom phases.

In the case of chitinase, where non-crystallographic symmetry is absent from the crystals (7), both the R-factor and free R-factor resulting from refinement of the pseudo-atoms are less convincing than in the case of OTCase, where non-crystallographic symmetry is present: the starting free R-factor and conventional R-factor were 41.2 and

41.4%, respectively. After pseudo-atom refinement, these figures were 46.1 and 27.2%. Such an increase in the free R-factor suggests a phase degradation. This is confirmed by the phase difference to the true phases (Table 3): the initial solvent flattening phases are 51.9^{o} away from the true phases. Pseudo-atom introduction followed by refinement leads to a rise of this figure to 58.0^{o}, whereas iterative skeletonization degrades the phases further, since the phase difference to the true phases obtained after this procedure is 63.5^{o}. It must be noted that in this case where no non-crystallographic redundancy is present, omitting the MSR stage leads to better results. These observations clearly indicate that, in cases where non-crystallographic symmetry is present, there is a strong relationship in reciprocal space between structure factors (G-function coupling, 21, 22), which leads to the simultaneous decrease of the conventional R-factor and of the free R-factor. This confirms that non-crystallographic symmetry leads to an effect on the free R-factor (23), which becomes a less sensitive indicator of phase correctness since the reflections which are removed for the free R-factor calculations are coupled to the reflections used for the conventional R-factor by this G-function coupling effect. The results also indicate that the pseudo-atom refinement method will be far more effective when non-crystallographic symmetry is present in the crystals under investigation.

TABLE 3. Phase differences between the different phase sets obtained and the phases from the refined model of chitinase

solvent flattening	51.9^{o}
skeletonization	63.5^{o}
before refinement	64.1^{o}
refinement	58.0^{o}

The phase set solvent flattening is that obtained by the iterative procedure of solvent flattening used in the structure determination of this chitinase (7). The skeletonization phase set is that obtained by combining the skeletonization phases with the solvent flattening phases. The phase set noted « before refinement » corresponds to the phases obtained from the pseudo-atom coordinates before refinement. The refinement phase set is that obtained by refinement of the coordinates of the pseudo-atoms resulting from the run of the program DARTH. Otherwise, as in the legend to table 1.

Examination of the electron density maps computed with pseudo-atom phases or with combined phases suggests the presence of several minor errors, such as the misplacement of side chains, in the 3.0 Å resolution refined structure of *Pseudomonas aeruginosa* ornithine transcarbamoylase (3).

Figure 1.

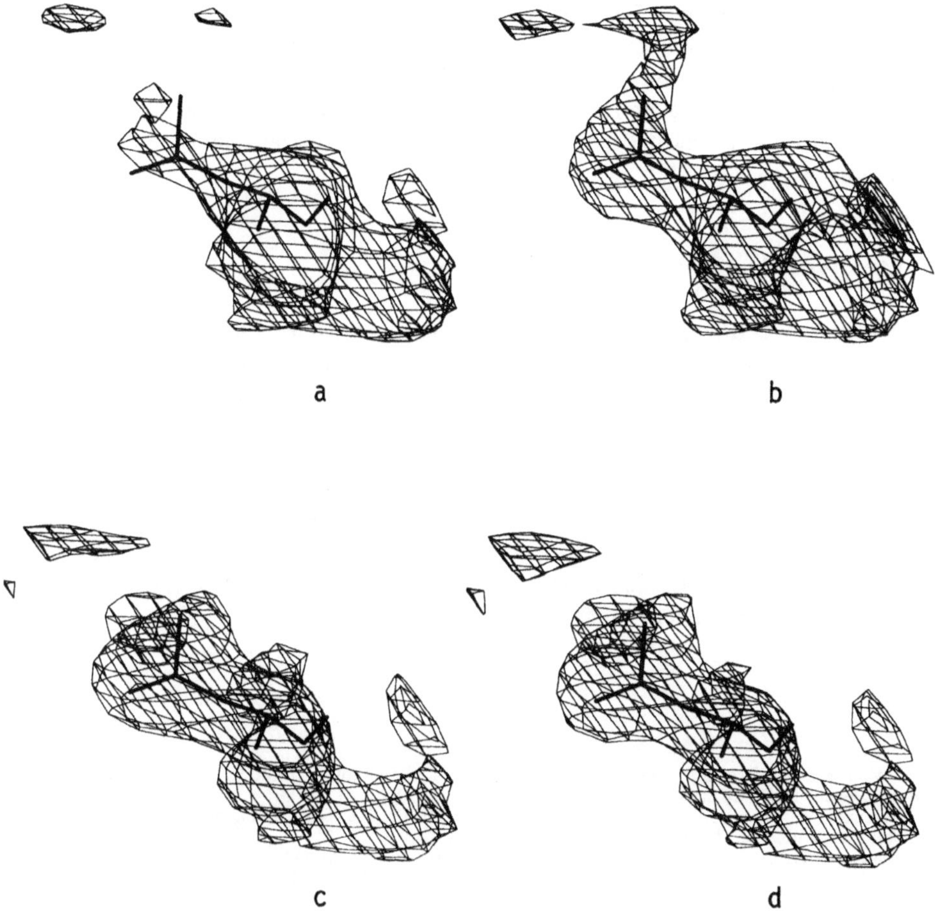

3.0 Å resolution electron densities corresponding to residue Leu 24 of *Pseudomonas aeruginosa* ornithine transcarbamoylase.
(a) electron density resulting from the procedure of iterative 12-fold averaging and solvent flattening (3). The map was calculated using mSim*|Fo|*exp(i*phi-calc);
(b) electron density resulting from 20 cycles of iterative skeletonization with DM (1). This map was calculated using coefficients of the form wSim*|Fo|*exp(i*phi-comb);
© SIGMAA electron density computed with phases resulting from the refinement of pseudo-atoms. This map was calculated with coefficients of the form 3m|Fo|-2D|Fc|*exp(i*phi-calc) for acentric reflections and 2m|Fo|-D|Fc|*exp(i*phi-calc) for centric reflections (15; 20);
(d) SIGMAA electron density computed with combined pseudo-atom refinement and density modification phases, as described in Results. This map was calculated with coefficients of the form 3m|Fo|-2D|Fc|*exp(i*phi-comb) for acentric reflections and 2m|Fo|-D|Fc|*exp(i*phi-comb) for centric reflections. All maps are contoured at a 1.0 sigma level, and were displayed with the program O (19).

In summary, these results indicate that, at the medium resolution (3.0 Å) where the comparison was performed, pseudo-atom introduction into an electron density map followed by refinement performs much better than iterative map skeletonization (at least in the case of OTCase, where non-crystallographic symmetry is present), with a phase improvement of 15.3^{o} over the initial phases. These results probably apply as well to other procedures which rely on the introduction of pseudo-atoms in an electron density map, followed by refinement of these atoms (namely the ARP and wARP procedures). It has also not escaped my attention that this procedure of pseudo-atom introduction may be modified to automatically introduce and delete water oxygen atoms in difference Fourier maps.

5. Acknowledgements

I thank Prof. A.T. Brünger (Yale University) for having provided a new crystallographic program system which includes an implementation of Pannu and Read's maximum likelihood refinement. I also thank Dr. O. Dideberg (IBS/LCM) and Dr. V. Villeret (University of Gent) for having provided the data of OTCase, and Dr. A. Perrakis (NKI, Amsterdam) for the chitinase data. Part of this work has been carried out during a short working visit at the Weizmann Institute of Science, Rehovot, Israel, in 1993 (with thanks due to Prof. J.L. Sussman). The DEMON/ANGEL program suite is now distributed under the terms of the Free Software Foundation's General Public License. Financial support of the Commissariat à l'Energie Atomique and of the Centre National de la Recherche Scientifique is gratefully acknowledged.

6. References

1. Cowtan, K.D. (1994) 'dm': an automated procedure for phase improvement by density modification, *Joint CCP4 and ESF-EACBM Newsletter on Protein Crystallography*, **31**, 34-38.
2. Vellieux, F.M.D, Hunt, J.F., Roy, S. & Read, R.J. (1995) DEMON/ANGEL: a suite of programms to carry out density modification, *J. Appl. Cryst.* **28**, 347-351.
3. Villeret, V., Tricot, C., Stalon, V. & Dideberg, O. (1995) Crystal structure of *Pseudomonas aeruginosa* catabolic ornithine transcarbamoylase at 3.0-Å resolution: a different oligomeric organization in the transcarbamoylase family, *Proc. Natl. Acad. Sci. USA* **92**, 10762-10766.
4. Wilson, C. & Agard, D.A. (1993) PRISM: automated crystallographic phase refinement by iterative skeletonization, *Acta Cryst.* **A49**, 97-104.
5. Lamzin, V.S. & Wilson, K.S. (1993) Automated refinement of protein models, *Acta Cryst.* **D49**, 129-147.
6. Perrakis, A., Sixma, T.K., Wilson, K.S. & Lamzin, V.S. (1997) wARP: improvement and extension of crystallographic phases by weighted averaging of multiple-refined dummy atom models, *Acta Cryst.* **D53**, 448-455.
7. Perrakis, A., Tews, I., Dauter, Z., Oppenheim, A.B., Chet, I., Wilson, K.S. & Vorgias, C.E. (1994) Crystal structure of a bacterial chitinase at 2.3 Å resolution, *Structure* **2**, 1169-1180.
8. Swanson, S.M. (1994) Core tracing: depicting connections between features in electron density, *Acta Cryst.* **D50**, 695-708.
9. Greer, J. (1985) Computer skeletonization and automatic electron density map analysis, in H.W. Wyckoff, C.H.W. Hirs and S.N. Timasheff (eds.), *Methods in Enzymology,* Academic Press, Orlando, **115**, pp. 206-226.
10. Konnert, J.H. & Hendrickson, W.A. (1980) A restrained-parameter thermal-factor refinement procedure, *Acta Cryst* **A36**, 344-350.
11. Pannu, N.S. & Read, R.J. (1996) Improved structure refinement through maximum likelihood, *Acta Cryst.* **A52**, 659-668.

12. Adams, P., Pannu, N.S., Read, R.J. & Brünger, A.T. (1997) Cross-validated maximum likelihood enhances crystallographic simulated annealing refinement, *Proc. Natl. Acad. Sci. USA* **94**, 5018-5023.
13. Sim, G.A. (1959) The distribution of phase angles for structures containing heavy atoms. II. A modification of the normal heavy atom method for non-centrosymmetrical structures, *Acta Cryst.* **12**, 813-815.
14. Sim, G.A. (1960) A note on the heavy-atom method, *Acta Cryst.* **13**, 511-512.
15. Read, R.J. (1986) Improved Fourier coefficients for maps using phases from partial structures with errors, *Acta Cryst.* **A42**, 140-149.
16. Vellieux, F.M.D. (1997) A systematic analysis of weighting schemes for structure factors in macromolecular crystallography, *J. Appl. Cryst.*, **30**, 400-401.
17. Agarwal, R.C. & Isaacs, N.W. (1977) Method for obtaining a high resolution protein map starting from a low resolution map, *Proc. Natl. Acad. Sci. USA* **74**, 2835-2839.
18. Brünger, A.T. (1992) Free R value: a novel statistical quantity for assessing the accuracy of crystal structures, *Nature* **355**, 472-474
19. Jones, T.A., Zou, J.Y., Cowan, S.W. & Kjeldgaard, M. (1991) Improved methods for building protein models in electron density maps and the location of errors in these models, *Acta Cryst.* **A47**, 110-119.
20. Vellieux, F.M.D. & Dijkstra, B.W. (1997) Computation of Bhat's OMIT maps with different coefficients, *J. Appl. Cryst.* **30**, 396-399.
21. Rossmann, M.G. & Blow, D.M. (1962) The detection of sub-units within the crystallographic asymmetric unit, *Acta Cryst.* **15**, 24-31.
22. Vellieux, F.M.D. & Read, R.J. (1997) Noncrystallographic symmetry averaging in phase refinement and extension, in R.M. Sweet and C.W. Carter (Eds) *Methods in Enzymology*, Academic Press, Orlando, **277**, pp. 18-53.
23. Kleywegt, G.J. & Brünger, A.T. (1996) Checking your imagination: applications of the free R value, *Structure* **4**, 897-904.

REAL-SPACE REFINEMENT USING RSREF

ERIC BLANC, ZHI CHEN & MICHAEL S. CHAPMAN[†]
Department of Chemistry & Institute of Molecular Biophysics, Florida State University, Tallahassee, FL 32306-3015, USA.

Abstract: Real-space methods for the refinement of macromolecular structures are briefly presented with regard to their most common application, when "experimental" phases are accurate. Recent improvements extend their use to initial stages of protein refinements, when phases are poor. RSRef can also be used interactively to hasten model (re)building and improve the starting point for conventional refinement.

1. Introduction

Although real-space methods of refinement were successfully applied to some of the first protein structures ([11], [18] for example), macromolecular structures are now usually refined by reciprocal space methods ([4] and [13]), as they are independent of the experimental phases, that are poorly determined. By contrast, real-space methods rely implicitly on phase information, as they minimize a residual based on the squared difference between observed and calculated electron densities: the observed electron density is calculated using experimental phases (from isomorphous replacement, symmetry averaging, *etc.*) and/or phases from a preliminary atomic model, in which case refinement may be biased towards the preliminary model.

Nevertheless, there are several niches for which real-space refinement is well-suited:

1. Manual optimization: A lot of time is spent manually optimizing a structure before or between rounds of reciprocal-space refinement. Because real-space refinement is a local method, *i.e.* a small part of the structure can be refined independently from the rest, it is a very fast and therefore very efficient way to keep a good fit between the model and the observed map while interactively

[†] To whom correspondence should be addressed.

S. Fortier (ed.), Direct Methods for Solving Macromolecular Structures, 513–519.

remodeling. For this reason, real-space refinement has been implemented within interactive molecular modeling programs (see [2] and [15] for example).

2. Accurate phases: When accurate phases are available (from MAD or averaging for example) Rees and Lewis [16], and Arnold and Rossmann [1] have shown that reciprocal-space refinement was improved through the addition of explicit phase restraints. These phase-restrained reciprocal-space refinements are near-equivalents of the real-space methods described.
3. Viruses: The very high non-crystallographic symmetry (ncs) present in icosahedral viruses has made real-space refinement the method of choice for optimizing such structures. The benefits of ncs are two fold. First, phases are of unusual quality because of the high order averaging (typically 15 to 60 fold) applied to the observed map (see accompanying articles [8] and [9]). As already pointed out, refinement is improved by the addition of phase information, if accurate. Secondly, real-space refinement is much faster than reciprocal-space methods, partly because it is carried out on a non-crystallographic asymmetric unit only, which contains typically 15 to 60 fold fewer atoms than the crystallographic asymmetric unit used by reciprocal space methods. It should also be noted that real-space refinement fits the model to the complete diffraction data set at once, whereas it is common to improve the efficiency of reciprocal-space virus structure refinement by using only alternating subsets of the *ca.* 1 million reflections on each cycle.

For conventional protein refinement, for example a protein structure with poor "observed" phases from MIR or molecular replacement, real-space methods are not widely used, even though it has been recently shown [7] that refinements could be improved by alternating real- and reciprocal-space methods. New techniques are currently under development to increase the power of real-space methods during the initial stages of refinement. These new methods, briefly described in section 3, take advantage of the large convergence radius of real-space refinement [12], of the much shorter range of interdependence of atoms, and of the relatively low computing requirements of real-space refinement.

2. Theory

The real-space refinement minimizes a least-squares residual S with respect to a set of atomic parameters $\{p_n\}$, where n runs over the refined atoms. The residual depends on atomic positions ξ_n, displacement parameters B_n and occupancies O_n; all these individual atomic parameters will be generically referred to as p_n. S is thus given by:

$$S = \int_V \left[\rho_{\text{obs}}(\mathbf{x}) - \rho_{\text{calc}}(\mathbf{x};\{p_n\})\right]^2 dx\,, \tag{1}$$

where ρ_{obs} and ρ_{calc} are respectively the observed and calculated electron densities (put on an absolute scale) evaluated at any point **x** in the crystal. The integration domain V depends on the positions of the refined atoms: V is the volume occupied by spheres of given radius R_n centered on atom position ξ_n. It must be noted that this dependency is not taken into account during the minimization process.

The calculated electronic density is assumed to be a linear superposition of individual atomic contributions:

$$\rho_{\mathrm{calc}}(\mathbf{x};\{p_n\}) = \sum_m O_m \rho_m(\|\mathbf{x} - \xi_m\|; B_m) \ , \qquad (2)$$

where $\rho_m(r)$ is the electron density of atom m at a distance r from its center. The individual atomic densities ρ_m are weighted by an atomic occupancy factor O_m and depend upon a displacement parameter B_m, which is usually assumed isotropic. In principle, all atoms in the crystal contribute to $\rho_{calc}(\mathbf{x})$, but as their distance to the point **x** increases, their contribution to the sum becomes vanishingly small. In practice, atoms for which $\|\mathbf{x} - \xi_m\|$ is larger than a cut-off radius R_m are not used in (2). It implies that unrefined atoms contribute to (2) if they are close to the refined region, or more precisely if their distance to a grid point in V is smaller than R_m.

Unlike previous implementations of real-space refinement, RsRef [6] explicitly accounts for resolution limits of the experimental data: ρ_m is calculated by the Fourier transform of the apparent atomic form factor smeared by the "temperature" factor $\exp(-B_m \|\mathbf{h}\|^2/4)$ (f_m), and restricted to the resolution limits d^*_{low} and d^*_{high}. With $h = \|\mathbf{h}\|$, ρ_m is given by

$$\rho_m(r) = \frac{2}{r}\int_{d^*_{\mathrm{low}}}^{d^*_{\mathrm{high}}} h\, f_m(h; B_m)\, \sin(2\pi h r)\, dh \qquad (3)$$

for r smaller than R_m, usually independent of the atom. For larger r, ρ_m is set to 0.

The resolution limits are responsible for the truncation ripple seen in electronic density maps, even for very accurate experimental or model phases. Figure 1a shows the comparison of the apparent density of a carbon atom "seen" at infinite resolution or only between 10^{-1} and 3^{-1} Å^{-1}. The difference between these two formulae is enhanced when the density is weighted by r^2 (fig. 1b), proportional to the surface of a spherical shell of radius r, and according to the total contribution to refinement of all grid points at a given r.

For macromolecular refinement at medium resolution, real-space refinement must be used with stereochemical information, that can be provided in several ways. RsRef is written as a module of the TNT package [17], using TNT's stereochemical restraints and minimizer.

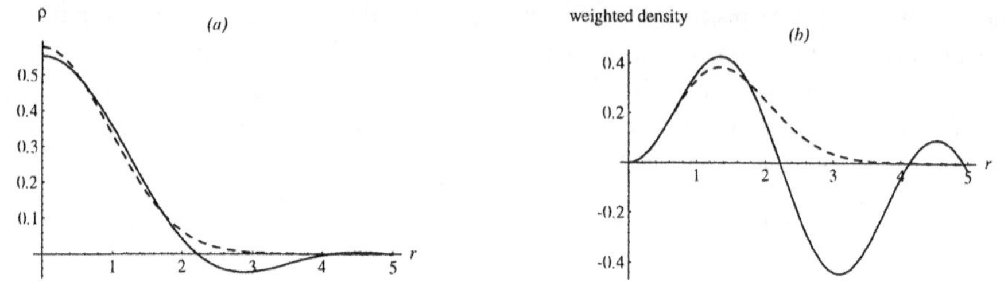

Figure 1: Effects of resolution limits on the apparent atomic electronic density. (a) Apparent electronic density of a carbon atom smeared by a displacement factor of B = 15.0, as a function of the distance from the atomic center r in Å. The continuous line shows the density for data collected between 10^{-1} and 3^{-1} Å^{-1}, the dashed line shows the apparent density at infinite resolution (sum of gaussians). (b) shows the densities of (a) weighted by the square of the radius, which is proportional to the contribution of each spherical shell to the total density.

3. Recent Developments

3.1. INITIAL REFINEMENT

When phases are poor, real-space methods suffer from their dependence to phases. While unlikely to replace reciprocal-space methods, it has recently been shown [7] that real-space methods complement the more usual methods effectively. Starting with an unrefined initial model and using a poor MIR-phased map, alternating real- and reciprocal-space refinements lead to an improvement of the model (monitored by R_{free}) nearly as great as can be realized through many hours of labor-intensive rounds of interactive model building and re-refinement. One efficient protocol for initial refinement iterates real- and reciprocal-space refinements with $2F_o$-F_c map calculation. It demonstrates the complementarity of real- and reciprocal space refinements, as the improvement realized by either method alone is smaller.

During the initial refinement, real-space methods likely improve the conditioning of the optimization by increasing the data to parameter ratio with phase information. Conditioning is also improved by restricting the interdependencies of atoms to local interactions, leading to improved convergence. Reduced interdependence also helps avoid overfitting (frequently observed in reciprocal-space refinements) by forcing a good *local* fit of the model and the experimental data. In a complementary way, by uncoupling atoms from phases, and reducing phase bias, the reciprocal-space refinement benefits real-space refinement.

Eventually however, progress of real-space methods is limited by the bias of the calculated phases. Several methods are under development to overcome this limitation.

During initial stages of model building, the calculation of phases from a partial or preliminary model has long been a method of improving experimental phases. It was soon recognized that these calculated phases have the potential to bias electron density maps towards this model, parts of which might be incorrect. Omit maps have been used to reduce such bias, but remaining bias has been the cause of model errors in several structures [3]. The source of bias in omit maps is refinement, which can cause atoms outside the omit region to adjust away from their correct positions to compensate for errors elsewhere and minimize the overall residual [14]. Real-space local refinements are thus well suited for reducing remaining bias, because mutually compensating adjustments of atoms from different regions of the structure are no longer possible. Implementations of such protocols are undergoing preliminary tests.

Complementing the efforts above, real-space refinement has been implemented with simulated annealing molecular dynamics replacing least-squares optimization. Preliminary tests indicate wide convergence radius for this refinement. Moreover, the implicit phase restraints of real-space refinement should decrease the risk of finding an incorrect structure with a low crystallographic residual when using powerful molecular dynamics searches. It also offers the possibility of fast local refinements.

3.2. DETERMINING THE QUALITY OF REAL-SPACE REFINED MODELS

Although standard free *R*-factor [5] have been used to date, recently it has become apparent that they do not show real-space refinement in fair light. Real-space methods are very sensitive to missing data. The common practice of removing part of the data set for cross-validation purposes is much more detrimental to refinement in real-space than in reciprocal-space, because missing reflections are effectively set to zero in the map used for real-space refinement, whereas they are simply ignored by reciprocal space methods. The real-space refined model is then optimized to reproduce amplitudes of zero for reflections in the cross-validation test set. This has deleterious effects on the model, but even for models of equal quality, the agreement between test and calculated amplitudes is artifactually reduced [10]. The harmful effects are mostly mitigated by using for map calculations, approximations to the test-set amplitudes that are independent of the actual observed value. Effective approximations are the average amplitudes of working set reflections at similar resolution. This procedure almost eliminates most of the systematic discrepancy between R_{free} calculated after real- and reciprocal-space refinements.

4. Conclusion

Although most appropriate when experimental phases are accurate, real-space refinement is a useful tool for many purposes, even during the initial stages of protein refinement, when phases are usually poor. The main reason for this usefulness comes from the uncoupling of the atoms belonging to different regions of the structure: because real-space refinement is local, overfitting is reduced, as is model bias in omit maps. Methodological developments are ongoing to efficiently use real-space refinement in the early stages of macromolecular structure determination.

This local nature of real-space methods results in relatively expeditious local refinements, which makes them well-suited for interactive modeling. The tutorial on real-space methods will focus on this aspect, namely by showing how RsRef [6] is used in conjunction with O [15] for model rebuilding in various map qualities.

5. Acknowledgements & References

This work was supported by the National Science Foundation.

1. Arnold, E. and Rossmann, M.G. (1988). The Use of Molecular-Replacement Phases for the Refinement of the Human Rhinovirus 14 Structure, *Acta Crystallographica* A**44**, 270-282.

2. Blanc, E. and Chapman, M.S. (1997). RsRef: Interactive Real-Space Stereochemically Restrained Refinement during Model Building, **accepted for publication in** *Journal of Applied Crystallography*.

3. Brändén, C.I. and Jones, T.A. (1990). Between Objectivity and Subjectivity, *Nature* **343**, 687-690.

4. Brünger, A.T. (1992). *X-PLOR A System for X-ray Crystallography and NMR*, Yale University, New Haven.

5. Brünger, A.T. (1992). Free R Value: a Novel Statistical Quantity for Assessing the Quality of Crystal Structures, *Nature* **355**, 472-475.

6. Chapman, M.S. (1995). Restrained Real-Space Macromolecular Atomic Refinement using a New Resolution-Dependent Electron-Density Function, *Acta Crystallographica* A**51**, 69-80.

7. Chapman, M.S. and Blanc, E. (1997). Potential Use of Real-Space Refinement in Protein Structure Determination, *Acta Crystallographica* D**53**, 203-206.

8. Chapman, M.S. (1997). Introduction to the Use of Non-Crystallographic Symmetry in Phasing, in Fortier, S. (ed.), *NATO Advanced Study Institute on Direct Methods for Solving Macromolecular Structures*, Kluwer Academic Publishers, Dordrecht, The Netherlands

9. Chapman, M.S., Blanc, E., McKenna, R., Munshi, R., Rossmann, M.G. and Tsao, J. (1997). Use of Non-Crystallographic Symmetry for *ab initio* Phasing of Virus Structures, in Fortier, S. (ed.), *NATO Advanced Study Institute on Direct Methods for Solving Macromolecular Structures*, Kluwer, Academic Publishers, Dordrecht, The Netherlands

10. Chen, Z., Blanc, E. and Chapman, M.S. (1997). Improved free R Factors for Cross-Validation of Macromolecular Structure, **submitted to** *Acta Crystallographica*.

11. Diamond, R. (1974). Real-space Refinement of the Structure of Hen Egg-white Lysozyme, *Journal of Molecular Biology* **82**, 371-391.

12. Diamond, R. (1985). Real-Space Refinement, *Methods in Enzymology* **115**, 237-252.

13. Hendrickson, W.W. (1985). Stereochemically Restrained Refinement of Macromolecular Structures, *Methods in Enzymology* **115**, 252-270.

14. Hodel, A., Kim, S.H. and Brünger, A.T. (1992). Model Bias in Macromolecular Crystal Structures, *Acta Crystallographica* A**48**, 851-858.

15. Jones, T.A., Zou, J.Y., Cowan, S.W., Kjeldgaard, M. (1991), Improved Methods for Building Protein Models in Electron Density Maps and the Location of Errors in these Models, *Acta Crystallographica* A**47**, 110-119.

16. Rees, D.C. and Lewis, M. (1983). Incorporation of Experimental Phases in a Restrained Least-Squares Refinement, *Acta Crystallographica* A**39**, 94-97.

17. Tronrud, D.E., Ten Eyck, L.F. and Matthews, B.W. (1987). An Efficient General-Purpose Least-Squares Refinement Program for Macromolecular Structures, *Acta Crystallographica* A**43**, 489-501.

18. Watenpaugh, K. D., Sieker, L.C., Herriot, J.R. and Jensen, L.H. (1973). Refinement of the Model of a Protein: Rubredoxin at 1.5 Å Resolution, *Acta Crystallographica* B**29**, 943-956.

TOWARDS THE ENANTIOMORPH SPECIFIC PROBABILISTIC THEORY OF THE STRUCTURE INVARIANTS

HERBERT A. HAUPTMAN
Hauptman-Woodward Medical Research Institute, Inc.,
73 High St., Buffalo, NY 14203 USA

Abstract

It is now generally recognized that the availability of single wavelength anomalous scattering (SAS) diffraction data greatly increases the power of the direct method. This enhancement of the direct method is undoubtedly due to the fact that, in contrast to the usual case, the observed data in the SAS case serve to fix the enantiomorph. It is now proposed to reach the same goal, even when SAS data are not available, by a suitable probabilistic approach which yields relationships among the structure invariants which respect the enantiomorph, that is to say, which are valid for fixed choice of enantiomorph. Specifically, this approach will derive relationships among the sines of the triplets, the value of any one of which is not uniquely determined by the diffraction intensities (in sharp contrast to the cosines) but is determined by the crystal structure, i.e. by either one of the two enantiomorphs permitted by the measured intensities. Only a brief summary of the final results is given here.

I. Definitions

Reciprocal lattice vectors H, K, G satisfy:

$$H \neq 0, \quad K \neq 0, \quad G \neq 0 \tag{I.1}$$

$$H \pm K \neq 0, \quad K \pm G \neq 0, \quad G \pm H \neq 0, \tag{I.2}$$

$$H + K + G \neq 0 \tag{I.3}$$

N is the number of atoms, presumed identical, in the unit cell.

$$A_{\mathbf{HK}} = \frac{2}{N^{1/2}}\left|E_{\mathbf{H}}E_{\mathbf{K}}E_{\mathbf{H+K}}\right|; \; A_{\mathbf{KG}} = \frac{2}{N^{1/2}}\left|E_{\mathbf{K}}E_{\mathbf{G}}E_{\mathbf{K+G}}\right|. \tag{I.4}$$

$$B_{\mathbf{HG}} = \frac{2}{N}\left|E_{\mathbf{H}}E_{\mathbf{G}}E_{\mathbf{H+K}}E_{\mathbf{G+K}}\right|\left(\left|E_{\mathbf{H-G}}\right|^2 + \left|E_{\mathbf{H+K+G}}\right|^2 - 1\right). \tag{I.5}$$

$$\psi_{\mathbf{HK}} = \phi_{\mathbf{H}} + \phi_{\mathbf{K}} + \phi_{\mathbf{-H-K}}, \; \psi_{\mathbf{KG}} = \phi_{\mathbf{K}} + \phi_{\mathbf{G}} + \phi_{\mathbf{-K-G}}. \tag{I.6}$$

Observe that the triplets $(\psi_{\mathbf{HK}}, \psi_{\mathbf{KG}})$ have the phase $\phi_{\mathbf{K}}$ in common.

S. Fortier (ed.), Direct Methods for Solving Macromolecular Structures, 521–523.

II. The Conditional Probability Distribution, $Q_{2/7}(\Psi_{HK}, \Psi_{KG}/|E|)$, of the Pair of Triplets (ψ_{HK}, ψ_{KG}), Given the Seven Magnitudes

$$|E_H|, |E_K|, |E_G|;\ |E_{H+K}|, |E_{K+G}|;\ |E_{H-G}|, |E_{H+K+G}| \qquad \text{(II.1)}$$

$$Q_{2/7}(\Psi_{HK}, \Psi_{KG}/|E|) =$$

$$\frac{1}{C'}\exp\{A_{HK}\cos\Psi_{HK} + A_{KG}\cos\Psi_{KG} + B_{HG}\cos(\Psi_{HK} - \Psi_{KG})\} \qquad \text{(II.2)}$$

where C' is an irrelevant normalizing parameter not needed for the present purpose.

III. The Conditional Probability Distribution $Q_{1/1,7}(\Psi_{KG}/\psi_{HK},|E|)$, of the Triplet ψ_{KG}, Given the Triplet ψ_{HK} and the Seven Magnitudes (II.1)

$$Q_{1/1,7}(\Psi_{KG}/\psi_{HK},|E|) =$$

$$\frac{1}{C}\exp\{A_{KG}\cos\Psi_{KG} + B_{HG}\cos(\Psi_{KG} - \psi_{HK})\} \qquad \text{(III.1)}$$

where

$$C = 2\pi I_0(Y_{KG}) \qquad \text{(III.2)}$$

and

$$Y_{KG} = \{A^2_{KG} + 2A_{KG}B_{HG}\cos\psi_{HK} + B^2_{HG}\}^{1/2} \qquad \text{(III.3)}$$

$$= \{(A_{KG}B_{HG}\cos\psi_{HK})^2 + B^2_{HG}\sin^2\psi_{HK}\}^{1/2} \qquad \text{(III.4)}$$

IV. The Conditional Expected Value, $\varepsilon(\cos\Psi_{KG}/\psi_{HK},|E|)$ of $\cos\Psi_{KG}$, Given the Triplet ψ_{HK}, Having a Phase in Common with ψ_{KG}, and the Seven Magnitudes (II.1)

$$\varepsilon(\cos\Psi_{KG}/\psi_{HK},|E|) = \frac{I_1(Y_{KG})(A_{KG} + B_{HG}\cos\psi_{HK})}{Y_{KG}I_0(Y_{KG})} \qquad \text{(IV.1)}$$

V. The Conditional Expected Value, $\varepsilon(\sin\Psi_{KG}/\psi_{HK},|E|)$ of $\sin\Psi_{KG}$, Given the Triplet ψ_{HK}, Having a Phase in Common with ψ_{KG}, and the Seven Magnitudes (II.1)

$$\varepsilon\left(\sin\Psi_{KG}/\psi_{HK},|E|\right)=\frac{B_{HG}I_1\left(Y_{KG}\right)\sin\psi_{HK}}{Y_{KG}I_0\left(Y_{KG}\right)} \tag{V.1}$$

VI. The Enantiomorph Specific Tangent Formula

Eqs. (IV.1) and (V.1) lead in the obvious way to the enantiomorph specific tangent formula:

$$\tan\psi_{HK}=\frac{\sum_G Y_{KG}\sin\psi_{KG}}{\sum_G Y_{KG}\cos\psi_{KG}-\sum_G\frac{A_{KG}I_1\left(Y_{KG}\right)}{I_0\left(Y_{KG}\right)}}, \tag{VI.1}$$

where the signs of $\sin\psi_{HK}$ and $\cos\psi_{HK}$ are determined by

$$\sin\psi_{HK}\left\{\frac{\sum_G\frac{B_{HG}I_1\left(Y_{KG}\right)}{I_0\left(Y_{KG}\right)}}{\sum_G Y_{KG}}\right\}=\frac{\sum_G Y_{KG}\sin\psi_{KG}}{\sum_G Y_{KG}} \tag{VI.2}$$

$$\cos\psi_{HK}\left\{\frac{\sum_G\frac{B_{HG}I_1\left(Y_{KG}\right)}{I_0\left(Y_{KG}\right)}}{\sum_G Y_{KG}}\right\}=\frac{\sum_G Y_{KG}\cos\psi_{KG}-\sum_G\frac{A_{KG}I_1\left(Y_{KG}\right)}{I_0\left(Y_{KG}\right)}}{\sum_G Y_{KG}} \tag{VI.3}$$

Research supported by the National Institutes of Health Grant No. GM46733.

** Please note: In many cases the author has indexed to the first page of their paper, as the keyword is a concept throughout.

GPSR Compliance

The European Union's (EU) General Product Safety Regulation (GPSR) is a set of rules that requires consumer products to be safe and our obligations to ensure this.

If you have any concerns about our products, you can contact us on ProductSafety@springernature.com

In case Publisher is established outside the EU, the EU authorized representative is:

Springer Nature Customer Service Center GmbH
Europaplatz 3
69115 Heidelberg, Germany

Batch number: 10809436

Printed by Printforce, the Netherlands